W9-CAT-528

FISH PHYSIOLOGY

VOLUME X
Gills

Part B
Ion and Water Transfer

CONTRIBUTORS

LIANA BOLIS
MICHEL BORNANCIN
P. S. DAVIE
C. DAXBOECK
GUY DE RENZIS
SUZANNE DUNEL-ERB
A. G. ELLIS
DAVID H. EVANS
J. P. GIRARD
JACQUES ISAIA
PIERRE LAURENT
N. MAYER-GOSTAN
THOMAS P. MOMMSEN
P. PAYAN
S. F. PERRY
W. T. W. POTTS
J. C. RANKIN
D. G. SMITH
J. A. ZADUNAISKY

FISH PHYSIOLOGY

Edited by

W. S. HOAR

DEPARTMENT OF ZOOLOGY
UNIVERSITY OF BRITISH COLUMBIA
VANCOUVER, CANADA

D. J. RANDALL

DEPARTMENT OF ZOOLOGY
UNIVERSITY OF BRITISH COLUMBIA
VANCOUVER, CANADA

VOLUME X

Gills

Part B

Ion and Water Transfer

1984

ACADEMIC PRESS, INC.

(Harcourt Brace Jovanovich, Publishers)
Orlando San Diego San Francisco New York London
Toronto Montreal Sydney Tokyo São Paulo

ACADEMIC PRESS, INC.
Orlando, Florida 32887

United Kingdom Edition published by
ACADEMIC PRESS, INC. (LONDON) LTD.
24/28 Oval Road, London NW1 7DX

Library of Congress Cataloging in Publication Data

Hoar, William Stewart, Date
Fish physiology.

Vols. 8- edited by W. S. Hoar, D. J. Randall, and J. R. Brett.
Includes bibliographies and indexes.
CONTENTS: v. 1 Excretion, ionic regulation, and metabolism.--v. 2. The endocrine system.--[etc.]--v. 10. Gills. pt. A. Anatomy, gas transfer, and acid-base regulation. pt. B. Ion and water transfer (2 v.)
1. Fishes--Physiology--Collected works. I. Randall, D. J., joint author. II. Conte, Frank P. III. Brett, J. R. IV. Title.
QL639.1.H6 597'.01 76-84233
ISBN 0-12-350430-9 (v. 10A)
ISBN 0-12-350432-5 (v. 10B)

PRINTED IN THE UNITED STATES OF AMERICA

84 85 86 87 9 8 7 6 5 4 3 2 1

CONTENTS

CONTRIBUTORS

Numbers in parentheses indicate the pages on which the authors' contributions begin.

LIANA BOLIS (177), *Istituto di Fisiologia Generale, 98100 Messina, Italy*

MICHEL BORNANCIN (65), *Laboratoire de Physiologie Cellulaire et Comparée, Université de Nice, France*

P. S. DAVIE (325), *Department of Physiology and Anatomy, Massey University, Palmerston North, New Zealand*

C. DAXBOECK (325), *Pacific Gamefish Foundation, Kailua-Kona, Hawaii 96745*

GUY DE RENZIS (65), *Laboratoire de Physiologie Cellulaire et Comparée, Université de Nice, France*

SUZANNE DUNEL-ERB (285), *Laboratoire de Physiologie Comparée des Régulations, Centre National de la Recherche Scientifique, 67037 Strasbourg, France*

A. G. ELLIS (325), *Department of Zoology, University of Melbourne, Parkville, Victoria 3052, Australia*

DAVID H. EVANS (239), *Department of Zoology, University of Florida, Gainesville, Florida 32611, and Mt. Desert Island Biological Laboratory, Salsbury Cove, Maine 04672*

J. P. GIRARD (39), *Laboratoire de Physiologie Cellulaire et Comparée, Université de Nice, 06034 Nice, France*

JACQUES ISAIA (1), *Laboratoire de Physiologie Cellulaire et Comparée, Université de Nice, 06034 Nice, France*

PIERRE LAURENT (285), *Laboratoire de Morphologie Fonctionnelle et Ultrastructurale des Adaptations, Centre National de la Recherche Scientifique, 67037 Strasbourg, France*

N. MAYER-GOSTAN (39), *Laboratoire Jean Maetz, Groupe de Biologie du C.E.A., 06230 Villefrance-sur-Mer, France*

THOMAS P. MOMMSEN (203), *Department of Biology, Dalhousie University, Halifax, Nova Scotia B3H 4J1, Canada*

P. PAYAN (39), *Laboratoire de Physiologie Cellulaire et Comparée, Université de Nice, 06034 Nice, France*

S. F. PERRY (325), *Department of Biology, University of Ottawa, Ottawa, Ontario K1N 6N5, Canada*

W. T. W. POTTS (105), *Department of Biological Sciences, University of Lancaster, Lancashire, England*

J. C. RANKIN (177), *School of Animal Biology, University College of North Wales, Bangor, Gwynedd LL57 2UW, United Kingdom*

D. G. SMITH (325), *Department of Zoology, University of Melbourne, Parkville, Victoria 3052, Australia*

J. A. ZADUNAISKY (129), *Department of Physiology and Biophysics, School of Medicine, New York University Medical Center, New York, New York 10016*

PREFACE

The fish gill is an intriguing tissue because of its multifunctional nature. Gills are involved in ion and water transfer as well as oxygen, carbon dioxide, acid, and ammonia exchange, and there are many interactions between these processes. These interactions lead to exchange of information among groups of scientists who otherwise might not meet, because they are brought together by their mutual interest in gills. Many aspects of gill structure and function have been studied, and our understanding of these different systems is reasonably well advanced. This is not to say that there are not vast gaps in our knowledge. Only a relatively few species have been studied in detail, goldfish, eels, and trout being most prominent because of availability and ease of handling. Freshwater fish are much more studied than marine species for the same reasons. The range of freshwater is very broad, and we are only at the beginning of understanding the differences in gill structure and function that, for example, allow some fish to flourish in soft water but restrict others to hard waters.

This volume attempts to review the structure and function of fish gills and also makes some attempt, particularly in the final chapter, to review some of the methodology used in studying gills. The terminology concerned with the gills has grown with the studies and is often confusing. We have attempted, with the help of Drs. Hughes, Laurent, Nilsson, and many others, to arrive at some terms and abbreviations to be used with fish gills. It is always difficult to change habits but some uniformity is always an aid to understanding, particularly for those just entering the field. We would urge the adoption of these terms and abbreviations, even though we cannot persuade all contributors to do the same.

Many people advised and helped us in editing this text; in particular, the chapters were reviewed by many people, and we are most grateful for all help given. We hope the end result is a work on gills that will be of use to those interested in this fascinating organ for many years to come.

W. S. HOAR
D. J. RANDALL

TERMS AND ABBREVIATIONS

There is a profusion of terms associated with fish gills; we have attempted herein to standardize the terminology and abbreviations, in the hope that these will be generally accepted and used by all in the field.

Term	Abbreviation	Synonym/Notes
Branchial (Gill) arch	B	Gill bar refers to arch
Filament	F	Primary lamella; gill rod refers to filament
Lamella (s)	L	Secondary lamella
Lamellae (pl)		
Proximal lamellae		Lamellae proximal to arch
Distal lamellae		Lamellae distal to arch
Interlamellar space/water		Space/water between lamellae
Inhalent water	I	Do not refer to afferent and
Exhalent	E	efferent water.
Prelamellar water		Water that has not passed over lamellae
Postlamellar water		Water that has passed over lamellae
Pillar cell		
Epithelial cell		
Interstitial cell		Avoid terms like stem cell, pillar cell II etc. unless better evidence indicates function of these cells.
Erythrocyte	RBC	
Interstitial space	IS	Not lymphatic space
Ventral aorta	VA	A = artery
Dorsal aorta	DA	
Suprabranchial artery	SBA	
Carotid artery	CA	
Coeliacomesenteric artery	CMA	
Afferent branchial artery	af.BA	af. = afferent
Afferent filament artery	af.FA	Afferent primary (lamella) artery
Afferent lamellar arteriole	af.La	Afferent secondary (lamella) arteriole a = arteriole
Efferent lamellar arteriole	ef.La	Efferent secondary (lamellar) arteriole ef. = efferent

(continued)

Continued

Term	Abbreviation	Synonym/Notes
Efferent filament artery	ef. FA	Efferent primary (lamellar) artery
Efferent branchial artery	ef. BA	
Branchial vein	BV	V = vein
Dorsal branchial vein	DBV	
Ventral branchial vein	VBV	
Inferior jugular vein	IJV	
Anterior cardinal vein	ACV	
Posterior cardinal vein	PCV	
Central venous sinus	CVS	Not filament sinus or *veno-lymphatic* sinus, not lymphatic.
Anterior-Venous anastomosis	AVas	Avoid shunt because this implies function, as = anastomosis
Efferent side	ef. AVas	
Afferent side	af. AVas	
Lamellar *basal* blood channel		Blood channel at base of lamella
Lamellar *marginal* blood channel		Blood channel in free edge of lamella
Filament epithelium		Primary (lamellar) epithelium
Lamellar epithelium		Secondary (lamellar) epithelium = respiratory epithelium
Leading edge[a] (refers to water)		Filament efferent side (refers to blood)
Trailing edge[a] (refers to water)		Filament afferent side (refers to blood)

[a]Leading and trailing edge refer to the direction of water flow over the gill arch, filament, or lamella.

CONTENTS OF OTHER VOLUMES

Volume III

Volume IV

Volume V

Volume VI

Volume VII

Volume VIII

Volume IXA

Volume IXB

Volume XA

1

WATER AND NONELECTROLYTE PERMEATION

JACQUES ISAIA

Laboratoire de Physiologie Cellulaire et Comparée
Université de Nice
Nice, France

I. INTRODUCTION

The first model concerning the osmoregulatory function of fish gills was proposed by Smith (1932). Since that time, branchial ionic permeability has been studied intensively, and at least some of the osmoregulatory mechanisms are now well known. In contrast, nonelectrolyte branchial permeability has not been studied as much, essential data having been primarily concerned with water and gas exchanges. The major difficulty arises from the fact that permeability measurements of various nonelectrolytes possessing a large range of physicochemical properties requires control of various plasma

ISBN 0-12-350432-5

parameters; for example, it has been shown that adrenaline and stress play a role in water exchanges (Lahlou and Giordan, 1970; Pic *et al.*, 1974). For these reasons, studies of branchial permeability are usually carried out on an *in vitro* preparation (perfused filament, gill arch, or isolated head). Various *in vitro* preparations have been used on certain euryhaline teleosts (eels or trout), whereas data are missing for other fish groups. Furthermore, the functional vascularization of the gills has now been described, and only physiological data appearing after this date allow for the separation of the fluxes across the nonrespiratory epithelium (i.e., chloride cells) and the respiratory epithelium (i.e., secondary lamellae).

The study of nonelectrolyte gill exchanges is important because it permits the elucidation of various phenomena. For example, one can (*a*) compare the branchial permeability coefficients with those obtained from other epithelia, which permits the classification of the gills in relation to other similar structures in the animal kingdom, (*b*) define the hydro- or lipophilic characteristics of the various pathways, (*c*) describe the correlation between membrane permeability and cellular structure as a function of environmental conditions, and finally, (*d*) show that membrane permeability is asymmetrical, a consequence of absorption and excretion phenomena.

The permeability of the gills to noncharged substances for which no specific transport system exists is of interest because transfer is unaffected by changes in transepithelial potential, the only driving force being the chemical gradient. This chapter reviews information concerning branchial nonelectrolyte permeation of teleost gills, and especially the data on branchial permeability of euryhaline teleosts.

II. GENERALITIES ON GILL PERMEABILITY

The fish gill can be considered as the sum of two epithelia: the respiratory and nonrespiratory (Laurent and Dunel, 1980). This complex structure does not allow for the treatment of transbranchial diffusion as a transfer of molecules across a homogeneous membrane. However, to simplify the analysis of permeation, the following pathways may be considered: (*a*) the membranes of respiratory cells (i.e., secondary lamellae respiratory epithelium), which represent the major part (96%) of the exchange surface of the gills (Hughes, 1972; Girard and Payan, 1980), (*b*) the membranes of chloride cells, which represent the exchange surface of the nonrespiratory epithelium (4% of the total area), and finally, (*c*) the intercellular junctions.

Morphological investigations have yielded two important results for branchial permeation studies.

1. A comprehensive account of the vascular organization of the gills in representative groups of fish has been presented (Dunel and Laurent, 1980;

Laurent and Dunel, 1980) indicating that the nonrespiratory epithelium that surrounds the primary lamellae has a close relationship with the venous compartment, whereas the respiratory epithelium that covers the free part of the secondary lamellae has an exclusive relationship with the arterio-arterial vasculature. These new anatomic results now permit, with an appropriate *in vitro* preparation, the separate calculation of transbranchial flux across (*a*) the respiratory cells and (*b*) the chloride cells (Isaia, 1982).

2. The comparison of branchial morphology of fishes adapted to fresh water or seawater pointed out structural modifications of the last two pathways. The number of chloride cells increases in seawater by a factor of about three (Shirai and Utida, 1970; Sargent *et al.*, 1977), and loose junctions, permeable to lanthanum ions, appear between chloride cells on the apical side (Sardet *et al.*, 1979). These ultrastructural modifications do not change the relative area of the nonrespiratory epithelium compared to the total exchange surface. In contrast, the permeability of this epithelium is increased by the appearance of leaky junctions.

III. THE VARIOUS GILL PATHWAYS AND THEIR CHARACTERISTICS

A. General Characteristics of Branchial Permeability

The values of total branchial permeability (P_s) to various nonelectrolytes and the permeability of various epithelia and the erythrocyte membrane are reported in Table I. Except for human skin, the gills are among the most impermeable epithelia especially with respect to small lipo- and hydrophilic solutes. Thus, P_{water} and P_{urea} of seawater-adapted fishes (an epithelium with leaky junctions) are lower than those of the toad urinary bladder, which is an epithelium with tight intercellular junctions and high transepithelial electrical resistance (Bindslev and Wright, 1976). This comparison illustrates the complexity of the relationship between structure and function in such an epithelium. The complexity of the analysis can be seen in Fig. 1. When freshwater-adapted euryhaline fishes (eels or trout) are transferred to seawater, P_s decreases, increases, or remains constant as a function of the properties of the diffusing molecules. These results can be explained by different pathways across the gills (Isaia *et al.*, 1978a; Isaia, 1979). As shown in Fig. 2, the separate calculation of nonrespiratory and respiratory fluxes has demonstrated the existence of various transbranchial routes as a function of the characteristics of the diffusing molecules (Isaia, 1982). There appear to be a transcellular pathway for small lipo- and hydrophilic molecules across the secondary lamellae (respiratory pathways), a transcellular pathway for

Table I

Permeability Coefficients to Various Nonelectrolytes of Various Epithelia and the Erythrocyte Membrane[a]

Molecule[b]	Nonelectrolyte permeability coefficient (cm sec^{-1} × 10^6)								References
	Ethyl acetate	Butanol	Antipyrine	Ethanol	Water	Urea	Mannitol	Inuline	
K_{oil}	2.5	0.44	3.2×10^{-2}	2.6×10^{-2}	7×10^{-4}	1.5×10^{-4}	1.2×10^{-6}	Insoluble	
Biological material									
Fish gills									
Salmo gairdneri									
FW	—	—	—	—	—	—	0.17	0.02	Kirschner (1980)
SW	—	—	—	—	—	—	0.12	0.02	Kirschner (1980)
FW	42.8	19.3	—	9.5	10.0	1.2	0.16	0.92	Isaia (1979, 1982)
SW	28.8	6.4	—	5.6	4.4	2.7	0.98	—	Isaia (1979, 1982)
Anguilla anguilla									
SW	—	—	7.2	14.6	14.6	2.2	—	—	Steen and Sray-Pedersen (1975)
FW	—	—	—	—	25[c]	0.12[d]	1.3[e]	0.29	
SW	—	—	—	—	18[c]	0.35[d]	—	0.29	
Scyliorhinus canicula									
SW	—	—	—	—	5.6	0.004	—	—	Payan and Maetz (1970)
Other epithelia									
Toad urinary bladder	—	2000	10	96	>130	0.6	0.09	—	Wright and Pietras (1974)
Frog choroid plexus	—	—	24.2	—	68	12	—	—	Bindslev and Wright (1976)
Rabbit gallbladder	—	—	76	—	202	89	—	—	
Human skin	—	0.37	—	—	0.08	—	—	—	Scheuplein and Blanck (1971)
Human erythrocyte	—	41	—	88	9150	239	—	—	Naccache and Sha'afi (1973)

[a]See text for further details.
[b]K_{oil}, partition coefficient; FW, freshwater; SW, seawater.
[c]Reference: Motais *et al.* (1969).
[d]Reference: Masoni and Payan (1974).
[e]Reference: Masoni and Isaia (1973).

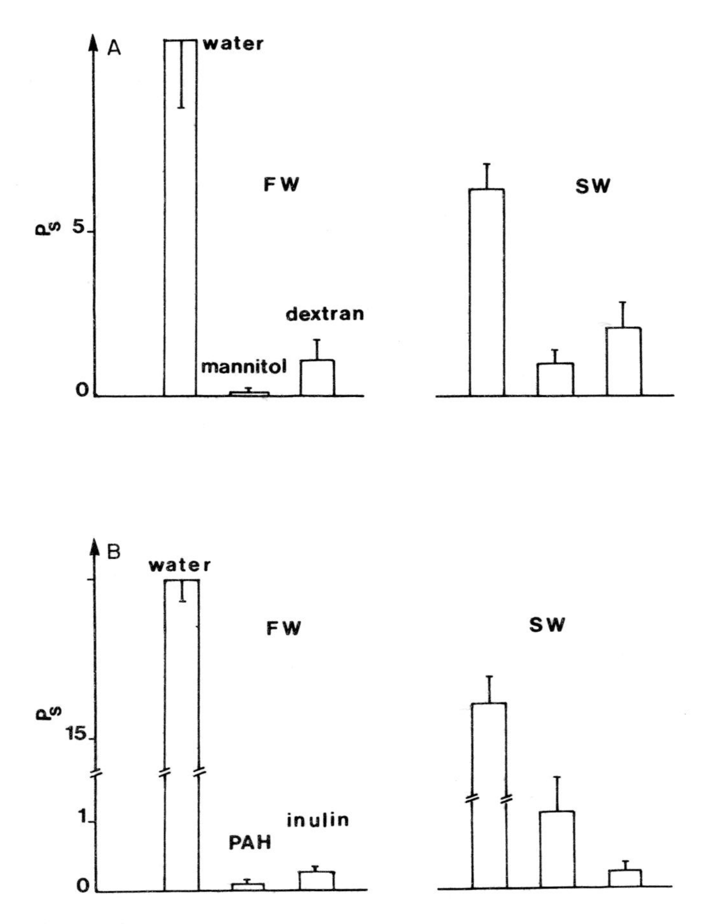

Fig. 1. The nonelectrolyte permeability of gills in (A) trout (*Salmo gairdneri*) and (B) eel (*Anguilla anguilla*) as a function of the adaptation medium. FW, fresh water; SW, seawater. Ordinate represents the branchial permeability calculated from the total efflux (serosal–mucosal direction), P_s in cm $sec^{-1} \times 10^6$. Note that P_{water} is decreased, $P_{mannitol}$ and P_{PAH} are increased, and P_{inulin} and $P_{dextran}$ are not significantly modified. (Data from Motais *et al.*, 1969; Masoni and Payan, 1974; Isaia *et al.*, 1978a, Isaia, 1981.)

hydrophilic macromolecules across the chloride cells (nonrespiratory pathway), and finally, a predominantly paracellular pathway for mannitol.

B. Hydrophilic Characteristics of the Respiratory Pathway

The analysis of membrane permeability as a function of the olive oil–water partition coefficient (K_{oil}) of nonelectrolytes is used to obtain informa-

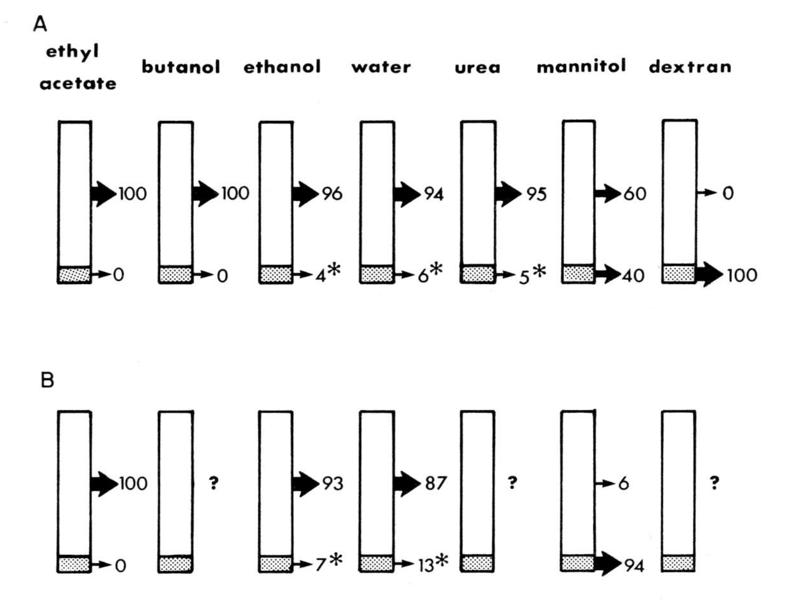

Fig. 2. Percentage of the nonelectrolyte fluxes crossing the lamellae (open columns) and the nonlamellar pathway (dotted columns) in freshwater (A) and seawater (B) trout. For small lipo- and hydrophilic molecules the fluxes crossing the nonlamellar pathway are not significant. The contrary is observed for macromolecules in freshwater trout and in saltwater eel, where Masoni and Isaia (1980) have shown that dextran is concentrated and excreted only by the chloride cells. Mannitol crosses mainly the chloride cells in saltwater fish, while in freshwater fish it crosses both epithelia (filament and lamellar). * Not significant. (Data from Girard, 1979; Isaia, 1982.)

tion concerning the selectivity and the hydro- or lipophilic nature of the pathways across biological membranes. The theory of membrane permeation of nonelectrolytes established by Diamond and Katz (1974) leads to a simple equation if the membrane is homogeneous and symmetrical, and if the intramembrane resistance is the factor limiting diffusion. The simplified equation, $P = KD/x$, where P is the membrane permeability, K the local nonelectrolyte partition coefficient, D the diffusion coefficient, and x the membrane thickness, shows that P_s depends essentially on K and D for different substances. A good correlation between P_s and K suggests that permeation of nonelectrolytes across a membrane is accounted for by solubility/diffusion processes in lipid of the plasmalemma (Stein, 1967). Studies of numerous lipophilic solutes have shown the existence of such correlations in permeation processes occurring in plants (*Nitella;* Collander, 1954) and animals (rabbits and pig gallbladder, Hingson and Diamond, 1972; Frog choroid plexus and toad urinary bladder, Wright and Pietras, 1974; bovine erythrocytes, Lieb and Stein, 1969). Figure 3 shows a good correlation

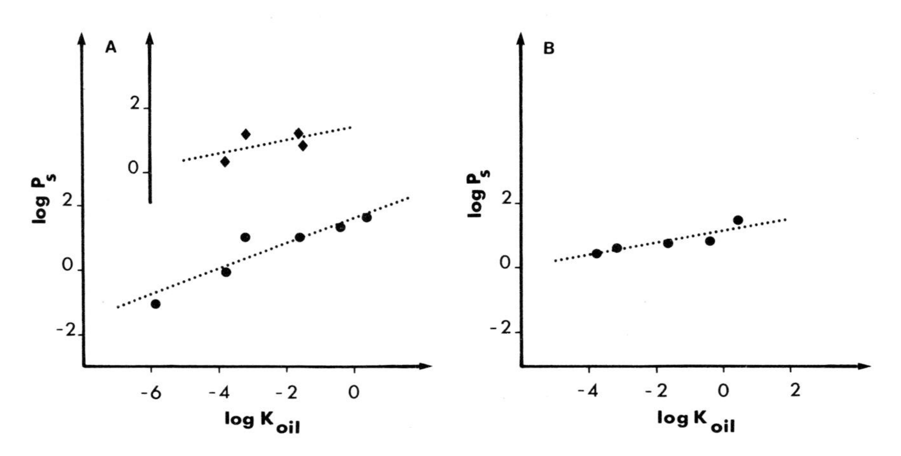

Fig. 3. Branchial permeabilities of the lamellar pathway to nonelectrolytes as a function of their partition coefficient between water and olive oil (K_{oil}). (A) Fresh water; ◆ eel; ● trout. (B) Seawater; ● trout. Permeability coefficients (P_s in cm $sec^{-1} \times 10^6$) are calculated from the total branchial efflux except for mannitol in freshwater (FW) trout, where P_s was calculated from 60% of the total efflux, a fraction corresponding to the flux taking place in respiratory cells. The linear regression curves of log P_s as a function of log K_{oil} are the following:

$$\text{FW trout} = \log P_s = 0.33 \pm 0.07 \log K_{oil} + 1.58 \pm 0.23 \ (n = 6,\ r = .94)$$
$$\text{FW eel} = \log P_s = 0.19 \pm 0.21 \log K_{oil} + 1.37 \pm 0.57 \ (n = 4,\ r = .53)$$
$$\text{SW trout} = \log P_s = 0.19 \pm 0.06 \log K_{oil} = 1.4 \pm 0.14 \ (n = 5,\ r = .87)$$

These curves have been traced for nonelectrolytes permeating especially through respiratory cells. (Data from Steen and Stray-Petersen, 1975; Isaia *et al.*, 1978a; Isaia, 1982.)

between P_s and K for the small molecules crossing the respiratory cells, although the slopes of these curves (log P_s/log K_{oil}) are low (from 0.38 to 0.19). Furthermore, the slope is decreased in seawater. This low selectivity of the gill could be interpreted as the result of various mechanisms: (*a*) a large flux between the cells through intercellular nonselective junctions (Wright and Pietras, 1974), (*b*) a transmembrane passage in aqueous channels (the self-diffusion coefficients in water of nonelectrolytes are not different in comparison to their partition coefficients, e.g., $D_{ethanol}$ and D_{urea} = 1.2 and 1.4 × 10^{-5} cm^2 sec^{-1}, respectively, while the ratio of their K_{oil} = 173), and (*c*) the presence of a lipid plasma membrane able to form numerous hydrogen bonds with the diffusing compound (Wright and Prather, 1970). The first mechanism can be rejected because the respiratory epithelium is one of the most impermeant (see Section III,A). Furthermore, for water permeation, it has been demonstrated that an inner epithelial barrier, possibly, the basal membrane, is the limiting barrier for diffusion (Isaia *et al.*, 1978b). Two arguments favor the last mechanism as an explanation of the low slope of P_s to K (Fig. 3).

1. Zwingelstein (1981) has compiled results concerning the amphiphilic phospholipids of the gills. He shows that phospholipids of fish gill cells are not different from those of other cell membranes. However, in the course of seawater adaptation, the percentage of sphingomyelin doubles. This is significant when taken in conjunction with the finding that permeability of molecules decreases with increasing sphingomyelin content (Barenholz and Thompson, 1980.

2. The selective increase in permeability of small lipophilic molecules under adrenaline action in freshwater-adapted fish indicates an effect of the catecholamine on the lipid mobility of plasmalemma (Isaia, 1979).

C. Properties of the Nonrespiratory Pathway

The branchial permeability to sugars presents several paradoxical properties:

1. At a trace level ($10^{-6}M$), $P_{glucose}$ is close to zero (Furspan and Isaia, 1983), whereas hexose monomers or polymers such as inulin and dextrans are permeant (Lam, 1969; Masoni and Isaia, 1973; Masoni and Payan, 1974; Isaia *et al.*, 1978a; Girard, 1979; Kirschner, 1980; Isaia, 1982) (see Fig. 4).
2. There is a large rectification of the membrane permeability of the branchial epithelium to dextran and glucose (when fluxes are high), the serosal–mucosal permeability being about 10-fold higher than in the reverse direction (Isaia, 1982; Furspan and Isaia, 1983).
3. Hexose polymers are concentrated and excreted primarily by the chloride cells (Masoni and Garcia-Romeu, 1972; Masoni and Isaia, 1980).
4. P_s to macromolecules is very high compared to P_s to small molecules (e.g., P_{inulin} is 2.5-fold higher than P_{urea} in the seawater-adapted eel (Masoni and Payan, 1974).

These apparently contradictory data probably are a result of the properties of the chloride cell. As suggested by Furspan and Isaia (1983), the tubulovesicular reticulum of the chloride cell is continuous with the cell's basal and lateral plasmalemma and brings the latter into intimate associations with mitochondria and other cytoplasmic inclusions (see review of Philpott, 1980). The fluid in the lumen is continuous with the fluid bathing the basal and lateral surfaces of the chloride cells, that is, the plasma. The tubulovesicular reticulum is connected to the low-pressure central venous sinus (Dunel and Laurent, 1977), and this system could constitute a mechanism for cellular excretion. It is suggested that a plasma filtrate enters the

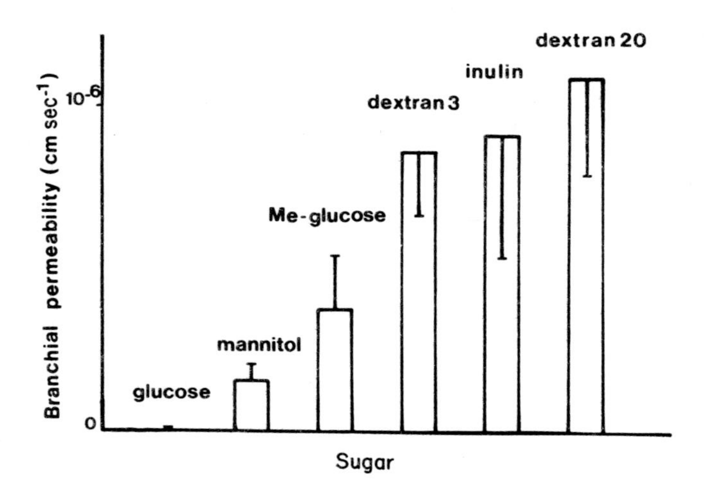

Fig. 4. Branchial permeability of various sugars in fresh-water-adapted trout (serosal–mucosal direction, in cm sec^{-1}). In contrast to others, glucose permeability is not significantly different from zero at "trace" and even normal plasma concentrations. (Data from Isaia *et al.*, 1978a; Isaia, 1982; Furspan and Isaia, 1983.)

tubule. Vital substances such as ions, glucose, and amino acids would be reabsorbed, the driving force being active ionic transport. Above a threshold of saturation of the reabsorption mechanism, these substances are excreted passively. The nonvital substances also are excreted passively. This model is supported by various experimental results:

1. Chloride cells display a maximal rate of glucose transport. The pharmacological agents that inhibit the sodium-dependent glucose transport (e.g., phlorizin) produce, at normal glycemia, a rate of branchial excretion of glucose similar to that observed with a nontransported glucose substitute (Me-glucose) (Furspan and Isaia, 1983).
2. In seawater-adapted fish, a continuity exists between the basal and apical sides of the chloride cell via the tubulovesicular reticulum. The plasma is virtually in contact with external medium through leaky junctions permeant to lanthanum and dextran of MW 20,000 (Sardet *et al.*, 1979; Masoni and Isaia, 1980). Only an efficient reabsorption of vital substances can explain the homeostasis of the "internal milieu."
3. The transport-associated (Na^+,K^+)-ATPase, isolated, purified, and characterized from fish gills (Sargent and Thomson, 1974), and the presence of active salt transport in chloride cells (Girard and Payan, 1977a,b, 1980) are compatible with ionic-dependent transport of glucose or amino acids in chloride cells.

4. Ultrastructural localization of D-glucose and homolog substitutes (Me-glucose, Au-S-glucose in the presence or absence of phlorizin) (Isaia *et al.*, 1984) support this model: in the presence of phlorizin, glucose is concentrated and excreted mainly by the tubulovesicular reticulum of chloride cells, the respiratory cells making no significant contribution.

IV. WATER PERMEABILITY OF THE GILLS

The problem of water exchange across the gills is of physiological importance. Indeed, water constitutes 70–80% of the internal medium of the fish, which is either hypo- or hyperosmotic to the external medium; this osmotic gradient generates water movements that are compensated by appropriate effector organs. Furthermore, the adaptation of fish to various salinities requires a large regulatory capacity of mechanisms that are basic to the functions of cells.

A. Water Diffusion through the Branchial Epithelium

1. Kinetic Studies of Diffusional Water Exchange

It has been shown *in vivo* that fish gills are the major site (more than 90%) of diffusional water exchanges (Motais *et al.*, 1969). *In vitro* preparations have confirmed this conclusion (Haywood *et al.*, 1977). It is established that water fluxes occur primarily through the respiratory cells (Isaia, 1982). *In vitro* preparations have permitted studies of the lamellar epithelium as a membrane mounted between two leucite chambers (Girard and Payan, 1977a,b). Two kinds of kinetic experiments can be performed to obtain essential information on water permeation across the gills: (1) unloading experiments following isotopic equilibrium of the lamellar compartment, in which radioactivity is followed kinetically in the external and vascular media, and (2) loading experiments, where the isotope is added to the external medium and the radioactivity followed in the vascular compartment. The equations describing the variations of the radioactivity in each compartment of such a system have been compiled by Girard and Payan (1977a,b). These equations permit the calculation of transfer coefficients and fluxes, respectively, across the two barriers separating the intermediate compartment of external and vascular media.

Two points are shown clearly from the data of loading and unloading experiments performed in *Salmo gairdneri* (Table II) (Isaia *et al.*, 1978b):

Table II

Transfer Coefficients and Unidirectional Flux of Water through the Gills of *Salmo gairdneri*[a,b]

Transfer coefficient ($\%$ min^{-1})		Unidirectional fluxes (ml min^{-1} 100 g^{-1})			
		Transepithelial		Transmembranous	
K_{21}	K_{23}	*In vivo*	*In vitro*	J_{21}	J_{23}
98 ± 7	14 ± 2	0.29 ± 0.02	0.17 ± 0.03	1.40 ± 0.37	0.18 ± 0.03

[a]According to Isaia *et al.* (1978b).
[b]Indices 1, 2, and 3 refer, respectively, to external, lamellar, and internal compartments.

1. An intermediate compartment, probably consisting of intracellular water, exists for water exchange. This compartment is limited by two barriers: the internal and external. Indeed, in the absence of an intermediate compartment, the curve showing the appearance of THO in the vascular compartment (loading experiments) should be linear from time zero, contrary to the actual findings, which are initially curvilinear. The fact that the flux across the internal barrier is comparable to the transflux measured by this technique or during *in vivo* experiments (Table II) demonstrates that the intermediate compartment may be equated to the lamellae, keeping in mind that vascular fluid collected via the dorsal aorta bathed only the lamellar epithelium of the gills.

2. Two barriers exist that water molecules must successfully pass in order to cross the epithelium. These are the external and the internal membranes. The basal barrier is about eight times less permeable and thus is the limiting factor for water diffusion as well as ionic exchanges in the gills (Girard and Payan, 1977a,b).

The size of the pool (140 μl 100 g fresh weight^{-1} (Isaia *et al.*, 1978b) that participates in transepithelial diffusional flux is only 13% of the total intracellular water content of the gills (1300 μl 100 g fresh weight^{-1}). This low percentage, however, is greater than that of the corresponding ionic pool, which represents only 0.3–2% of the intracellular ion content.

These kinetic data permit discussion concerning the role of unstirred water layers as a limiting factor in diffusional water flux. Indeed, the water diffusional permeability coefficient calculations have been questioned by Dainty and House (1966a,b). They found that vigorous stirring of the medium bathing a membrane (frog skin) produced an increase in diffusional fluxes proportional to the rate of stirring. This increase in water flux is associated with a decrease in thickness of the unstirred water layers. This thickness of unstirred layers of water has been measured on various epithelia and varies from 200 to 900 $\times 10^{-6}$ m (Parisi and Piccini, 1973; Wright and Pietras, 1974). Several theoretical and experimental considerations exclude any interference between unstirred water layers and nonelectrolyte branchial permeability determinations: (1) The thickness of the respiratory epithelium is about 5×10^{-6} m. The diffusion coefficient of water in trout gills is 3×10^{-9} $cm^2\ sec^{-1}$ (Isaia *et al.*, 1978b). The self-diffusion of HTO being 1.5×10^{-5} $cm^2\ sec^{-1}$ (Wang *et al.*, 1953), the thickness of limiting unstirred layers of water should be 2.5 cm. This is impossible, considering the lamellar blood flow and the external ventilation of the gills (Hughes, 1980). (2) The apical barrier is more permeable than the basal limiting barrier, thereby excluding limiting factors on the mucosal side. (3) Adrenaline selectively increases diffusional water flux as a function of its concentration in perfusion fluid, the effect being greater in freshwater-adapted fish (Isaia *et al.*, 1979). These arguments are also valid for nonelectrolyte permeation occurring across the secondary lamellae if $P_s < 10P_{water}$.

2. Structure of Water in Gills

Diffusional water transfer across the secondary lamellae is limited by a basal barrier. The cytoplasmic water, structured by cellular macromolecules in a number of tissues (House, 1974), should constitute the limiting factor to water diffusion taking into account the relatively small size of the transport pool. Studies of water contained in gill filaments using nmr show resonance signals of the same frequency but broader than that of tap water or seawater (Isaia and Isaia, 1978). The use of nmr experiments in biological systems was pioneered by Odeblad *et al.* (1956). The conclusion from this and following studies was that some intracellular water was relatively immobilized. Odeblad and his collaborators observed a broadening of the proton absorption signal in the nmr spectrum of cellular water relative to that in pure water. Similar broadening of the nmr spectrum of water was observed in fish muscle (Sussman and Chin, 1966). Numerous internal and external factors such as the nature of the material utilized, the homogeneity of the applied field, and the viscosity of the medium can produce an enlargement of the proton absorption signal (Emsley *et al.*, 1965). Paramagnetic impurities, such as hemoglobin-bound iron, can also produce a broadening of the absorption signal. Nevertheless, it has been shown in skeletal muscle, where red blood cells were not removed, that this factor is negligible (Hazlewood *et al.*, 1974). Finally, an enlargement of the nmr signal can be brought about by interactions of the hydrogen atoms of water with the surrounding molecules. These molecules restrict the movement of the water molecules to varying degrees and can sometimes lead to structural modifications of the water. Despite the numerous artifacts that could produce an enlargement of the resonance signal, only an important restriction of the movement of water molecules by the environmental macromolecules can produce such an enlargement (Cope, 1967; Vick *et al.*, 1973; Burke *et al.*, 1974; Hazlewood *et al.*, 1974; Raaphorst *et al.*, 1975; Fung *et al.*, 1975).

It is generally accepted that a certain fraction of cell water is structured by macromolecules, but the size of this fraction is controversial. Hazlewood's group (1974) interpret their data using a model (*a*), where the entire cellular environment is appreciably modified by the molecular environment. The resonance signal should represent the different pools of structured water. Other authors, such as Fung *et al.* (1975), propose a two-phase model (*b*) to describe the state of water in the cell. In this model, the relaxation time of hydrogen atoms of water (proportional to the full-width, half-maximum of the resonance signal) is determined by a minor pool that forms the hydration layer of macromolecules. The major part of the cellular water is not structured and exchanges rapidly with the minor fraction of structured water.

The percentage of water in gill filaments is 81.8 ± 0.66% ($n = 19$). Since the integration of the resonance signal (which represents the water detected by the spectrometer) is only 60.0 ± 2.0% ($n = 4$) of the integral of pure water made in similar experimental conditions, this percentage is significantly different ($p < .01$) than the water content of the gill filaments (Isaia and Isaia, 1978). This phenomenon can be interpreted according to model (*a*), which suggests that 60% of cell water is structured yet visible to the spectrometer, whereas the remainder is much more highly structured thereby producing a signal so broad that it is detected by high-resolution scan. The results can also be interpreted according to model (*b*), where the nmr signal represents the cellular nonstructured water but is broadened by rapid exchange with a minor fraction of water molecules in both the hydration layer of gill macromolecules and in mucus, this fraction not being detected by the spectrometer. It seems unlikely that all the filament water, including water contained in arterial and venous sinuses, is structured. Furthermore, the addition of water to the nmr tube containing gill filaments produces a shrinkage of resonance signal and an increase of the integral signal proportional to the quantity of water added (J. Isaia and A. Isaia, unpublished results). It seems that the biphasic model (*b*) more accurately describes the state in the gill epithelium in which a fraction of the water (25% minimum) is structured by environmental macromolecules.

The use of deuterated water (undetectable by the spectrometer) has shown that cellular water exchange occurs at the same rate as isotopic branchial exchange (Isaia and Isaia, 1978). The destruction of cellular membranes increases the isotopic exchange rate and the exchange rate of structured water, without changing the width of the resonance signal. These experiments have demonstrated that structured cellular water cannot constitute a limiting factor for the diffusional process. Thus, it is concluded that a basal barrier is the major resistance to water permeation.

3. Activation Energy of Water Transport

One way to investigate the mechanisms of diffusion is to examine the temperature dependence of this process. Such data are expressed in form of an Arrhenius plot where the diffusional coefficient is plotted against the reciprocal of absolute temperature. When the slope of the graph is linear one can say that a single process is occurring and that it is characterized by an apparent activation energy, E_a (kcal mol^{-1}). The temperature dependence of water permeability also can be expressed by means of the well-known Q_{10}; at 20°C, Q_{10} is approximately equal to $\exp(0.056\ E_a)$ (House, 1974). Numerous studies on artificial and natural membranes have shown that E_a often is higher than the self-diffusion coefficient of water (see review by House, 1974), that is, 4.5 kcal mol^{-1} (Wang *et al.*, 1953), suggesting diffusion in a highly structured molecular pathway. For example, the study of Gary-Bobo and Solomon (1971) on cellulose acetate membranes has demonstrated that high E_a values for water diffusion (and filtration) occur in nonporous membranes that have relatively low water permeabilities, and where diffusion seems to be the only mechanism of water transport. According to Stein (1967), E_a for water permeation across lipid membranes should be considerably larger than 4.5 kcal mol^{-1}. He considers that E_a should lie between 13.5 and 18 kcal mol^{-1}. Indeed Price and Thompson (1969) have found that E_a lies in the range 12.7–13.1 kcal mol^{-1} for permeation through a homogeneous lipid barrier.

The values of E_a obtained on various species of fish are listed in Table III. These values are significantly higher than 4.5 kcal mol^{-1}. Some of these coefficients have been obtained *in vivo* from the variation of the water turnover rate of the whole animal as a function of temperature. However, results presented by Loretz (1979) challenge the calculation of permeability coefficients derived from turnover experiments. The major argument is that hypoxia (50% that of normoxia) produces abrupt increases in the rate of branchial water turnover (+56%). He utilized small *Carassius auratus* (1–4 g) in which the branchial water clearance can be estimated at about 100%. Under these experimental conditions, where the behavior of the fish appeared normal, although with the qualification that ventilation rate was increased and that increased lamellar perfusion occurs in the trout during hypoxia (Holeton and Randall, 1967; Booth, 1979), the results emphasize

Table III

Activation Energies (E_a) for Water Diffusion in Various Fish Species

Species	E_a (kcal mol^{-1})	References
Carassius auratus	13.5	Evans (1969)
Phoxinus phoxinus	10.2	Evans (1969)
Platichthys platessa	10.6	Evans (1969)
Anguilla anguilla		
FW	21.5	Motais and Isaia (1972a)
SW	17.7	Motais and Isaia (1972a)
Serranus scriba, Serranus cabrilla	11.3	Isaia (1972)
Carassius auratus	15.6	Isaia (1972)
Carassius auratus	16.2	Loretz (1979)
Salmo gairdneri		
FW	8.4	Isaia (1979)
SW	9.0	Isaia (1979)

that changing surface area and blood perfusion patterns can be major factors influencing the observed fluxes.

These results, however, can be interpreted differently. When a fish is stressed, its plasma catecholamine concentration varies (Nakano and Tomlinson, 1967); in addition, *in vivo* studies have shown that intraperitoneal injections of adrenaline in small *Mugil capito* (5–41 g) increase the diffusional and osmotic water fluxes, as well as branchial cyclic AMP content (Pic *et al.*, 1974). Since it has been shown *in vitro* that adrenaline (10^{-6} *M*) increases the diffusional water permeability by about 100% in freshwater trout independently of hemodynamic effects (Haywood *et al.*, 1977) or other limiting factors such as unstirred layers of water (Isaia *et al.*, 1978b), the increased turnover rate of water observed by Loretz (1979) may be a consequence of hypoxic stress. Indeed various results have indicated that nonelectrolyte permeability measurements obtained with fish weighing 100–200 g are not affected by cardiac output, unstirred layers, or variations in gill area consecutive to lamellar recruitment or swelling of secondary lamellae (Haywood *et al.*, 1977; Isaia *et al.*, 1978a,b; Isaia, 1979). Meanwhile, if the results obtained in the isolated gill arches under adrenaline action may be interpreted purely as changes in permeability (Haywood *et al.*, 1977; Isaia, 1979), the increased fluxes in the perfused head represent a combination of altered permeability and altered surface area (Haywood *et al.*, 1977). The high values of E_a for water diffusion through the gills can be interpreted by permeation in highly structured media such as crystalline water (the diffusion of water molecule through an ice lattice is characterized by an E_a of 13.5 kcal mol^{-1} (Dengel and Riehl, 1963) or hydrogen bond donor lipids. Keeping in mind the following experimental results—that is, that (1) water diffusion is limited by the basal membrane of respiratory cells, (2) this membrane is one of the less permeable to small hydrophilic molecules, (3) this membrane is not selective, and (4) water diffusion occurs in a highly structured medium—one can conclude that water diffusion proceeds through hydrophilic lipid plasmalemma, a conclusion reached in Section III,B.

B. Osmotic Permeability of the Gills

1. Osmotic Net Flux Measurement

Until now, it has not been possible to measure directly and accurately the osmotic net flux occurring in the whole animal. The majority of results have been obtained indirectly. The algebraic sum of drinking rate and urine flow is taken as a measure of the osmotic net flux. These experiments, however, are open to criticism, because the osmotic net flow is determined indirectly from two independent series of measurements (urine flow and drinking rate), while the fish are assumed to be in a steady state with respect to water balance. Some of these measurements have been made on handled fish, which may result in weight loss in saltwater fish and weight gain in freshwater fish (Stevens, 1972). These factors may lead to an underestimate of the net osmotic flow. The amount of water swallowed also is assumed to be completely absorbed. While this is probably true in freshwater fish, it is not the case in saltwater fish, in which only 60–80% of the water is absorbed (Smith, 1930; Hickman, 1968; Oide and Utida, 1968; Shehadeh and Gordon, 1969; Pickering and Morris, 1970). The urine flow also may be overestimated, as catheterization may prevent the bladder from completing its absorptive function.

Meanwhile, *in vivo* techniques creating nonstressful experimental conditions over long time periods have been used on silver eels to study the kinetics of water and ion exchange (Kirsch, 1972a,b; Kirsch *et al.*, 1975). Besides the *in vivo* techniques, the net flux can be measured directly by weighing excised and ligatured gills. This last method can be used only for a short time because the tissue is not perfused by blood and the ionic content of the isolated gill varies in a range between 20 and 50% of the total ionic content after 30 min of incubation. These variations produce a decrease in the osmotic gradient and an attenuation of osmotic net flux (Kamiya, 1967; Isaia and Hirano, 1975). This technique nevertheless has the double advantage of precision and lack of stress.

Table IV compares the results obtained with the different methods.

Table IV

Comparison of Osmotic Water Fluxes Measured by *in Vivo* and *in Vitro* Techniques

Species	Adaptation[a]	Temperature (°C)	Osmotic net flux (μl hr^{-1} 100 g^{-1})	References
In vivo				
Anguilla rostrata	SW	20	−236	Smith (1930)
Anguilla anguilla	FW	12	+68	Gaitskell and Chester-Jones (1971)
Anguilla anguilla	FW	15	+221	Motais and Isaia (1972a)
	SW	15	−105	
Anguilla anguilla	FW	12	+29	Kirsch *et al.* (1975)
	SW	12	−50	
Salmo gairdneri	SW	20	−259	Oide and Utida (1968)
Anguilla japonica	SW	20	−246	
Salmo trutta	FW	15	+336	Oduleye (1975)
In vitro				
Anguilla japonica	FW	20	+168	Ogawa (1974)
	SW	20	−236	
Anguilla dieffenbachi	FW	17	+27	Suttleworth and Freeman (1974)
	SW	17	−53	
Anguilla anguilla	FW	15	+444	Isaia and Hirano (1975)
	SW	15	−48	
Salmo gairdneri	FW	15	+178	Isaia *et al.* (1979)
	SW	15	−489	
Platichthys flesus	FW	10	+185	T. J. Shuttleworth (personal communication)
	SW	10	−249	

[a]FW, freshwater; SW, seawater.

There are large variations with the technique used for the same species of fish, up to a factor of about eight. These variations cannot be explained by the difference in size of animals or the temperature of experiments. Thus, one cannot discuss absolute values; only relative modifications of osmotic permeability may be discussed, using the same species of fish and the same experimental protocol.

2. Ionic Dependence of the Osmotic Pressure Gradient

The driving force of osmotic water flux is the "real" osmotic pressure gradient existing through the gill epithelium. The theoretical osmotic gradient is equal to the difference between the osmotic pressure of plasma and the external medium; its value is about 0.3 and 0.8 Osm liter^{-1} in fish living in fresh water and seawater, respectively.

In freshwater fish one can assume that theoretical and "real" osmotic gradients are identical, because in this medium osmotic water flux is high compared to ionic fluxes. For example, in *Salmo gairdneri*, the sum of sodium and chloride influxes is about 50 μM hr^{-1} 100 g^{-1} (Payan, 1978; Payan *et al.*, 1978), whereas the net water influx is about 10^4 μM hr^{-1} 100 g^{-1} (Isaia *et al.*, 1979). In other words, 200 water molecules enter when 1 ion each of Na^+ and Cl^- is absorbed. Even if there are common exchange sites between ions and water, the degree of interaction should be weak considering that water and solute movements occur in the same direction.

In contrast, in seawater the gills are the site of large sodium and chloride exchanges (Mullins, 1950; Motais, 1961, 1967; Motais *et al.*, 1966; Potts and Evans, 1967). Unidirectional fluxes are 10-fold higher than net excretion of salt through the branchial epithelium, and 25–60% of the total salt content of the fish is exchanged every hour (Maetz, 1974). Thus, in the eel the influx of sodium or chloride is about 1 m mol hr^{-1} 100 g^{-1} (Motais, 1967; Motais and Isaia, 1972b), compared to 2.5 mmol hr^{-1} 100 g^{-1} for water (Isaia and Hirano, 1975). Thus, the influx of ions and the efflux of water are approximately equal in seawater. If the interactions between water and ionic movements exist, they may significantly modify water flux. In fact, the monovalent ions are responsible for the osmotic gradient. Replacement of ions by an equimolar solution of a less permeant solute (e.g., mannitol), is followed by physiological and morphological disturbances. Thus, when a seawater-adapted eel is transferred to a 1 Osm liter^{-1} solution of mannitol, its survival time does not exceed 2 hr. In this medium the hydromineral equilibrium is altered, especially osmotic water fluxes, which are increased by 100%. The ultrastructure of chloride cells is affected (Masoni and Isaia, 1973): a great number of mitochondria are destroyed; a swelling and a subsidence toward the apical side of the cell is observed, the erythrocytes have an echinocyte conformation, a sign of intense dehydration. The respiratory cells are apparently not affected. These modifications are the consequence of the substitution of saltwater ions by an isoosmotic mannitol solution. This sugar is much less permeant than ions and increases the hypertonicity of the external medium. The selective histolysis of chloride cells is not the result of a variation in salt concentration, because transfer to fresh water produces a swelling of the tubular reticulum only after a considerable period of time. It is worth noting that (*a*) any large osmotic flux variation (serosal–mucosal direction in the course of the transfer of a saltwater fish to a mannitol solution, reverse direction in the course of the transfer of a saltwater fish to fresh water) is followed by an increase in the diameter of the tubules of the ionocyte reticulum, and (*b*) that an increase in osmotic flux is accompanied by a decrease in ionic exchange (Masoni and Isaia, 1973; Isaia and Masoni, 1976). There are strong arguments for the localization of ionic pumps in the tubular reticulum of ionocytes (Shirai, 1972; Masoni and Garcia-Romeu, 1973; Sargent *et al.*, 1975; Karnaky *et al.*, 1976; Thomson and Sargent, 1977; Sardet *et al.*, 1979; Philpott, 1980; Girard and Payan, 1980).

Interference between ionic and osmotic water movements may occur in one of two ways: (1) as the result of a common transfer site, so that the hypertonicity produced by mannitol is followed by a partial destruction of ionocytes and a simultaneous increase of water movement and a decrease in ionic exchange, or (2) because of the existence of a connection between ionic and water movements, the former reducing the latter by an attenuation of the osmotic gradient. The fact that osmotic permeability is less than diffusional permeability in several experimental conditions (see Section IV,C) suggests an association of this type.

In conclusion, one might speculate that ionic movements cancel the osmotic gradient in part, thereby preventing dehydration in marine fish.

3. Rectification Mechanism of Osmotic Permeability

The variations of osmotic permeability (P_{os}) of the gills of euryhaline teleosts, as a function of the salinity of the ambient medium, present notable differences if they are studied after an acute transfer or following the adaptation of a given salinity. Although a rapid change in external salinity has no effect on diffusional permeability (P_{dif}), the instantaneous transfer of saltwater fish to fresh water or vice versa is accompanied by acute changes in P_{os}.

The intensity of net fluxes measured in the serosal–mucosal direction (negative net flux) is less than those measured in the reverse direction (positive net flux) (Isaia and Hirano, 1975; Isaia *et al.*, 1979). This phenomenon, called rectification, was observed on various epithelia including amphibian bladder (Bentley, 1961, 1964; Earley *et al.*, 1962), amphibian intestine (Loeschke *et al.*, 1970), fish intestine (Skadhauge, 1969), insect rectal cuticle (Phillips and Beaumont, 1971), and erythrocyte membrane (Farmer and Macey, 1970). Concerning the gills of euryhaline fish, this may be explained by the superimposition of three effects:

1. One effect is related to the different degrees of semipermeability of the gill epithelium (see preceding paragraph). The great porosity of saltwater fish gills (at level of the leaky junctional complex of chloride cells) compared to freshwater fish gills explains the great influx of water when saltwater fish enter fresh water.

2. An effect on permeability is mediated by divalent ions (Potts and

Fleming, 1971) as follows: in seawater, the calcium concentration is 10-fold higher than in fresh water (Wiggins, 1971) and helps to maintain the cohesion of intercellular "cements." This may explain why a reduction in P_{os} is observed after a transfer from fresh to seawater and the opposite effect following the transfer from seawater to fresh water.

3. A mechanical effect on permeability is linked to the structure of the tubulovesicular reticulum of chloride cells. Such an effect is illustrated by Diamond (1966) in experiments on rabbit gallbladder. The results obtained on this epithelium indicate the presence of aqueous channels in the cell membrane that behave as osmometers, shrinking in concentrated solutions and thereby increasing membrane resistance to water flow. In contrast, one can expect a swelling of a channel in dilute solutions. Indeed, the tubules of ionocytes have been shown to swell in the course of transfer from seawater to fresh water (Masoni and Isaia, 1973). This phenomenon is of physiological interest when considering euryhaline fishes such as the eel (catadromous migration for reproduction), salmon or flounder (anadromous migration for reproduction), *Mugil* (frequent changes of salinity when searching for food in estuaries that encounter changes in salinity. Occurring instantaneously, the rectification attenuates the stress caused by dehydration in seawater.

C. Comparison of Osmotic and Diffusional Permeability

Values of P_{os}, P_{dif}, and the ratio P_{os}/P_{dif} are listed in Table V. The ratio P_{os}/P_{dif} has been calculated assuming on the one hand that the reflection coefficient of salt is near 1 (gills acting as a semipermeable membrane) and on the other hand that osmotic and diffusional fluxes take place in the same pathway.

The ratio P_{os}/P_{dif} is greater than 1 in freshwater fish. The difference between the two coefficients can be explained by physical considerations. Diffusion presents a maximal interaction between molecule and membrane and obeys Fick's law. In contrast, osmosis takes place preferably by way of water-filled channels and bulk flow, and obeys fluid mechanisms (Poiseuille's law). Koefoed-Johnsen and Ussing (1953) have postulated the existence of transmembrane aqueous channels in frog skin on the basis of such considerations. This hypothesis is supported by the effect of neurohypophyseal hormones, which greatly increase P_{os} without affecting P_{dif}, and by solvent-drag experiments. This hypothesis has been used to explain the discrepancy between osmotic and diffusional fluxes in various epithelia (rabbit gallbladder, Diamond, 1966; frog skin, Parisi and Piccini, 1973; rabbit renal tubule, Al-

Table V

Comparison of Osmotic (P_{os}) and Diffusional Permeability (P_{dif}) Coefficients on Various Species of Fish

Species	Medium[a]	Permeability coefficients (cm sec^{-1} × 10^6)			References
		P_{os}	P_{dif}	P_{os}/P_{dif}	
Teleost					
Xisphister atropurpureus	SW/10	—	—	1.6	Evans (1967a,b)
	SW	—	—	0.15	
Serranus scriba	SW	16	12	1.3	Isaia (1972)
Carassius auratus	FW	—	—	3.3	Lahlou and Giordan (1970)
Salmo salar	SW	—	—	1.5	Potts *et al.* (1970)
Plactichthys flesus	SW	14	17	0.8	Motais *et al.* (1969)
	FW	70	27	2.6	
Platichthys flesus	SW	—	—	1.1	T. J. Shuttleworth (personal communication)
	FW	—	—	2.6	
Anguilla anguilla	SW	6.4	18	0.3	Motais and Isaia (1972a)
	FW	38	23	2.1	
Salmo gairdneri	SW	30	4.4	7	Isaia *et al.* (1979)
	FW	25	7.7	3	
Elasmobranch					
Scyliorhinus canicula	SW	8.1	0.56	14.5	Payan and Maetz (1971)
Raja erinacea *Raja eglanteria*	SW	0.84	0.21	4.0	Payan *et al.* (1973)

[a]FW, fresh water; SW, seawater; SW/10, seawater diluted 10-fold.

Zahid *et al.*, 1977). Although the presence of transmembrane aqueous channels has been challenged by Dainty and House (1966a,b), the unstirred water layers cannot explain the variation of the ratio P_{os}/P_{dif} (Parisi and Piccini, 1973). Indeed, it has been shown (Bourguet *et al.*, 1976; Parisi *et al.*, 1979a,b; Chevalier *et al.*, 1979) that formation of transmembrane channels can be induced by hormonal action and by an assembling of aggregates of intramembrane particles. It can be assumed that water-filled channels extend across the branchial epithelium of freshwater fish.

In saltwater fish, the measured values of P_{os}/P_{dif} are near 1 and often are less than 1, which makes no sense physically (0.15 at 5°C for the eel, Motais and Isaia, 1972a; 0.15 for *Xiphister*, Evans, 1967a). Indeed, the theoretical considerations might lead one to expect a value much greater than unity. In these experimental conditions, the osmotic gradient may be underestimated, and consequently P_{os} and P_{os}/P_{dif} are underestimated. It is impossible to determine the reflection coefficient of ions; the biggest molecule used (dextran, MW 20,000) crosses the gills, and it is probable that higher molecular weight compounds are also permeant. The fact that P_{os} is less than P_{dif} in seawater probably is attributable to an underestimation of the "real" osmotic gradient, the theoretical osmotic gradient being attenuated by a high salt permeability. It is suggested that net movements of sodium, chloride, and water take place in the same pathway. Morphological studies have demonstrated the presence of 500-Å diameter tubules (Sardet *et al.*, 1979; Masoni and Isaia, 1980) that cross the chloride cell and that may be involved in water transfer by bulk flow.

These conclusions have been reached assuming a common pathway for osmotic and diffusional water fluxes. In reality, various experimental results suggest that osmotic phenomena take place in chloride cells while diffusional processes occur in respiratory cells:

1. There is a tight degree of interaction between ionic and osmotic exchanges when isotonic mannitol solution replaces seawater, while the diffusional exchanges of water remain the same (Masoni and Isaia, 1973).
2. The reduction of divalent ions in seawater is followed by a different increase in diffusional (+33%) and osmotic (+80%) fluxes, and an inhibition of ionic "pumps" (Isaia and Masoni, 1976).
3. An acute change in ambient salinity is followed by a rectification of P_{os} without a change in P_{dif} (Isaia and Hirano, 1975; Isaia *et al.*, 1979).
4. Variations of net water fluxes are accompanied by structural variations in the tubular reticulum of chloride cells without changes in intercellular spaces or morphological aspects of respiratory cells, sites essential for diffusion (Masoni and Isaia, 1973; Isaia and Masoni, 1976; Isaia, 1982).
5. The comparative modification of water fluxes as a function of temperature changes show a different response of P_{os} and P_{dif} (Motais and Isaia, 1972a).

These morphological and physiological data are only consistent with some separation of pathways for water fluxes. In fact, P_{os}/P_{dif} is greatly underestimated if one takes into account that the external exchange surface of respiratory cells is about 24-fold larger than that of chloride cells. In reality, P_{os} may be $\gg P_{dif}$, which is compatible with the lower permeability of the secondary lamellae and the great porosity of chloride cells and associated structures in marine fish.

D. Model to Explain the Osmoregulatory Mechanism of the Gill of Seawater-Adapted Euryhaline Fishes

In the course of seawater adaptation osmoregulatory parameters apparently evolve in opposite ways. When salinity increases, the number of chloride cells increases and the intercellular junctions between them open as evidenced by the permeation of La^{3+} (Sardet *et al.*, 1979). This "hyperosmotic response" is accompanied by hypertrophy of the chloride cells and amplification of the tubular reticulum (Philpott, 1980). At the same time, one observes variations in salt and water exchanges. Exchange of Na^+ and Cl^- across the gills increases so that 25–60% of the total salt content of the fish is exchanged per hour in seawater (Maetz, 1974) and the branchial diffusional water permeability and osmotic water permeability are also reduced in seawater-adapted animals (see preceding paragraphs). The osmotic water permeability decreases despite the fact that the osmotic gradient between internal and external milieus is threefold more important in seawater than in fresh water and the intercellular junctions become more permeable.

It remains to be explained how the reduction of net water flux is obtained in the presence of changes in structure and function of the intercellular junctions which should result in an increase in permeability. Only one model has been proposed by Motais and Garcia-Romeu (1972) to explain a reduction of osmotic permeability in chloride cells. This model involves recycling ions in the tubular reticulum. The morphological basis for this model has been brought into question by the observation that the tubular reticulum communicates with the intercellular spaces and is thus potentially open to the external medium.

A new model is proposed that integrates the new morphological and physiological data. Schematically, a tubule is represented vertically open to the external medium (Fig. 5). The Na^+ and Cl^- ions of the external medium diffuse toward the blood along their concentration gradient. In contrast, water moves in the opposite direction along its osmotic gradient. The decrease and even the cancellation of the effects of the osmotic gradient between plasma and seawater are obtained by the increased transcellular ion permeability of the chloride cells (aperture of cellular junctions). The increase of the ionic permeability of the cell produces a decrease in its semipermeability, that is, the decrease of the relfection coefficient of the membranes toward these ions. The expected accumulation of ions in the plasma, however, is not observed. In consequence, Na^+ and Cl^- ions entering the intercellular spaces could be exchanged across the lateral membranes for intracellular monovalent ions. A high concentration (e.g., K^+ and B^-) in the intercellular spaces (Fig. 5) would then permit the passive diffusion of the

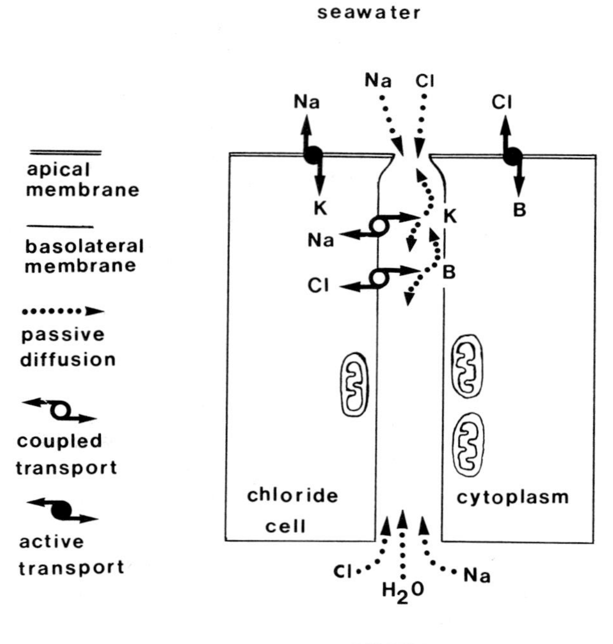

Fig. 5. Schematic representation of the attenuation of an osmotic pressure gradient established across the chloride cells by monovalent ions. A tubule of a chloride cell is shown vertically open to the external medium. The aperture of the junction provokes the decrease of semipermeability of the junctional complex. The apical and basolateral membranes are impermeable to water and passive ionic fluxes. Sodium and chloride of seawater enter the tubule along the chemical gradient through the leaky junction between chloride cells. They are exchanged against counterions that can diffuse toward seawater and plasma. This reverses the ionic movement and avoids the inrush of external Na^+ and Cl^- into the plasma. The portion of counterions that diffuses towards the plasma could be exchanged against Na^+ and Cl^- at the basal part of tubule in such a way that the final Na^+ and Cl^- concentrations in the tubule are the same as plasmatic concentrations. Cytoplasmic K^+ against tubular Na^+ and cytoplasmic B against tubular Cl^- exchanges can function with plasma Na^+ and Cl^- ions.

exchanged ions into seawater. The part of potassium and base that diffuse toward the plasma could be exchanged by means of Na^+_{ext}/K^+_{int} and $Cl^-_{ext}/$ $base_{int}$ at the basal part of the tubular reticulum. For example, the concentrations of K^+ in seawater and blood are 10 and 4 m*M*. It is sufficient that tubular K^+ concentration exceeds 10 m*M* for this ion to diffuse into seawater. The same arguments can be applied to Cl^- ions which could be exchanged against a base, such as HCO_3^-. Near the apical sides, ions that have diffused in the tubular reticulum can be absorbed toward the cytoplasm. The resulting decrease in sodium and chloride concentration can create a diffusion chemical gradient along the tubule. This permits a net extrusion of plasma ions to compensate for their intestinal absorption (Hickman, 1968). This model requires that cellular membranes (apical and basolateral) show reduced or nonexistent passive exchanges of ions and water.

This scheme, although disputable, explains how an epithelium possessing junctions permeable to substances of high molecular weight (Lam, 1969; Kirschner, 1980; Masoni and Isaia, 1980; Isaia, 1982) prevents a disequilibrium of the hydromineral balance and extrudes plasma salts against their electrochemical gradient. Concerning sodium, several laboratories have approached the question of its distribution between the internal and external media on the grounds that it results passively from the transepithelial membrane potential (see review by Evans, 1980). The two following arguments suggest that an active transport of sodium exists in seawater-adapted gills: (1) In *Salmo gairdneri*, for example, the measured potential (10 mV) cannot explain the theoretical sodium equilibrium potential (23 mV) (Kirschner *et al.*, 1974). (2) Marine teleosts ingest seawater. Monovalent ions (Na^+ and Cl^-) absorbed across the intestine (Skadhauge, 1969, 1974) are extruded independently from each other via the gills (see review by Maetz, 1974).

Various experimental results support this model. Na_{int}/K_{ext} exchange at the apex of chloride cells could explain competition between external potassium and ouabain observed when the glycoside is present in the external medium (Motais and Isaia, 1972b). The ionic exchanges at the apex of chloride cells account for net extrusion of monovalent ions. Actually, Girard and Payan (1980) have demonstrated that the filamental epithelium is the site of net excretion of sodium and chloride and recently, Foskett and Scheffey (1982) using the vibrating-probe technique, have established definitively that chloride cells are the sites of active chloride secretion and high ionic permeability. This scheme can explain the variations observed in gill ion flux and osmotic permeability following rapid transfer experiments. For example, when a seawater-adapted fish is transfered to fresh water, there is an instantaneous increase in net inward water flux (Masoni and Isaia, 1973; Isaia and Hirano, 1975) and a passive loss of plasma salt (Maetz, 1971) owing to the presence of leaky junctions. After adaptation to fresh water, the filamental epithelium becomes one of the most impermeant structures ever found (Isaia *et al.*, 1978a). Net water influx is reduced (Isaia and Hirano, 1975), while the leaky loss of ions is suppressed (Girard and Payan, 1977b). In the reverse transfer, from fresh water to seawater, the observed reduction of osmotic permeability takes place following the opening of intercellular junctions and development of ionic exchanges (Kirsch and Mayer-Gostan, 1973).

This model is open to criticism because (1) it supposes the same exchange mechanisms operating in opposite directions at the apical and basolateral membranes of chloride cells (2). Furthermore, while (Na^+,K^+)-ATPase activity has not been revealed in apical membranes, considerable body of evidence shows the localization of this enzyme in the tubular reticulum (Karnacky *et al.*, 1976; Sargent *et al.*, 1980; Philpott, 1980). The exchange of cytoplasmic sodium against tubular potassium could permit a recycling of the potassium ions that diffuse towards the plasma. However, if these exchanges only are taken into account, they cannot explain the maintenance of hydromineral balance in seawater fish. (3) The recycling of monovalent ions necessitates an important consumption of energy for active transport. Recent investigations (Furspan and Isaia, 1983) have shown in perfused heads of *Salmo gairdneri* that glucose breakdown at normal glycemia is about 35 μmol hr^{-1} 100 g^{-1}. The glucose consumed in the arterial circulation (lamellar epithelium) represents 5% of this figure against 95% of the venous circulation (filament epithelium). This results reveals a high metabolism in chloride cells.

The separation of apical and basolateral membranes of chloride cells and the determination of their enzymatic activity could provide decisive information in regard to this problem.

V. ADRENERGIC CONTROL OF NONELECTROLYTE PERMEABILITY

A. Separation of the Hemodynamic and Permeability Effects of Adrenaline

Adrenaline produces important vascular effects. Numerous authors have shown that this hormone provokes a decrease in the resistance of gill vessels to blood flow (Keys and Bateman, 1932; Richards and Fromm, 1969; Wood, 1974). Two hypotheses have been suggested to explain this effect. Steen and Kruysse (1964) proposed that adrenaline causes a vasodilation of the gill secondary lamellae, whereas Hughes (1972) postulated the existence of a mechanism of recruitment that would increase the number of lamellae perfused in the presence of the catecholamine. These two hypotheses describe the effect of adrenaline in terms of an increase in the functional exchange surface between the gill and the external medium (Bergman *et al.*, 1974). In the perfused head of trout (constant pressure perfusion), the addition of adrenaline (10^{-6} *M*) causes an increase of about 100% in the urea and the water effluxes, and an increase of about 30% in the perfusion flow (Haywood *et al.*, 1977).

Although the perfusion flow should be 5- to 10-fold larger than the water and urea branchial clearance and cannot constitute a limiting factor, this kind of experiment does not rule out the possibility of interaction between perfusion flow and efflux variations. Nevertheless, a drastic mechanical reduction of blood by 80% (from 130 to 26.6 ml hr^{-1} 100 g^{-1}) only reduces water and urea effluxes by 45 and 37%, respectively (for water, from 26.9 to 14.8 ml hr^{-1} 100 g^{-1}). We can consider that perfusion flow becomes a limiting factor when it represents less than twofold the branchial clearance of product. Indeed, results obtained on isolated gill arch experiments support the conclusion that there is a direct effect of the catecholamine on membrane permeability (Haywood *et al.*, 1977). Further experiments showing the selective role of adrenaline on nonelectrolyte permeability according to the physicochemical properties of the molecules and the ambient salinity of the adaptation medium (see Section V,B) have supported this hypothesis.

B. Adrenergic Effects on Nonelectrolyte Permeability

As shown in Fig. 6, adrenaline increases branchial permeability to small molecules (oxygen, butanol, ethanol, water, and urea) without affecting the branchial permeability to mannitol and dextran. The effect of adrenaline on ethyl acetate (>25 ± 8%, $n = 6$), although significant, is not seen in freshwater trout, because for this substance the rate of diffusion is high and the blood clearance (59 ± 6%, $n = 6$ in the absence of adrenaline; Isaia, 1981) shows that perfusion flow becomes a limiting factor (see Section V,A). The branchial permeability to ethyl acetate and the effect of adrenaline are certainly underestimated. The increase of P_s by hormonal action is bidirectional (Isaia, 1981) and is a function of (*a*) the chemical nature of the diffusing molecule and (*b*) the salinity of the ambient adaptative medium. The adrenaline effect is more marked in the gills of freshwater fish than in saltwater fish. Furthermore, the increase in P_s in freshwater fish is related to the lipid solubility of the molecule (Isaia *et al.*, 1978a). This phenomenon has been confirmed on isolated gills in experiments involving tritiated and ^{14}C isotopes together (for example, 10^{-6} *M* adrenaline increases $P_{butanol}$ and P_{water} by 160 and 100%, respectively, the difference being significant; +60 ± 18%, $n = 47$; $p < .001$, Isaia, 1979). Furthermore, the transfer coefficient of oxygen is increased by about 400% (Randall *et al.*, 1967) in trout during moderate swimming, which is indicative of a much greater catecholamine action on highly liposoluble molecules. This hormone acts primarily on the more selective epithelium (gills of freshwater fish) and preferentially on permeability to lipophilic molecules. This mode of action is in part similar to that described by Svelto *et al.* (1977) on frog skin; in this epithelium, nor-

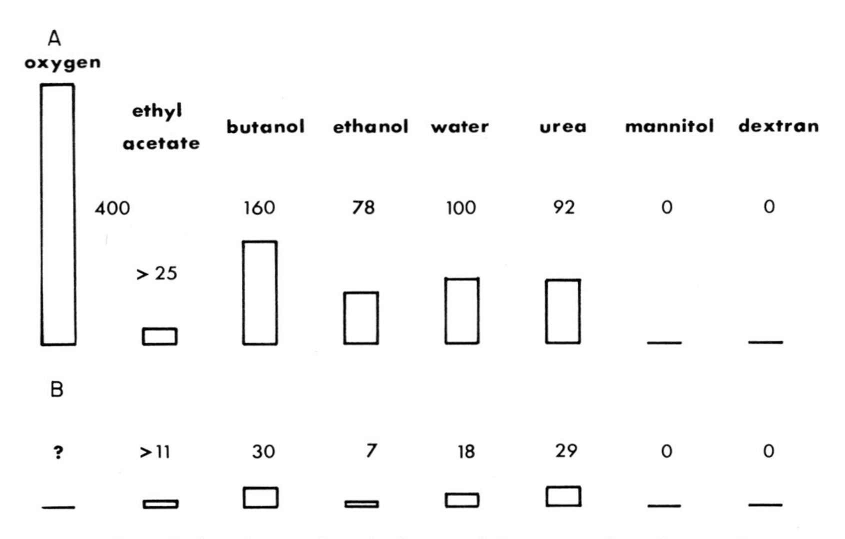

Fig. 6. Effect of adrenaline on branchial permeability to nonelectrolytes. Values are expressed as the percentage of the control efflux. Note that there is no effect on sugars (mannitol and dextran). There is a marked effect in freshwater trout (A), contrary to a small effect in saltwater trout (B). The action of adrenaline is a function of the liposolubility of the nonelectrolyte in freshwater fish. The low percentage obtained for ethyl acetate flux is an underestimate, because the branchial clearance becomes a limiting factor for this highly permeant molecule. (See Randall *et al.*, 1967; Isaia, 1979, 1981.)

adrenaline produces a double stimulation: (1) an increase in bidirectional permeability to the solute independent of its chemical nature and (2) an increase in excretion of the nonelectrolyte related to the liposolubility of the compound.

It seems logical to interpret the effect of adrenaline by an action on the lipid phase of the plasmalemma of respiratory cells, resulting in a facilitation of molecular transfer. This facilitation can occur by an effect on the structure (decrease in enthalpy of activation, $\Delta H\dagger$) or on the fluidity (increase in entropy of activation, $\Delta S\dagger$) of the plasmalemma. The first mechanism is illustrated by the work of Bourguet *et al.* (1976) and Parisi *et al.* (1979b) with respect of the effect of antidiuretic hormone (ADH) on water permeability of frog urinary bladder. This hormone induces the formation of intramembrane particle aggregates. These aggregates constitute specific channels that become operational in an intermittent manner. In the absence of ADH, water diffusion occurs at lower rates than in the presence of hormone. The second mechanism involves increase in the disorder of the lipid membrane, which produces a facilitation in molecular transfer. A third hypothesis was proposed by Hays *et al.* (1971) to explain the effect of neurohypophyseal hormones on water permeability through the amphibian urinary bladder and urinary bladder and was used again by Al-Zahid *et al.* (1977) to explain the effect of ADH on rabbit renal collecting tubules. According to these authors, the hormone may induce an increase in the number of transfer sites without changing their characteristics.

The study of P_s variations as a function of temperature for various nonelectrolytes has shown that adrenaline action is followed by a decrease in $\Delta S^\dagger$ in the case of substances whose permeability is under adrenergic control (butanol and water in freshwater trout) without any change in $\Delta H^\dagger$ (Isaia, 1979). These results argue in favor of an effect of the catecholamine on lipid mobility of plasma membrane. It is of special interest to consider the effects of adrenaline in fish water balance. This hormone has different effects on osmotic and diffusional permeabilities. Concerning osmotic permeability, adrenaline abolishes the rectification phenomena (Isaia *et al.*, 1979), while it increases water diffusional permeability. These conjugate effects may explain the rise in the plasma osmolarity (Pickford *et al.*, 1971; Pic, 1972) and the body weight loss (Pic *et al.*, 1974) observed following the injection of adrenaline.

This double effect of adrenaline on water fluxes reflects a disequilibrium of the water balance, greater in freshwater fish. This disequilibrium may only be secondary to the necessary increased permeabilization of the membrane to gases, which permits a better oxygenation of the blood when fish are stressed.

C. Effects of Drugs

The hemodynamic effects of catecholamines are well known. Their agonistic effects occur via hormonal sites of fixation; for adrenaline the α and β receptors perform this function (Ahlquist, 1948, 1966, 1968, 1973). These receptors are revealed by using pharmacological substances that compete with adrenaline for the fixation sites. For example, propranolol and phentolamine are β- and α-receptor-blocking agents, respectively. Figure 7 shows the separate or simultaneous effect of the presence of these blocking agents in the perfusion medium. In accordance with the findings of Payan and Girard (1977), these results show that blockade of the β-adrenergic sites results in decreased vascular flow under adrenaline stimulation, whereas the converse is apparent when the α-adrenergic sites are blocked in the presence of adrenaline. These hemodynamic changes are associated with a decrease and an increase in effluxes of water and urea, respectively. When only propranolol is used, it produces a nonsignificant decrease in water and urea fluxes that can be related to a drastic reduction of flow. On the contrary,

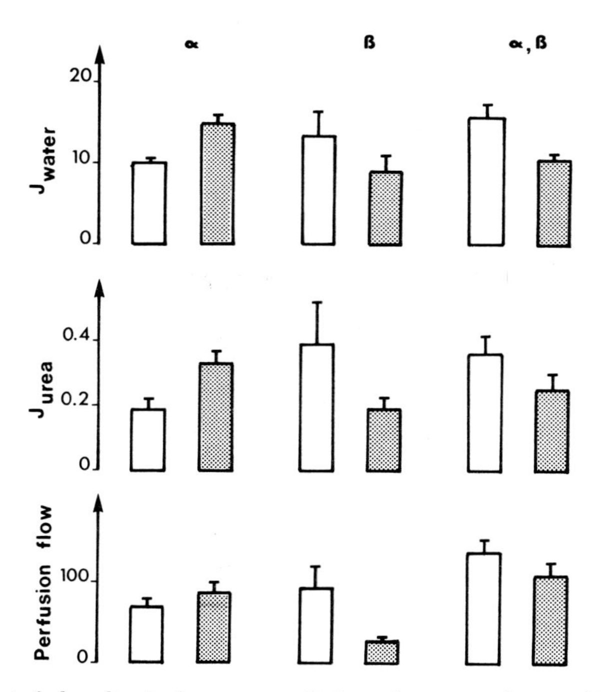

Fig. 7. Effect of adrenaline in the presence of independent or simultaneous blocking agents on water and urea fluxes and perfusion flow. Open columns; control values; dotted columns, preparation under adrenaline action. In presence of α blocker (β-site stimulated), there is small increase in perfusion flow and a rise in permeability to water and urea. When the β site is blocked the contrary is observed. With the simultaneous presence of both blockers the perfusion flow does not fall significantly, while the branchial permeability is decreased. (Data from Haywood *et al.*, 1977.)

phentolamine produces a significant increase in the urea and water fluxes. The simultaneous use of the blocking agents cancels the adrenaline effect, showing the role played by β receptors. It also has been shown that propranolol inhibits the increase of P_{butanol} during adrenaline treatment.

The increase in permeability as a result of β-adrenergic receptor stimulation is supported by the work of Rajerison *et al.* (1972) on water permeability of frog skin. Payan (1978) also suggested that stimulation of β receptors results in an increase in the passive diffusion of ammonia (a liposoluble substance) across the basal membrane of respiratory cells, involving an increase in NH_4^+/Na^+ exchange. The effect of adrenaline on the branchial epithelium is illustrated in the following scheme:

Adrenaline → β-Adrenergic receptors → Fluidity of lipid plasmalemma

In conclusion, the specific effect of adrenaline on membrane permeability to small molecules, especially in freshwater-adapted fish, is an efficient tool that confirms the presence and hydrophilic nature of various pathways across the gills, especially in seawater-adapted animals. The physiological interest in this hormonal action resides in the selective increase in permeability to more lipophilic substances (oxygen), which may result in (*a*) the facilitation of gaseous exchange and (*b*) an increase in the excretion of products of protein catabolism. Furthermore, this hormone reduces flow into the central venous sinus, and thus presumably also enhance flow in the arterial circulation (Girard and Payan, 1976), and it increases branchial ventilation (Steen and Kruysse, 1964). The specific effect on lipid membrane fluidity results in an increase in the respiratory capacity, while reducing the disequilibrium of the hydromineral balance in marine fish. In freshwater fish, catecholamines, while augmenting oxygen transfer, also increase water uptake across the gills.

REFERENCES

Ahlquist, R. P. (1948). A study of the adrenotropic receptors. *Am. J. Physiol.* **153,** 586–600.

Ahlquist, R. P. (1966). The adrenergic receptor. *J. Pharm. Sci.* **55,** 359–367.

Ahlquist, R. P. (1968). Adrenergic β-receptor blocking agents. *Annu. Rev. Pharmacol.* **8,** 259–272.

Ahlquist, R. P. (1973). Adrenergic receptors: A personal and practical view. *Perspect. Biol. Med.* **17,** 119–122.

Al-Zahid, G., Schafer, J. A., Troutman, S. L., and Andreoli, T. E. (1977). Effect of antidiuretic hormone on water and solute permeation, and the activation energies for these processes, in mammalian cortical collecting tubules: Evidence for parallel ADH- sensitive pathways for water and solute diffusion in luminal plasma membranes. *J. Membr. Biol.* **31,** 103–130.

Barenholz, Y., and Thompson, T. E. (1980). Sphingomyelins in bilayers and biological membranes. *Biochim. Biophys. Acta* **604,** 129–158.

Bentley, P. J. (1961). Directional differences in the permeability to water of the isolated urinary bladder of the toad, *Bufo marinus*. *J. Endocrinol.* **22,** 95–100.

Bentley, P. J. (1964). Physiological properties of the isolated frog bladder in hyperosmotic solutions. *Comp. Biochem. Physiol.* **12,** 233–239.

Bergman, H. L., Olson, K. R., and Fromm, P. O. (1974). The effects of vasoactive agents on the functional surface area of isolated-perfused gills of rainbow trout. *J. Comp. Physiol.* **94,** 267–286.

Bindslev, N., and Wright, E. M. (1976). Effect of temperature on nonelectrolyte permeation across the toad urinary bladder. *J. Membr. Biol.* **29,** 265–288.

Booth, J. H. (1979). The effects of oxygen supply, epinephrine and acetylcholine on the distribution of blood flow in trout gills. *J. Exp. Biol.* **83,** 31–39.

Bourguet, J., Chevalier, J., and Hugon, J. S. (1976). Alterations in membrane associated particle distribution during antidiuretic challenge in frog urinary bladder epithelium. *Biophys. J.* **16,** 627–639.

Burke, M. J., Bryant, R. C., and Weiser, C. J. (1974). RMN of water in cold acclimating red osier dogwood stem. *Plant Physiol.* **54,** 392–398.

Chevalier, J., Bourguet, J., and Parisi, M. (1979). New evidence on the role of intramembranous particle aggregates as the ADH induced water pathways: The effect of a low HLB surfactant, cemulsol NP-E06. *INSERM* **85,** 147–158.

Collander, R. (1954). The permeability of Nitella cells to non electrolytes. *Physiol. Plant.* **7,** 420–445.

Cope, F. W. (1967). NMR evidence for complexing of Na^+ in muscle, kidney, and brain, and by actomyosin. The relation of cellular complexing of Na^+ to water structure and to transport kinetics. *J. Gen. Physiol.* **50,** 1353–1375.

Dainty, J., and House, C. R. (1966a). 'Unstirred layers' in frog skin. *J. Physiol.* (*London*) **182,** 66–78.

Dainty, J., and House, C. R. (1966b). An examination of the evidence for membrane pores in frog skin. *J. Physiol.* (*London*) **185,** 172–184.

Dengel, O., and Riehl, N. (1963). Diffusion von protonen (tritonen) in eiskristallen. *Phys. Kondens. Mater.* **1,** 191–196.

Diamond, J. M. (1966). Non-linear osmosis. *J. Physiol.* (*London*) **183,** 58–82.

Diamond, J. M., and Katz, Y. (1974). Interpretation of nonelectrolyte partition coefficients between dimyristoyl lecithin and water. *J. Membr. Biol.* **17,** 121–154.

Dunel, S., and Laurent, P. (1977). La vascularisation branchiale chez l'anguille: Action de l'acétylcholine et de l'adrénaline sur la répartition d'une résine polymérisable dans les différents compartiments vasculaires. *C. R. Hebd. Seances Acad. Sci.* **284,** 2011–2014.

Dunel, S., and Laurent, P. (1980). Functional organization of the gill vasculature in different classes of fish. *In* "Epithelial Transport in the Lower Vertebrates" (B. Lahlou, ed.), pp. 37–58. Cambridge Univ. Press, London and New York.

Earley, L. E., Sidel, V. W., and Orloff, J. (1962). Factors influencing permeability of vasopressin-sensitive membrane. *Fed. Proc., Fed. Am. Soc. Exp. Biol.* **21,** 145.

Emsley, J. W., Feeney, J., and Sutcliffe, L. H. (1965). "High Resolution Nuclear Magnetic Resonance Spectroscopy." Pergamon, Oxford.

Evans, D. H. (1967a). Sodium, chloride and water balance of the intertidal teleost, *Xiphister atropurpureus*. I. Regulation of plasma concentration and body water content. *J. Exp. Biol.* **47,** 513–517.

Evans, D. H. (1967b). Sodium, chloride and water balance of the intertidal teleost, *Xiphister atropurpureus*. II. The role of the kidney and the gut. *J. Exp. Biol.* **47,** 519–524.

Evans, D. H. (1969). Studies on the permeability to water of selected marine freshwater and euryhaline teleost. *J. Exp. Biol.* **50,** 689–703.

Evans. D. H. (1980). Kinetic studies of ion transport by fish gill epithelium. *Am. J. Physiol.* **238,** R224–R230.

Farmer, R. E. L., and Macey, R. I. (1970). Perturbation of red cell volume: Rectification of osmotic flow. *Biochim. Biophys. Acta* **196,** 53–65.

Foskett, J. K., and Scheffey, C. (1982). The chloride cell: definitive identification as the salt-secretory cell in teleosts. *Science* **215,** 164–166.

Fung, B. M., Durham, D. L., and Wassil, D. A. (1975). The state of water in biological systems as studied by proton and deuterium relaxation. *Biochim. Biophys. Acta* **399,** 191–202.

Furspan, P., and Isaia, J. (1983). Glucose transport across the gill of the rainbow trout, *Salmo gairdneri. Comp. Biochem. Physiol.* **75 A,** 401–406.

Gaitskell, R. E., and Chester Jones, I. (1971). Drinking and urine production in the european eel (*Anguilla anguilla* L.). *Gen. Comp. Endocrinol.* **16,** 478–483.

Gary-Bobo, C. M., and Solomon, A. K. (1971). Effect of geometrical and chemical constraints on water flux across artificial membranes. *J. Gen. Physiol.* **57,** 610–622.

Girard, J. P. (1979). Etude du rôle respectif des cellules "à chlorure" et des cellules respiratoires dans les échanges ioniques branchiaux chez la truite adaptée à l'eau de mer *Salmo gairdneri:* Mise en évidence de la séparation spatiale de ces deux types cellulaires. Thèse de Doctorat ès Sciences, Université de Nice.

Girard, J. P., and Payan, P. (1976). Effect of epinephrine on vascular space of gills and head of rainbow trout. *Am. J. Physiol.* **230,** 1555–1560.

Girard, J. P., and Payan, P. (1977a). Kinetic analysis and partitioning of sodium and chloride influxes across the gills of sea water adapted trout. *J. Physiol.* (*London*) **267,** 519–536.

Girard, J. P., and Payan, P. (1977b). Kinetic analysis of sodium and chloride influxes across the gill of the trout in fresh water. *J. Physiol.* (*London*) **273,** 195–209.

Girard, J. P., and Payan, P. (1980). Ion exchanges through respiratory and chloride cells in freshwater- and seawater- adapted teleosteans. *Am. J. Physiol.* **238,** R260–R268.

Hays, R. M., Franki, N., and Soberman, R. (1971). Activation energy for water diffusion across the toad bladder: Evidence against the pore enlargement hypothesis. *J. Clin. Invest.* **50,** 1016–1019.

Haywood, G. P., Isaia, J., and Maetz, J. (1977). Epinephrine effects on branchial water and urea flux in rainbow trout. *Am. J. Physiol.* **232,** R110–R115.

Hazlewood, C. F., Chang, D. C., Nichols, B. L., and Woessner, D. E. (1974). Nuclear magnetic resonance transverse relaxation times of water protons in skeletal muscle. *Biophys. J.* **14,** 583–606.

Hickman, C. P. (1968). Ingestion, intestinal absorption, and elimination of seawater and salts in the southern flounder, *Paralichthys lethostigma. Can. J. Zool.* **46,** 457–466.

Hingson, D. J., and Diamond, J. M. (1972). Comparison of nonelectrolyte permeability patterns in several epithealia. *J. Membr. Biol.* **10,** 93–135.

Holeton, G. F., and Randall, D. J. (1967). The effect of hypoxia upon the partial pressure of gases in the blood and water afferent and efferent to the gills of rainbow trout. *J. Exp. Biol.* **46,** 317–327.

House, C. R. (1974). "Water Transport in Cells and Tissues." Arnold, London.

Hughes, G. M. (1972). Morphometrics of fish gills. *Respir. Physiol.* **14,** 1–25.

Hughes, G. M. (1980). Functional morphology of fish gills. *In* "Epithelial Transport in the Lower Vertebrates" (B. Lahlou, ed.), pp. 15–36. Cambridge Univ. Press, London and New York.

Isaia, J. (1972). Comparative effects of temperature on the sodium and water permeabilities of the gills of a stenohaline freshwater fish (*Carassius auratus*) and a stenohaline marine fish (*Serranus scriba, Serranus cabrilla*). *J. Exp. Biol.* **57,** 359–366.

Isaia, J. (1979). Non-electrolyte permeability of trout gills. Effect of temperature and adrenaline. *J. Physiol.* (*London*) **286,** 361–373.

Isaia, J. (1981). Etude de la perméabilité branchiale de quelques téléostéens à l'eau et aux nonelectrolytes. Contrôle adrénergique. Thèse de Doctorat d'Etat es Sciences, Université de Nice.

Isaia, J. (1982). Effects of environmental salinity on branchial permeability of rainbow trout, *Salmo gairdneri. J. Physiol.* (*London*) **326,** 297–307.

Isaia, J., and Hirano, T. (1975). Effect of environmental salinity change on osmotic permeability of the isolated gill of the eel, *Anguilla anguilla* L. *J. Physiol.* (*Paris*) **70,** 737–747.

Isaia, J., and Isaia, A. (1978). Measurements of water exchanges in the eel gill compared by nuclear magnetic resonance and isotopic techniques. *J. Comp. Physiol.* **124,** 137–142.

Isaia, J., and Masoni, A. (1976). The effects of calcium and magnesium on water and ionic permeabilities in the sea water adapted eel, *Anguilla anguilla* L. *J. Comp. Physiol.* **109,** 221–233.

Isaia, J., Maetz, J., and Haywood, G. P. (1978a). Effects of epinephrine on branchial nonelectrolyte permeability in rainbow trout. *J. Exp. Biol.* **74,** 227–237.

Isaia, J., Girard, J. P., and Payan, P. (1978b). Kinetic study of gill epithelium permeability to water diffusion in the fresh water trout, *Salmo gairdneri:* effect of adrenaline. *J. Membr. Biol.* **41,** 337–347.

Isaia, J., Payan, P., and Girard, J. P. (1979). A study of water permeability of trout gills (*Salmo gairdneri*) adapted to fresh water and to sea water. Mode of action of epinephrine. *Physiol. Zool.* **52,** 269–279.

Isaia, J., Masoni, A., and Furspan, P. (1984). In preparation.

Kamiya, M. (1967). Changes in ion and water transport in isolated gills of the cultured eel during the course of salt adaptation. *Annot. Zool. Jpn.* **40,** 123–129.

Karnaky, K. J., Kinter, L. B., Kinter, W. B., and Stirling, C. E. (1976). Teleost chloride cell. II. Autoradiography localization of gill Na, K-ATPase in killifish *Fundulus heteroclitus* adapted to low and high salinity environments. *J. Cell Biol.* **70,** 157–177.

Keys, A. B., and Bateman, J. B. (1932). Branchial response to adrenaline and to pitressin in the eel. *Biol. Bull.* (*Woods Hole, Mass.*) **63,** 327–336.

Kirsch, R. (1972a). Plasma chloride and sodium, and chloride space in the European eel, *Anguilla anguilla* L. *J. Exp. Biol.* **57,** 113–131.

Kirsch, R. (1972b). The kinetics of peripheral exchanges of water and electrolytes in the silver eels (*Anguilla anguilla* L.) in fresh water and in sea water. *J. Exp. Biol.* **57,** 489–512.

Kirsch, R., and Mayer-Gostan, N. (1973). Kinetics of water and chloride exchanges during adaptation of the European eel to sea water. *J. Exp. Biol.* **58,** 105–121.

Kirsch, R., Guinier, D., and Meens, R. (1975). L'équilibre hydrique de l'anguille européenne (*Anguilla anguilla* L.). Etude du rôle de l'oesophage dans l'utilisation de l'eau de boisson et étude de la perméabilité osmotique branchiale. *J. Physiol.* (*Paris*) **70,** 605–626.

Kirschner, L. B. (1980). Uses and limitations of inulin and mannitol for monitoring gill permeability changes in the rainbow trout. *J. Exp. Biol.* **85,** 203–211.

Kirschner, L. B., Greenwald, L., and Sanders, M. (1974). On the mechanism of sodium extrusion across the gill of sea water adapted rainbow trout (*Salmo gairdneri*) *J. Gen. Physiol.* **64,** 148–165.

Koefoed-Johnsen, V., and Ussing, H. H. (1953). The contributions of diffusion and flow to the passage of D_2O through living membranes. Effect of neurohypophyseal hormone on isolated anuran skin. *Acta Physiol. Scand.* **28,** 60–76.

Lahlou, B., and Giordan, A. (1970). Le contrôle hormonal des échanges et de la balance de l'eau chez le téléostéen d'eau douce *Carassius auratus,* intact et hypophysectomisé. *Gen. Comp. Endocrinol.* **14,** 491–509.

Lam, T. J. (1969). Effect of prolactin on loss of solutes via the head region of the early-winter marine threespine stick-black (*Gasterosterus aculeatus L.*, form trachurus) in fresh water. *Can. J. Zool.* **47,** 865–869.

Laurent, P., and Dunel, S. (1980). Morphology of gill epithelia in fish. *Am. J. Physiol.* **238,** R147–R159.

Lieb, W. R., and Stein, W. D. (1969). Biological membranes behave as non-perous polymeric shetts with respect to the diffusion of non-electrolytes. *Nature* (*London*) **224,** 240–243.

Loeschke, K., Bentzel, C. J., and Csaky, T. Z. (1970). Asymmetry of osmotic flow in frog intestine: Functional and structural correlation. *Am. J. Physiol.* **218,** 1723–1731.

Loretz, C. L. (1979). Water exchange across fish gills: The significance of tritiated water flux measurements. *J. Exp. Biol.* **79,** 147–162.

Maetz, J. (1971). Fish gills: mechanisms of salt transfer in fresh water and sea water. *Philos. Trans. R. Soc. Lond. B.* **262,** 209–249.

Maetz, J. (1974). Aspects of adaptation to hypo-osmotic and hyperosmotic environments. *In* "Biochemical and Biophysical Perspectives in Marine Biology" (D. C. Malins and J. R. Sargent, eds.), Vol. 1, pp. 1–166. Academic Press, New York.

Masoni, A., and Garcia-Romeu, F. (1972). Accumulation et excrétion de substances organiques par les cellules à chlorure de la branchie d'*Anguilla anguilla* L. adaptée à l'eau de mer. *Z. Zellforsch. Mikrosk. Anat.* **133,** 389–398.

Masoni, A., and Garcia-Romeu, F. (1973). Localisation autoradiographique des ions Cl^- et Na^+ dans les cellules à chlorure de la branchie d'anguille (*Anguilla anguilla* L.) adaptée à l'eau de mer. *Z. Zellforsch. Mikrosk. Anat.* **141,** 575–578.

Masoni, A., and Isaia, J. (1973). Influence du mannitol et de la salinité externe sur l'équilibre hydrique et l'aspect morphologique de la branchie d'anguille adaptée à l'eau de mer. *Arch. Anat. Microsc. Morphol. Exp.* **62,** 293–306.

Masoni, A., and Isaia, J. (1980). Mise en évidence du rôle des cellules à chlorure dans le transport des macromolécules organiques. *In* "Epithelial Transport in the Lower Vertebrates" (B. Lahlou, ed.), pp. 71–80. Cambridge Univ. Press, London and New York.

Masoni, A., and Payan, P. (1974). Urea, inulin and para-amino-hippuric acid (PAH) excretion by the gills of the eel, *Anguilla anguilla* L. *Comp. Biochem. Physiol.* **47,** 1241–1244.

Motais, R. (1961). Les échanges de sodium chez un Téléostéen euryhalin, *Platichthys flesus.* Cinétique de ces èchanges lors des passages d'eau de mer en eau douce et d'eau couce en eau de mer. *C. R. Hebd. Seances Acad. Sci.* **253,** 724–726.

Motais, R. (1967). Les mécanismes d'échanges ioniques branchiaux chez les Téléostéens. Ann. Inst. Oceanogr. (*Paris*) **45,** 1–84.

Motais, R., and Garcia-Romeu, F. (1972). Transport mechanisms in the Teleostean gill and Amphibian skin. *Ann. Rev. Physiol.* **34,** 141–176.

Motais, R., and Isaia, J. (1972a). Temperature-dependence of permeability to water and to sodium of the gill epithelium of the eel *Anguilla anguilla. J. Exp. Biol.* **56,** 587–600.

Motais, R., and Isaia, J. (1972b). Evidence for an effect of ouabain on the branchial sodium-excreting pump of marine teleosts: interaction between the inhibitor and external Na and K. *J. Exp. Biol.* **57,** 367–373.

Motais, R., Garcia-Romeu, F., and Maetz, J. (1966). Exchange diffusion effect and euryhalinity in Teleosts. *J. Gen. Physiol.* **50,** 391–422.

Motais, R., Isaia, J., Rankin, J. C., and Maetz, J. (1969). Adaptative changes of the water permeability of the teleostean gill epithelium in relation to external salinity. *J. Exp. Biol.* **51,** 529–546.

Mullins, L. J. (1950). Osmotic regulation in fish as studied with radioisotopes. *Acta Physiol. Scand.* **21,** 303–314.

Naccache, N., and Sha'afi, R. I. (1973). Patterns of nonelectrolyte permeability in human red blood cell membrane. *J. Gen. Physiol.* **62,** 714–736.

Nakano, T., and Tomlinson, N. (1967). Catecholamine and carbohydrate concentrations in rainbow trout (*Salmo gairdneri*) in relation to physical disturbance. *J. Fish. Res. Board Can.* **241,** 1701–1715.

Odeblad, E., Bhar, B. N., and Lindström, G. (1956). Proton magnetic resonance of human red blood cells in heavy-water exchange experiments. *Arch. Biochem. Biophys.* **63,** 221–225.

Oduleye, S. O. (1975). The effects of calcium on water balance of the brown trout *Salmo trutta. J. Exp. Biol.* **63,** 343–356.

Ogawa, M. (1974). The effects of bovine prolactin, sea water and environmental calcium on water influx in isolated gills of the euryhaline teleosts, *Anguilla japonica* and *Salmo gairdneri. Comp. Biochem. Physiol.* **49,** 545–553.

Oide, H., and Utida, S. (1968). Changes in intestinal absorption and renal excretion of water during adaptation to sea-water in the japanese eel. *Mar. Biol.* (*Berlin*) **1,** 172–177.

Parisi, M., and Piccinni, Z. F. (1973). The penetration of water into the epithelium of toad urinary bladder and its modification by oxytocin. *J. Membr. Biol.* **12,** 227–246.

Parisi, M., Ripoche, P., Chevalier, J., and Bourguet, J. (1979a). A low HLB surfactant (NP-

E06) differently modifies water, sodium, urea and nicotinamide permeation in frog urinary bladder. *INSERM* **85,** 289–300.

Parisi, M., Bourguet, J., Ripoche, P., and Chevalier, J. (1979b). Simultaneous minute by minute determination of unidirectional and net water fluxes in frog urinary bladder: A reexamination of the two barriers in series hypothesis. *Biochim. Biophys. Acta* **556,** 509–523.

Payan, P. (1978). Mise en oeuvre d'une technique de perfusion de la tête isolée pour l'étude de l'hémodynamique et de l'échange Na^+/NH_4^+ au niveau de la branchie de truite *Salmo gairdneri.* Contrôle adrénergique. Thèse de Doctorat d'Etat, Université de Nice.

Payan, P., and Girard, J. P. (1977). Adrenergic receptors regulating patterns of blood flow through the gills of trout. *Am. J. Physiol.* **232,** H18–H23.

Payan, P., and Maetz, J. (1970). Balance hydrique et minérale chez les élasmobranches: Arguments en faveur d'un contrôle endocrinien. *Bull. Inf. Sci. Tech. Commis. Energ. At. (Fr.)* **146,** 77–96.

Payan, P., and Maetz, J. (1971). Balance hydrique chez les élasmobranches: Arguments en faveur d'un contrôle endocrinien. *Gen. Comp. Endocrinol.* **16,** 535–554.

Payan, P., Goldstein, L., and Forster, R. P. (1973). Gills and kidneys in ureosmotic regulation in euryhaline skates. *Am. J. Physiol.* **224,** 367–372.

Payan, P., Pic, P., and De Renzis, G. (1978). Comparison des échanges branchiaux de Cl^- en eau douce chez *Salmo gairdneri in vivo* et *in vitro* sur la tête isolée et perfusée. *J. Fish. Res. Board Can.* **35,** 477–479.

Phillips, J. E., and Beaumont, C. (1971). Symmetry and non-linearity of osmotic flow across rectal cuticle of the desert locust. *J. Exp. Biol.* **54,** 317–328.

Philpott, C. W. (1980). Tubular system membranes of teleost chloride cells: Osmotic response and transport sites. *Am. J. Physiol.* **238,** R171–R184.

Pic, P. (1972). Hypernatrémie consécutive à l'administration d'adrénaline chez *Mugil capito* (Téléostéen mugilide adapté à l'eau de mer). *C. R. Seances Soc. Biol. Ses Fil.* **166,** 131–136.

Pic, P., Mayer-Gostan, N., and Maetz, J. (1974). Branchial effects of epinephrine in the seawater-adapted mullet. I. Water permeability. *Am. J. Physiol.* **226,** 698–702.

Pickering, A. D., and Morris, R. (1970). Osmoregulation of *Lampetra fluviatis* L. and *Petromyzon marinus* (Cyclostomata) in hyperosmotic solution. *J. Exp. Biol.* **53,** 231–243.

Pickford, G. E., Srivastava, A. M., Slicher, A. M., and Pang, P. K. T. (1971). The stress response in the abundance of circulating leucocytes in the killifish *Fundulus heteroclitus.* II. The role of catecholamines. *J. Exp. Zool.* **17,** 97–108.

Potts, W. T. W., and Evans, D. H. (1967). Sodium and chloride balance in the killifish *Fundulus heteroclitus. Biol. Bull. (Woods Hole, Mass.)* **133,** 411–425.

Potts, W. T. W., and Fleming, W. R. (1971). The effect of environmental calcium and ovine prolactin on sodium balance in *Fundulus kansae. J. Exp. Biol.* **55,** 63–76.

Potts, W. T. W., Foster, M. A., and Stather, J. W. (1970). Salt and water balance in salmon smolts. *J. Exp. Biol.* **52,** 553–564.

Price, H. D., and Thompson, T. E. (1969). Properties of liquid bilayer membranes separating two aqueous phases; temperature dependance of water permeability. *J. Mol. Biol.* **41,** 443–457.

Raaphorst, G. P., Kruuv, J., and Pintar, M. M. (1975). Nuclear magnetic resonance study of mammalian cell water. Influence of water content and ionic environment. *Biophys. J.* **15,** 191–402.

Rajerison, R. M., Montegut, M., Jard, S., and Morel, F. (1972). The isolated frog skin epithelium: Presence of α and β adrenergic receptors regulating active sodium transport and water permeability. *Pfluegers Arch.* **332,** 313–331.

Randall, D. J., Holeton, G. F., and Stevens, E. D. (1967). The exchange of oxygen and carbon dioxide across the gills of rainbow trout. *J. Exp. Biol.* **46,** 339–348.

Richards, B. D., and Fromm, P. O. (1969). Patterns of blood flow through filaments and lamellae of isolated-perfused rainbow trout (*Salmo gairdneri*) gills. *Comp. Biochem. Physiol.* **29,** 1063–1070.

Sardet, C., Pisam, M., and Maetz, J. (1979). The surface epithelium of Teleostean fish gills. Cellular and junctional adaptations of chloride cell in relation to salt adaptation. *J. Cell Biol.* **80,** 96–117.

Sargent, J. R., and Thomson, A. J. (1974). The nature and properties of the inducible sodium-plus-potassium ion-dependent adenosine triphosphatase in the gills of eels (*Anguilla anguilla*) adapted to fresh water and sea water. *Biochem. J.* **144,** 69–75.

Sargent, J. R., Thomson, A. J., and Bornancin, M. (1975). Activities and localization of succinic deshydrogenase and Na^+/K^+ activated adenosine triphosphatase in the gills of fresh water and sea water eels (*Anguilla anguilla*). *Comp. Biochem. Physiol. B* **51B,** 75–79.

Sargent, J. R., Pirie, B. J. S., Thomson, A. J., and George, S. G. (1977). Structure and junction of chloride cells in the gills of *Anguilla anguilla. In* "Physiology and Behaviour of Marine Organisms" (D. S. McLusky and A. J. Berry, eds.), pp. 123–134. Pergamon, Oxford.

Sargent, J. R., Bell, M. V., and Kelly, K. F. (1980). The nature and properties of sodium ion plus potassium ion-activated adenosine triphosphatase and its role in marine salt secreting epithelia. *In* "Epithalial Transport in the Lower Vertebrates." (B. Lahlou, ed.), pp. 251–267. Cambridge Univ. Press, London and New York.

Scheuplein, R. J., and Blank, I. H. (1971). Permeability of the skin. *Physiol. Rev.* **51,** 702–747.

Shehadeh, Z., and Gordon, M. S. (1969). The role of the intestine in salinity adaptation of the rainbow trout (*Salmo gairdneri*). *Comp. Biochem. Physiol.* **30,** 397–418.

Shirai, N. (1972). Electron-microscope localization of sodium ions and adenosine triphosphatase in chloride cells of the Japanese eel, *Anguilla japonica. J. Fac. Sci., Univ. Tokyo, Sect. 4* **12,** 385–403.

Shirai, N., and Utida, S. (1970). Development and degeneration of the chloride cell during seawater and freshwater adaptation of the Japanese eel, *Anguilla japonica. Z. Zellforsch. Mikrosk. Anat.* **103,** 247–264.

Shuttleworth, T. J., and Freeman, R. F. H. (1974). Factors affecting the net fluxes of ions in the isolated perfused gills of freshwater *Anguilla dieffenbachii. J. Comp. Physiol.* **94,** 297–307.

Skadhauge, E. (1969). The mechanism of salt and water absorption in the intestine of the eel (*Anguilla anguilla*) adapted to waters of various salinities. *J. Physiol.* (*London*) **204,** 135–158.

Skadhauge, E. (1974). Coupling of transmural flows of NaCl and water in the intestine of the eel (*Anguilla anguilla*). *J. Exp. Biol.* **60,** 535–546.

Smith, H. W. (1930). The absorption and excretion of water and salts by marine Teleosts. *Am. J. Physiol.* **93,** 480–505.

Smith, H. W. (1932). Water regulation and its evolution in the fishes. *Q. Rev. Biol.* **7,** 1–26.

Steen, J. B., and Kruysse, A. (1964). The respiratory function of teleostean gills. *Comp. Biochem. Physiol.* **12,** 127–142.

Steen, J. B., and Stray-Pedersen, S. (1975). The permeability of fish gills with comments on the osmotic behaviour of cellular membranes. *Acta Physiol. Scand.* **95,** 6–20.

Stein, W. D., ed. (1967). "The Movement of Molecules across Cell Membranes." Academic Press, New York.

Stevens, E. D. (1972). Change in body weight caused by handling and exercise in fish. *J. Fish. Res. Board Can.* **29,** 202–203.

Sussman, M. W., and Chin, L. (1966). Liquid water in frozen tissue; study by nuclear magnetic resonance. *Science* **151,** 324–325.

Svelto, M., Perrini, M. C. R., and Lippe, C. (1977). Noradrenaline induced secretion of non electrolytes through frog skin. *J. Membr. Biol.* **36,** 1–11.

Thomson, A. J., and Sargent, J. R. (1977). Changes in the levels of chloride cells and ($Na^+ + K^+$) -dependent ATPase in the gills of yellow and silver eels adapting to seawater. *J. Exp. Zool.* **200,** 33–40.

Vick, R. L., Chang, D. C., Nichols, B. L., Hazlewood, C. F., and Harvey, M. C. (1973). Sodium, potassium and water in cardiac tissues. *Ann. N.Y. Acad. Sci.* **204,** 575–592.

Wang, J. W., Robinson, C. V., and Edelman, I. S. (1953). Self-diffusion and structure of liquid water. III. Measurement of the self-diffusion of liquid water with H^2, H^3, O^{18} as tracers. *J. Am. Chem. Soc.* **75,** 466–470.

Wiggins, P. H. (1971). Water structure as a determinant of ion distribution in living tissue. *J. Theor. Biol.* **32,** 131–146.

Wood, C. M. (1974). A critical examination of the physical and adrenergic factors affecting blood flow through the gills of the rainbow trout. *J. Exp. Biol.* **60,** 241–265.

Wright, E. M., and Pietras, R. J. (1974). Routes of nonelectrolyte permeation across epithelial membranes. *J. Membr. Biol.* **17,** 293–312.

Wright, E. M., and Prather, J. W. (1970). The permeability of the frog choroid plexus to nonelectrolytes. *J. Membr. Biol.* **2,** 127–149.

Zwingelstein, G. (1981). Les lipides amphiphiles du poisson. *CNRS Symp. Ser. Colloq.* pp. 249–260.

2

BRANCHIAL ION MOVEMENTS IN TELEOSTS: THE ROLES OF RESPIRATORY AND CHLORIDE CELLS

P. PAYAN AND J. P. GIRARD
Laboratoire de Physiologie Cellulaire et Comparée
Université de Nice et ERA CNRS 943
06034 Nice, France

N. MAYER-GOSTAN
Groupe de Biologie Marine du Département de Biologie du CEA
Station Marine 06230 Villefranche-sur-Mer, France

I. INTRODUCTION

The interest that is shown in studies of transport and permeability in epithelial cells is justified by the diversity and importance of the physiologi-

FISH PHYSIOLOGY, VOL. XB

ISBN 0-12-350432-5

cal functions carried out by these cells. In this connection the fish gill affords a fascinating example of a multifunction structure, the epithelial cells assuming respiratory, osmoregulatory, and excretory roles. It also has a remarkable capacity for integrating these various functions and adjusting them to the needs of the organism.

Since excellent reviews on ionic exchanges in fish have already been published (Evans, 1980; Evans *et al.*, 1982), the present chapter will be limited to a consideration of the respective roles of the various types of branchial cells in ionic exchanges in teleosts.

The branchial epithelium is an extremely complex tissue in both histological structure (there are several cell types: mitochondria-rich cells, mucous cells, respiratory cells) and blood circulation. In fish the branchial and systemic circulations are in series (Randall, 1970), the heart driving the blood at a sufficiently high pressure to maintain an adequate perfusion of the entire system in spite of the flow resistance encountered in the gills. The blood pressure in the afferent branchial artery (~30 mm Hg) is one of the highest recorded in vertebrates (Johansen, 1972) and causes passive water and ion movements across the branchial epithelium, which is thus continually called on to compromise between the respiratory and osmoregulatory needs of the organism.

By 1930 Homer Smith had described the main features of osmoregulation in fish, particularly stressing the importance of the branchial epithelium. In freshwater teleosts the gills are the site of active NaCl absorption compensating renal loss, whereas in saltwater teleosts the gills actively excrete the salt absorbed through the intestines. The direction of active salt transport is thus reversed when a fish passes from fresh water to seawater. Many subsequent workers have shown that in fresh water, NaCl absorption is mediated by coupled Na^+/NH_4^+, Na^+/H^+, and Cl^-/HCO_3^- exchanges (Krogh, 1939; Maetz and Garcia-Romeu, 1964; Maetz *et al.*, 1976; De Vooys, 1968; Kerstetter *et al.*, 1970; Maetz, 1973; Kirschner *et al.*, 1973; De Renzis, 1978; Payan, 1978; Kerstetter and Keeler, 1976; Claiborne *et al.*, 1982).

In seawater, in contrast, various and often contradictory mechanisms have successively been proposed. Figure 1 illustrates several such models.

In 1932 Keys and Willmer described mitochondria-rich cells in the branchial epithelium of seawater-adapted eels and called them "chloride-secreting cells" by analogy with the oxyntic cells of the stomach. All subsequent models concerning the mechanisms of NaCl absorption in fresh water and NaCl excretion in seawater have sited the salt exchanges in the branchial chloride cells. Only subsequent work on branchial irrigation (Laurent and Dunel, 1976, 1978; Dunel and Laurent, 1977, 1980; Vogel *et al.*, 1973, 1974, 1976) and on isolated fish head under perfusion (Payan and Matty, 1975) has

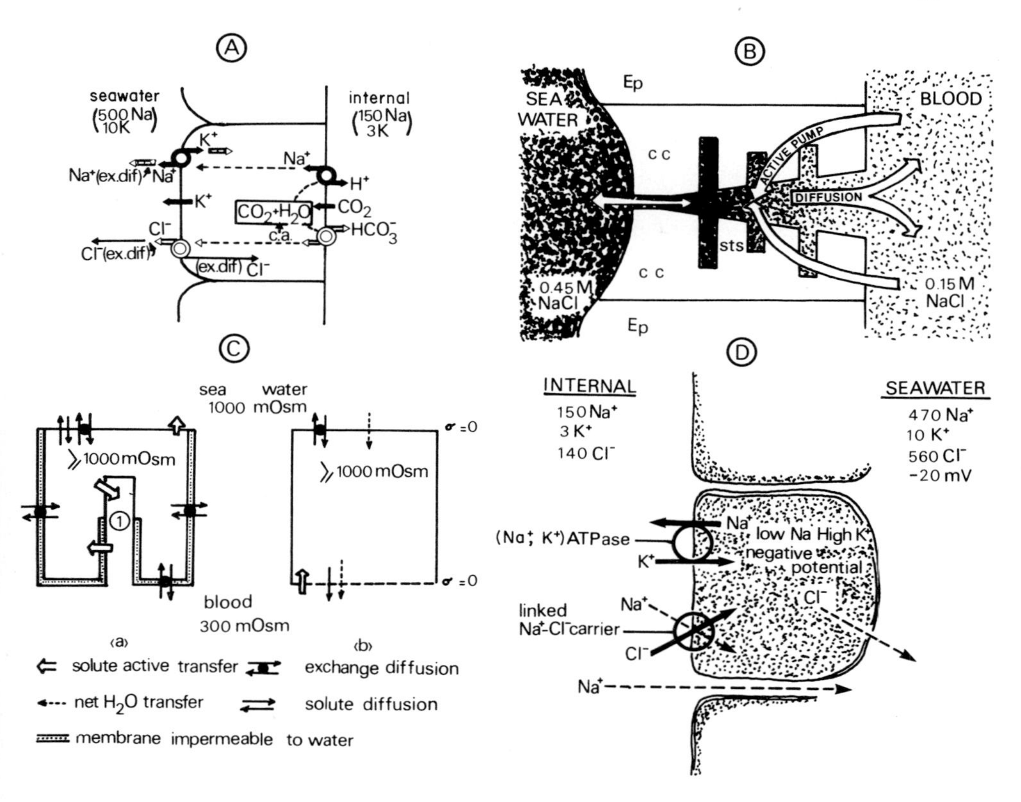

Fig. 1. Some of the numerous hypothetical models showing how seawater teleosts excrete NaCl through "chloride cells." Different symbols are used to indicate active transport and exchange diffusion in (A) and (C). The explanations of symbols in (C) are applicable to (C) only. Values for Na^+, Cl^-, and K^+ are in milliequivalents. cc, chloride cell; Ep, epithelium; ex. dif, exchange diffusion; sts, smooth tubular system. [References: (A) Maetz, 1971; (B) Sargent *et al.*, 1977; (C) Motais and Garcia-Romeu, 1972 (reproduced, with permission, from the Annual Review of Physiology, Volume 34. © 1972 by Annual Reviews, Inc.); (D) Silva *et al.*, 1977.].

thrown fresh light on the relative importance of chloride cells and respiratory cells in branchial ion transport in teleosts.

The complexity of the branchial epithelium with its two dominant cell types makes it very difficult to study ion transport in either cell type independently. In 1972 Karnaky found that the opercular membrane of certain fish had a high density of mitochondria-rich cells histologically similar to branchial chloride cells. The membrane can be mounted in an Ussing chamber, and such preparations have proved invaluable in the study of active NaCl excretion by chloride cells in marine teleosts (see Chapter 5). It now appears that this opercular membrane plays a significant role in the total ionic exchanges in certain marine teleosts. Thus, a comparison of Na^+ fluxes in *Fundulus heteroclitus in vivo* (Mayer, 1970) and *in vitro* in the opercular membrane (Mayer Gostan and Maetz, 1980) gives a value of 10 to 15% for the contribution of the opercular membrane to total ionic fluxes in these species.

The present chapter will discuss the respective roles of respiratory cells and chloride cells in the ionic exchanges of freshwater and saltwater teleosts.

II. THE VASCULARIZATION AND ULTRASTRUCTURE OF THE BRANCHIAL EPITHELIUM

A brief review of the vascular structure of the teleostean gill is necessary for an understanding of the isotopic kinetic studies carried out on perfused isolated heads. The following description of the branchial vascular system is drawn from the works of Steen and Kruysse (1964), Newstead (1967), Richards and Fromm (1969), Hughes and Morgan (1973), Morgan and Tovell (1973), Vogel *et al.* (1973, 1974, 1976), Laurent and Dunel (1976, 1978), and Dunel and Laurent (1977, 1980).

A. Branchial Vascularization

The gill is composed of three structural subunits: (1) the gill arches supporting (2) the filaments, which themselves support (3) the lamellae. The filaments are branched in parallel on the gill arch, and their irrigation is assured by three blood vessels: two arteries (afferent and efferent) and a vein (Fig. 2).

The filament contains two distinct vascular compartments: (1) the lamellar compartment located between the afferent and efferent arteries, and (2) the central venous sinus, forming a separate vascular compartment inside the filament. In the trout this sinus is irrigated from the efferent filamental artery and drained by the branchial vein (Fig. 2). The central venous sinus thus forms an alternative pathway for the blood flowing through the lamella, which, instead of taking the arterial route to the dorsal aorta, may be recycled via the heart into the branchial circulation.

B. Ultrastructure of the Gill Epithelium

The branchial epithelium is made up of a mosaic of respiratory, mucous, and mitochondria-rich cells, the latter being the "chloride cells" of Keys and

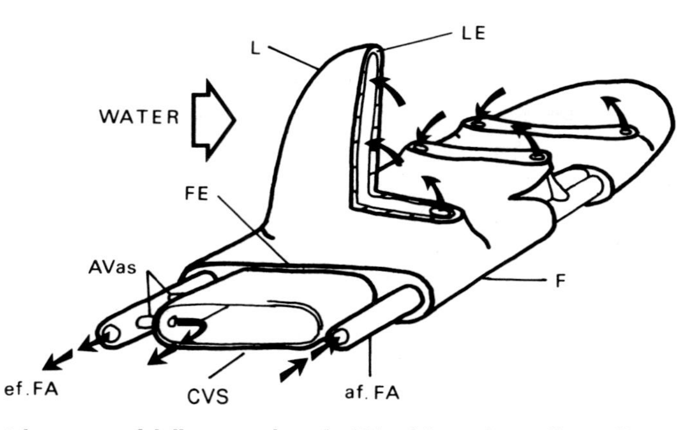

Fig. 2. Schematic model illustrating branchial blood flow. af.FA, afferent filamental artery; AVas, arteriovenous anastomosis; ef.FA, efferent filamental artery; F, filament; FE, filamental epithelium; CVS, central venous sinus; L, lamella; LE, lamellar epithelium. (From Girard and Payan, 1980.)

Willmer (1932). The respiratory cells form the greater part of the surface of the lamella epithelium, while the chloride cells are found mainly in the epithelium of the filament (Fig. 3). However, Laurent and Dunel (1978) considered that in all cases investigated the chloride cells, also called ionocytes, are functionally linked with the venous compartment.

During adaptation of a freshwater euryhaline teleost to seawater, the gill epithelium undergoes the following modifications (Fig. 3):

1. There is an increase in the number of chloride cells (Newstead, 1967).
2. The chloride cells show a fundamental structural reorganization (Sardet *et al.*, 1979). "Young" chloride cells extend fingers into the mature chloride cells and in so doing create intercellular spaces that communicate directly with the external medium. Thus, the filamental epithelium in seawater-adapted animals has the morphological characteristics of a leaky membrane.

The structural characteristics of the respiratory cells, in contrast, are not modified by adaptation to seawater; they retain their tight junctions, characteristic of poorly permeable epithelia.

The gills, constituting the principal surface of contact with the environment, can thus be considered as a combination of two epithelia distinguished by their irrigation (arterio-arterial for the lamellar epithelium and arteriovenous for the filamental epithelium) and by their ultrastructure.

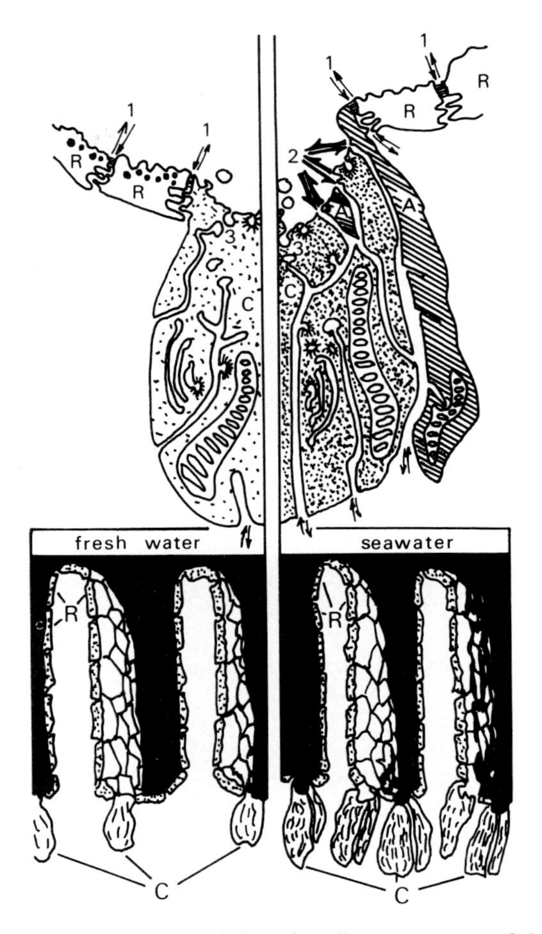

Fig. 3. Details of the organization of chloride cells in seawater and fresh water. Three possible routes across the epithelium are shown (1, 2, and 3). In fresh water only respiratory cells (R) and a chloride cell (C) are shown linked by tight junctions (1). In seawater the presence of the developing chloride cell and its arm (A) allows the tubular reticulum and extracellular space to communicate with the external milieu via leaky junctions (2). We propose that they could also communicate via a tubulovesicular system (3) present in both seawater- and fresh-water-adapted fish. (From Sardet *et al.*, 1980.)

"Leaky" epithelia are also found in the small intestine, gallbladder, proximal kidney tubule, and choroid plexus; "tight" epithelia are found in various urinary bladders, amphibian skin, rabbit large intestine, and distal kidney tubule. The gill is the only organ in which a juxtaposition of these two different types of epithelia is known to occur.

III. LOCATION OF IONIC EXCHANGES IN THE FILAMENTAL AND LAMELLAR EPITHELIUM

The technique of perfusion of the isolated trout head (Girard and Payan, 1977) has enabled workers to study separately the ionic fluxes of the filamental epithelium and lamellar epithelium (Girard and Payan, 1980).

The gills are perfused by a constant volume from the ventral aorta, and the efferent arterial and venous liquids are collected separately after passing over lamellar and filamental epithelia, respectively (Fig. 4). Isotopes (^{24}Na, ^{36}Cl, ^{47}Ca) added to the external medium may be measured in the outflowing arterial and venous liquids, and the unidirectional influxes of each ion across the lamellar epithelium and the filamental epithelium (i.e., by the respiratory cells and chloride cells) may thus be individually calculated. This preparation is considered to be functional as long as ionic fluxes are comparable to those found *in vivo*. The ionic exchanges in this epithelium are generally studied for 1 hr, a period during which a net absorption of NaCl can be measured in fresh water.

A. Ionic Movements through the Filamental Epithelium and Lamellar Epithelium

The ratios of the radioactive concentrations of venous and arterial efferent liquids (V/A) at isotopic equilibrium were used to determine whether, after passing over the lamellar epithelium, the perfusion liquid increased in radioactivity on passing through the venous sinus. These data from measurements on freshwater- and seawater-adapted trout are given in Table I. In freshwater trout the V/A ratios for sodium and chloride are not significantly different from 1, indicating that the Na^+ and Cl^- influxes are entirely carried out by the lamellar epithelium, that is, by the respiratory cells. The V/A ratio for Ca^{2+}, in contrast, is 38, implying that calcium is transported across the branchial epithelium mainly by the chloride cells. This last result is illustrated in Fig. 5. It can be seen that even with different flow rates (5 ml min^{-1} for the lamellar epithelium and 1 ml min^{-1} for the filamental epithelium), the influx through the filamental epithelium is 10 times greater than through the lamellar epithelium.

In seawater-adapted trout the V/A ratios for Na^+ and Cl^- are significantly greater than 1 (Table I), showing that ions are also entering through the filamental epithelium—presumably through the chloride cells.

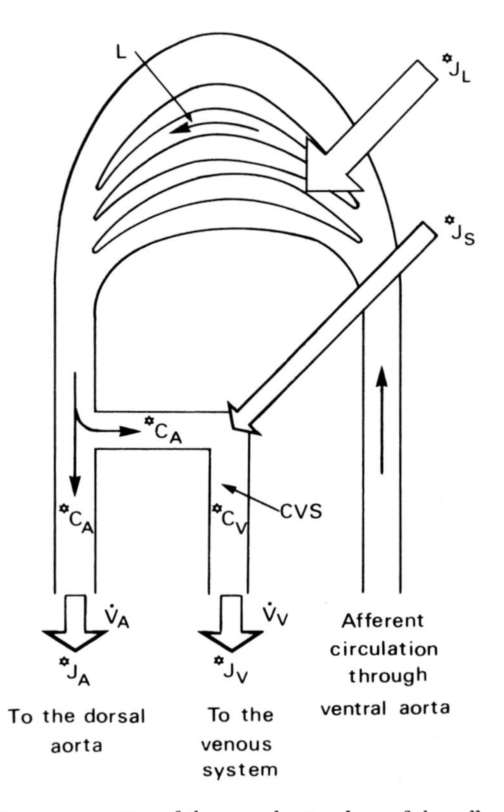

Fig. 4. Schematic representation of the vascular topology of the gill apparatus. The wide arrows indicate the two routes of entry of the ions—lamellar ($*J_L$) and sinusal ($*J_S$)—as well as their repartition into the vascular gill system—dorsal aorta ($*J_A$) and venous system ($*J_V$). The concentration of the isotope originating from $*J_L$ is $*C_A$ found in the efferent artery and in the arteriovenous anastomoses. As a result of $*J_S$, $*C_A$ is increased to $*C_V$. $\dot{V}_A$ and $\dot{V}_V$ represent the flow rate through the two efferent routes. CVS, central venous sinus; L, lamellae. (From Girard and Payan, 1977.)

B. Unidirectional Fluxes

In freshwater teleosts Na^+ and Cl^- ionic exchanges with the external medium take place through the lamellar epithelium (Fig. 6), whereas Ca^{2+} movements are localized through the filamental epithelium. Previous work demonstrated that excretions of NH_4^+ or H^+ and HCO_3^- are linked, respectively, to the absorption of Na^+ and Cl^- (Maetz and Garcia Romeu, 1964; De Renzis, 1978; Payan, 1978). The results obtained with the perfused head technique bring a new point of view for such mechanisms. Those

Table I

Ratio between Radioactive Concentrations of Venous and Arterial Fluid during Influx Measurements in Trout[a,b]

Environmental salinity	Na^+	Cl^-	Ca^{2+}
Fresh water	1 ± 0.13 (8)	0.9 ± 0.09 (8)	38* ± 7.3 (4)
Seawater	1.9 ± 0.32 (9)	1.6 ± 0.18 (9)	—

[a]Values taken from Girard and Payan (1980), except * from Payan *et al.* (1981).

[b]Numbers in parentheses are numbers of experiments. Data x̄ ± SEM.

exchanges were considered to be situated in the mitochondria-rich cells, and there is now evidence that they are located in the lamellar epithelium (i.e., in the respiratory cells; Girard and Payan, 1980), the role of which is not limited to gaseous exchanges.

In seawater-adapted teleosts it has proved possible to calculate the components of the effluxes across the filamental and the lamellar epithelia. By using a Ringer solution traced with ^{36}Cl and ^{24}Na with a low specific radio-

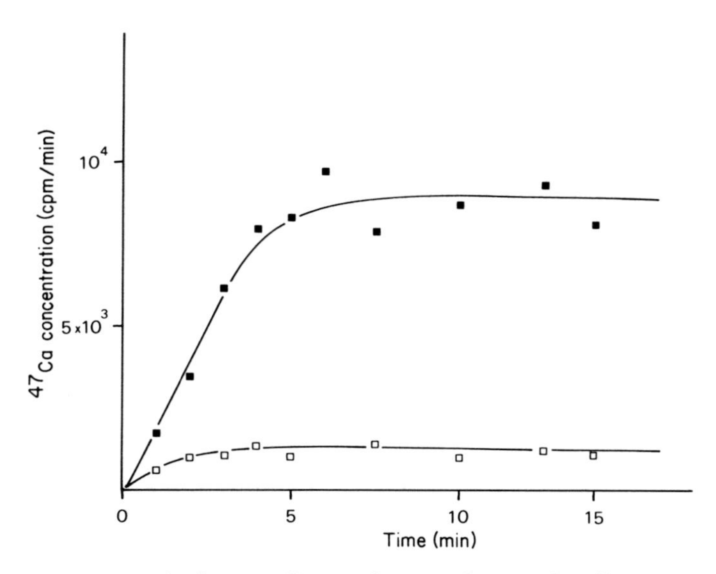

Fig. 5. Appearance of radioactive ^{47}Ca as a function of time in the efferent arterial (□) and venous fluid (■). (From Payan *et al.*, 1981.)

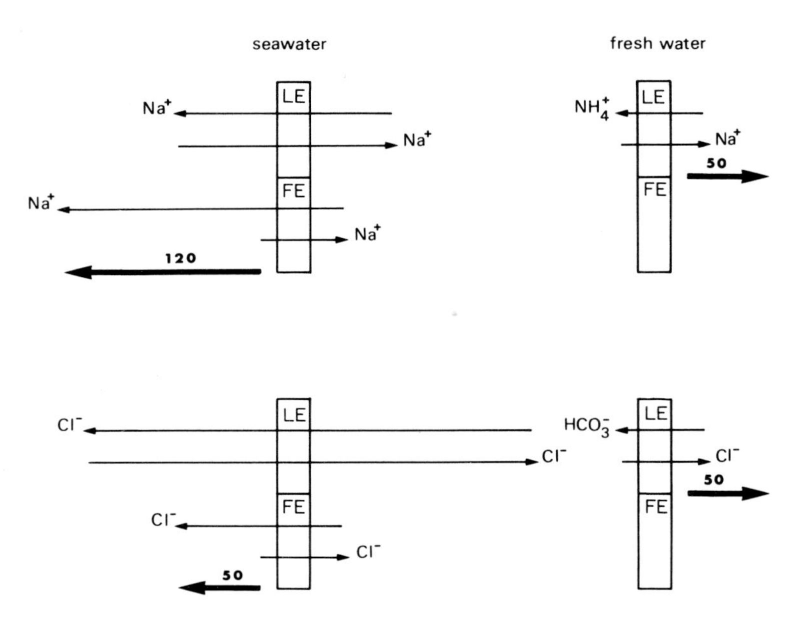

Fig. 6. Na^+ and Cl^- fluxes through filamental epithelium (FE) and lamellar epithelium (LE) in freshwater- and seawater-adapted trout. The arrows represent the approximate magnitude of the fluxes. Large arrows represent the net fluxes expressed as μEq hr^{-1} 100 g^{-1}. (From Girard and Payan, 1980.)

activity, effluxes through the lamellar epithelium can be measured by dorsoventral arterial clearance. The difference between the total efflux as measured by the appearance of radioactivity in the external medium, and this lamellar efflux allows a calculation of the filamental efflux (Fig. 6).

The difference between the total Na^+ and Cl^- influxes and effluxes shows that there is a net excretion of Na^+ and Cl^- through the seawater-adapted trout gill.

The influxes and effluxes of Na^+ and Cl^- through the lamellar epithelium are not significantly different: there is thus no excretory net flux through the respiratory cells. Evans (1980) suggested that the thinness of the lamellar epithelium may be related to passive ionic exchanges across this barrier, but no experimental evidence for his hypothesis was presented.

However, the effluxes are significantly higher than the influxes, indicating that a net excretory flux of Na^+ and Cl^- occurs through the filamental epithelium. This confirms that the mitochondria-rich cells, which are bathed by the branchial venous system, are responsible for the salt excretion.

The (Na^+,K^+)-ATPase localized in the mitochondria-rich cells (Maetz and Bornancin, 1975; Karnaky *et al.*, 1976) is considered to be the biochemical mediator for the active Na^+ extrusion. One way of investigating further the nature of the salt net fluxes is to compare the magnitude of the Na^+ and Cl^- net fluxes with the results of experiments in which the Na^+ effluxes are stimulated by addition of K^+ to the external medium during rapid transfer of fish to fresh water. This technique, called the "K test" (Maetz, 1969) and used by several workers on various euryhaline teleosts (see review by Maetz and Bornancin, 1975), makes possible an evaluation of the active fraction of the Na^+ efflux. Table II shows the effect of addition of external potassium (5 mM K_2SO_4) on the Na^+ and Cl^- effluxes after rapid transfer into fresh water. The intensity of the Na^+ efflux related to the presence of external K^+ (FWK-FW) is comparable to that of the Na^+ net flux through the filamental epithelium (Fig. 6). In the case of Cl^-, stimulation of the efflux by external K^+, described by Epstein *et al.* (1973) in *Anguilla anguilla,* was also observed by Maetz and Pic (1975) in *Mugil capito,* suggesting that (Na^+,K^+)-ATPase participates in active Cl^- excretion. This is confirmed by the observations of Foskett and Scheffey (1982). However, the value of the increased Cl^- efflux and that of the Cl^- net flux through the filamental epithelium (Fig. 6) do not correspond as well as for Na^+, suggesting that a part of the chloride efflux is not related to the (Na^+,K^+)-ATPase.

C. Localization of Ammonium Excretion

In both freshwater- and seawater-adapted fish, ammonium is excreted by the lamellar epithelium. Dorsoventral arterial clearance measurements show that in both media the quantity of ammonium extracted from the blood

Table II

Components of Sodium and Chloride Effluxes Measured by Rapid-Transfer Experiment in Trout[a,b]

Ion	Environment[c]			
	SW	FW	FWK	FWK-FW
Na^+	298 ± 32.2 (8)	127 ± 18.3 (8)	297 ± 42.8 (8)	170 ± 50.2 (8)
Cl^-	373 ± 49.8 (5)	177 ± 56.3 (5)	320 ± 61.2 (5)	143 ± 54.3 (5)

[a]Results from Girard (1976).

[b]Ionic fluxes are expressed as μEq hr^{-1} 100 g^{-1}. Numbers in parentheses are numbers of experiments. Data $\bar{x}$ ± SEM.

[c]SW, seawater; FW, fresh water; FWK, fresh water containing 10 mEq of K^+; FWK-FW, intensity of Na^+ efflux related to presence of external K^+.

Table III

Ammonium Excretion by Freshwater (FW) and Saltwater (SW) Trout Gills Measured by Appearance in the External Medium and by Dorsoventral Clearance through Lamellar Epithelium[a,b]

Ammonium excretion	FW	SW
Appearance in external medium	50.9 ± 4.0 (6)	51.7 ± 6.5 (5)
Dorsoventral clearance	49.2 ± 3.3 (6)	49.6 ± 6.0 (5)

[a]Data for FW gills from Payan and Pic (1977); for SW gills, from Girard (1979).

[b]Ammonium concentrations in perfusing fluid: FW, 0.6 m*M*; SW, 0.4 m*M*. Fluxes are expressed in μEq hr^{-1} 100 g^{-1}. Numbers in parentheses are numbers of experiments. Data x̄ ± SEM.

during its passage through the lamellar epithelium is equal to the amount of NH_4^+ appearing in the external medium (Table III).

Since Krogh's study (1939), there has been controversy regarding whether ammonium crosses the gill epithelium as NH_3 or NH_4^+. In experimental conditions the ammonia concentration in the fluid was varied, and in this case NH_4^+ was the main form of ammonia. Figure 7 shows clearly that

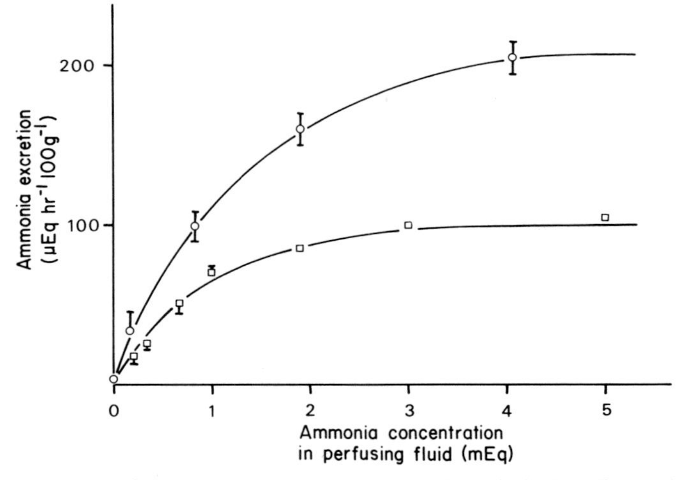

Fig. 7. Relationship between the ammonia excretion through the lamellar epithelium and the concentration of ammonia in the perfusing fluid in freshwater- (□) and seawater-adapted (○) trout. Means ± SE of five experiments. In these experiments external medium was continually renewed in order to avoid an external accumulation of ammonium. Ammonium fluxes were calculated by dorsoventral clearance between afferent and efferent arterial fluids. (From Girard, 1979; Payan, 1978.)

in freshwater- and seawater-adapted trout, ammonium excretion measured by dorsoventral clearance as a function of the ammonium concentration in the perfusion liquid presents saturable kinetics. This result favors the hypothesis of ammonium excretion by means of a Na^+/NH_4^+ exchange localized in the respiratory cells. However, some authors have shown that NH_3 may diffuse across the gill epithelium (see review by Maetz *et al.*, 1976).

IV. KINETIC CHARACTERISTICS OF THE APICAL AND BASAL MEMBRANES OF THE RESPIRATORY AND CHLORIDE CELLS

In the "loading–unloading radioactivity experiments" using radioactive isotopes, the opercular and branchial epithelia can be considered as systems with three compartments in series (Fig. 8; Girard and Payan, 1977a,b; Mayer-Gostan *et al.*, in preparation). In the freshwater trout the intermediate

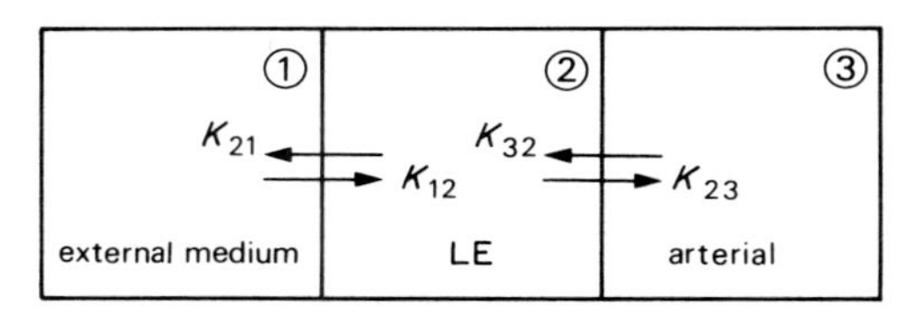

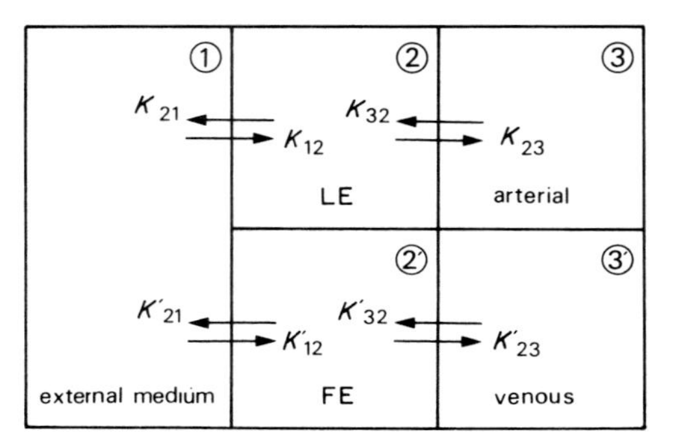

Fig. 8. Schematic representation of ionic transporting compartments in gill of freshwater- and seawater-adapted trout. FE, filamental epithelium; LE, lamellar epithelium. When one considers the operculum, the two chambers and the membrane constitute three serial compartments. In the case of the perfused head the three serial compartments are represented by the

compartment (the lamella), consisting mostly of respiratory cells, is sandwiched between the external medium and the arterial circulation (Fig. 8). The opercular membrane also forms a serial three-component system with an intermediate compartment rich in chloride cells (Fig. 8).

In the gill epithelium of seawater-adapted trout, the intermediate com-

---

external medium, either the lamellar or filamental epithelium, and the arterial or venous blood. The transfer coefficient K of an ion from one compartment to another is a specific characteristic of the barrier for the considered ion. K is the rate constant representing a fraction of the compartment exchanged per unit of time. The cumulative $K = K_{21} + K_{23}$ will represent the constant rate of the ion from compartment 2 (the cells) across the basal and serosal barriers of this compartment.

In loading experiments the isotope is introduced in compartment 1, and its appearance in (3) is monitored. The specific radioactivity in (1) has to be large enough to be considered as a constant during the experiment, and backflux from (3) toward (2) has to be negligible. Under these two conditions the amount of radioactivity in (3) is given by the equation

$$z = x_0 \frac{K_{21}\,K_{23}}{K_{21} + K_{23}} \left[t + \frac{1}{K_{21} + K_{23}} e^{-(K_{21} + K_{23})t} - \frac{1}{K_{21} + K_{23}} \right]$$

and if t is sufficiently large the equation becomes

$$z = x_0 \frac{K_{21}\,K_{23}}{K_{21} + K_{23}} \left[t - \frac{1}{K_{21} + K_{23}} \right]$$

for $z = 0$. $t = \theta$ (point of intersection of the curve with the t axis). The value of θ can be graphically evaluated from this last equation:

$$\theta = \frac{1}{K_{21} + K_{23}}$$

In unloading experiments the compartments 2 and 2′ are loaded, and the appearance of the radioactivity is monitored in compartments 1 and 3. These compartments may be considered infinitely large compared with compartments 2 and 2′, so that radioactive backfluxes from compartments 1 and 3 toward (2) may be neglected. The variation y of tracers in (2) as a function of time will be $dy/dt = -(K_{21} + K_{23})y$, the integral of which is an exponential function.

$$y = y_0 e^{-(K_{21}+K_{23})t}$$

where y_0 is the quantity of isotopes in compartment 2 at time zero.

The rates of appearance of radioactivity x in (1) and z in (3) are given by the following equations:

$$\frac{dx}{dt} = K_{21} y_0 e^{-(K_{21}+K_{23})t} \text{ and } \frac{dz}{dt} = K_{23} y_0 e^{-(K_{21}+K_{23})t}$$

The exponential functions were either estimated graphically or calculated with a computer (Hewlett Packard, HP 30) using an exponential adjustment program. From the two last equations K_{23} and K_{21} can be estimated, and the ratio of K_{23} and K_{21} gives r:

$$r = \frac{dx/dt}{dz/dt} = \frac{K_{23}}{K_{21}}$$

partment is composite. It is either the respiratory cells of the lamellar epithelium (Fig. 8, compartment **2**), drained on their internal faces by the arterial blood (**3**), or the chloride cells of the filamental epithelium (**2′**), drained on their internal faces by the venous system (**3′**) (Fig. 8).

A. Respiratory Cells of Freshwater- and Seawater-Adapted Gills

An example of a radioactive (^{24}Na) loading curve of the internal compartment (arterio-arterial circuit bathing the respiratory cells of freshwater- or seawater-adapted trout gills) is given in Fig. 9. From such experiments the turnover value (K_{21} + K_{23}) of a given ion transported in the intermediate compartment can be calculated. The kinetic characteristics of the apical (K_{21}) and basal (K_{23}) membranes of the respiratory cells are summarized in Table IV. The values of the transfer coefficients provide information on the relative permeabilities to Na^+ and Cl^- ions of the basal and apical barriers. They indicate that in both fresh water and seawater it is the basal membrane that limits ion movements. It is worth noticing that adaptation to a marine environment does not significantly modify the kinetic characteristics of the apical membrane of the respiratory cells despite a change of salinity from 0.2 to 550 m*M* NaCl. Nevertheless, the unidirectional fluxes across the apical face of these cells are much higher in seawater than in fresh water (Table IV). These results suggest that in the respiratory cells lies an ionic "pool" involved in the exchanges, and that this pool is much larger in seawater than in fresh water (Girard and Payan, 1977a,b).

The shape of the respiratory cells (Sardet *et al.*, 1979) would suggest that the area of internal and external surfaces of contact are similar. The transfer coefficients as measured for the basal and apical membranes of the respiratory cells are therefore truly representative of the permeabilities of these barriers.

B. Chloride Cells of the Seawater Gill and of the Opercular Membrane

An example of ^{24}Na unloading in the opercular chloride cells of *Fundulus* is given in Fig. 10.

The hypothesis of a three-compartment system will be valid only if the intermediary compartment is a homogeneous unit. In this case the exponential increments measured on each side during unloading should be identical (Whittam and Guinnebault, 1960). This is true in the present experiments (Fig. 10), because no significant differences between serosal and mucosal exponential increments were observed.

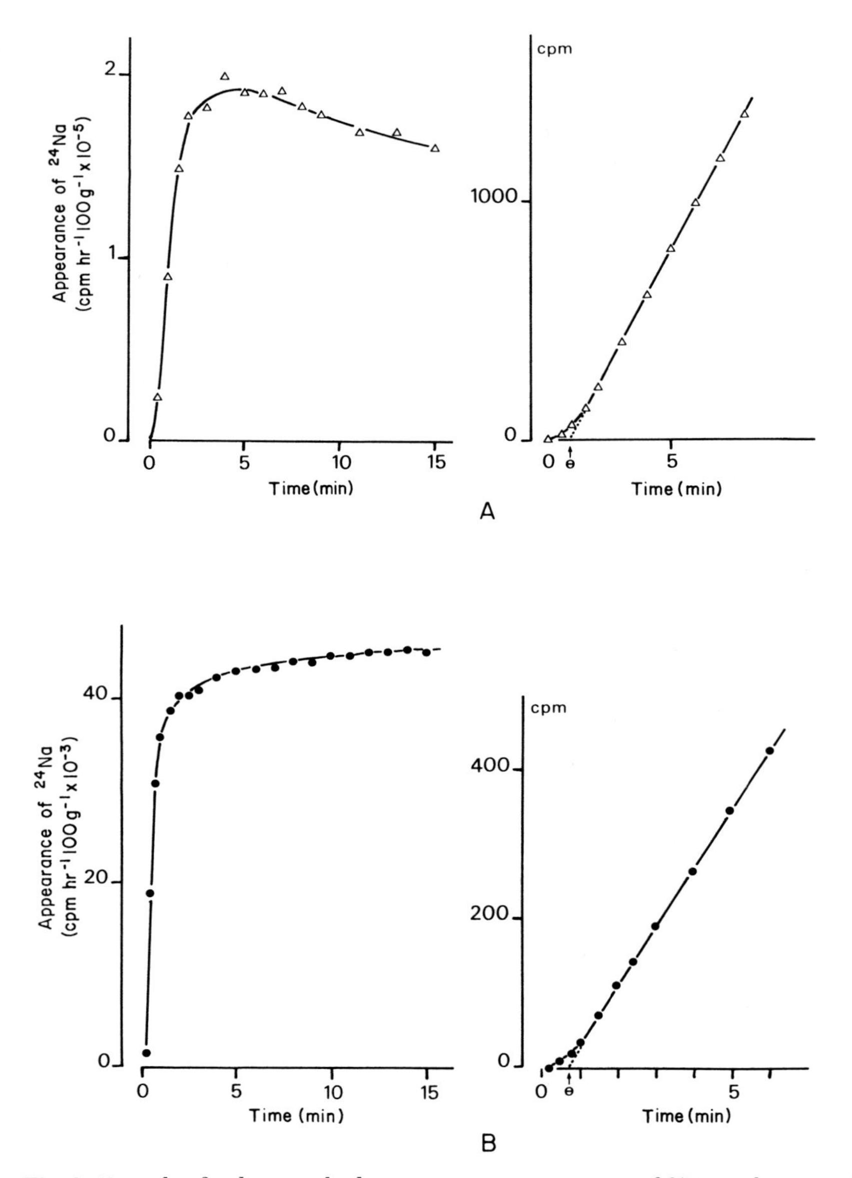

Fig. 9. Example of radioactive loading experiments: appearance of ^{24}Na in the internal lamellar compartment of the gill of (A) seawater- and (B) freshwater-adapted trout. Loading curves allow for the calculation of θ (see legend of Fig. 8).

Table IV

Transfer Coefficients (K) and Fluxes (F) for Na^+ and Cl^- through Apical (21) and Basal (23) Membranes of Respiratory Cells in Freshwater (FW) and Saltwater (SW) Trout Gills[a,b]

		K_{21}	K_{23}	F_{21}	F_{23}
FW	Na^+	101 ± 5.1	11 ± 2.8	78 ± 13.7	8 ± 1.7
(n = 8)	Cl^-	91 ± 7.2	14 ± 3.1	61 ± 11.0	8 ± 2.0
SW	Na^+	206 ± 73.0	6 ± 0.7	4961 ± 763	148 ± 23
(n = 5)	Cl^-	89 ± 2.3	2 ± 2.2	12,440 ± 2010	276 ± 45

[a] Data for FW gills from Girard and Payan (1977b); for SW gills, from Girard and Payan (1977b) and Girard (1979).

[b] K expressed as % min^{-1} and F as μEq hr^{-1} 100 g^{-1}. n, number of experiments. Data $\bar{x}$ ± SEM.

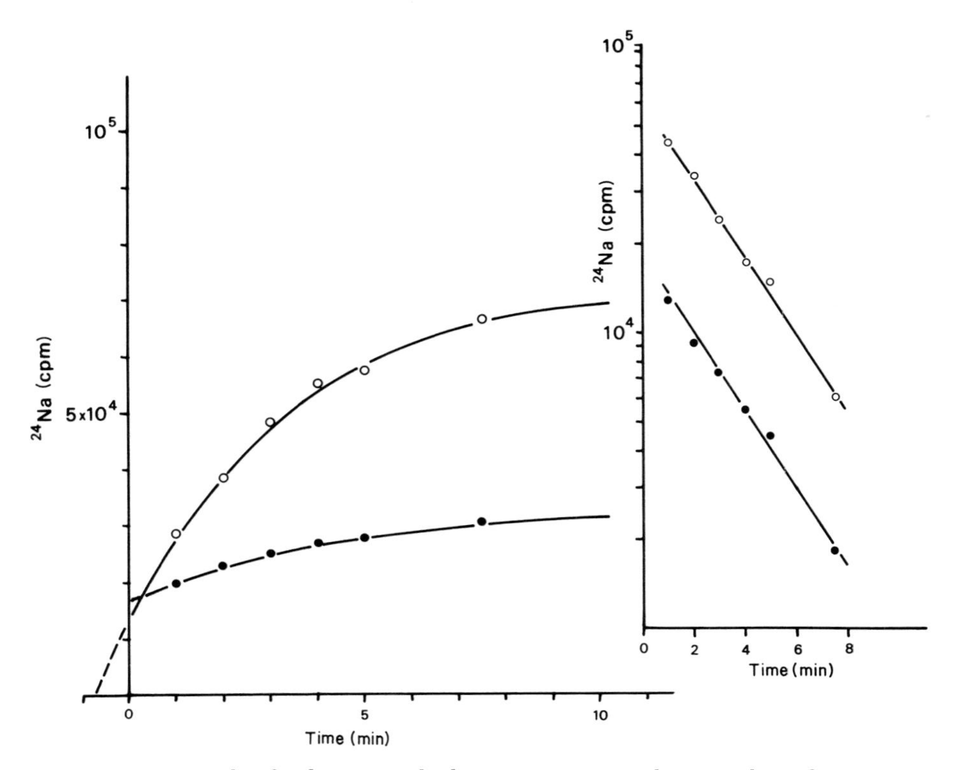

Fig. 10. Example of radioactive unloading experiments in the operculum of seawater-adapted *Fundulus heteroclitus*. Appearance of ^{24}Na in the serosal (○) and mucosal (●) chambers. Graphic analysis of the unloading curves (see legend of Fig. 8).

Table V

Transfer Coefficients (*K*) and Fluxes (*F*) for Na^+ and Cl^- through Apical (21) and Basal (23) Membranes of Chloride Cells of Seawater-Adapted Trout Gill and Opercular Membrane of *Fundulus*[a,b]

		K_{21}	K_{23}	F_{21}	F_{23}
Gill of seawater-adapted trout (n = 9)	Na^+	24 ± 1.2	2 ± 0.8	831 ± 201	54 ± 18
	Cl^-	23 ± 2.1	2 ± 0.2	498 ± 112	56 ± 12
Opercular membrane of *Fundulus*	Na^+ (n = 8)	5.9 ± 0.5	16.7 ± 1.4	—	—
	Cl^- (n = 13)	5.2 ± 0.5	16.3 ± 1.6	—	—

[a]Data for trout from Girard (1979); for *Fundulus*, from Mayer-Gostan *et al.* (in preparation).

[b]K is expressed as % min^{-1}; F is in $\mu Eq\ hr^{-1}\ 100\ g^{-1}$ for gill (and $\mu Eq\ cm^2\ hr^{-1}$ for opercula). n, number of experiments. Data $\bar{x}$ ± SEM.

In Table V the kinetic characteristics of gill and opercular chloride cells are compared. The total renewal rates of Na^+ and of Cl^- in the chloride cell compartment of the gill and that of the operculum are very similar (K_{21} + K_{23} not significantly different). Nevertheless, the chloride cells of the operculum differ from those of the gill in that their apical membranes represent a limiting factor in ionic transport. This difference may be a result of the experimental conditions employed. Thus, the opercular membrane was studied in the absence of an electrochemical gradient: membrane mounted in a Ussing chamber with Ringer solution on both sides and short-circuited. Further studies would be necessary to evaluate the importance of membrane potential and chemical gradients.

In the case of the chloride cells the interpretation of the transfer coefficients in terms of permeability must be made with caution in view of the histological structure of these cells. Their area of contact with the external medium (pit) is probably smaller than that of the basal membrane in contact with the blood. Calculating the ratio of apical to basal surface is complicated by the presence of a complex multicellular system of chloride cells (Sardet *et al.*, 1979), and the intercellular spaces may be derived from either the serous side or the mucous side.

C. Relationship of Flux to Chemical Gradient

Figure 11 gives the flux values across the apical and basal membranes as functions of the ionic (Na^+ and Cl^-) concentration differences between the

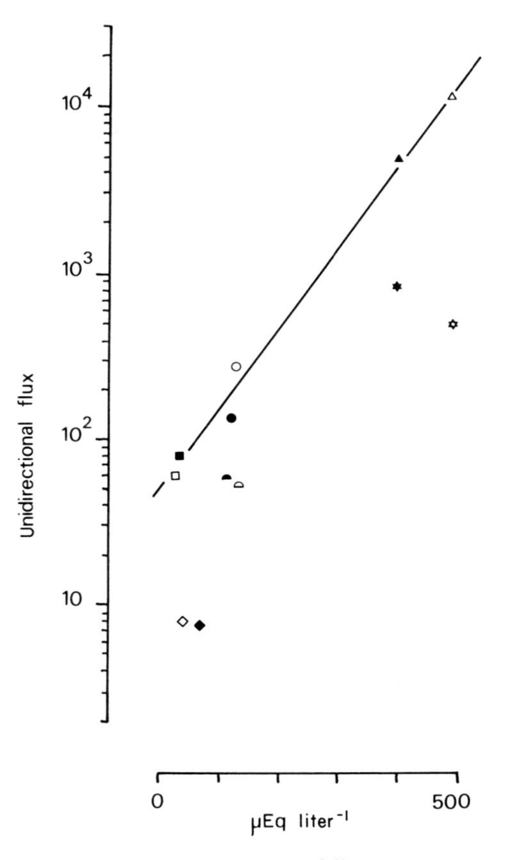

Fig. 11. Relationship between ionic concentration difference across apical (21) and basal (23) barriers of the lamelar and filamental epithelia and corresponding unidirectional flux (F) in freshwater- and seawater-adapted trout. ($y = 0.0049 \pm 0.001x \pm 0.44$; $n = 6$; $r = 0.996$; $p < .001$).

	Sodium	Chloride
Lamellar epithelium	▲ F_{21} seawater	△ F_{21} seawater
	● F_{23} seawater	○ F_{23} seawater
	■ F_{21} fresh water	□ F_{21} fresh water
	◆ F_{23} fresh water	◇ F_{23} fresh water
Filamental epithelium	★ F_{21} seawater	☆ F_{21} seawater
	◓ F_{23} seawater	◠ F_{23} seawater

two sides of the corresponding membranes (see Table IV). It can be seen that the fluxes through the lamellar epithelium of the seawater-adapted trout and those across the apical face of the lamellar epithelium in the freshwater-adapted trout are related to the same degree to the ionic concentration differences. This figure shows that fluxes through the lamellar epithelium are not saturable with regard to the osmotic gradient across this barrier, thus supporting the hypothesis that diffusional fluxes occur through this epithelium. If the membrane is symmetrical, it is difficult to see how the influxes and effluxes could be the result of simple passive diffusion, because this would not result in similar intensities. It is possible that a coupling mechanism of an exchange diffusion type occurs. However, the experimental conditions, described by Ussing (1960) and later by Motais (1973), under which an exchange diffusion can be characterized have not been applied, so this possibility must remain hypothetical.

The apical and basal fluxes through the filamental epithelium in seawater-adapted trout and fluxes through the basal membrane of the lamellar epithelium of freshwater-adapted trout (Payan, 1978) cannot be correlated with the ionic gradient (Fig. 11). This would seem to indicate that the exchanges under these conditions are the result of active ionic transport mechanisms: excretion of NaCl in seawater (Fig. 6) and active uptake in fresh water (Payan, 1978).

V. PHYSIOLOGICAL ROLES OF RESPIRATORY AND CHLORIDE CELLS

It would appear that in fresh water all ionic exchanges (except those of Ca^{2+}) occur through the lamellar epithelium. The respiratory cells are the site of coupled Na^+/NH_4^+, Na^+/H^+, and Cl^-/HCO_3^- exchanges (Girard and Payan, 1980). Payan (1978) and De Renzis (1978) have shown that the intensity of these exchanges is closely linked to the regulation of the internal acid–base equilibrium and the elimination of ammonium, the catabolic nitrogenous end product of ammoniotelic organisms.

This suggests that the respiratory cells, in addition to their well-known role in gas exchanges (Hughes, 1972), also carry out the important physiological role of acid and ammonium excretion, which in mammals is performed by the kidney. These conclusions are schematized in Fig. 12 in the model proposed by De Renzis (1978) for the freshwater teleost *Carassius auratus*.

We proposed the following model for seawater-adapted trout:

1. The lamellar epithelium is the site of Na^+/NH_4^+ exchanges and probably also Cl^-/HCO_3^- exchanges, and also of about 70% of the

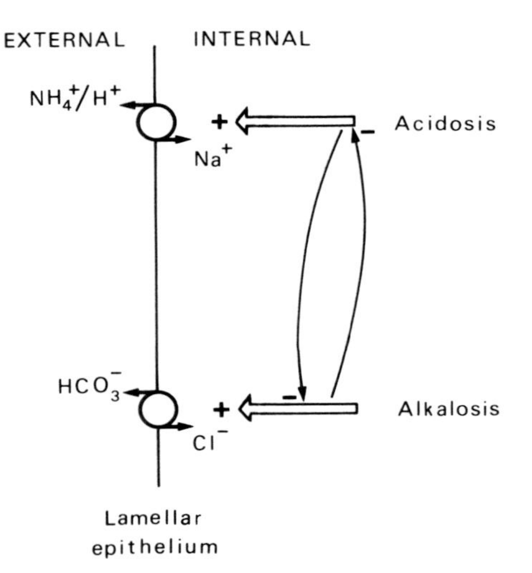

Fig. 12. Interrelations of Na^+/NH_4^+, or Na^+/H^+, and Cl^-/HCO_3^- exchange mechanisms in regulation of acid–base equilibrium and ammonia clearance from internal medium in a freshwater teleost, after De Renzis (1978). Acidosis, whether respiratory (Cameron, 1976) or metabolic due to absence of Na^+ in external medium (De Renzis and Maetz, 1973), stimulates Na^+/NH_4^+ exchange. Metabolic alkalosis caused by Cl^- lack in external medium (De Renzis and Maetz, 1973) intensifies Cl^-/HCO_3^- exchange.

total unidirectional fluxes of Na^+ and Cl^- assumed to be exchanged as Na^+/Na^+ and Cl^-/Cl^-.

2. The filamental epithelium is the site of net NaCl excretion corresponding to from 10 to 20% of the total fluxes. The predominance of chloride cells in this epithelium confirms the hypothesis of Keys and Willmer (1932), subsequently supported by numerous workers, that these cells are responsible for net NaCl excretion in marine teleosts. This model is schematized in Fig. 13.

The occurrence of chloride cells in an epithelium that is not irrigated by the arterial system implies that part of the blood that has bathed these cells is recycled in the branchial circulatory system instead of passing into the systemic circulation. Certain functional advantages may be proposed for this position in the primary gill lamella.

Fischbarg *et al.* (1977) have suggested that paracellular pathways of high permeability (particularly to substances of high molecular weight) are very probably the means by which nonosmotic water flux occurs. They maintain that in certain epithelia such as the gastric mucosa or the rumen epithelium (Moody and Durbin, 1969), hydraulic force determines water movement, which is related to epithelial permeability. Sardet *et al.* (1979) have de-

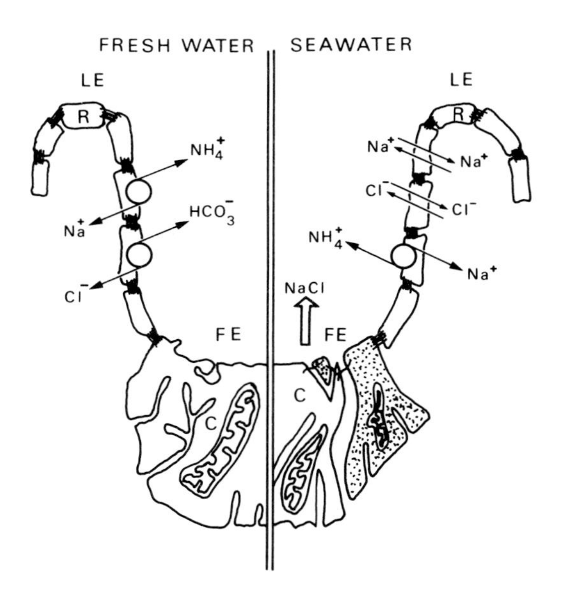

Fig. 13. Schematic illustration of the partition of Na^- and Cl^- exchanges across the lamellar and filamental epithelia (LE and FE) in fresh water- and seawater-adapted trout. C, chloride cell; R, respiratory cell.

scribed the chloride cells in seawater as bunches of cells linked together by leaky junctions forming a structure known as open. This observation is supported by conductance measurements of the opercular membrane of *Fundulus heteroclitus* (Degnan *et al.*, 1977; Mayer-Gostan and Maetz, 1980), which was found to be one of the most "leaky" membranes known. The work of Girard (1979) also showed that the filamental epithelia were much more permeable to nonelectrolytes (passing by a paracellular path) than was the lamellar epithelium. In the filamental epithelium we have a structure of high permeability, but because of its position in the arteriovenous circulatory system where hydraulic pressure is absent, it is not the site of a significant water flux. Thus, the gill strikes a compromise between its need to possess chloride cells with open junctions of the salt-excretory type described by Ernst and Mills (1977) and Kirschner (1977), and its need to avoid loss of water from the paracellular system of these cells.

REFERENCES

Cameron, J. N. (1976). Branchial ion uptake in arctic grayling: resting values and effects of acid–base disturbance. *J. Exp. Biol.* **64,** 711–725.

Claiborne, J. B., Evans, D. H., and Goldstein, L. (1982). Fish branchial Na^+/NH_4^+ exchange is via basolateral Na^+-K^+ activated ATPase. *J. Exp. Biol.* **96,** 431–434.

Degnan, K. J., Karnaky, Jr., J., and Zadunaisky, J. A. (1977). Active chloride transport in the *in vitro* opercular skin of a teleost (*Fundulus heteroclitus*), a gill-like epithelium rich in chloride cells. *J. Physiol.* (*London*) **271,** 155–191.

De Renzis, G. (1978). Etude des caractéristiques de l'absorption branchiale de Cl^- et de l'échange Cl^-/HCO_3^- chez *Carassius auratus:* aspect physiologique et biochimique. Relation avec la régulation de l'équilibre acidobasique du milieu intérieur. Thèse Doctorat, Université de Nice, France.

De Renzis, G., and Maetz, J. (1973). Studies on the mechanism of chloride absorption by the goldfish gill: relation with acid–base regulation. *J. Exp. Biol.* **59,** 339–358.

De Vooys, G. G. (1968). Formation and excretion of ammonia in teleostei. I: Excretion of ammonia through the gills. *Arch. Intern. Physiol. Biochim.* **76,** 268–272.

Dunel, S., and Laurent, P. (1977). La vascularisation branchiale chez l'anguille: action de l'acétylcholine et de l'adrénaline sur la répartition d'une résine polymérisable dans les différents compartiments vasculaires. *C. R. Hebd. Seances Acad. Sci.* **284,** 2011–2014.

Dunel, S., and Laurent, P. (1980). Functional organisation of the gill vasculature in different classes of fish. *In* "Epithelial Transport in the Lower Vertebrates" pp. 37–58. (B. Lahlou, ed.), Cambridge Univ. Press, London and New York.

Epstein, F. H., Maetz, J., and De Renzis, G. (1973). Active transport of chloride by the teleost gill: inhibition by thiocyanate. *Am. J. Physiol.* **224,** 1295–1299.

Ernst, S. A., and Mills, J. W. (1977). Basolateral plasma membrane localization of ouabaine sensitive transport sites in the secretory epithelium of the avian salt gland. *J. Cell Biol.* **75,** 74–94.

Evans, D. H. (1980). Osmotic and ionic regulation by freshwater and marine fishes. *In* "Environmental Physiology of Fishes" (M. A. Ali, ed.), pp. 93–122. Plenum Press, New York.

Evans, D., Clairborne, J. B., Farmer, L., Mallery, C., and Krasny, J. C. (1982). Fish gill ionic transport: methods and models. *Biol. Bull.* **163,** 108–130.

Fischbarg, J., Warshavsky, C. R., and Lim, J. J. (1977). Pathways for hydraulically and osmotically induced water flows across epithelia. *Nature* (*London*) **266,** 71–74.

Foskett, J. K., and Scheffey, C. (1982). The chloride cell: definitive identification as the salt secretory cell in teleosts. *Science* **215,** 164–166.

Girard, J. P. (1976). Salt excretion by the perfused head of trout adapted to seawater and its inhibition by adrenaline. *J. Comp. Physiol.* **111,** 77–91.

Girard, J. P. (1979). Etude du rôle respectif des cellules "à chlorure" et des cellules respiratoires dans les échanges ioniques branchiaux chez la truite adaptée à l'eau de mer (Salmo gairdneri): Mise en évidence de la séparation spatiale de ces deux types cellulaires. Thèse Doctorat, Université de Nice, France.

Girard, J. P., and Payan, P. (1977a). Kinetic analysis of sodium and chloride influxes across the gills of the trout in freshwater. *J. Physiol.* (*London*) **273,** 195–209.

Girard, J. P., and Payan, P. (1977b). Kinetic analysis and partitioning of sodium and chloride influxes across the gills of seawater adapted trout. *J. Physiol.* (*London*) **267,** 519–536.

Girard, J. P., and Payan, P. (1980). Ionic exchanges through respiratory and chloride cells in freshwater and seawater adapted teleosteans: adrenergic control *Am. J. Physiol.* **238,** R260–R268.

Hughes, G. M. (1972). Morphometrics of fish gills. *Respir. Physiol.* **14,** 1–25.

Hughes, G. M., and Morgan, M. (1973). The structure of fish gills in relation to their respiratory function. *Biol. Rev. Cambridge Philos. Soc.* **48,** 419–475.

Johansen, K. (1972). Heart and circulation in gill, skin and lung breathing. *Respir. Physiol.* **14,** 193–210.

Karnaky, K. J. (1972). A system to study the teleost chloride cell with the Ussing chamber. *Bull. Mt. Desert Isl. Biol. Lab.* **12,** 60–61.

Karnaky, K. J., Kinter, L. B., Kinter, W. B., and Stirling, C. E. (1976). II. Autoradiographic localization of gill Na, K-ATPase in killifish *Fundulus heteroclitus* adapted to low and high salinity environments. *J. Cell Biol.* **70,** 157–177.

Kerstetter, T. H., and Keeler, M. (1976). On the interaction of NH_4^+ and Na^+ fluxes in the isolated trout gills. *J. Exp. Biol.* **64,** 517–527.

Kerstetter, T. H., Kirschner, L. B., and Rafuse, D. (1970). On the mechanisms of sodium ion transport by the irrigated gills of rainbow trout (*Salmo gairdneri*) *J. Gen. Physiol.* **56,** 342–359.

Keys, A., and Willmer, E. N. (1932). "Chloride secreting cells" in the gills of fishes, with special reference to the common eel. *J. Physiol.* (*London*) **76,** 368–378.

Kirschner, L. B. (1977). The sodium chloride excreting cells in marine Vertebrates. *In* "Transport of Ions and Water in Animals" (B. L. Gupta, R. B. Moreton, J. L. Oschman, B. J. Wall, eds.), pp. 427–452. Academic Press, London.

Kirschner, L. B., Greenwald, L., and Kerstetter, T. H. (1973). Effect of amiloride on sodium transport across body surfaces of freshwater animals. *Am. J. Physiol.* **224,** 832–837.

Krogh, A. (1939). "Osmotic Regulation in Aquatic Animals." Cambridge University Press.

Laurent, P., and Dunel, S. (1976). Functional organization of the teleost gill. I. Blood pathways. *Acta Zool.* (*Stockholm*) **57,** 189–209.

Laurent, P., and Dunel, S. (1978). Relations anatomiques des ionocytes (cellules à chlorure) avec le compartiment veineux branchial: définition de deux types d'épithélium de la branchie des Poissons. *C. R. Hebd. Seances Acad. Sci.* **286,** 1447–1450.

Maetz, J. (1969). Seawater teleosts: evidence for a sodium-potassium exchange in the branchial sodium-excreting pump. *Science* **166,** 631–615.

Maetz, J. (1971). Fish gills: mechanisms of salt transfer in fresh water and sea water. *Philos. Trans. R. Soc. London Ser. B* **262,** 109–249.

Maetz, J. (1973). Na^+/NH_4^+, Na^+/H^+ Exchanges and NH_3 movement across the gill of *Carassius auratus. J. Exp. Biol.* **58,** 255–275.

Maetz, J., and Bornancin, M. (1975). Biochemical and biophysical aspects of salt excretion by chloride cells in teleosts. *Fortschr. Zool.* **23,** 322–362.

Maetz, J., and Garcia-Romeu, F. (1964). The mechanism of sodium and chloride uptake by the gills of a freshwater fish, *Carassius auratus.* II. Evidence for NH_4^+/Na^+ and HCO_3^-/Cl^- exchanges. *J. Gen. Physiol.* **47,** 1209–1227.

Maetz, J., and Pic, P. (1975). New evidence for a Na/K and Na/Na exchange carrier linked with the Cl^- pump in the gill of *Mugil capito* in sea water. *J. Comp. Physiol.* **102,** 85–100.

Maetz, J., Payan, P., and De Renzis, G. (1976). Controversial aspects of ionic uptake in freshwater animals. *In* "Perspectives in Experimental Biology," Vol. I (S. Davies, ed.), pp. 77–92. Pergamon Press, New York.

Mayer, N. (1970). Contrôle endocrinien de l'osmorégulation chez les téléostéens, Rôle de l'axe hypophyse-interrenal et de la prolactine. *Bull Inf. Sci. Tech. Commis. Energ. At.* (*Fr.*) **86,** 11–70.

Mayer-Gostan, N., and Maetz, J. (1980). Ionic exchanges in the opercular membrane of *Fundulus heteroclitus* adapted to sea water. *In* "Transport in Lower vertebrates," (B. Lahlou, ed.), pp. 233–250. Cambridge University Press, London and New York.

Moody, F. G., and Durbin, R. P. (1969). Water flow induced by osmotic and hydrostatic pressure in the stomach. *Am. J. Physiol.* **217,** 255–262.

Morgan, M., and Tovell, P. W. A. (1973). The structure of the gill of the trout, *Salmo gairdneri. Z. Zellforsch. Mikrosk. Anat.* **142,** 147–162.

Motais, R. (1973). Sodium movements in high-sodium beef red cells: properties of a ouabain-insensitive exchange diffusion. *J. Physiol.* **233,** 395–422.

Motais, R., and Garcia-Romeu, F. (1972). Transport mechanisms in the teleostean gill and amphibian skin. *Annu. Rev. Physiol.* **34,** 141–176.

Newstead, J. D. (1967). Fine structure of the respiratory lamellae of teleostean gills. *Z. Zellforsch. Mikrosk. Anat.* **79,** 396–428.

Payan, P. (1978). A study of the Na^+/NH_4^+ exchange across the gill of the perfused head of the trout (*Salmo gairdneri*) *J. Comp. Physiol.* **124,** 181–188.

Payan, P., and Matty, A. J. (1975). The characteristics of ammonia excretion by a perfused isolated head of trout (*Salmo gairdneri*): Effect of temperature and CO_2-free ringer. *J. Comp. Physiol.* **96,** 167–184.

Payan, P., and Pic, P. (1977). Origine de l'ammonium excrété par les branchies chez la truite (*Salmo gairdneri*). *C. R. Hebd. Seances Acad. Sci.* **284,** 2519–2522.

Payan, P., Mayer-Gostan, N., and Pang, P. K. T. (1981). Site of calcium uptake in the fresh water trout gill. *J. Exp. Zool.* **216,** 345–347.

Randall, D. J. (1970). The circulatory system. *In* "Fish Physiology," Vol. IV (W. S. Hoar and D. J. Randall, eds.), pp. 133–168. Academic Press, New York.

Richards, B. D., and Fromm, P. O. (1969). Patterns of blood flow through filaments and lamellae of isolated perfused rainbow trout gills. *Am. Zool.* **8,** 766–771.

Sardet, C., Pisam, M., and Maetz, J. (1979). The surface epithelium of teleostean fish gills. Cellular and junctional adaptations of the chloride cell in relation to salt adaptation. *J. Cell Biol.* **80,** 96–117.

Sardet, C., Pisam, M., and Maetz, J. (1980). Structure and function of gill epithelium of euryhaline teleost fish. *In* "Epithelial Transport in Lower Vertebrates" pp. 59–68. (B. Lahlou, ed.), Cambridge University Press, London and New York.

Sargent, J. R., Pirie, B. J. S., Thomson, A. J., and George, S. G. (1977). Structure and function of chloride cells in the gills of *Anguilla anguilla. In* "Physiology and Behaviour of Marine Organisms" (M. McLusky, and A. J. Berry, eds.), pp. 123–132. Pergamon Press, New York.

Silva, P., Solomon, R., Spokes, K., and Epstein, F. H. (1977). Ouabain inhibition of gill Na/K ATPase: relationship to active chloride transport. *J. Exp. Zool.* **199,** 419–426.

Smith, H. (1930). The absorption and excretion of water and salts by marine teleosts. *Am. J. Physiol.* **93,** 480–505.

Smith, H. W. (1932). Water regulation and its evolution in fishes. *Quart. Rev. Biol.* **7,** 1–26.

Steen, J. B., and Kruysse, A. (1964). The respiratory function of teleostean gills. *Comp. Biochem. Physiol.* **12,** 127–142.

Ussing, H. H. (1960). The frog skin potential. *J. Gen. Physiol.* **43,** 135–147.

Vogel, W., Vogel, V., and Kremers, H. (1973). New aspects of the intrafilamental vascular system in gills of a euryhaline teleost *Tilapia mossambica. Z. Zellforsch. Mikrosk. Anat.* **144,** 573–583.

Vogel, W., Vogel, V., and Schlote, W. (1974). Ultrastructural study of arteriovenous anastomoses in gill filaments of *Tilapia mossambica. Cell Tissue Res.* **155,** 491–512.

Vogel, W., Vogel, V., and Pfautsch, M. (1976). Arterio-venous anastomoses in rainbow trout gill filaments. *Cell Tissue Res.* **167,** 373–385.

Whittam, R., Guinnebault, M. (1960). The efflux of potassium from electroplaques of electric eels. *J. Gen. Physiol.* **43,** 1171–1191.

3

ION TRANSPORT AND GILL ATPASES*

GUY DE RENZIS AND MICHEL BORNANCIN

Laboratoire de Physiologie Cellulaire et Comparée
Université de Nice
Nice, France

I. INTRODUCTION

Since Smith's pioneering work (1930), successive studies on fish gills have shown not only their important role in osmoregulation but also their role in the regulation of acid–base balance. The gill epithelium is located between two liquid compartments of very different ionic composition. The gills are the site not only of entry for selected ions essential to life (Na^+, K^+,

*The authors acknowledge support from the Centre National de la Recherche Scientifique (ERA 943) and the Commissariat à l'Energie Atomique, Département de Biologie.

ISBN 0-12-350432-5

Cl^-), but also for extrusion of other ions such as HCO_3^-, NH_4^+, and H^+, which are the ionic forms of metabolic by-products.

The magnitude of the unidirectional fluxes, as well as the degree and direction of the net fluxes, is dependent on the equilibrium established between the fish and its environment. Knowledge of the ionic composition of the external as well as the internal medium of the fish is essential, therefore, to explain the ionoregulatory capacities of the gill epithelium.

It is within marine fishes that the transbranchial fluxes of sodium chloride are the greatest. The gills actively excrete ions, which necessitates a significant energy expenditure to compensate for the influx of ions resulting from the difference in ionic composition between the internal and external environments. Even though the chemical gradient is smaller for freshwater fish, extraction of salts from a dilute environment requires mechanisms that themselves are energy dependent.

The only known source of energy for ionic transfer across membranes is ATP; certain ions such as Na^+, K^+, and Ca^{2+} are transported by molecular carriers that hydrolyze ATP as a source of energy. The enzymes that effect the utilization of stored energy from ATP are the ATPases (EC 3.6.1.4). Among those that play a role in ionic transport are (Na^+,K^+)-ATPase, Ca^{2+}-ATPase, and K^+-ATPase. In the gills, the (Na^+,K^+)-ATPase and Ca^{2+}-ATPase have been identified and studied. Some others such as the anion-dependent ATPase have been described in the gill. The presence of an ATPase in epithelial membranes, however, does not infer that it has a role in transepithelial ionic transport. Although it has been demonstrated on vesicular plasma membranes and liposomes as well as on whole cells that (Na^+,K^+)-ATPase and Ca^{2+}-ATPase are ionic pumps, the presence of these enzymes in the gills does not necessarily imply these molecules in transbranchial ionic transfers. We will see that our knowledge on this subject is still too fragmentary to define exactly the role of ATPases in transbranchial ionic transport.

II. ATPASES: GENERAL CONSIDERATIONS

A. Identification

ATPases catalyze the release of phosphate from ATP in the presence of Mg^{2+} and an adequate ionic environment containing, respectively, Na^+ and K^+ for (Na^+,K^+)-ATPase, Ca^{2+} for Ca^{2+}-ATPase, certain anions for anion-ATPase, and so on. This hydrolysis reaction may be inhibited by cardiac glycosides like ouabain in the case of (Na^+,K^+)-ATPase or by such substances as vanadate [(Na^+,K^+)-ATPase], ruthenium red (Ca^{2+}-ATPase), oligomycin, or carboxyatractyloside (anion-ATPase).

These enzymes do not display a strict specificity for ATP, since other triphosphate nucleosides may be broken down as well. However the affinity for the latter substances is lower than for ATP. Although Mg^{2+} may be replaced with cobalt or manganese, the true substrate in the reaction is the ATP–Mg^{2+} complex. In the presence of this complex, the hydrolytic activity of ATPase is described as Mg^{2+} dependent or as residual activity. This activity is enhanced (up to 20 times) on addition of certain ions. There is no absolute requirement concerning the ionic species involved. Thus, while Na^+ is needed for the expression of (Na^+,K^+)-ATPase, K^+ may be replaced by other cations with more or less efficiency, the latter decreasing in the following order: $Tl^+ > Rb^+ > K^+ > NH_4^+ > Cs^+ > Li^+$, the affinity for thallium being 10 times that for potassium. With regard to Ca^{2+}-ATPase, it was observed that Ca^{2+} may be replaced with cobalt or strontium. Anion-ATPase is strongly stimulated by HCO_3^-, but also by many other ions, whether inorganic (SO_3^{2-}, SO_4^{2-}, Cl^-, etc.) or organic (acetate, citrate, etc.).

B. Purification, Structure, and Reconstitution

Most commonly, membranes are isolated as a microsomal fraction, which is made up of several components. In salt-transporting epithelia, plasma membranes account for the major part of this fraction.

1. Purification

Two protocols are generally used to separate (Na^+,K^+)-ATPase from other protein components of the membranes:

1. Membranes are gently treated with ionic detergents such as deoxycholate, dodecyl sulfate, or chaotropic agents (e.g., NaI) in order to release proteic constituents that are less strongly associated with phospholipids, (Na^+,K^+)-ATPase remaining in situ. A classical example is the purification of kidney (Na^+,K^+)-ATPase as described by Jørgensen (1975). Fractions enriched in enzyme activity are then separated on sucrose gradient, the maximum specific activity attaining 1800 μmol P_i hr^{-1} mg $protein^{-1}$.
2. Nonionic detergents are used to solubilize gross fractions containing the enzyme. The extract is submitted subsequently to classical techniques of protein purification. This second method was utilized by Hokin (1977) for the electrical organ of the electric eel, yielding a maximum specific activity of 900 μmol P_i hr^{-1} mg $protein^{-1}$.

Intermediate methods have also been devised. Thus, Esmann *et al.* (1979) purified the enzyme by using deoxycholate, then solubilized it with octoethylene glycol–dodecylether ($C_{12}E_8$).

Ca^{2+}-ATPase also meets criteria of intrinsic membrane proteins and therefore must be extracted from the lipid matrix by employing detergents. Both of the approaches just described have been used for this enzyme. Ikemoto *et al.* (1972) purified it by detergent treatment followed by chromatography on Sephadex. Specific activity reaches 900 μmol P_i hr^{-1} mg $protein^{-1}$. Warren *et al.* (1974) chose the second approach, consisting of deoxycholate elimination of reticulum extrinsic protein followed by fractionation on sucrose gradient. The activity obtained was 720–900 μmol P_i hr^{-1} mg $protein^{-1}$.

Anion-sensitive ATPase was solubilized with Triton X-100, a nonionic detergent (Wiebelhauss *et al.*, 1971). This method yielded an increase in specific activity by a factor of three to seven. Because no other type of separation was attempted, no structural study has been made available for this enzyme.

2. Structure

Although virtually pure preparations of (Na^+,K^+)-ATPase allowed for the characterization of its structural elements, it is not possible at the moment to propose a stoichiometry or a subunit arrangement of the functional transporting unit.

All purified (Na^+,K^+)-ATPases are made up of two polypeptide chains (α and β) of different molecular weights: 95,000 and 45,000–57,000 (depending on the source).

1. The α subunit is phosphorylated during ATP hydrolysis. It bears the binding site of ouabain. Because this site is on the outer membrane side whereas the phosphorylation site is on the inner face, this peptide obviously spans the entire thickness of the membrane.
2. The β subunit is heavily glycosylated. The quantity and chemical nature of the glycopeptides depend on the biological material used for extraction. Although this peptide copurifies with the subunits, no functional role has yet been ascribed to it.

There is no agreement regarding α:β stoichiometry, but the ratio 1:1 is the most frequently quoted (Esmann *et al.*, 1979). Several authors have also described the presence of a γ subunit of low molecular weight (12,000).

Attempts to determine molecular weight of (Na^+,K^+)-ATPase by gel filtration proved disappointing, proposed values ranging between 280,000 and 500,000. The most reliable result was obtained by Kepner and Macey (1968) by using irradiation inactivation technique. According to these authors, a 280,000 MW unit is necessary for the expression of enzyme activity.

Analysis of results on electron microscopy of purified (Na^+,K^+)-ATPase preparations after negative staining or freeze-fracturing suggests a symmetrical, probably tetrameric structure. Altogether, available structural data are in favor of an α_2, β_2-subunit arrangement (Maunsbach *et al.*, 1979; Esmann *et al.*, 1979).

Highly purified fractions of Ca^{2+}-ATPase submitted to SDS–polyacrylamide gel electrophoresis reveal the presence of a major band corresponding to an apparent molecular weight of 102,000. There is a remarkable similarity in amino acid sequence of this subunit and of the 100,000 MW subunit of (Na^+,K^+)-ATPase.

By irradiation inactivation technique, Vegh *et al.* (1968) determined a molecular weight of 190,000 for the unit responsible for Ca^{2+}-ATPase activity. Murphy (1976) and Le Marie *et al.* (1976) indicated a molecular weight of 400,000 after using other methods (phenanthroleine–Cu^{2+} reticulation; sedimentation in nonionic detergent).

3. Reconstitution

Lipid-free (Na^+,K^+)-ATPase is completely inactive. One can regenerate the catalytic activity by incubating the enzyme protein fraction with increasing concentration of phospholipids. Ottolenghi (1979) obtained maximum regeneration of activity when 72 molecules of dioleylphosphatidylethanolamine and 102 molecules of dioleylphosphatidylcholine were associated with 1 mol of the enzyme.

Several investigators have now demonstrated that purified (Na^+,K^+)-ATPase is able to catalyze ATP-dependent and ouabain-inhibitable coupled Na^+/K^+ transport when reconstituted into phospholipid vesicles. These reconstitution experiments have provided the first evidence that (Na^+,K^+)-ATPase is in fact the sodium pump. Using a highly efficient *in vitro* transport system, Goldin (1977) has shown that the active transport of Na^+ and K^+ is coupled to ATP hydrolysis in approximately a 3:2:1 ratio.

Ca^{2+}-ATPase is inhibited irreversibly by complete extraction of the phospholids participating in its structure. Warren *et al.* (1974) succeeded in regenerating 100% of the original activity after incubation of partially delipidized enzyme in the presence of various phospholipid solutions. The fixation stoichiometry was 90 mol of dioleylecithine per mole of completely active enzyme. Cholesterol was totally excluded from the lipids surrounding Ca^{2+}-ATPase.

It is possible to reconstitute a system pumping Ca^{2+} actively by using Ca^{2+}-ATPase highly purified by sonication in the presence of a phospholipid–cholate mixture, followed by dialysis (Racker and Eytam, 1973). The technique is identical to that described for (Na^+,K^+)-ATPase. This system can accumulate Ca^{2+} against an electrochemical gradient and utilizes ATP as energy source. It posseses the characteristics of the Ca^{2+} pump acting *in vivo*.

C. Universality and Main Functions

The (Na^+,K^+)-dependent ATPase is an enzyme that has been described since 1960 (see Skou, 1965). It can be considered as an electrogenic cationic pump (Hilden and Hokin, 1975) that actively transports sodium out of and potassium into animal cells through the plasma membranes. Its main function is to maintain the separation of ions across the plasma membrane, assuring the constancy of the ionic concentrations in the intracellular environment and the osmotic equilibrium with the external milieu. Even though it is present in the membranes of all vertebrate cells, the (Na^+,K^+)-dependent ATPase displays a greater activity in some tissues like the electric organ from electric eels, or the epithelia of effector organs involved in hydromineral and osmotic control (e.g., the kidneys of higher vertebrates, rectal glands in marine elasmobranchs, salt glands of reptiles and birds, gills of marine teleosts). In these latter examples the membrane (Na^+,K^+)-ATPase not only contributes to the physiological integrity of epithelial cells, but because of the particular cellular structure and polarity, it also participates in the transepithelial transport of NaCl.

Calcium is involved in the regulation of a large number of cellular functions whose expression depends on the cytoplasmic concentration of ionized calcium. A low concentration of cytoplasmic calcium is, in fact, an essential condition to cellular life. In vertebrates, the plasma membranes maintain a large difference in concentration between the intracellular and extracellular fluids; for example, in erythrocytes the intracytoplasmic concentration of Ca^{2+} is less than 1 μM, being more than 1000 times less than that of plasma. Despite a low permeability of the membrane to Ca^{2+}, an active transport of Ca^{2+} toward the exterior of the cell is necessary to maintain this low intracellular concentration of Ca^{2+}. The mechanism of this transport is not known, except for muscle cells and red blood cells, where it is accomplished by a calcium pump dependent on ATP. The erythrocyte membrane contains a calmodulin-dependent Ca^{2+}-ATPase, with maximum activation by calcium occurring at a concentration of 10 μM (Scharff, 1980). The enzymatic activity is dependent on other cations, particularly K^+, which in a number of tissues modulate the transport of Ca^{2+} (Adams and Duggan, 1981).

III. (Na^+,K^+)-DEPENDENT ATPASE

The presence of a (Na^+,K^+)-dependent ATPase in the gills of fish was demonstrated for the first time simultaneously, by two different research groups: Epstein *et al.* (1967) in the gills and pseudobranchs of *Fundulus heteroclitus*, and Kamiya and Utida (1968) in the gills of Japanese eels (*Anguilla japonica*) adapted to fresh water or seawater. Because the activity of the enzyme was strongly enhanced in the gills of those fish after their adaptation to seawater, the (Na^+,K^+)-ATPase was implicated in the excretion of salt from marine fish.

These first results were confirmed during the following 5 years in numerous euryhaline species where a (Na^+,K^+)-dependent ATPase activity was measured in the gills. The activity in freshwater fish gills was, however, significantly less than in marine forms. Among the most mentioned examples were *Anguilla rostrata* (Jampol and Epstein, 1970; Butler and Carmichael, 1972), *Anguilla anguilla* (Motais, 1970; Milne *et all*, 1971; Bornancin and de Renzis, 1972), *Salmo gairdneri* (Kamiya and Utida, 1969; Pfeiler and Kirschner, 1972), and several different salmonid species (Zaugg and McLain, 1970, 1972; Zaugg *et al.*, 1972) (Table I).

In stenohaline fish the (Na^+,K^+)-ATPase activities are 4–10 times greater in marine species compared to those in fresh water (Kamiya and Utida, 1969; Jampol and Epstein, 1970). This general observation has rare exceptions as Lasserre (1971) demonstrated in *Chelon labrosus* and *Dicentrarchus labrax* (Table I).

A. Branchial Localization

1. Biochemical Studies

The first investigations to localize the (Na^+,K^+)-dependent ATPase in gill membranes were conducted according to fractionating methods permitting a separation of different cell populations based on their size and density. Kamiya (1972) used an elastase to dissociate the gill filaments of *Anguilla japonica*, and Sargent *et al.* (1975) caused mechanical disaggregation of the branchial epithelium of *Anguilla anguilla* in a medium without Ca^{2+} to obtain preparations of isolated cells. It is possible to separate, by density gradient centrifugation, some populations of isolated cells enriched either with respiratory or chloride cells, which can be easily identified with a light microscope. The cells obtained have lost their original epithelial orientation, and their structure is altered. These cells cannot be used for physiological measurements, but biochemical studies are possible.

It is possible to identify the mitochondria-rich chloride cells by their high content of succinic dehydrogenase (SDH) or (Na^+,K^+)-ATPase. Table II shows the measured activities on isolated cells from the gills of Japanese and European eels. This table illustrates the differences in the activity of the (Na^+,K^+)-dependent ATPase between the gills of animals adapted to fresh water or seawater.

Table I

(Na⁺,K⁺)-Dependent ATPase Activities in Homogenate and Microsomes from Gills of Some Teleosts

Species and acclimation medium[a]	Preparation procedures	Conditions of measurement: Temperature (°C)	pH	Na^+,K^+ (m*M*)	Homogenate activities[b]: Residual	Na^+,K^+	Microsomes: (g min^{-1})	Activities[b]: Residual	Na^+,K^+	Reference
Anguilla rostrata										
FW	Deoxycholate 1‰	37	7.8	Na^+ = 100 K^+ = 20	11.5 ± 0.6	6 ± 0.6	—	—	—	Jampol and Epstein (1970)
FW	Deoxycholate 1‰	37	7.8	Na^+ = 100	9.4 ± 5.6	4.6 ± 1.2	—	—	—	Epstein *et al.* (1971)
SW				K^+ = 20	8.5 ± 2.6	6.8 ± 1.4				
Anguilla japonica										
FW	—	25	7.5	Na^+ = 100	1.5	0	2×10^5<	1.7	0.9	Kamiya and Utida (1968)
SW	—			K^+ = 20	1.7	0.9	<3×10^6	7.6	3.1	
FW	NaI 2 *M*	25	7.5	—	—	—	1×10^4<	—	8.8 ± 1.0	
SW							<3×10^6	—	45.2 ± 2.8	
FW	—	25	7.0	Na^+ = 100 K^+ = 20	—	2.5 ± 0.3	H[c] 1.5×10^5< <1.1×10^6	—	H 14.4 ± 1.2 L 5.3 ± 0.7	Ho and Chan (1980a)
SW					—	6.1 ± 0.6	L[c] 1.1×10^6< <6×10^7	—	H 32.3 ± 3.4 L 11.0 ± 3.3	
Anguilla anguilla										
FW	Deoxycholate 1‰	37	7.8	Na^+ = 100	9.3 ± 0.7	5.5 ± 0.6	—	—	—	Motais (1970)
SW				K^+ = 20	12.4 ± 0.9	10.8 ± 0.8				
FW	Deoxycholate 1‰	30	7.0	—	3.0	3.1	2.2×10^5<	5.0	10.1	Sargent and Thomson (1974)
SW					4.3	7.9	<6×10^6	6.8	37.3	
FW	Deoxycholate 0.6‰	26	7.2	Na^+ = 50	2.5 ± 0.3	0.9 ± 0.2	7×10^5<	3.8 ± 0.4	1.7 ± 0.7	Maetz and Bornancin (1975)
SW				K^+ = 20	6.6 ± 0.7	6.5 ± 0.8	<5×10^6	8.2 ± 0.5	9.4 ± 1.2	
Carassius auratus 25°C	—	25	7.5	Na^+ = 100 K^+ = 20	—	—	4×10^5< <3×10^6	12.2 ± 0.9	2.1 ± 0.2	Murphy and Houston (1974)
Chelon labrosus										
FW	—	37	7.8	Na^+ = 8 to 508	—	—	1×10^4<	22.3 ± 2.9	7.7 ± 1.1	Lasserre (1971)
SW				K^+ = 20			<1.8×10^6	23.1 ± 3.4	3.3 ± 0.9	

FW	—	37	7.8	Na^+ = 100	—	—	$4.5 \times 10^4<$	32.3 ± 3.8	22.6 ± 1.8	Gallis and Bourdichon (1976)
SW				K^+ = 20			$<1.44 \times 10^6$	30.1 ± 2.5	10.4 ± 0.4	
Cyprinodon salinus										
FW 25°C	EDTA 70 n*M*	37	7.2	Na^+ = 240	55.2	28.0 ± 1.3	—	—	—	Stuenkel and Hillyard (1980)
SW	Osmotic concentration 810 mOs			K^+ = 120	51.5	25.3 ± 1.2				
Cyprinodon variegatus										
50% SW	—	37	7.2	Na^+ = 60	20.6 ± 1.1	4.1 ± 0.5	—	—	—	Karnaky *et al.* (1976a)
SW				K^+ = 5	23.5 ± 1	6.5 ± 0.8				
200% SW					26.0 ± 0.8	25.5 ± 0.8				
Fundulus heteroclitus										
FW	Deoxycholate 1‰	37	7.8	Na^+ = 100	14.6	3.6	$3 \times 10^5<$	23.6	12.4	Epstein *et al.* (1967)
SW				K^+ = 20	18.6	20.1	$<6.3 \times 10^6$	31.5	84.4	
SW	—	37	7.8	Na^+ = 100 K^+ = 20	24.8	11.1	—	—	—	Jampol and Epstein (1970)
10% SW	—	37	7.2	Na^+ = 60	—	3.5 ± 0.3	—	—	—	Karnaky *et al.* (1976b)
SW						5.7 ± 0.7				
200% SW				K^+ = 5		12.1 ± 0.9				
Oncorhynchus kisutch										
FW	—	37	7.8	Na^+ = 100 K^+ = 20	—	—	$6 \times 10^4<$ $<4.4 \times 10^6$		Seasonal variations 3.8 ± 0.5 to 27.8 ± 4.0	Lasserre *et al.* (1978)
SW 2 months	—	37	7.8	Na^+ = 100 K^+ = 20	—	—	—	14.3 ± 2	32 ± 2	Boeuf *et al.* (1978)
Salmo gairdneri										
FW	NaI 2 *M*	13	7.1	Na^+ = 100	—	—	$9.7 \pm 10^4<$	10.4 ± 0.9	1.3 ± 0.8	Pfeiler and Kirschner (1972)
SW				K^+ = 20			$<2.1 \times 10^6$	6.8 ± 0.7	3.4 ± 0.5	
FW	NaI 2 *M*	37	7.1	Na^+ = 100	—	—		7.2 ± 0.9	12.6 ± 1.7	
SW				K^+ = 20				4.2 ± 0.8	23.6 ± 2.1	
Tilapia mossambica										
FW	Deoxycholate 1‰	26	7.2	Na = 60	2.9 ± 0.3	0.6 ± 0.1	$7 \times 10^5<$	6 ± 0.8	1.5 ± 0.1	Dharmamba *et al.* (1975)
SW				K = 20	5.8 ± 0.3	1.6 ± 0.3	$<5 \times 10^6$	11.4 ± 0.6	3.5 ± 0.8	

[a] FW, fresh water; SW, seawater.

[b] ATPase activities in μmol P_i hr^{-1} mg $protein^{-1}$.

[c] H, heavy; L, light.

Table II

(Na^+,K^+)-ATPase and Mg^{2+}-ATPase Activities in Cell Fractions Isolated by Density Gradient Centrifugation from Disrupted Gill Suspensions

		Anguilla japonica (Kamiya, 1972)		*Anguilla anguilla* (Sargent *et al.*, 1975)		*Lagodon rhomboides* (Hootman and Philpott, 1978)	
Acclimatization medium	Cell fraction[b]	Residual ATPase	(Na^+,K^+)-ATPase	Residual ATPase	(Na^+,K^+)-ATPase	Residual ATPase	(Na^+,K^+)-ATPase
Fresh water	ZRC	2.4 ($n = 2$)	−0.1 ($n = 2$)	3.15 ± 0.6 ($n = 4$)	1.9 ± 0.4 ($n = 4$)	—	—
	ZCC	3.35	−1	3.25 ± 1.3	4.9 ± 1.1	—	—
Seawater 33.3%	ZRC	—	—	—	—	13.5 ± 2.1 ($n = 4$)	2.3 ± 0.9 ($n = 4$)
	ZCC	—	—	—	—	24.4 ± 2.1	10.3 ± 0.3
Seawater	ZRC	6.55 ($n = 3$)	0.65 ($n = 3$)	8.25 ± 0.6 ($n = 4$)	1.1 ± 0.4 ($n = 4$)	17.9 ± 5.7 ($n = 4$)	5.7 ± 3.1 ($n = 4$)
	ZCC	15.9	6.9	13.0 ± 3	38.9 ± 6.5	32.2 ± 4.1	37.2 ± 8.8

[a]Activities in μmol P_i hr^{-1} mg $protein^{-1}$.

[b]ZRC, zone containing light cells, principally respiratory cells; ZCC, zone enriched with chloride cells.

The most remarkable observation is that, in seawater-adapted fish, the enzyme is 10–30 times more active in cellular fractions enriched with chloride cells compared with fractions containing respiratory cells. Hootman and Philpott (1978) subsequently obtained some comparable results (Table II) on *Lagodon rhomboides*, confirming the presence of (Na^+,K^+)-ATPase in the chloride cells from the gills of fish acclimated to seawater.

This enzyme activity is thus not only related to the importance of branchial ionic fluxes in fish adapted to seawater, but also to the particular cell types that differentiate in the course of adaptation of euryhaline fish species to seawater. This was demonstrated by the work of Thomson and Sargent (1977), who showed, using dissociated gill cells, that the number of chloride cells and the activity of (Na^+,K^+)-ATPase in the gills of immature yellow eels increased in parallel during the course of adaptation to seawater, a maximum being reached after 2 weeks in seawater (Fig. 1).

2. Autoradiographic, Histological, and Histochemical Studies

These studies, more direct than the preceding ones, tend to use certain characteristics of the (Na^+,K^+)-dependent ATPase to locate the enzyme in

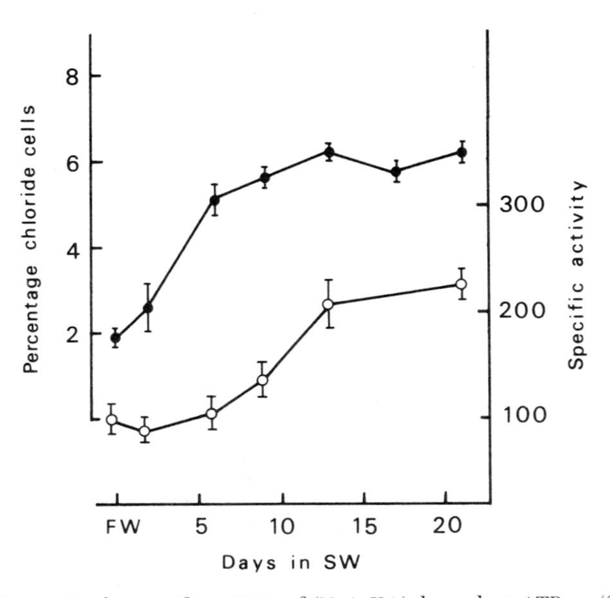

Fig. 1. Changes in the specific activity of (Na^+,K^+)-dependent ATPase (○) and in the percentage of chloride cells in total cells released from the gills of yellow eels adapting to seawater (●). Specific activity, 100% = 26.13 μmol P_i hr^{-1} mg protein^{-1}. (From Thomson and Sargent, 1977.)

situ and hence to localize the cells and intracellular structures that contain it. In order to clarify the discussion, we will begin with the autoradiographical studies, even though studies using histochemical methods of localization were conducted first.

a. Autoradiographic Studies. Ouabain is known to bind selectively with and inhibit (Na^+,K^+)-ATPase. This inhibition is potassium sensitive and only occurs on the side of the enzyme opposite to the site of ATP hydrolysis. The number of sites sensitive to potassium is approximately equal to the number of phosphorylation sites (Wallick *et al.*, 1979). The use of tritiated ouabain is therefore an appropriate technique to study the localization of the enzyme by autoradiography, especially because one can alter ouabain binding by changing the concentration of potassium in the medium. On gills excised from marine eels, Masoni and Bornancin (see Maetz and Bornancin, 1975) showed that after a 2-min exposure to tritiated ouabain ($5 \times 10^{-7} M$) in seawater without K^+ and a rinse with unlabeled ouabain, the chloride cells were specifically tagged. The same experiment in seawater with potassium at 10 mM did not allow ouabain binding. This demonstration shows, consequently, that ouabain enters the cells from the external medium.

More complete studies done by Stirling *et al.* (1973) and Karnaky *et al.* (1976b) on *Fundulus heteroclitus* adapted to different salinities (10% seawater, full-strength seawater, and 200% seawater) obtained more precise results on the enzyme's localization. Fish were perfused through an intracardiac catheter with tritiated ouabain ($5 \times 10^{-5} M$) for a period of 45 min, followed by a 30-min rinse; autoradiography showed that tritiated ouabain was bound to chloride cells, distributing itself over the entire surface except the apical crypt, indicating enzyme localization on the basolateral membrane.

b. Histological and Histochemical Studies. The preceding studies do not permit the precise intracellular localization of the (Na^+,K^+)-ATPase, because they are limited in resolution and sensitivity by use of the light microscope. The combination of histochemical and ultrastructural studies has allowed for a greater knowledge of the structure of chloride cells and for localization of the (Na^+,K^+)-ATPase, and it suggests a possible function of this cell type.

The first studies using cytochemical reactions of ATPase were conducted by Laurent *et al.* (1968) on the pseudobranch, then by Ernst and Philpott (1970) on the gill of *Fundulus grandis* adapted to seawater; in the latter case the ATPase seemed to be localized in the mitochondria. Shirai (1972) showed a deposit on the endoplasmic reticulum and in the apical area of chloride cells of the gills of the Japanese eel; unfortunately, the technique employed was not specific for the (Na^+,K^+)-ATPase, since the sensitivity to ouabain was lost during the binding. A similar but more complete study was done by Hootman and Philpott (1979), providing evidence for a K^+-dependent phosphatase activity that is part of the (Na^+,K^+)-ATPase. A derivation of Ernst's technique (1975) was applied to the gill of *Lagodon rhomboides*, and they found a cytochemically good labeling of the (Na^+,K^+)-ATPase. With the use of a medium combining formaldehyde and glutaraldehyde at low temperature for a short time, they were able to preserve cellular structures without inhibiting phosphatase activity. This activity was identified by deposits of Sr^{2+} phosphate; the strontium was replaced by lead, which is more electron dense. To ensure the specificity of this reaction some experimental controls were conducted in the absence of K^+ and in the presence of ouabain. Then an inhibitor of alkaline phosphatase (1-*p*-bromotetramizole oxalate) eliminated the possibility of a reaction due to this enzyme.

For this fish species adapted to seawater, the main localization of the enzyme is the side of the basolateral tubules in the chloride cells; in addition, some deposits were observed on the apical membrane and on the cytoplasmic surfaces of most of the apical vesicles.

Detection of (Na^+,K^+)-dependent ATPase on the sides of the cytoplasmic tubular system allows for interpretation of other purely structural observations such as the presence of asymmetrical repeating particles on the inside of the tubular walls (Philpott, 1980; Sardet, 1980). These particles can be interpreted as functional ATPases (Maunsbach *et al.*, 1979).

B. Properties and Characteristics of the Branchial (Na^+,K^+)-ATPase

1. Enzymatic Properties

The enzymatic reaction takes place in the presence of the substrate ATP. The ATPase is phosphorylated in the presence of Na^+ and Mg^{2+}, and the intermediate phosphoenzyme formed is dephosphorylated in the presence of K^+. In fact, Na^+ and K^+ are required to obtain a complete reaction and the speed of the reaction depends on the concentration of these ions in the reaction medium. With K^+ alone, the enzyme possesses phosphatase activity that can be measured via the hydrolysis of *p*-nitrophenylphosphate.

A study of the principal characteristics of the enzyme reaction, therefore, will comprise determination of the affinity constants for ATP, Mg^{2+}, Na^+, and K^+ as well as some conditions for maximal activation of the enzyme (concentrations of the substrates and cosubstrates, temperature, pH). This analysis can be performed on subcellular fractions from which the purification is variable; it is difficult to compare results directly because of differences in the preparative procedures, especially when supplementary treatments (detergents, activators, etc.) are used. We have presented in

Table III

Kinetic Parameters Observed for the Branchial (Na^+,K^+)-ATPase from Some Teleosts

Species and acclimation medium	Enzyme preparation	Kinetic parameters	Na^+ (mM)	K^+ (mM)	Mg^{2+} (mM)	pH_{max}	K_i Ouabain (M)	Experimental particularities	Reference
Anguilla anguilla									
FW	Crude homogenate	V_m	40 100	20	—	—	—	Temperature = 37°C	Motais (1970)
		K_m	= 7.4	3.4					
SW		V_m	40 100	10	—	—	—		
		K_m	= 9.4	2					
Anguilla japonica									
SW	NaI-treated extract	V_m	100	20	—	7	1.5×10^{-6}	—	Kamiya and Utida (1968)
		K_m		1.3					
Anguilla japonica									
FW	Deoxycholate-treated microsomes	V_m	100	20	—	7.3	2.7×10^{-6}	Enzyme preincubated 10 min at 25°C	Ho and Chan (1980a)
		K_m	8	1.9					
Oncorhynchus tshawytscha	Deoxycholate-treated microsomes	V_m	200	50	—	7.2	6×10^{-6}	Temperature = 37°C Na_2ATP = 10 mM $MgCl_2$ = 20 mM NaCl = 240 mM KCl = 120 mM	Johnson *et al.* (1977)
		K_m	26	14					
Salmo gairdneri									
SW	NaI-treated microsomes	V_m	100	20	5	7.1	1.3×10^{-5}	Enzyme preincubated 30 min at 37°C; temperature measure = 13°C	Pfeiler and Kirschner (1972)
		K_m	7	0.8	0.1				
Salmo gairdneri									
SW	NaI-treated microsomes	V_m	200					Enzyme preincubated 30 min at 37°C; temperature measure = 37°C	Pfeiler (1978)
		K_m	36						

Table III the kinetic parameters for a certain number of branchial enzymatic preparations; two species have been studied particularly: the trout and the eel. This table demonstrates that no single exhaustive study has been completed on the kinetic parameters of this enzyme; only the sensitivity to Na^+, K^+, pH optimum, and the affinity of the enzyme for its specific inhibitor ouabain have been studied. In most cases the maximum activation occurred at a pH of about 7.2, the concentrations of Na^+ and K^+ being 100 and 20 m*M*, respectively. The affinity constant of the enzyme for these ions varies between 7 and 10 m*M* for Na^+ and is around 1 m*M* for K^+. The different results obtained by Pfeiler (1978) on *Salmo gairdneri* and Johnson *et al.* (1977) on *Oncorhynchus tshawytscha* could have been due to peculiar experimental conditions.

From these observations it appears difficult to compare the gill (Na^+,K^+)-dependent ATPases from diverse origins by their kinetic characteristics. In addition, when a comparative study is done on the same euryhaline species adapted to fresh water or seawater the results are quite similar.

As on ATPases from other biological material (see Skou, 1975), it has been shown on the eel by Bell *et al.* (1977) that a number of cations are able to replace K^+ in the following order of affinity: $Tl^+ > Rb^+ > K^+ > NH_4^+ > Cs^+ > Li^+$.

In conclusion, examination of the enzymatic properties of the (Na^+,K^+)-dependent ATPases from fish gills shows no differences from those studied in other tissues.

2. Purification and Chemical Characteristics

A detailed analysis of the chemical characteristics of gill (Na^+,K^+)-ATPase was done by a team directed by J. Sargent in Aberdeen, Scotland. They purified many hundreds of times the enzyme extracted from the gills of eels (*Anguilla anguilla*) adapted to seawater with an extraction efficiency of 70% (Bell and Sargent, 1979). An extract enriched in chloride cells has been prepared that contained at least 90% of the (Na^+,K^+)-ATPase from gill scrapings. This crude cellular extraction is subjected to an enrichment and purification procedure (see Fig. 2). The final preparation has a specific activity of 392 ± 22 μmol P_i hr^{-1} mg $protein^{-1}$, which is the highest obtained for gills. The enzymatic activity measured at 30°C cannot be directly compared to activities measured at 37°C on enzymes from other biological materials.

The purity of the final preparation is estimated to be 56%. Like the ATPases of other tissues (Skou, 1975), the (Na^+,K^+)-ATPase of the eel gill is made up of two parts: a protein of 90,000 MW and a glycoprotein of 45,000 MW (Sargent and Thomson, 1974).

The totality of these investigations shows clearly that the structure and

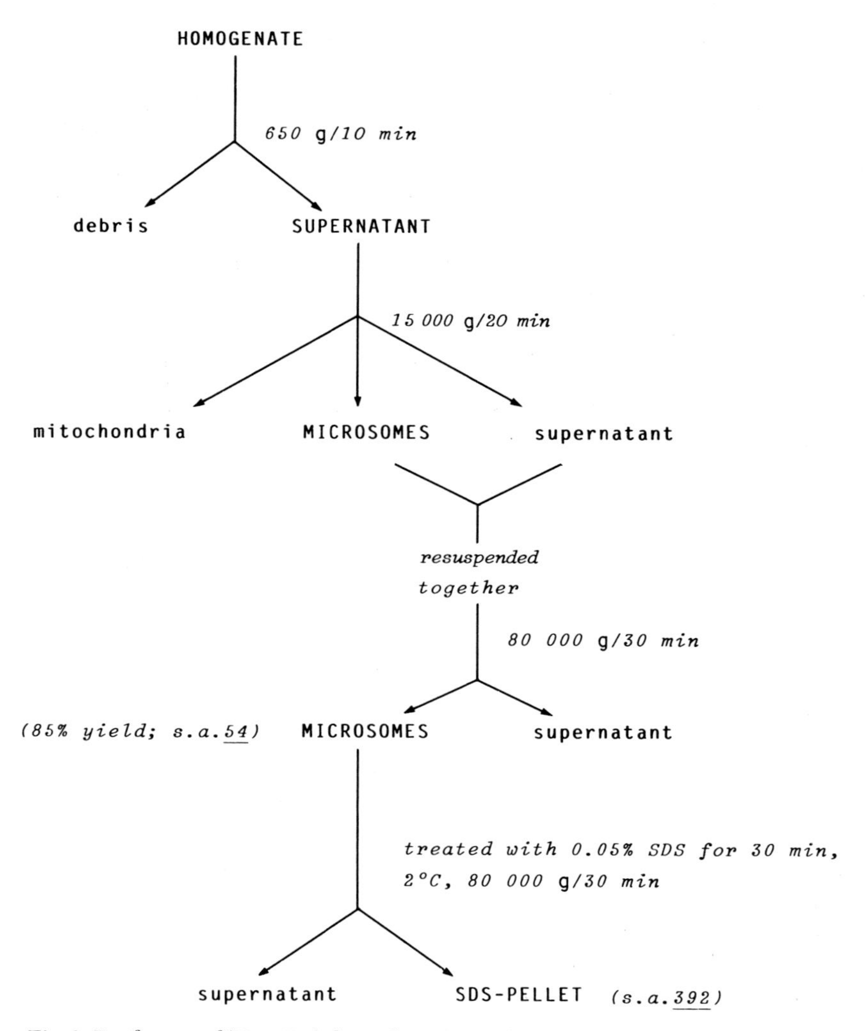

Fig. 2. Purification of (Na^+,K^+)-dependent ATPase from the gills of *Anguilla anguilla*. s.a., specific activity (μmol P_i hr^{-1} mg $protein^{-1}$). (From Sargent *et al.*, 1980.)

properties of the branchial enzyme (binding sites, kinetic parameters, etc.) are identical with those demonstrated on purified ATPases from other tissues. Also, there does not appear to be any difference between the characteristics of the (Na^+,K^+)-ATPases extracted from the gills of eels adapted to fresh water or seawater. This last point therefore directs research elsewhere than the enzymatic properties to explain the differences in the intensity and direction of the ionic fluxes observed in these fish in relation to the adaptation medium.

C. (Na^+,K^+)-Dependent ATPase and Branchial Ion Transport

1. Marine Teleosts

Generally, the gills of marine stenohaline teleosts contain a greater amount of ATPase activity than freshwater stenohaline species. With some exceptions, this rule applies when ATPase activities from euryhaline species adapted to fresh water and seawater are compared; the total activity of branchial (Na^+,K^+)-ATPase reported to a similar level of protein is greater in animals acclimated to seawater (Table I).

In seawater the ionic exchanges, particularly through the digestive mucous membranes and gill epithelia, are larger compared with those observed in fresh water (Maetz, 1974; Maetz and Bornancin, 1975). In marine fish, large bidirectional transepithelial exchanges and net excretion of sodium and chloride take place in the gill. The parallel that can be established between the direct branchial excretion and the activity of (Na^+,K^+)-ATPase leads one to consider the enzyme responsible for the active excretion of NaCl. We will describe the experimental arguments that are in favor of this hypothesis.

The first argument concerns the particular role of potassium in the branchial excretion of sodium. Maetz (1969) demonstrated that the branchial sodium and chloride effluxes are dependent on the concentration of external potassium in marine stenohaline or euryhaline teleosts: the transfer of a marine teleost to seawater without potassium is accompanied by a drop in sodium efflux and an increase in plasma Na^+ concentration. These results have been confirmed by Motais and Isaia (1972) and Evans *et al.* (1973). A short-term transfer (a few minutes) of a marine fish to fresh water is accompanied by a rapid drop in the branchial efflux of Na^+ that can be restored (from 50 to 100%) by addition of potassium to the external medium. The Na^+ efflux depends on the concentration of potassium added (Maetz, 1969; confirmed by numerous authors, see the review by Evans, 1979). These observations can be interpreted as proof of the direct role of (Na^+,K^+)-ATPase in the active transport of sodium. These results can also be explained by taking into consideration the effects caused by manipulating the composition of the external environment and in particular the influence of K^+ on the transepithelial branchial potential. The reader will be able to refer to some previous reviews (Maetz and Bornancin, 1975; Evans, 1979; Kirschner, 1980) and also to Chapter 4 of this volume.

A second series of arguments are derived from the use of ouabain to inhibit branchial fluxes. Motais and Isaia (1972) showed that ouabain (10^{-4} *M*) added to the external medium provoked a reduction in sodium efflux in *Anguilla anguilla* whether the external medium was seawater or fresh water plus KCl. These observations have been confirmed by Evans *et al.* (1973) on *Dormitator maculatus* but raise difficulties in interpretation because of the behavior of ouabain at the level of the gills. In effect, when the gills of *A. anguilla* are exposed for short periods (2–4 min) to seawater containing ouabain, it enters the epithelia cells, particularly the chloride cells (Maetz and Bornancin, 1975). If the exposure is longer (1 hr), ouabain passes through the epithelial barrier and enters the blood (Silva *et al.*, 1977). These same investigators have shown on *Anguilla rostrata* that the injection of this inhibitor is accompanied by reductions in gill (Na^+,K^+)-ATPase activity as well as sodium and chloride fluxes. These latter results are in complete contradiction with those obtained by Girard *et al.* (1977) on *Salmo gairdneri* adapted to seawater; in this case the internal application of ouabain provoked an increase in the efflux of both Na^+ and Cl^-. Sargent *et al.* (1980) have discussed the effect of ouabain and orthovanadate, another inhibitor of ATPase (Josephson and Cantley, 1977) by showing the hemodynamic effect of these substances that alter branchial vasoconstriction and modify the perfusion pressure. Since these consequences are directly linked to the concentration of inhibitors in the plasma, it is very difficult to compare the effects of inhibitors from one experiment to another (Sargent *et al.*, 1980; Kelly *et al.*, 1979). Thus, injections of ouabain or a very small amount of vanadate have shown the importance of the blood flow in branchial ionic excretion; they do not provide any direct proof for the role of (Na^+,K^+)-ATPase in this function.

2. Adaptation to Variations of the External Medium

a. Modifications of Salinity. Euryhaline species are an attractive model to compare branchial physiology from the adaptation medium and to relate physiological and biochemical aspects of ionic transport. In addition to the comparisons of animals fully adapted to different salinities (Table I), some work has been devoted to fish in the course of adaptation after an abrupt change in the salinity of their environment. As early as 1968, Kamiya and Utida showed with *Anguilla japonica* that adaptation to seawater was accompanied by an activation of gill (Na^+,K^+)-ATPase. In the yellow eel, *Anguilla anguilla*, Bornancin and de Renzis (1972) corroborated this observation showing a good correlation between the development of ATPase activity and the magnitude of ionic fluxes, in particular the Na^+ efflux dependent on external K^+. This last result was verified by Evans and Mallery (1975) on

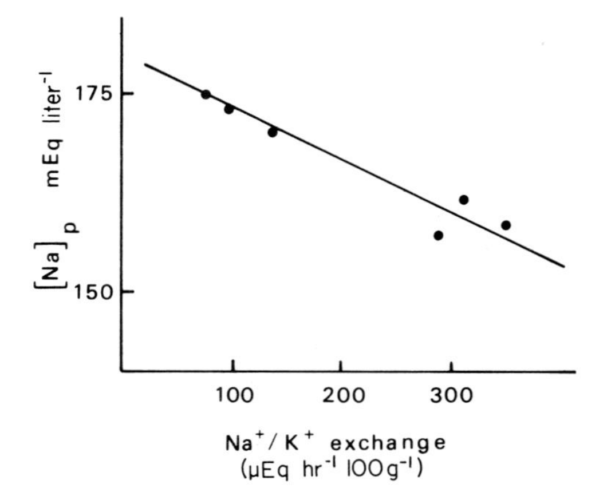

Fig. 3. Correlation between plasma sodium concentration and Na^+/K^+ exchange during adaptation of *Anguilla anguilla* to seawater ($r = 0.953$; $p < .001$). (From Bornancin, 1978; see Bornancin and de Renzis, 1972.)

Dormitator maculatus. As shown in Fig. 3, plasma sodium concentration seems to be regulated by this Na^+/K^+ exchange. Other work by Forrest *et al.* (1973) on *Anguilla rostrata* as well as Thomson and Sargent (1977) on *A. anguilla* corroborate the general scheme of saltwater adaptation (Fig. 4). Utida *et al.* (1971) have described the seawater adaptation of the Japanese

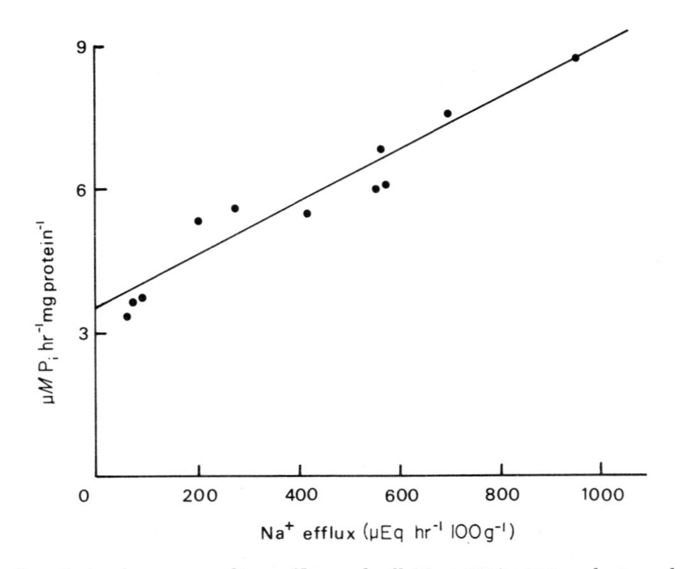

Fig. 4. Correlation between sodium efflux and gill (Na^+,K^+)-ATPase during adaptation to seawater ($r = 0.963$; $p < .001$). (From Epstein *et al.*, 1980.)

eel as a proliferation of the functional chloride cells and a parallel increase in (Na^+,K^+)-ATPase activity. A comparable study done by Thomson and Sargent (1977) established a clear correlation between the number of chloride cells in the branchial epithelium and the degree of saltwater adaptation. This study was unique in that the authors were able to determine the processes of marine adaptation by the sexual maturity of the eels. In immature yellow eels, transfer from fresh water to seawater is followed by an increase, first in the number of chloride cells and then in the activity of branchial (Na^+,K^+)-ATPase over a period of 5 to 10 days (Fig. 1). In the mature silver eel, the preparation for catadromous migration occurs involving the structural and biochemical modifications of the branchial epithelium before the fish makes contact with seawater.

Augmentation of branchial (Na^+,K^+)-ATPase in relation to the migratory behavior of teleosts had already been observed by Zaugg and McLain (1970, 1972) and Zaugg and Wagner (1973) on *Oncorhynchus kisutch*, *Oncorhynchus tshawytscha*, and *Salmo gairdneri;* these authors described seasonal variations of ATPase activity linked to photoperiod. These results have been confirmed by a systematic study conducted by Lassere *et al.* (1978) and Boeuf *et al.* (1978) on coho salmon. In this teleost, two seasonal activations (spring and fall) of gill (Na^+,K^+)-ATPase were observed that prepared the animal for its eventual entry into seawater. In contrast, Hart *et al.* (1981), working on *O. tshawytscha* to establish a relationship between the high level of gill (Na^+,K^+)-ATPase and migratory tendencies, raised doubt about the use of a single factor such as salinity as an indicator of parr–smolt transformation. All of the published results on salmonids indicate that other factors than salinity (temperature of the external medium, animal size, etc.) should be considered to explain changes in ATPase activity.

The influence of salinity changes of the external environment on branchial ATPase activity does not allow establishment of a single rule to include all species. In *Fundulus heteroclitus*, transfer from fresh water to seawater provokes a large increase in gill (Na^+,K^+)-dependent ATPase, but this transformation occurs after an extremely short delay on entry into seawater: 1 hr (Towle *et al.*, 1977). In *Chelon labrosus,* activation of ATPase is secondary to the progressive transfer from seawater to fresh water (Gallis *et al.*, 1979).

b. Modifications of Temperature and Pressure. Some authors (Zaugg and Wagner, 1973; Zaugg and McLain, 1976; Lassere *et al.*, 1978) have observed the limiting effects of temperature elevation on the seasonal activation of branchial (Na^+,K^+)-ATPase. Although the very great sensitivity of this ATPase to temperature was noted by Gilles and Vanstone (1976), it is difficult to establish clear relationships between the fishes' acclimation temperature, the *in vitro* enzymatic assay temperature, and measured activity.

Should one choose assay temperatures that of acclimation or should one compare the activities measured *in vitro* at the same temperature but corresponding to different acclimation temperatures? The dilemma facing investigators when they determine the experimental conditions for *in vitro* measurements and when they interpret the results is particularly clear in this case; this is why we have chosen to emphasize it. This could be extended to all measurements of ATPase activity that are carried out in experimental conditions often very far from *in vivo* conditions, poorly understood for an enzyme that functions between the intracellular and extracellular media of different physicochemical characteristics. The correlations established between the measured enzyme activities and physiological processes must be interpreted carefully.

Cossins and Prosser (1978) have shown that prolonged thermal acclimation modified the lipid composition of the cell membrane and its fluidity; these factors influence ATPase activity and its dependence on temperature (Skou, 1975; Wallick *et al.*, 1979). In addition, the contradictory observations of Murphy and Houston (1974), McCarty and Houston (1977), and Stuenkel and Hillyard (1980) indicate a complex relationship linking the temperature and the activity of branchial (Na^+,K^+)-ATPase.

The effects of variations of another parameter in the fishes' environment, pressure, have also been studied. Variations of small amplitude do not affect the activity of branchial (Na^+,K^+)-ATPase. The effect of large changes in pressure are reported and discussed in an article by Pequeux (1980).

3. Concluding Remarks

Euryhaline fish adapted to seawater develop a particular, highly efficient branchial ultrastructure to eliminate significant quantities of salt against a large chemical gradient, and at the same time to reduce water loss across the gills. This multicellular structure effectively extrudes NaCl (Foskett and Scheffey, 1982) and contains in quantity an enzyme known to exchange intracellular sodium with potassium. Sargent *et al.* (1980) have calculated in *Anguilla anguilla* adapted to seawater the intensity of Na^+ effluxes related to the quantity of (Na^+,K^+)-ATPase in the gills. This estimate is a good approximation of the measured fluxes in seawater-adapted eels (Bornancin and de Renzis, 1972). The (Na^+,K^+)-ATPase can then be considered as responsible for the transport of NaCl by the gills. This allows for a number of possibilities to be considered in the real function of the enzyme and its role in branchial ionic transport, which have already been discussed (Maetz and Bornancin, 1975; Potts, 1977; Kirschner, 1977, 1980; Evans, 1979, 1980; Epstein *et al.*, 1980; Sargent *et al.*, 1980), as well as some functional models. These models favor peculiar experimental results or certain observations, but they cannot account for the complexity of the macrostructure made up of the total of the intracellular basolateral tubules network. The repeating particles of molecules of (Na^+,K^+)-ATPase on these tubules are some of the essential elements in the function of these structures but probably the only ones.

The existence of an electrochemical gradient unfavorable for Cl^- transport indicates that it is actively transported, whereas the distribution of Na^+ could be passive (Silva *et al.*, 1977; Epstein *et al.*, 1980). According to these authors the ATPase could be indirectly supplying the energy for active chloride transport against the electrochemical gradient. Furthermore, they present the following arguments showing in certain conditions that there could be dissociation between the intensity of ionic fluxes and ATPase activity.

1. The initial activation of ATPase in the gills of eels by cortisol on transfer to seawater does not result in the immediate establishment of elevated ionic fluxes after transfer; a delay of a few days in seawater is necessary.
2. During the return to fresh water of seawater-adapted eels for short periods (2 hr), a large decrease of ionic fluxes occurs, even though the activity of (Na^+,K^+)-ATPase remains high.

These two observations do not constitute major objections regarding the role of ATPase in the branchial transport of salts. In these two cases the enzyme is present, but the structures permitting its function are not yet developed or are destroyed. In the first case, the channels allowing establishment of this flux do not exist in the branchial structure; they progressively appear during the course of seawater adaptation associated with the gradual development of fluxes (Bornancin and de Renzis, 1972). In the second case, it has been demonstrated that the return to fresh water for a euryhaline fish adapted to seawater can only be done for extremely brief periods (a few minutes and not 2 hr) without loss of the seawater adaptation (Motais, 1967) and destruction of intracellular structures and chloride cells (Masoni and Isaia, 1973). This is why the model presented by Sargent *et al.* (1980) seems to us closer to the real situation; it has the merit of attempting to include results from biochemical, physiological, and ultrastructural studies. This model permits explanation of the existence of intense ionic fluxes (by exchange diffusion Na^+/Na^+ and by Na^+/K^+ exchange) as well as the low branchial osmotic permeability of saltwater fish (Motais *et al.*, 1969). This last property could result from the development of the chloride cell tubular network, which is continuous with the extracellular spaces and which probably builds up high concentrations of ions within these spaces; the salt is thus able to flow toward the external milieu via the leaky junction (Sardet *et al.*, 1979).

This mechanism reduces the amplitude of the osmotic gradient and then the branchial osmotic permeability (Bornancin, 1978).

IV. CA^{2+}-DEPENDENT ATPASE

The Ca^{2+}-ATPase is also a membrane ATPase whose role in the cell is essential and evident but that could also play a role in transepithelial transport and be implicated in the regulation of plasma Ca^{2+} concentration. In fish, the Ca^{2+}-ATPase has been principally studied in freshwater fish and in euryhaline fish adapted to fresh water or seawater.

A. Enzymatic Characteristics

Ca^{2+}-ATPase has been studied on microsomal fractions from the gills of different fish. The enzymatic characteristics are quite similar irrespective of the fish studied (Table IV), except for the results obtained on *Fundulus* in our laboratory, which show lower Ca^{2+} concentration requirements for the K_m and maximal activation.

B. Branchial Ca^{2+}-ATPase and Adaptation Medium

All the studies performed to date try to establish a correlation between enzymatic activity and adaptation of fish to media of variable calcium concentrations. Ma *et al.* (1974) did not find any difference between trout adapted to fresh water or seawater. These results were not confirmed by the studies conducted with *Anguilla rostrata* and *Anguilla japonica*. In the former case, it is the gills of fish adapted to fresh water that contain maximum Ca^{2+}-ATPase activity (Fenwick, 1979). In the latter case, adaptation to seawater is followed by an increase in Ca^{2+}-ATPase activity (Ho and Chan, 1980b).

Similar changes have been found in studies on *Fundulus heteroclitus*. Burdick *et al.* (1976) found a positive correlation between the activity of branchial Ca^{2+}-ATPase and the quantity of Ca^{2+} whether the fish are adapted to fresh water or seawater. In the same species, Mayer-Gostan *et al.* (1983) could not reproduce this result; in contrast, Shephard (1981) described for *Rutilus rutilus* in fresh water a parallelism between the quantity of external Ca^{2+} and the activity of Ca^{2+}-dependent ATPase. These contradictory results need additional studied to interpret the actual results and understand the role played by branchial Ca^{2+}-ATPase. The most urgent work should consist of a reevaluation of the biochemical characteristics of the

Table IV

Kinetic Parameters Observed for the Branchial Ca^{2+}-ATPase from Some Teleosts

Species and acclimatization medium	Enzyme preparation	Ca^{2+} activation[a]	pH	Temperature (°C)	Reference
Salmo gairdneri					
FW	Microsomes	K_m = 0.33 mM	7.8–8.1	12.5	Ma *et al.* (1974)
		Optimal conc. = 4 mM			
Rutilus rutilus					
FW	Heavy microsomes	Optimal conc. = 2 mM	7.8	—	Shephard and Simkiss (1978)
Anguilla anguilla					
FW	Microsomes	K_m = 0.40 mM	7.8	12	Fenwick (1979)
SW		Optimal conc. = 4 mM			
Anguilla japonica					
FW	Heavy microsomes or homogenate	K_m = 0.44 mM	8.2	25	Ho and Chan (1980b)
		Optimal conc. = 3 mM			
SW		K_m = 0.66 mM			
		Optimal conc. = 3 mM			
Fundulus heteroclitus					
FW	Microsomes	K_m = 0.035 mM	8.0	25	Mayer-Gostan *et al.* (1983)
SW		Optimal conc. = 0.25 mM			

[a]Optimal conc.: Ca^{2+} concentration for which the maximal activation of the enzyme is observed.

enzyme from purified preparations by controlling the concentration of free Ca^{2+} by utilization of appropriate buffers.

V. ANION-DEPENDENT ATPase

Absorption of chloride from the external medium in freshwater teleosts is probably the result of an active transport, since it is accomplished against an electrochemical gradient. On the one hand, the plasma chloride concentration is many hundred times greater than that of the ambient environment, and on the other hand, determinations of differences in transepithelial potential in freshwater fish (Kerstetter *et al.*, 1970; Kerstetter and Kirschner, 1972; Maetz, 1974; Eddy, 1975; de Renzis, 1978) indicate that it is on the order of 10 to 30 mV negative inside. The chloride is absorbed through the gills independently from sodium in exchange for an endogenous ion, HCO_3^- (Krogh, 1939; Maetz and Garcia-Romeu, 1964; Kerstetter and Kirschner, 1972; de Renzis, 1975; Haswell *et al.*, 1980; Perry *et al.*, 1981). Since the intensity of this exchange is more closely linked to the state of acid–base balance of the internal medium rather than that of mineral balance (de Renzis and Maetz, 1973), de Renzis (1978) suggests that in freshwater teleosts the Cl^-/HCO_3^- exchange plays a more important role in the regulation of internal pH than in the maintenance of hydromineral balance. In addition, the mechanisms responsible for active chloride absorption in fresh water can be inhibited by thiocyanate (Epstein *et al.*, 1973; Kerstetter and Kirschner, 1974; de Renzis, 1975).

Although many studies have been directed to its investigation, the mechanism of chloride transfer through the gills of saltwater fish is still poorly understood. It is generally admitted that chloride is excreted by an active transport mechanism against a chemical and electrical gradient. The chemical gradient is less than in freshwater fish but at the same time unfavorable; the plasma chloride concentration is always lower than 200 m*M*, whereas in seawater it is about 500 m*M*. Concerning the electrical gradient, Kirschner (1977, 1979) and Evans (1979, 1980), in assembling all the values of potential difference measured by different authors through the branchial epithelium of saltwater fish, have clearly shown that all the fish studied maintain a transepithelial potential very different from the equilibrium potential of chloride.

The net transepithelial movements of chloride are effected in both freshwater and saltwater fish by mechanisms of active transport that require energy. Thus far ATP is the only molecule known to supply energy for membrane ion transport. The discovery by Kasbekar and Durbin (1965) of an ATPase in the gastric mucosa of frog whose activity is modulated by anions

Table V

Experimental Conditions Adopted for the Measurement of Anion-Dependent ATPase in Fish Gills

Species and acclimatization medium	Buffer	pH	Temperature (°C)	ATP (m*M*)	ATP/Mg^{2+}	Experimental particularities	Reference
Salmo gairdneri							
FW	30 m*M* Tris-Hepes	8.3	25	3.5	0.7	Preincubation at 37°C in 0.2% Triton X-100	Kerstetter and Kirchner (1974)
SW	or 30 m*M* Tris-acetate	8.3					
Salmo gairdneri							
FW	25 m*M* Hepes-Tris	8.0	27	1	1	—	Bornancin *et al.* (1980)
	or 50 m*M* glycine-Tris	8.0					
	or 25 m*M* glycine-Tris	7.5					
Salmo irideus	100 m*M* Tris-acid corresponding to the added salt	8.4	37	2	1	—	Van Amelsvoort *et al.* (1977)

Salmo gairdneri							
FW	40 m*M* tris-acetate	7.5	25	3	0.6	0.1% Sodium deoxycholate	McCarty and Houston (1977)
Carassius auratus							
FW	50 m*M* Hepes-Tris or 5 m*M* bicarbonate	8.2 8.0	27	0.5	1	—	de Renzis and Bornancin (1977)
Anguilla anguilla							
FW SW	25 m*M* Hepes-Tris or 50 m*M* glycine-Tris	8.0 8.0	27	0.5	1	—	Naon *et al.* (1981)
Anguilla anguilla							
FW	50 m*M* Hepes-Tris or 5 m*M* bicarbonate	8.0 8.0	27	1	1	—	Bornancin *et al.* (1977)
Anguilla rostrata							
FW SW	10 m*M* Imidazole	7.8	37	6	1	0.1% Sodium deoxycholate	Solomon *et al.* (1975)
Anguilla japonica							
FW SW	30 m*M* Tris-acetate	—	25	3	1	Preincubation at 30°C in 0.2% Triton X-100	Ho and Chan (1980c)

caused a number of authors to look for a similar ATPase in other tissues such as the gills of fish. The involvement of an anion-dependent ATPase in active Cl^- transport by a mechanism analogous to Na^+ and K^+ transports was considered.

A. Demonstration of an Anion-Dependent ATPase Activity in Teleostean Gills

The presence of ATPase activity sensitive to Mg^{2+} and inhibited by thiocyanate was mentioned for the first time in 1973 by Solomon *et al.* in the gills of eel (*Anguilla rostrata*) adapted to fresh water and seawater. In 1974, Kerstetter and Kirschner studied trout (*Salmo gairdneri*) adapted to fresh water and seawater and reported an ATPase activated by HCO_3^- that is inhibited by thiocyanate. These pioneer works were followed by some other studies, sometimes more detailed, but still few in number. Table V shows the species studied (only four) and the principal experimental conditions in which the anion-ATPase of fish gills has been studied. Note that certain characteristics have been described by all the investigators, namely, sensitivity of this ATPase to Mg^{2+}, stimulation by anions—in particular the oxyanions—and nonspecific inhibition of thiocyanate. Furthermore, absence of a specific inhibitor leads to problems of identification of this enzyme and determination of its activation by anions. The activity of (Na^+,K^+)-ATPase is determined as being the difference between the measured hydrolysis of ATP in the presence of sodium and potassium, with or without ouabain. It is different for the ATPase activated by anions. According to some authors, the activity of anion-sensitive ATPase is estimated by (1) the difference of activity measured in the presence of Mg^{2+} with or without thiocyanate (Solomon *et al.*, 1975; Ho and Chan, 1980c), (2) the difference of activity measured in the presence or absence of the anion whose concentration varies between 5 and 30 m*M* depending on the study (Kerstetter and Kirschner, 1974; de Renzis and Bornancin, 1977; Bornancin *et al.*, 1977, 1980; Naon *et al.*, 1981; Ho and Chan, 1980c; McCarty and Houston, 1977), (3) the difference of activity measured in the presence of anion with or without thiocyanate (de Renzis and Bornancin, 1977; Bornancin *et al.*, 1980), or (4) the difference of activity measured in the presence of chloride or the presence of other anions put in place of chloride (Van Amelsvoort *et al.*, 1977).

Whatever the method of assay and the selected experimental conditions, all studies show that an enzyme exists in the gills of fish that hydrolyzes ATP—that is to say, it is capable of catalyzing a reaction freeing energy and whose activity is modulated in particular by oxyanions.

B. Enzymatic Properties

1. Intracellular Distribution

The studies of anion-dependent ATPase have led to some contradictory conclusions regarding the intracellular localization of this enzyme, and because of this, its participation in ionic transports. According to the investigators and the methods of separation, this ATPase activity is linked either to the mitochondrial fraction or to the mitochondrial and microsomal fractions. The microsomal fraction contains plasma membranes in which should be located the enzymatic activities directly implicated in transepithelial ionic exchanges.

The intracellular localization of the anion-sensitive ATPase has generally been studied by measuring the activity of the enzyme in the fractions obtained after differential centrifugation. After separation of the nuclei and cellular debris by centrifugation at 10^4 *g* min, a fraction containing the heavy mitochondria is obtained by centrifugation at 10^5 *g* min, a fraction containing the light mitochondria and heavy microsomes is obtained at $7–12 \times 10^5$ *g* min, and finally the microsomal fraction is obtained after centrifugation at $4–6 \times 10^6$ *g* min. Most experimenters have studied the problem of the intracellular distribution of the anion-ATPase by comparing the distribution of this enzymatic activity with mitochondrial enzyme markers such as succinate dehydrogenase and cytochrome *c* oxidase or membrane markers like (Na^+,K^+)-ATPase or 5′-nucleotidase. A mitochondrial localization has been postulated by Ho and Chan (1980c) in the eel, whereas Kerstetter and Kirschner (1974) in the trout, de Renzis and Bornancin (1977) in the goldfish, Bornancin *et al.* (1980) in trout, and Naon *et al.* (1981) in the eel have concluded that there is ATPase activity specifically linked to the plasma membranes.

To resolve the problem of the localization of the anion-ATPase with less doubt, some mitochondria and microsomal fractions have been separated by sucrose density gradient centrifugation. This method has also led to contradictory results: Van Amelsvoort *et al.* (1977) decided there was contamination of the membranes by mitochondrial fragments, whereas de Renzis *et al.* (1980) and Naon *et al.* (1981) concluded from the results presented in Fig. 5 that there was anion-ATPase activity specifically linked to the plasma membranes.

The problem of the intracellular localization of the anion-ATPase in the gills of fish is still open; it is unlikely that it will be resolved easily in this tissue because of difficulties in preparing purified fractions from an organ with such a complex structure and cellular heterogeneity.

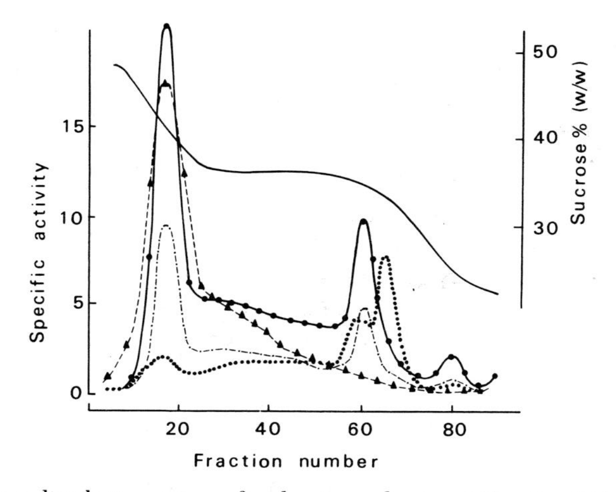

Fig. 5. Enzyme distribution pattern after density gradient centrifugation (16 hr at 100,000 *g*) of a fraction (>50,000 *g* min^{-1}) containing light mitochondria and plasma membrane of a saltwater eel gill extract. (●——●) HCO_3^- activation; (·-·) Cl^- activation; (▲--▲) cytochrome *c* oxidase activity; (···) (Na^+,K^+)-ATPase activity; (———) distribution of sucrose. Specific activities are given directly for HCO_3^- and Cl^- activations in μmol P_i hr^{-1} mg $protein^{-1}$ and divided by 20 for cytochrome *c* oxidase activity in nmol cytochrome *c* oxidized min^{-1} mg $protein^{-1}$. (From Naon *et al.*, 1981.)

2. Sensitivity to Anions

Considering the possible physiological role of the anion-ATPase, bicarbonate is the ion that has been most studied. This ion always stimulates the ATPase. In the microsomal fraction, maximal activation, obtained for concentrations ranging from 10 to 50 m*M*, is between 4 μ*M* P_i hr^{-1} mg $protein^{-1}$ in the freshwater eel and 40 μ*M* P_i hr^{-1} mg $protein^{-1}$ in the trout. Half-maximal activation is obtained for bicarbonate concentrations ranging from 2 to 17 m*M* (Kerstetter and Kirschner, 1974; Bornancin *et al.*, 1980; Naon *et al.*, 1981). When greater than optimal concentrations of bicarbonate are used, an inhibition of activity has sometimes been observed (Kerstetter and Kirschner, 1974; Bornancin *et al.*, 1977).

Bicarbonate can be replaced by other anions, such as sulfate or sulfite. The half-maximal stimulating concentration of sulfite is lower than that of bicarbonate in rainbow trout gill (2.8 vs 12 mEq, Bornancin *et al.*, 1980) or equal in freshwater eel gill (2.9 vs 2.1 mEq, de Renzis *et al.*, 1980); sulfate is higher (9.2 vs 2.5 mEq) in goldfish.

Among other anions studied on branchial preparations, acetate stimulates the anion-dependent ATPase activity but to a lesser degree than bicarbonate (Kerstetter and Kirschner, 1974; de Renzis and Bornancin, 1977), and nitrate has an inhibitory effect, its action being proportional to its concentration (de Renzis and Bornancin, 1977).

Participation of the anion-ATPase in active transport has often been doubted because of its intracellular localization and because clear stimulation of this enzyme by chloride has not been measured in epithelia actively transporting this ion, like the gills.

By modifying the experimental assay conditions, that is to say by using MgATP as substrate and a glycine–Tris buffer without any other anion in the medium, Bornancin *et al.* (1980) and Naon *et al.* (1981) showed stimulation of the anion-ATPase by chloride. The K_m of this stimulation is about the same in goldfish, trout, and eel, being 3.5 mEq. Enzyme activity was always less than one-third of that observed with bicarbonate at similar concentration. When buffers containing sulfonate groups (Pipes, Hepes) were used, stimulation by chloride could not be shown. It is the same when the medium contains high concentrations of anions, in particular the buffers Tris-HCl, imidazole HCl, and Tris-acetate.

The anion-sensitive ATPase present in the gills of fish can generally be inhibited by the lipophilic anion thiocyanate. The concentration necessary to obtain maximal inhibition is about 10 m*M*. Activation by HCO_3^- or Cl^- ions is competitively inhibited by thiocyanate (Naon *et al.*, 1981; Bornancin *et al.*, 1980). This suggests the existence of a single site on the enzyme with different affinities for various anions. The relative affinity for bicarbonate and thiocyanate is 3.3 in eel adapted to fresh water (Naon *et al.*, 1981); it is 10 in goldfish (de Renzis and Bornancin, 1977). The relative affinity for chloride and thiocyanate is 18 in goldfish (de Renzis and Bornancin, 1977); it is 35 in freshwater trout (Bornancin *et al.*, 1980). Thiocyanate does not affect residual ATPase activity. Inhibition of this activity observed by some authors (Solomon *et al.*, 1975; Ho and Chan, 1980c) probably results from inhibition of the activity dependent on the anion in the buffer. Since the value of maximum activity and/or the affinity constant are dependent on each anion, the sensitivity of this branchial anion-ATPase is therefore a specific reaction independent from Mg^{2+}-ATPase.

3. Other Properties of the Enzyme

The maximal activation of anion-ATPase by anions is observed at a concentration of ATP between 0.5 and 2 m*M*, and a ratio of ATP:Mg^{2+} equal to 1.

The effect of pH is not clearly elucidated, probably because the study requires a large number of precautions (Bornancin *et al.*, 1980). The pH optimum seems, however, to be between 7.0 and 8.0 (de Renzis and Bornancin, 1977; Bornancin *et al.*, 1980; Ho and Chan, 1980c).

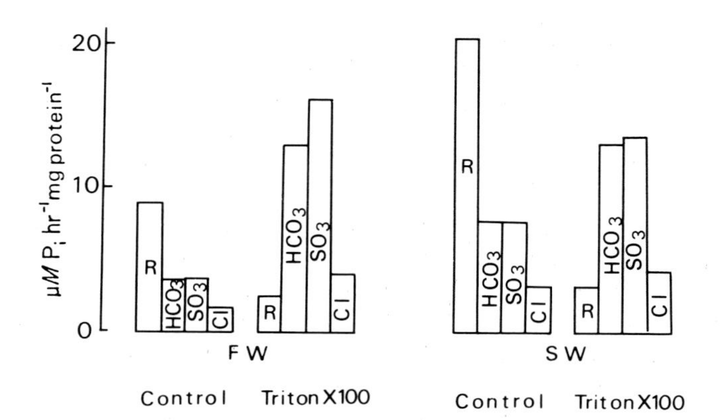

Fig. 6. Effect of Triton X-100 on residual (Mg^{2+}) and anion-ATPase activities of the plasma membrane fraction. Experiments were done on gill extracts from fresh water- (FW) or seawater (SW)-adapted eels. R, residual activity measured in the presence of ATP-Mg^{2+} 0.5 mM and ouabain 10^{-4} M; HCO_3^- activation measured in the presence of $NaHCO_3$ 20 mM; Cl^- activation measured in the presence of NaCl 10 mM; SO_3^{2-} activation measured in the presence of Na_2SO_3 5 mM. (From Naon *et al.*, 1981.)

ATPase activity is not modified by the replacement of Na^+ by K^+ or NH_4^+ (Kerstetter and Kirschner, 1974; de Renzis and Bornancin, 1977). Replacement of Mg^{2+} by Ca^{2+} causes, first, an increase in residual activity which suggests the presence of a Ca^{2+}-sensitive ATPase, and second, a decrease in the stimulation by anions, which indicates a requirement of Mg^{2+} in order to obtain this effect.

The anion-sensitive ATPase can be solubilized by Triton X-100; the optimal ratio of detergent:protein is 3:1. Figure 6 shows that the solubilization produces an increase in anion-sensitive activity by a factor of two to three and a significant reduction in residual activity. The two-way reaction to detergent permits one to distinguish the anion-ATPase from the residual Mg^{2+}-dependent ATPase.

C. Physiological Considerations

The possibility that a branchial anion-ATPase is involved in the branchial transport of chloride in fish is based on arguments that consider (1) the intracellular localization of the enzyme, (2) activation by the transported anions, and (3) the action of inhibitor, which affects both the enzymatic activity and anionic transport.

1. As we have seen previously, many results favor the existence of a membranous anion-ATPase. However, the controversy that exists at this

time should be taken into consideration. Fish gills do not seem to be the biological material to provide an end to the conflict. Furthermore, in cells without mitochondria such as mammalian erythrocytes, the presence of an anion-ATPase could not be demonstrated (unpublished personal results).

2. Activation by anions, and especially by Cl^-, has been clearly demonstrated (Bornancin *et al.*, 1980).

3. Thiocyanate inhibits both Cl^-/HCO_3^- exchange in freshwater fish and anion-sensitive ATPase activity. However, this parallelism could be coincidental. In effect, as De Pont and Bonting (1981) remarked, SCN^- is a lipophilic anion that can inhibit the transport without necessarily causing an inhibition of ATPase.

The hypothesis of a participation of the branchial anion-ATPase in Cl^-/HCO_3^- exchange must be discussed in the context of branchial physiology, taking into account the differences of osmoregulatory functions of this epithelium in fish acclimated to fresh water or seawater. The anion-sensitive ATPase activity has been measured on branchial extracts from fish adapted to these two media (Kerstetter and Kirschner, 1974; Ho and Chan, 1980c; Naon *et al.*, 1981). No significant difference in the kinetic characteristics of the enzyme was observed. This and the fact that small differences measured are abolished after solubilization by Triton X-100 (Fig. 6) make it difficult to incorporate this enzyme into a functional model of branchial ionic exchanges. There is nothing comparable to (Na^+,K^+)-ATPase, whose activity varies with salinity in parallel with modifications of transbranchial sodium fluxes. For this enzyme, the differences are accentuated after solubilization with deoxycholate. The different behavior of the two enzymes in response to variations of salinity suggests the possibility that the anion-dependent ATPase is implicated in regulation of acid–base balance rather than in osmoregulation. This hypothesis is corroborated by the results of Bornancin *et al.* (1977), who observed in freshwater eel that the induction of respiratory acidosis is accompanied by an increase in internal bicarbonate concentration and the capacity to absorb Cl^-. This is associated with an increase in the affinity of the branchial ATPase for this anion. This modification could permit the animal, under normal environmental conditions, to eliminate its excess plasma bicarbonate through a facilitated Cl^-/HCO_3^- exchange.

The energy supplied by this anion-ATPase could energize an external Cl^-–internal HCO_3^- exchange. Such an exchange occurs in freshwater fish and may possibly be retained in seawater by euryhaline species, similar to that observed for the external Na^+–internal NH_4^+ exchange across the freshwater trout gill when adapted to seawater (Payan and Girard, 1978). Alternatively, the energy may be utilized for proton transport, the allosteric control of which would depend on the Cl^- and HCO_3^- concentrations, as suggested by Kinne-Saffran and Kinne (1979).

ACKNOWLEDGMENTS

We wish to thank the editors for help in English translation.

REFERENCES

Adams, M., and Duggan, P. F. (1981). ATP-dependent membrane-located calcium transport systems are widely distributed and activated by potassium ions. *Comp. Biochem. Physiol. B* **70B**, 85–91.

Bell, M. V., and Sargent, J. R. (1979). Purification and properties of the (Na^+ + K^+)-dependent ATPase from the gills of *Anguilla anguilla*. *Biochem. J.* **179**, 431–438.

Bell, M. V., Tondeur, F., and Sargent, J. R. (1977). The activation of sodium-plus-potassium ion-dependent adenosine triphosphatase from marine teleost gills by univalent cations. *Biochem. J.* **163**, 185–188.

Boeuf, G., Lasserre, P., and Harache, Y. (1978). Osmotic adaptation of *Oncorhynchus kisutch* walbaum. II. Plasmaosmotic and ionic variations and gill Na^+-K^+ ATPase activity of yearling coho salmon transferred to sea water. *Aquaculture* **15**, 35–52.

Bornancin, M. (1978). Echanges ioniques et activités ATPasiques cation et anion dépendantes dans les branchies de l'anguille en relation avec la salinité du milieu d'adaptation. Thèse de Doctorat es Sciences, Université de Nice, France.

Bornancin, M., and de Renzis, G. (1972). Evolution of the branchial sodium outflux and its components especially the Na/K exchange and the Na-K dependent ATPase activity during adaptation to sea water in *Anguilla anguilla*. *Comp. Biochem. Physiol. A* **43A**, 577–591.

Bornancin, M., de Renzis, G., and Maetz, J. (1977). Branchial Cl transport anion-stimulated ATPase and acid-base balance in *Anguilla anguilla* adapted to fresh-water: Effects of hyperoxia. *J. Comp. Physiol.* **117**, 313–322.

Bornancin, M., de Renzis, G., and Naon, R. (1980). Cl^- -HCO_3^- ATPase in gills of the rainbow trout: Evidence for its microsomal localization. *Am. J. Physiol.* **238**, R251–R259.

Burdick, C. J., Mendlinger, S., Pang, R. K., and Pang, P. K. T. (1976). Effects of environmental calcium & stannius corpuscule extracts on Ca-ATPase in killifish, *Fundulus heteroclitus*. *Am. Zool.* **16**, 224.

Butler, D. G., and Carmichael, F. J. (1972). (Na^+-K^+)- ATPase activity in eel (*Anguilla rostrata*) gills in relation to changes in environmental salinity: Role of adrenocortical steroids. *Gen. Comp. Endocrinol.* **19**, 421–427.

Cossins, A. R., and Prosser, C. L. (1978). Evolutionary adaptation of membranes to temperature. *Proc. Natl. Acad. Sci. U.S.A.* **75**, 2040–2043.

De Pont, J. J. H. H. M., and Bonting, S. L. (1981). Anion-sensitive ATPase and (K^++H^+)-ATPase. *In* "Membrane Transport" (S. L. Bonting and J. J. H. H. M. De Pont, eds.), pp. 209–234. Elsevier, Amsterdam.

de Renzis, G. (1975). The branchial chloride pump in the goldfish *Carassius auratus:* Relationship between Cl^-/HCO_3^- and Cl^-/Cl^- exchanges and the effect of thiocyanate. *J. Exp. Biol.* **63**, 587–602.

de Renzis, G. (1978). Etude des caractéristiques de l'absorption branchiale de chlore et de l'échange Cl^-/HCO_3^- chez *Carassius auratus:* Aspects physiologique et biochimique. Relation avec la régulation de l'équilibre acido-basique du milieu intérieur. Thèse de Doctorat, Université de Nice, France.

de Renzis, G., and Bornancin, M. (1977). A Cl^-/HCO_3^- ATPase in the gills of *Carassius auratus*. Its inhibition by thiocyanate. *Biochim. Biophys. Acta* **467**, 192–207.

de Renzis, G., and Maetz, J. (1973). Studies on the mechanism of chloride absorption by the goldfish gill. Relation with acid-base regulation. *J. Exp. Biol.* **59,** 339–358.

de Renzis, G., Naon, R., and Bornancin, M. (1980). Localisation intracellulaire et caractérisation d'une activité ATPasique sensible aux anions présente dans les branchies de téléostéens. *In* "Epithelial Transport in the Lower Vetebrates" (B. Lahlou, ed.), pp. 297–315. Cambridge Univ. Press, London and New York.

Dharmanba, M., Bornancin, M., and Maetz, J. (1975). Enviromental salinity and sodium and chloride exchanges across the gill of *Tilapia mossambica. J. Physiol. (Paris)* **70,** 627–636.

Eddy, F. B. (1975). The effect of calcium on gill potentials and on sodium and chloride fluxes in the goldfish, *Carassius auratus. J. Comp. Physiol.* **96,** 131–142.

Epstein, F. H., Katz, A. I., and Pickford, G. E. (1967). Sodium and Potassium-activated adenosine triphosphatase of gills: Role in adaptation of teleosts to salt water. *Science* **156,** 1245–1247.

Epstein, F. H., Cynamon, M., and McKay, W. (1971). Endocrine control of Na-K-ATPase and seawater adaptation in *Anguilla rostrata. Gen. Comp. Endocrinol.* **16,** 323–328.

Epstein, F. H., Maetz, J., and de Renzis, G. (1973). Active transport of chloride by the teleost gill: inhibition by thiocyanate. *Am. J. Physiol.* **224,** 1295–1299.

Epstein, F. H., Silva, P., and Kormanik, G. (1980). Role of Na-K-ATPase in chloride cell function. *Am. J. Physiol.* **238,** 246–250.

Ernst, S. A. (1975). Transport ATPase cytochemistry: Ultrastructural localization of potassium-dependent and potassium-independent phosphatase activities in rat kidney cortex. *J. Cell Biol.* **66,** 586–608.

Ernst, S. A., and Philpott, C. W. (1970). Preservation of Na-K-activated and Mg-activated adenosine triphosphatase activities of avian salt gland and teleost gill with formaldehyde as fixative. *J. Histochem. Cytochem.* **18,** 251–263.

Esmann, M., Skou, J. C., and Christiansen, C. (1979). Solubilization and molecular weight determination of Na, K-ATPase from rectal gland of *Squalus acanthias. In* "Na,K-ATPase, Structure and Kinetics" (J. C. Skou and J. G. Nørby, eds.), pp. 21–24. Academic Press, New York.

Evans, D. H. (1979). Fish. *In* "Comparative Physiology of Osmoregulation in Animals" (G. M. O. Maloiy, ed.), Vol. 1, pp. 305–390. Academic Press, New York.

Evans, D. H. (1980). Kinetic studies of ion transport by fish gill epithelium. *Am. J. Physiol.* **238,** 224–229.

Evans, D. H., and Mallery, C. H. (1975). Time course of sea water acclimation by the euryhaline teleost, *Dormitator maculatus:* Correlation between potassium stimulation of sodium efflux and Na/K activated ATPase activity. *J. Comp. Physiol.* **96,** 117–122.

Evans, D. H., Mallery, C. H., and Kravitz, L. (1973). Sodium extrusion by a fish acclimated to sea water: Physiological and biochemical description of a Na for K exchange system. *J. Exp. Biol.* **58,** 627–636.

Fenwick, J. C. (1979). Ca^{2+}-activated adenosinetriphosphatase activity in the gills of freshwater- and seawater-adapted eels (*Anguilla rostrata*). *Comp. Biochem. Physiol. B* **62B,** 67–70.

Forrest, J. N., Cohen, A. D., Jr., Schon, D. A., and Epstein, F. H. (1973). Na transport and Na-K ATPase in gills during adaptation to sea water: effects of cortisol. *Am. J. Physiol.* **224,** 709–713.

Foskett, J. K., and Scheffey, C. (1982). The chloride cell: Definitive identification as the salt-secretory cell in teleosts. *Science* **215,** 164–166.

Gallis, J. L., and Bourdichon, M. (1976). Changes of (Na^+-K^+) dependent ATPase activity in gills and kidneys of two mullet *Chelon labrossus* (Risso) and *Liza ramada* (Risso) during fresh water adaptation. *Biochemie* **58,** 625–627.

Gallis, J. L., Lasserre, P., and Belloc, F. (1979). Freshwater adaptation in the Euryhaline

Teleost *Chelon labrosus*. I. Effects of adaptation prolactin, cortisol and actinomycin D on plasma osmotic balance and (Na^+-K^+) ATPase in gill and kidney. *Gen. Comp. Endocrinol.* **38,** 1–10.

Gilles, M. A., and Vanstone, W. E. (1976). Changes in ouabaine-sensitive adenosine triphosphatase activity in gills of coho salmon (*Oncorhynchus kisutch*) during parr-smolt transformation. *J. Fish. Res. Board Can.* **25,** 2717–2720.

Girard, J. P., Sardet, C., Maetz, J., Thomson, A. J., and Sargent, J. R. (1977). Effet de l'ouabaine sur les branchies de truite adaptée à l'eau de mer. *Proc. Int. Union Physiol. Sci.* **13,** Abstr. No. 778.

Goldin, S. M. (1977). Active transport of sodium and potassium ions by sodium and potassium ion-activated adenosine triphosphatase from renal medulla. *J. Biol. Chem.* **252,** 5630–5642.

Hart, C. E., Concannon, G., Fustish, C. A., and Ewing, R. D. (1981). Seaward migration and gill (sodium, potassium) ATPase activity of spring chinook salmon in an artificial stream. *Trans. Am. Fish. Soc.* **110,** 44–50.

Haswell, M. S., Randall, D. J., and Perry, S. F. (1980). Fish gill carbonic anhydrase: Acid-base regulation or salt transport? *Am. J. Physiol.* **228,** 240–245.

Hilden, S., and Hokin, L. E. (1975). Active potassium transport coupled to active sodium transport in vesicles reconstituted from purified sodium and potassium ion-activated adenosine triphosphatase from the rectal gland of *Squalus acanthias*. *J. Biol. Chem.* **250,** 6296–6303.

Ho, S. M., and Chan, D. K. O. (1980a). Branchial ATPases and Ionic transport in the eel, *Anguilla Japonica*. I. Na^+-K^+-ATPase. *Comp. Biochem. Physiol. B* **66B,** 255–260.

Ho, S. M., and Chan, D. K. O. (1980b). Branchial ATPases and ionic transport in the eel *Anguilla Japonica*. II. Ca^{2+}-ATPase. *Comp. Biochem. Physiol. B* **67B,** 639–645.

Ho, S. M., and Chan, D. K. O. (1980c). Branchial ATPases and ionic transport in the eel *Anguilla japonica*. III. HCO_3-stimulated, SCN-inhibited Mg^{2+} ATPase. *Comp. Biochem. Physiol. B* **68B,** 113–117.

Hokin, L. E. (1977). Purification and properties of Na, K-ATPase from the rectal gland of *Squalus acanthias* and the electric organ of *Electrophorus electricus* and reconstitution of the Na^+-K^+ pump from the purified enzyme. *In* "Biochemistry of Membrane Transport" (G. Semenza and E. Carafoli, eds.), pp. 374–388. Springer-Verlag, Berlin and New York.

Hootman, S. R., and Philpott, C. W. (1978). Rapid isolation of chloride cell from pinfish gill. *Anat. Rec.* **190,** 687–702.

Hootman, S. R., and Philpott, C. W. (1979). Ultracytochemical localization of Na^+,K-activated ATPase in chloride cells from the gills of a euryhaline teleost. *Anat. Rec.* **193,** 99–129.

Ikemoto, N., Bhatnagar, G. M., Nagy, B., and Gergely, J. (1972). Interaction of divalent cations with the 55,000 dalton protein component of the sarcoplasmic reticulum. Fluorescence and circular dichroism. *J. Biol. Chem.* **247,** 7835–7837.

Jampol, L. M., and Epstein, F. H. (1970). Sodium potassium-activated adenosine triphosphatase and osmotic regulation by fishes. *Am. J. Physiol.* **218,** 607–611.

Johnson, S. L., Ewing, R. D., and Lichatowick, J. A. (1977). Characterization of gill (sodium, potassium ion) activated adenosine triphosphatase from chinook salmon, *Oncorhynchus tshawytscha*. *J. Exp. Zool.* **199,** 345–354.

Jørgensen, P. L. (1975). Isolation and characterization of the components of the sodium pump. *Q. Rev. Biophys.* **7,** 239–274.

Josephson, L., and Cantley, L. C. (1977). Isolation of a potent (Na-K) ATPase inhibitor from striated muscle. *Biochemistry* **16,** 4572–4578.

Kamiya, M. (1972). Sodium-potassium -activated adenosine triphosphatase in isolated chloride cells from eel gills. *Comp. Biochem. Physiol. B* **43B,** 611–617.

Kamiya, M., and Utida, S. (1968). Changes in activity of sodium-potassium-activated adenosinetriphosphatase in gills during adaptation of the japanese eel to sea water. *Comp. Biochem. Physiol.* **26,** 675–685.

Kamiya, M., and Utida, S. (1969). Sodium-potassium-activated adenosinetriphosphatase activity in gills of freshwater, marine and euryhaline teleosts. *Comp. Biochem. Physiol.* **31,** 671–674.

Karnaky, K. J., Ernst, S. A., and Philpott, C. W. (1976a). Teleost chloride cell. I. Response of gill Na, K-ATPase and chloride cell fine structure to various environments. *J. Cell Biol.* **70,** 144–156.

Karnaky, K. J., Kinter, L. B., Kinter, W. B., and Stirling, C. E. (1976b). Teleost chloride cell. II. Auroradiographic localization of gill Na-K-ATPase in killifish. *Fundulus heteroclitus* adapted to low and high salinity environments. *J. Cell Biol.* **70,** 157–177.

Kasbekar, D. K., and Durbin, R. P. (1965). An adenosine triphosphatase from frog gastric mucosa. *Biochim. Biophys. Acta* **105,** 472–482.

Kelly, K. F., Pirie, B. J. S., Bell, M. V., and Sargent, J. R. (1979). A light microscopic study of the action of sodium orthovanadate on the gills of fresh water and seawater eels (*Anguilla anguilla* L.). *J. Mar. Biol. Assoc. U.K.* **59,** 859–865.

Kepner, G. R., and Macey, R. I. (1968). Membrane enzyme systems: molecular size determination by radiation inactivation. *Biochim. Biophys. Acta* **163,** 188–203.

Kerstetter, T. H., and Kirschner, L. B. (1972). Active chloride transport by the gills of rainbow trout (*Salmo gairdneri*). *J. Exp. Biol.* **56,** 263–272.

Kerstetter, T. H., and Kirschner, L. B. (1974). HCO_3^--dependent ATPase activity in the gills of rainbow trout (*Salmo gairdneri*). *Comp. Biochem. Physiol. B* **48B,** 581–589.

Kerstetter, T. H., Kirschner, L. B., and Rafuse, D. D. (1970). On the mechanisms of sodium ion transport by the irrigated gills of rainbow trout (*Salmo gairdneri*). *J. Gen. Physiol.* **56,** 342–359.

Kinne-Saffran, E., and Kinne, R. (1979). Further evidence for the existence of an intrinsic bicarbonate-stimulated Mg^{2+} ATPase in brush border membranes isolated from rat kidney cortex. *J. Membr. Biol.* **49,** 235–251.

Kirschner, L. B. (1977). The sodium chloride excreting cells in marine vertebrates. *In* "Transport of Ions and Water in Animals" (B. L. Gupta, R. B. Moreton, J. L. Oschman, and B. J. Wall, eds.), pp. 427–452. Academic Press, New York.

Kirschner, L. B. (1979). Control mechanisms in crustaceans and fishes. *In* "Osmoregulation in Animals" (R. Gilles, ed.), pp. 157–222. Wiley (Interscience), New York.

Kirschner, L. B. (1980). Comparison of vertebrate salt-excreting organs. *Am. J. Physiol.* **238,** 219–223.

Krogh, A. (1939). "Osmotic Regulation in Aquatic Animals." Cambridge Univ. Press, London and New York.

Lassere, P. (1971). Increase of (Na^+ + K^+)-dependent ATPase activity in gills and kidneys of two euryhaline marine teleosts, *Crenimugil labrosus* and *Dicentrarchus labrax*, during adaptation to fresh water. *Life Sci.* **10,** 113–119.

Lassere, P., Boeuf, G., and Harache, Y. (1978). Osmotic adaptation of *Oncorhynchus kisutch* Walbaum. I. Seasonal variations of gill Na^+-K^+-ATPase activity in cohosalmon, O^+-age and yearling, reared in fresh water. *Aquaculture* **14,** 365–382.

Laurent, P., Dunel, S., and Barets, A. (1968). Tentative de localisation histochimique d'une ATPase Na^+ K^+ au niveau de l'épithélium pseudobranchial des Téléostéens. *Histochemie* **14,** 308–314.

Le Marie, M., Moeller, J. V., and Tanford, C. (1976). Retention of enzyme activity by detergent-solubilized sarcoplasmic calcium (2^+) ion activated ATPase. *Biochemistry* **15,** 2336–2342.

Ma, S. W. Y., Shami, Y., Messer, H. H., and Copp, D. H. (1974). Properties of Ca^{2+}-ATPase from the gill of rainbow trout (*Salmo gairdneri*). *Biochim. Biophys. Acta* **345,** 243–251.

McCarty, L. S., and Houston, A. H. (1977). Na^{+}:K^{+} and HCO_3^{-} stimulated ATPase activities in the gills and kidneys of thermally acclimated rainbow trout, *Salmo gairdneri*. *Can. J. Zool.* **55,** 704–712.

Maetz, J. (1969). Sea water teleosts: Evidence for a sodium-potassium exchange in the branchial sodium-excreting pump. *Science* **166,** 613–615.

Maetz, J. (1974). Aspects of adaptation to hypo-osmotic and hyper-osmotic environments. *Biochem. Biophys. Perspect. Mar. Biol.* **1,** 1–167.

Maetz, J., and Bornancin, M. (1975). Biochemical and biophysical aspects of salt excretion by chloride cells in teleosts. *Fortschr. Zool.* **23,** 322–362.

Maetz, J., and Garcia-Romeu, F. (1964). The mechanism of sodium and chloride uptake by the gills of a Fresh water fish, *Carassius auratus*. II. Evidence for NH_4^{+} / Na^{+} and HCO_3^{-}/Cl^{-} exchanges. *J. Gen. Physiol.* **47,** 1209–1227.

Masoni, A., and Isaia, J. (1973). Influence du mannitol et de la salinité externe sur l'équilibre hydrique et l'aspect morphologique de la branchie d'anguille adaptée à l'eau de mer. *Arch. Anat. Microsc. Morphol. Exp.* **62,** 293–306.

Maunsbach, A. B., Skiver, E., and Jorgensen, P. L. (1979). Ultrastructure of purified Na, K-ATPase membranes. *In* "Na, K-ATPase, Structure and Kinetics" (J. C. Skou and J. G. Nørby, eds.), pp. 3–13. Academic Press, New York.

Mayer-Gostan, N., Bornancin, M., de Renzis, G., Naon, R., Yee, R., Chew, J., and Pang, P. (1983). Extrainstetinal Ca^{2+} uptake in killyfish (*Fundulus heteroclitus*). *J. Exp. Zool.* **277,** 329–338.

Milne, K. P., Ball, J. N., and Chester Jones, I. (1971). Effects of salinity hypophysectomy and corticotrophin on branchial Na and K-activated ATPase in the eel, *Anguilla anguilla* L. *J. Endocrinol.* **49,** 177–178.

Motais, R. (1967). Les mécanismes d'échanges ioniques branchiaux chez les Téléostéens. *Ann. Inst. Oceanogr.* (*Monaco*) **45,** 1–84.

Motais, R. (1970). Effect of actinomycin D on branchial Na-K dependent ATPase activity in relation to sodium balance of the eel. *Comp. Biochem. Physiol.* **34,** 497–501.

Motais, R., and Isaia, J. (1972). Evidence for an effect of ouabain on the branchial sodium-excreting pump of marine teleosts: Interaction between the inhibitor and external Na and K. *J. Exp. Biol.* **57,** 367–373.

Motais, R., Isaia, J., Rankin, J. C., and Maetz, J. (1969). Adaptative changes of the water permeability of the teleostean gill epithelium in relation to external salinity. *J. Exp. Biol.* **51,** 529–546.

Murphy, A. J. (1976). Sulfhydryl group modification of sarcoplasmic reticulum membranes. *Biochemistry* **15,** 4492–4496.

Murphy, P. G., and Houston, A. H. (1974). Environmental temperature and the body fluid system of the fresh-water teleost. V. Plasma electrolyte levels and branchial microsomal (Na^{+}-K^{+}) ATPase activity in thermally acclimated goldfish (*Carassius auratus*). *Comp. Biochem. Physiol. B* **47B,** 563–570.

Naon, R., Bornancin, M., and de Renzis, G. (1981). Cl^{-}-HCO_3^{-} dependent ATPase in gills of freshwater and seawater eels (*Anguilla anguilla* L.). *Biochimie* **63,** 37–43.

Ottolenghi, P. (1979). The relipidation of delipidated (sodium, potassium)-ATPase (EC 3.6.1.3): An analysis of complex formation with dioleoylphosphatidylcholine and with dioleoylphosphatidylethamolamine. *Eur. J. Biochem.* **99,** 113–132.

Payan, P., and Girard, J. P. (1978). Mise en évidence d'un échange Na^{+}/NH_4^{+} dans la branchie de la truite adaptée à l'eau de mer: Contrôle adrénergique. *C. R. Hebd. Acad. Sci. Seances* **286,** 335–338.

Pequeux, A. (1980). Effects of high pressure on ion transport and osmoregulation. *In* "Environmental Physiology of Fishes" (M. A. Ali, ed.), pp. 163–200. Plenum, New York.

Perry, S. F., Haswell, M. S., Randall, D. J., and Farrel, A. P. (1981). Branchial ionic uptake and acid-base regulation in the rainbow trout, *Salmo gairdneri. J. Exp. Biol.* **92,** 289–303.

Pfeiler, E. (1978). Na-ATPase and NaK-ATPase activities in gills of Rainbow trout (*Salmo grairdneri*). *J. Comp. Physiol.* **124,** 97–104.

Pfeiler, E., and Kirschner, L. B. (1972). Studies on gill ATPase of rainbow trout (*Salmo gairdneri*). *Biochim. Biophys. Acta* **282,** 301–310.

Philpott, C. W. (1980). Tubular system membranes of teleost chloride cells: Osmotic response and transport sites. *Am. J. Physiol.* **238,** 171–184.

Potts, W. T. W. (1977). Fish gills. *In* "Transport and Water in Animals" (B. L. Gupta, R. G. Moreton, J. L. Oschman, and B. J. Wall, eds.), pp. 453–480. Academic Press, New York.

Racker, E., and Eytam, E. (1973). Reconstitution of an efficient calcium pump without detergents. *Biochem. Biophys. Res. Commun.* **55,** 174–178.

Sardet, C. (1980). Freeze fracture of the gill epithelium of euryhaline teleost fish. *Am. J. Physiol.* **238,** 207–212.

Sardet, C., Pisam, M., and Maetz, J. (1979). The surface epithelium of teleostean fish gills: Cellular and functional adaptations of the chloride cell in relation to salt adaptation. *J. Cell Biol.* **80,** 96–117.

Sargent, J. R., and Thomson, A. J. (1974). The nature and properties of the inducible sodium-plus-potassium ion-dependent adenosine triphosphatase in the gills of eels (*Anguilla anguilla*) adapted to freshwater and seawater. *Biochem. J.* **144,** 69–75.

Sargent, J. R., Thomson, A. J., and Bornancin, M. (1975). Activities and localisation of succinic dehydrogenase and Na^+/K^+-activated adenosine triphosphatase in the gill of eels (*Anguilla anguilla*) adapted to fresh water and seawater. *Comp. Biochem. Physiol. B* **51B,** 75–79.

Sargent, J. R., Bell, M. V., and Kelly, K. F. (1980). The nature and properties of sodium ion plus potassium ion-activated adenosine triphosphatase and its role in marine salt secreting epithelia. *In* "Epithelial Transport in the Lower Vertebrates" (B. Lahlou, ed.), pp. 251–267. Cambridge Univ. Press, London and New York.

Scharff, O. (1980). Kinetics of calcium-dependent membrane ATPase in human erythrocytes. *In* "Membrane Transport in Erythrocytes" (U. V. Lassen, H. H. Ussing, and J. O. Wieth, eds.), pp. 236–254. Munksgaard, Copenhagen.

Shephard, K. L. (1981). The activity and characteristics of the Ca^{2+} +ATPase of fish gills in relation to environmental calcium concentrations. *J. Exp. Biol.* **90,** 115–121.

Shephard, K. L., and Simkiss, K. (1978). The effects of heavy metal ions on Ca^{2+}ATPase extracted from fish gills. *Comp. Biochem. Physiol. B* **61B,** 69–72.

Shirai, H. (1972). Electron-microscope localization of sodium ions and adenosinetriphosphatase in chloride cells of the japanese eel, *Anguilla japonica. J. Fac. Sci., Univ. Tokyo, Sect. 4* **12,** 385–403.

Silva, P., Solomon, R., Spokes, K., and Epstein, F. H. (1977). Ouabain inhibition of gill Na-K-ATPase: Relationship to active chloride transport. *J. Exp. Zool.* **199,** 419–426.

Skou, J. C. (1965). Enzymatic basis for active transport of Na^+ and K^+ across cell membrane. *Physiol. Rev.* **45,** 596–617.

Skou, J. C. (1975). The $(Na^+ + K^+)$ activated enzyme system and its relationship to transport of sodium and potassium. *Q. Rev. Biophys.* **7,** 401–434.

Smith, H. W. (1930). The absorption and excretion of water and salts by marine teleost. *Am. J. Physiol.* **93,** 485–505.

Solomon, R. J., Silva, P., Bend, J. R., and Epstein, F. H. (1975). Thiocyanate inhibition of ATPase and its relationship to anion transport. *Am. J. Physiol.* **229,** 801–806.

Stirling, C. E., Karnaky, K. J., Kinter, L. B., and Kinter, W. B. (1973). Autoradiographic localization of ^{3}H ouabain binding by Na-K-ATPase in perfused gills of *Fundulus heteroclitus*. *Bull. Mt Desert Isl. Biol. Lab.* **13,** 117–120.

Stuenkel, E. L., and Hillyard, S. D. (1980). Effects of temperature and salinity on gill Na^+-K^+ATPase activity in the pupfish, *Cyprinodon salinus*. *Comp. Biochem. Physiol. A* **67A,** 179–182.

Thomson, A. J., and Sargent, J. R. (1977). Changes in the levels of chloride cells and ($Na^+ + K^+$) dependent ATPase in the gills of yellow and silver eels adapting to sea water. *J. Exp. Zool.* **200,** 33–40.

Towle, D. W., Gilmann, M. E., and Hempel, J. D. (1977). Rapid modulation of gill (sodium + potassium ion)-dependent ATPase activity during acclimatation of the killifish *Fundulus heteroclitus* to salinity change. *J. Exp. Zool.* **202,** 179–185.

Utida, S., Kamiya, M., and Shirai, N. (1971). Relationship between the activity of Na^+-K^+ activated adenosinetriphosphatase and the number of chloride cells in eel gills with special reference to sea-water adaptation. *Comp. Biochem. Physiol. A* **38A,** 443–447.

Van Amelswoort, J. M. M., De Pont, J. J. H. H. M., and Bonting, S. L. (1977). Is there a plasma membrane located anion-sensitive ATPase? *Biochim. Biophys. Acta* **466,** 283–301.

Vegh, K., Spiegler, P., Chamberlain, C., and Monmaerts, W. F. H. M. (1968). The molecular size of the calcium-transport ATPase of sarcotubular vesicles estimated from radiation inactivation. *Biochim. Biophys. Acta* **163,** 266–268.

Wallick, E. T., Lane, L. K., and Schwartz, A. (1979). Biochemical mechanism of the sodium pump. *Annu. Rev. Physiol.* **41,** 397–411.

Warren, G. B., Toon, P. A., Birdsall, N. J. M., Lee, A. G., and Metcalfe, F. C. (1974). Reconstitution of a calcium pump using defined membrane components. *Proc. Natl. Acad. Sci. U.S.A.* **71,** 622–626.

Wielbelhaus, V. D., Sung, C. P., Helander, H. F., Shah, G., Blum, A. L., and Sachs, G. (1971). Solubilization of anion ATPase from *Necturus* oxyntic cells. *Biochim. Biophys. Acta* **241,** 49–56.

Zaugg, W. S., and McLain, L. R. (1970). Adenosinetriphosphatase activity in gills of salmonids: Seasonal variations and salt water influence in coho salmon *Oncorhynchus kisutch*. *Comp. Biochem. Physiol.* **35,** 587–596.

Zaugg, W. S., and McLain, L. R. (1972). Changes in gill adenosinetriphosphatase activity associated with Parr-Smolt transformation in steelhead trout, Coho, and Spring chinooh salmon. *J. Fish. Res. Board Can.* **29,** 167–171.

Zaugg, W. S., and McLain, L. R. (1976). Influence of water temperature on gill sodium, potassium-stimulated ATPase activity in juvenile coho salmon (*Oncorhynchus kisutch*). *Comp. Biochem. Physiol. A* **54A,** 419–421.

Zaugg, W. S., and Wagner, H. H. (1973). Gill ATPase activity related to parr-smolt transformation and migration in steelhead trout (*Salmo gairdneri*): Influence of photoperiod and temperature. *Comp. Biochem. Physiol. B* **45B,** 955–965.

Zaugg, W. S., Adams, B. L., and McLain, L. R. (1972). Steelhead migration: Potential temperature effects as indicated by gill adenosine triphosphatase activities. *Science* **176,** 415–416.

4

TRANSEPITHELIAL POTENTIALS IN FISH GILLS

W. T. W. POTTS
Department of Biological Sciences
University of Lancaster
Lancaster, England

I. INTRODUCTION

A knowledge of the transepithelial potential is essential in the interpretation of the movements of ions across an epithelium. Although transepithelial potentials had been studied in other classes of vertebrates for more than 40 years, the first measurements of transepithelial potentials in teleost gills were made only in 1963, and a significant body of data has been collected only in the last decade.

The first attempts to measure gill potential adapted the techniques developed for intracellular recording. House (1963) inserted microelectrodes through the skin of a blenny on the probable but untested assumption that all extracellular fluids are at the same potential. Maetz and Campanini (1966) inserted microelectrodes through the gill epithelium of an eel, but this required the immobilization of the respiratory movements. The fragility of microelectrodes makes it almost essential to anesthetize the fish, which may affect both the potentials and ion fluxes.

ISBN 0-12-350432-5

Whole blood is a good electrical conductor, and therefore any potential gradients within the circulatory system should be small. The potential in the circulatory system can conveniently be monitored by an indwelling catheter, but indwelling catheters in fish blood vessels easily become blocked, leak, or interfere with the circulation. Comparison of the potentials in an artery and in the peritoneal fluid showed that the potentials were effectively identical (Potts and Eddy, 1973), and most recent measurements have been made with peritoneal catheters.

Most experiments have been performed *in vivo,* but experiments with whole fish suffer from two disadvantages. First, it is difficult to produce substantial changes in the composition of the body fluids, and the range of drug concentration is limited; second, most experiments have been carried out with euryhaline fishes, possibly because these are assumed to be more resistant to changes of the external medium. This may have given a distorted or limited picture of the gills of marine teleosts, as only a small proportion of species are euryhaline.

In vitro experiments using isolated perfused gills make it possible to study the effect of changes in the ionic composition of the internal medium on the transepithelial potentials and of the effects of the application of a limitless range of drugs on the serosal surface, but isolated perfused gills are inconvenient preparations. In particular it has proved difficult to maintain normal rates of flow at normal pressures and to recover all the perfusate. The discovery that some nonrespiratory epithelia are also rich in chloride cells (Karnaky *et al.*, 1977; Marshall, 1977) and can be used in Ussing chambers has opened up a new range of possibilities.

Fish gills are much more complex than many of the better known ion-transporting epithelia such as frog skin or toad bladder. The greater part of the surface of fish gills is covered by respiratory epithelial cells, which are situated mainly in the secondary lamellae and are irrigated by the afferent–efferent arterial system. The chloride or mitochondria-rich cells, covering only a small part of the gill area, are situated mainly on the primary lamellae, and their efferent blood supply leaves mainly by the venous drainage. In addition, mucous cells are distributed over both the primary and secondary lamellae.

In fish adapted to fresh water the major ion exchanges are sodium uptake in exchange for ammonium or hydrogen ions, and chloride uptake in exchange for bicarbonate ions. Studies of the distribution of labeled ions in the venous and arterial returns from the gill by Girard and Payan (1980) suggest that these exchanges take place mainly over the respiratory epithelia in the secondary lamellae. In freshwater fish both the respiratory epithelia and the chloride cells, which are found even in stenohaline freshwater fish, are joined by multistrand tight junctions consisting of between five and nine strands, (Sardet, 1980), and in consequence the gills have very low permeabilities to both ions and small organic molecules. The function of the mitochondria-rich cells in freshwater fishes, particularly in stenohaline species, is uncertain.

In marine teleosts the respiratory epithelial cells are very similar to those in freshwater fishes. They are connected by multistrand tight junctions, and they continue to exchange sodium ions for ammonium ions, even though the resulting salt uptake runs contrary to the major ion fluxes in these circumstances. This low-level exchange is probably part of the acid–base balancing mechanism (Evans, 1980). A variety of studies have shown that the chloride cell is the site of the active excretion of chloride ions in fish adapted to seawater. In these fish the chloride cells are linked only by single-strand junctions that are very permeable both to sodium ions and to small organic molecules. Girard and Payan (1980) estimate that in the sea trout the primary lamellar epithelium is 400 times as permeable per unit area as the secondary lamellar epithelium. The potentials generated by the active excretion of chloride ions induces a passive efflux of sodium ions through the leaky junctions. It will be shown later that many species of marine teleosts, possibly the great majority, do not generate sufficient potential to maintain passive sodium balance and must therefore be excreting sodium ions actively as well, but the site of sodium excretion is uncertain.

The chloride cells cover only a small part of the total area of the gill but are the site of the major chloride fluxes (Foskett and Scheffey, 1982). The occurrence of the simultaneous transport of the same ion species in opposite directions in the whole gill and the active transport of both cations and anions in parallel make the relatively simple analyses developed for more homogeneous tissues inapplicable. The situation is even more complicated *in vivo* where significant fluxes and electrical leaks may occur elsewhere through the skin and gut. In the special case of the epithelia consisting predominantly of chloride cells, such as are found on the inner sides of the operculi of *Fundulus* (Karnaky *et al.*, 1972) and very recently on *Blennius* (Nonette *et al.*, 1982), or the jaw of *Gallichthys* (Marshall, 1977) studies in Ussing chambers make it possible to compare ion fluxes and short-circuit current under a variety of conditions and to analyze the tissue in terms of active and passive fluxes and resistance (Degnan and Zadunaisky, 1980, 1982). This is not possible with *in vivo* studies but they are essential to analyze the functions of the ion transport systems in life, and studies of the gill potentials in a variety of solutions and comparison with appropriate physical models do provide valuable insights into the properties of the ion pumps. It should be noted that there are often interesting discrepancies

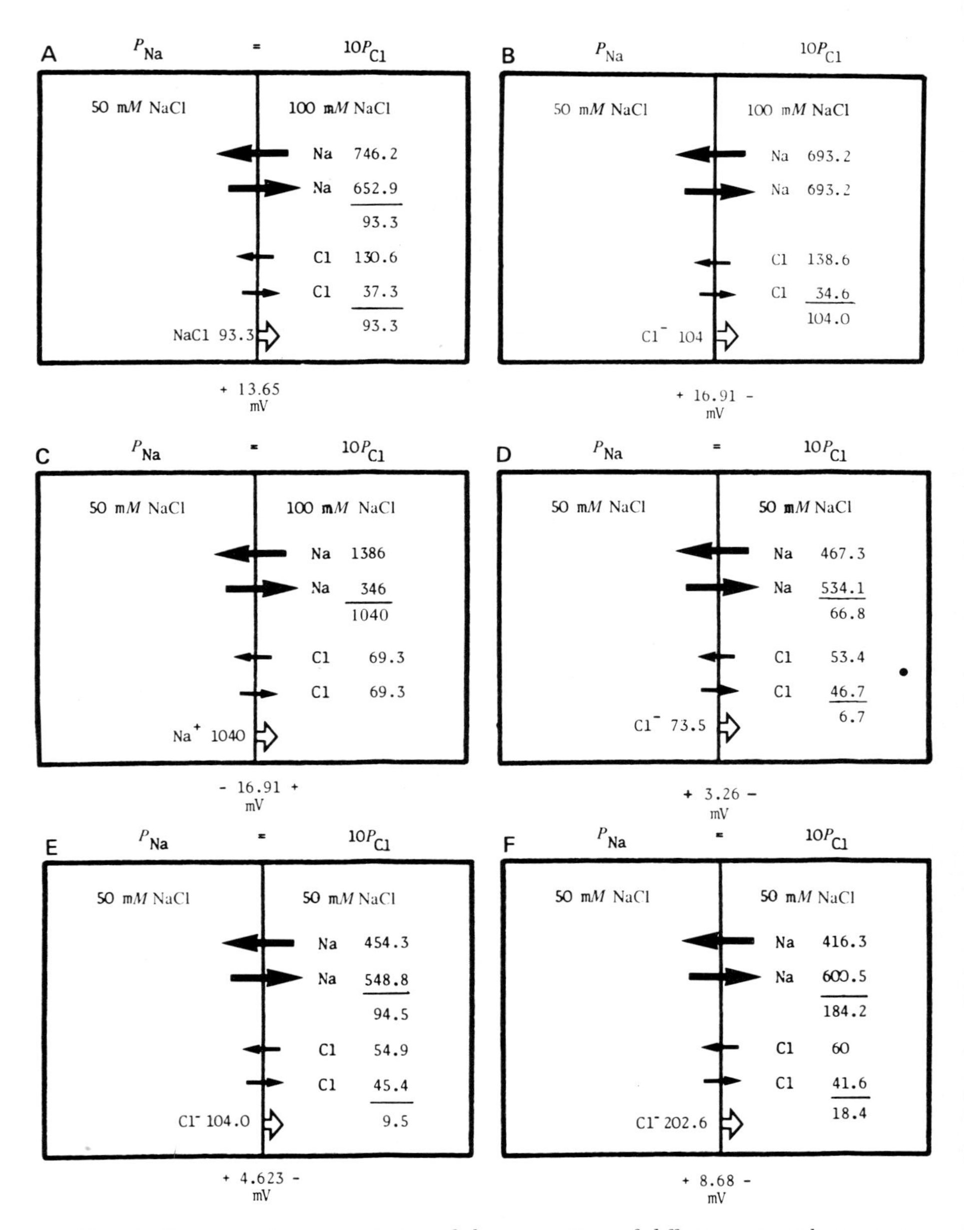

Fig. 1. Diagrammatic representation of the interactions of diffusion potentials across a cation-permeable membrane and active transport under various conditions. Solid arrows, passive transfer; open arrows, active transport. In all cases, $P_{Na} = 10P_{Cl}$. (A) Twofold concentration difference and a neutral NaCl pump. (B) Twofold concentration difference and an electrogenic chloride pump. (C) Twofold concentration difference and an electrogenic sodium pump. (D–F) Concentration in two compartments identical, electrogenic chloride pumps producing various potentials. For details see text.

between *in vivo* and *in vitro* studies. For example opercular epithelia from *Fundulus* adapted to fresh water still excrete chloride ions whereas *in vivo* active uptake must predominate.

II. THEORETICAL BACKGROUND

The transepithelial potential, $\Delta\psi$, observed in the gills of a teleost is generated by a diffusion potential, E_D, arising from the difference in composition between the blood plasma and the external medium and a metabolically maintained component, E_A, arising from the active transport of ions. Potentials, like pressures, are additive:

$$\Delta\psi = E_D + E_A$$

but the interactions of a diffusion potential and an electrogenic pump are complicated and not fully predictable, so some theoretical discussion may be advantageous.

A diffusion potential may arise whenever two solutions of electrolytes of different composition are in contact. In the simple case illustrated in Fig. 1A the passive net flux of chloride ions across the membrane must be equal to the passive net flux of sodium ions, because the cations must balance the anions in each compartment, except for the minute charge on the membrane itself. The potential at which the net fluxes are equal is given by

$$E_D = \frac{RT}{F} \ln \left(\frac{P_{Na}[Na_2] + P_{Cl}[Cl_1]}{P_{Na}[Na_1] + P_{Cl}[Cl_2]} \right) \tag{1}$$

where P_{Na} and P_{Cl} are the permeabilities of the membrane to sodium and chloride and $[Na_1]$, $[Na_2]$, and so on are the concentrations of sodium ions in compartments 1 and 2, and R, T, and F have their usual meanings. Differences in activity coefficients between the two compartments will be neglected. In the absence of any potential

$$\begin{matrix} J_{Cl}^{1\,2} \,\alpha\, P_{Cl}[Cl_1] & \text{and} & J_{Na}^{1\,2} \,\alpha\, P_{Na}[Na_1] \\ J_{Cl}^{2\,1} \,\alpha\, P_{Cl}[Cl_2] & & J_{Na}^{2\,1} \,\alpha\, P_{Na}[Na_2] \end{matrix} \tag{2}$$

where $J_{Cl}^{1\,2}$ is the chloride flux from compartment 1 to compartment 2, and so on. In the presence of a potential E, these fluxes become

$$J_{Cl}^{1\,2} \,\alpha\, P_{Cl}[Cl_2]\frac{EF/RT}{1 - \exp(EF/RT)}$$

When $E = E_D$ the results are illustrated in Fig. 1A.

Such a system could be maintained in equilibrium by a unidirectional ion pump transferring equal quantities of sodium and chloride ions against the concentration gradient, illustrated by the open arrow in Fig. 1A. As this pump is transferring equal quantities of cations and anions it would be electrically neutral, and if it were turned on and off it would have no instantaneous observable effect on the potential. If the activity of the neutral pump were increased, a net flux of equivalent quantities of both sodium and chloride ions—against the concentration gradient—would be generated, again without any instantaneous effect on potential.

The simple system illustrates two more subtle points. First, although electrically neutral this pump, in association with a membrane selectively permeable to ions, is in effect generating the potential. Second, although the whole pump is electrically neutral the energy required to transport the chloride ions against both the concentration and electrical gradients is much greater than that required to transport the sodium ion, where the potential assists the movement of the ion.

In Fig. 1A the energy required actively to transport 93 mEq sodium is only 0.304×10^{-3} J, while the energy required to transport the chloride is 2.851×10^{-3} J. As the permeability to the less permeant ion is reduced, the diffusion potential approaches the Nernst potential of the more permeant ion, but most of the energy is dissipated by the diffusion of the less permeant ion. Energy dissipation is approximately inversely proportional to the permeabilities of the membrane to the respective ions. If the concentration gradient were changed, the load on the chloride ion pump would change 10 times more than that on the sodium pump, and the active fluxes might no longer be equivalent. The pump might then become electrogenic.

The concentration gradient in Fig. 1A could also be maintained by an electrogenic pump. In Fig. 1B only the chloride ion is transported, the electrogenic pump raising the observed potential to the Nernst potential for sodium, $E_{Na} = (RT/F) \ln([Na_2]/[Na_1])$, at which there is no net sodium flux. If the chloride pump were turned off the potential would decline instantly by 3.26 mV to the diffusion potential illustrated in Fig. 1A, so that in the example illustrated in Fig. 1B $E_A = 3.26$ mV, $E_D = 13.65$, and $\Delta\psi = 16.91$. If the active chloride flux were increased to produce a net efflux of chloride from compartment 1, the potential would rise above the Nernst potential to maintain an equivalent net flux of sodium.

Equilibrium could also be maintained by an electrogenic sodium pump alone (Fig. 1C), although the transport of the more permeant ion is a prodigal exercise as a large flux has to be driven against both the potential and concentration gradients (Table I).

Were the permeability to sodium not much greater than that to chloride, the diffusion potential would be low. In these circumstances equilibrium

Table I

Energy Consumption in Systems Illustrated in Fig. 1A–F

Figure		Conditions	Potential (mV)	Active flux (mEq)	Energy consumed (J)
1A	Neutral pump	Anisosmotic	13.65	NaCl 93.3	1.093×10^{-3}
1B	Chloride ion pump	Anisosmotic	16.91	Cl^- 104	1.759×10^{-3}
1C	Sodium ion pump	Anisosmotic	16.91	Na^+ 1040	45.32×10^{-3}
1D	Chloride ion pump	Isosmotic	3.26	Cl^- 73.5	0.24×10^{-3}
1E	Chloride ion pump	Isosmotic	4.623	Cl^- 104	0.481×10^{-3}
1F	Chloride ion pump	Isosmotic	8.68	Cl^- 202.6	1.759×10^{-3}

could still be maintained by a pure chloride pump operating at the Nernst potential for sodium, but this again would be uneconomical. When the permeabilities are similar an economical system requires both a sodium and a chloride pump. Only when $P_{Na} \gg P_{Cl}$ does an electrogenic chloride pump approach a neutral sodium chloride pump in economy.

Equation (1) might seem to imply that E_A could be measured by eliminating E_D, when $E_A = \Delta\psi$. However, it is most unlikely that E_A would be independent of E_D. If compartment 2 were filled with a solution of 50 mM NaCl so that E_D were reduced to zero, it is unlikely that the pump would then continue to generate a potential of only 3.26 mV. No electrogenic pump can establish an equilibrium between two identical solutions, because any potential will drive sodium and chloride ions in opposite directions; however, it could generate a potential if it were to produce a net flux of ions across the membrane, as illustrated in Fig. 1D. In this model an electrogenic pump produces a potential of 3.26 mV (equivalent to the electrogenic component in Fig. 1B alone). In these circumstances the active flux, 73.5 mEq NaCl, is much smaller than in Fig. 1B and the energy required is less than a quarter of that required in the model illustrated in Fig. 1B (Table I).

If an electrogenic pump were operating so that the active flux of chloride was the same as in Fig. 1B, (namely 104 mEq Cl), a potential of 4.623 mV would be established (Fig. 1E), but again the work done would be less (Table I), and it is only at a potential of 8.68 mV and a chloride flux of 202.6 mEq (Fig. 1F) that the energy dissipation reaches that in Fig. 1B (Table I).

In biological systems the ionic pumps are likely to be under some form of nervous or hormonal control, but these calculations suggest that E_A is unlikely to be independent of E_D and that E_A is likely to increase when E_D is reduced or eliminated. An analysis of the sparse data available seems to confirm this.

Similar principles apply in both freshwater and marine conditions, but in seawater teleost fish need to maintain a net efflux of sodium and chloride across the gills to compensate for salts ingested in maintaining water balance, whereas fish in fresh water need to maintain a net influx to replace losses. Ignoring the problems that may arise from the active transport or diffusion of the minor ions in seawater, particularly potassium, calcium, magnesium, sulfate, and bicarbonate, it is apparent that when the transepithelial potential lies between the Nernst potentials of sodium and chloride both sodium and chloride ions must be actively transported outwards, although not necessarily in equivalent proportions. If chloride ions only are actively transported a net efflux of sodium ions can only be maintained if the transepithelial potential exceeds the Nernst potential of sodium. In fresh water the concentrations of sodium and chloride are so low that the Nernst potentials for these ions are usually more than 150 mV, much greater than the observed potentials, so it may be concluded that both ions are usually transported. A further distinction between the marine and freshwater condition may be relevant here. In fresh water the fluxes are so small that metabolically produced counterions (H^+, NH_4^+, OH^+, and HCO_3^-) are available in sufficient quantity that both the sodium and chloride pumps could be electrically neutral. In seawater the fluxes are usually so large that an electrically neutral pump could only operate by transporting equivalent quantities of both sodium and chloride ions.

A neutral pump is more economical than an electrogenic pump when the function is the movement of ions, although similar analyses show that where the primary object is the generation of potential an electrogenic pump is the more economical (Table I). The further the potential diverges from the diffusion potential—that is, the more electrogenic the pump becomes—the greater the energy required to maintain the same concentration difference (Potts *et al.*, 1973).

III. TRANSEPITHELIAL POTENTIALS IN SEAWATER

A. The Diffusion Potential

Transepithelial potentials have been examined in a number of species (Table II), but most of those scrutinized, such as the salmonids, the eel, the mullet, the blenny, and the flounder, are euryhaline fishes, and few stenohaline marine fishes have been studied in detail. The sample is so small that it is dangerous to generalize, but it may be significant that stenohaline species such as *Opsanus tau, Lagodon rhomboides, Achirus lineatus, Hippocampus erectus,* and *Gadus callarius* all maintain low potentials, whereas the euryhaline species all maintain higher potentials, closer to the Nernst potential for sodium. In euryhaline fishes more detailed experiments show that the potentials have many of the properties of diffusion potentials. The evidence that the transepithelial potentials are largely diffusional in origin can be summarized as follows. In the euryhaline fishes the potentials approximate to the Nernst potentials for sodium, depolarizing as the external concentration falls and even reversing in hypo-osmotic solutions (Table II). Substitution of sodium by the larger, less permeant organic ion choline also depolarizes the gills, whereas the substitution of organic anions for chloride hyperpolarizes them, but to a smaller extent. When the relative permeabilities to Na^+ and Cl^-, calculated from experiments of this type, are substituted in Eq. (2) the observed potential approximates to E_D in a wide

Table II

Transepithelial Potentials for Na^+ and Cl^- in Some Fish Species in Seawater[a]

Species	Na_{sw}	Cl_{sw}	Na_p	Cl_p	E_{Na}	E_{Cl}	$\Delta\psi$	References
Platichthys flessus	470	550	190	175	22	−27	19	Potts *et al.* (1973)
Platichthys flessus	520	610	170	155	28	−34	34	Macfarlane and Maetz (1975)
Blennius pholis	470	548	170	170	26	−29	23	House (1967)
Gillichthys mirabilis	480	560	164	—	27	—	21	Evans (1980)
Dormitator maculatus	500	583	—	—	—	—	17	Evans *et al.* (1974)
Anguilla anguilla	505	600	145	140	32	−38	23	House and Maetz (1974)
Sarotherodon mossambicus	540	615	164	146	31	−37	35	Dharmamba *et al.* (1975)
Serranus sp.	525	600	188	162	27	−35	25	Maetz and Bornancin (1975)
Mugil capito	520	608	170	155	27	36	20	Pic (1978)
Fundulus heteroclitus	520	610	175	140	28	38	18	Pic (1978)
Salmo gairdneri	450	520	172	142	23	32	10	Kirschner *et al.* (1974)
Lagodon rhomboides	500	583	172	—	28	—	5	Carrier and Evans (1976)
Gadus callarius	480	548	190	157	2	−3	3	Fletcher (1978a)
Hippocampus erectus	500	583	162	146	29	−36	−4	Evans and Cooper (1976)
Opsanus beta	455	530	159	146	26	32	−3	Evans (1980)
Achirus lineatus	500	583	—	—	—	—	−7	Evans and Cooper (1976)
Opsanus beta	500	583	159	154	30	−34	−8	Evans (1980)

[a] Na_{sw}, concentration of sodium in seawater (mM); Cl_{sw}, concentration of chloride in seawater (mM); Na_p, concentration of sodium in plasma (mM); Cl_p, concentration of chloride in plasma (mM); E_{Na}, Nernst potential of sodium ions across gill epithelium (mV); E_{Cl}, Nernst potential of chloride ions across gill epithelium (mV); $\Delta\psi$, observed potential (mV).

Table III

Relative Gill Permeabilities to Sodium and Chloride in Several Species of Fish

Species	P_{Cl}/P_{Na}	P_K/P_{Na}	Reference
Platichthys flessus	0.03	2.5	Potts and Eddy (1973)
Anguilla anguilla	0.11	34	Macfarlane and Maetz (1978)
Salmo gairdneri	0.3	—	Kirschner *et al.* (1974)
Mugil capito	0.14	—	Pic (1978)
Fundulus heteroclitus	0.23	—	Pic (1978)

range of solutions. The relative permeabilities of several gills to sodium and chloride calculated in this way are shown in Table III. In these early experiments the presence of an electrogenic component of the potential was not recognized or was ignored. This must have introduced minor systematic errors, but minor discrepancies were generally attributed to the uncertainties of the exact values of the activity coefficients, to changes in calcium concentration, or to experimental error (e.g., Potts and Eddy, 1973).

Little work has been carried out on those fish that display low or negative potentials in seawater, although they may prove to be more typical of marine fish. A detailed study of the cod, *Gadus callarias* (Fletcher, 1978b), showed that here the potential, +2.9 mV, is the combination of a slightly negative diffusion potential and a small positive electrogenic potential. The permeability to sodium is only slightly greater than that to chloride, but this is offset by the higher chloride concentration ratio across the gill. In all fish in which the transepithelial potential $\Delta\psi$ is small it seems likely that E_D is also low and that the permeabilities to both sodium and chloride ions will prove to be similar.

Although it is dangerous to generalize from such a small sample, it is possible that the selectively permeable gills found in the flounder, eel, mullet, and trout are specific adaptations to euryhalinity. In these gills the absolute permeabilities to chloride are low, so that on transfer to fresh water, chloride ions are conserved and the large diffusion potential created instantaneously retards sodium loss. A fish that is equally permeable to both ions must continue to lose large quantities of both sodium and chloride on transfer to fresh water, until the permeabilities are reduced.

Except in the case of the euryhaline fish, the observed potentials are far below E_{Na}. In these cases sodium as well as chloride ions must be actively transported across the epithelium. Since this was written, Nonette *et al.* (1982) have shown that the isolated opercular epithelium of *Blennius pholis*, which maintains a potential of 10.6 mV on circuit, maintains net effluxes of 1.1 μEq Na cm^{-2} hr^{-1} and 2.3 μEq cm^{-2} hr^{-1} on open circuit.

B. The Electrogenic Component of the Potential

The electrogenic component of the potential is easier to demonstrate than to measure. If the diffusion potential is eliminated, any remaining potential must be due to the active transport of ions, but as demonstrated earlier, in these circumstances both the fluxes and the potential are likely to be greater than when the ion pumps are operating against concentration differences and a diffusion potential. The presence of an electrogenic component can be demonstrated in several ways. First, the diffusion potential should be eliminated when the fish is immersed in an artificial solution similar in composition to the plasma. Any differences between the activity coefficients in the solution and in the protein-rich plasma will be small, so the diffusion potential should be negligible. Any remaining potential must be metabolically maintained. The mullet maintains a potential 5.0 mV under isotonic conditions (Pic and Maetz, 1975). Second, isolated gills can be perfused while immersed in a solution of identical composition to the perfusate. Experiments of this kind with flounder gills (Shuttleworth *et al.*, 1974; Shuttleworth, 1978) demonstrate the unambiguous presence of a potential of from 4 to 12 mV, inside positive, suggesting the active excretion of chloride ions. This potential is greatly reduced by anoxia or ouabain or by adrenaline.

Because E_D cannot exceed the Nernst potential for sodium, the potentials observed in *Oreochromis* [formerly *Sarotherodon* (formerly *Tilapia*)] and *Platichthys* (Macfarlane and Maetz, 1975) (Table II) are a priori evidence of the existence of electrogenic pumps, but the determination of E_A in the presence of E_D is much more difficult. E_A may be calculated from $\Delta\psi$ and E_D when the latter can be determined from P_{Na}, P_{Cl}, and P_K, but unfortunately these are rarely known with sufficient accuracy and are usually determined on the assumption that $\Delta\psi = E_D$.

Fletcher (1977) points out that E_D can be calculated from the rate of change of $\Delta\psi$ when the ambient seawater is concentrated or diluted, providing E_A remains constant. If $\delta\psi$ is the change in potential when the medium is concentrated or diluted x times,

$$\delta\psi = 1 - 2t_{Cl} \frac{RT}{F} \ln x$$

when t_{Cl} is the transport number of the chloride ion, or more exactly of all permeant anions. The transport number is the proportion of total charge carried by the ionic species in question, and the sum of all transport numbers is unity.

$$E_D = \Sigma\, t_i\, E_i$$

where t_i is the transport number of the ith permeant ion whose Nernst potential is E_i or, to a first approximation,

$$E_D = t_{Na} E_{Na} + t_{Cl} E_{Cl}$$

Unfortunately this is only strictly true if E_A is constant, but if $\delta\psi$ is kept small E_A is unlikely to change substantially. For example, in Fig. 1A where $\Delta\psi =$ 13.65 mV, if the 100 mM solution in compartment 2 were replaced with 90 mM NaCl and the pump remained neutral, then $\delta\psi = 2.03$ mV and t_{Cl} calculated by this method is 0.1055, close to the true value of $1/(10 + 1)$. If an electrogenic pump were involved, the method would again only be applicable if E_A remained almost constant, despite the reduction of the load on the pump.

A number of physiologists have investigated the effects of changes of the concentration of seawater on $\Delta\psi$, but not with the intention of determining E_A. When these are gleaned from the literature and E_D and E_A calculated (Table IV), the electrogenic component so derived is usually positive, implying that the major activity of the pump is the extrusion of chloride ions, and as expected, is smaller than the E_A measured in isotonic Ringer solution.

In *Salmo gairdneri* the diffusion potential calculated from the data of Kirschner *et al.* (1974) is greater than $\Delta\psi$, implying that E_A is negative, that is, sodium ions are being exported more energetically than chloride ions. This is a surprising conclusion but is apparently confirmed by the negative potential observed when the fish is in an approximately isotonic saline. This requires confirmation but is not impossible when taken in consideration with the fact that the fish with low potentials must actively extrude both sodium and chloride ions.

Inspection of the relationship between the potential across the isolated epithelium of *Fundulus* and the concentration of external seawater (Degnan and Zadunaisky, 1980) shows that the potential changes by about 56 mV for a 10-fold change of concentration; that is, it is almost impermeable to chloride, so that $E_D \simeq E_{Na}$ but the mean observed $\Delta\psi$ was 5.5 mV higher than E_{Na}. The stenohaline fish *Opsanus beta* maintains a potential of -3 mV with Ringer solution on both sides (Howe, 1980). This is probably to be interpreted as the active transport of both sodium and chloride, the former predominating in accordance with the low potential in seawater (Table II).

The mitochondria-rich cells on the gills of marine teleosts are commonly described as chloride cells. The term arose from Keys' experiments in 1930 when the active extrusion of ions from the fish was first demonstrated (Keys, 1931). It was quite fortuitous that because of the great accuracy of chloride assay by silver nitrate titration, chloride movements rather than sodium movements were first determined. The first gill potentials to be measured were positive with respect to the blood, which confirmed the occurrence of

Table IV

Electrogenic Component of Gill Potential (mV)

Species	t_{Na}	t_{Cl}	E_D	$\Delta\psi_{sw}$	E_A	$\Delta\psi_{iso}$	Reference
Artemia salina	0.65	0.35	+5.0	23.4	+18.4		Smith (1969)
Blennius pholis	0.747	0.253	+11.4	23.0	+11.6	<+20 but high	House (1963)
Salmo gairdneri	0.835	0.165	+12.9	10.1	−2.8	−5.0[b]	Kirschner *et al.* (1974)
Platichthys flessus	0.87	0.13	+15.6	19.0	+3.6	+8–12[c]	Potts and Eddy (1973)
Gadus callarius	0.527	0.475	−2.3	2.9	+5.2	—	Fletcher (1978b)
Fundulus heteroclitus	—	—	—	—	—	+7.2	Degnan and Zadunaisky (1982)
Mugil capito	—	—	—	—	—	+5.0	Pic and Maetz (1975)
Sarotherodon mossambicus	—	—	—	—	—	+21.5	Foskett *et al.* (1981)
Opsanus beta	—	—	—	—	—	−3	Howe (1980)
Gallichthys mirobilis	—	—	—	—	—	+14.5	Marshall and Bern (1980)

[a] Interpolation (House, 1963), (Fig. 2a).
[b] Interpolation (Kirschner *et al.* 1974) (Fig. 4).
[c] Shuttleworth (1978).

active chloride transport. The term chloride cell is now well established but is misleading because it ignores the active transport of sodium, which appears to be widespared in the gills of marine fishes.

It is not appropriate to discuss the details of ion transport in detail in a chapter on transepithelial potentials, but it may be observed that the Silva–Epstein model of chloride transport driven by the inward diffusion of sodium into the cell, together with a sodium chloride-linked carrier, does not require much modification to produce an active efflux of sodium. The driving force of the chloride cell is believed to be the active extrusion of sodium across the basolateral membranes. An active efflux of sodium could also take place if some of the extruded sodium were passed into the intercellular spaces outside the tight junctions or were sufficiently concentrated below the junctions to produce a net efflux. So far the only detailed descriptions of the intercellular junction in a marine teleost are those of euryhaline fish, particularly the mullet and the eel, two fish that operate close to the Nernst potential for sodium (Sargent, 1980; Sardet *et al.*, 1979).

Calculation shows that the efficiency of a cell actively extruding both sodium and chloride may be greater than a cell that extrudes chloride alone. Efficiency may have to be lowered to meet the peculiar requirements of the euryhaline species. It may also be significant that the gross fluxes of ions appear to be higher in euryhaline species that operate close to the Nernst potential for sodium. In the mullet, *Fundulus*, the blenny, and the flounder, the sodium fluxes are of the order of 20 to 25% hr^{-1} of total body salt, whereas in *Opsanus tau* and the cod they are only 10%. The total energy expended is mainly a function of the ionic fluxes.

C. Potentials in Freshwater Fishes

In fresh water transepithelial potentials again have many of the characteristics of diffusion potentials, but the absolute permeabilities of the gills of freshwater fish are much lower than those of marine fish and the reversal of the concentration gradient reverses the potentials so that most euryhaline fishes are negative in fresh water. The lined sole *Achirus* has a positive but low potential, and the sea horse *Hippocampus* is almost neutral, although neither lives naturally in fresh waters (Table V) (Evans and Cooper, 1976).

The term fresh water covers a much wider range of solutions than seawater, the major variables being sodium, chloride, hydrogen, and calcium ions, which are also the most important parameters in determining diffusion potentials. In acid waters the effects of hydrogen ions may be significant. The proton is hundreds or thousands of times more permeant than sodium or chloride ions, so that although the concentration of hydrogen ions in most

Table V

Transepithelial Potentials of Several Fish Species in Fresh Water

Species	Medium (m*M* liter^{-1})			$\Delta\psi$ (mV)	Reference
	Na^+	Cl^-	Ca^{2+}		
Carassius auratus	2	2	5	−10	Eddy (1975)
Mugil capito	0.1	—	1.5	−28	Pic (1978)
Salmo trutta	0.25	—	0.5	−6	McWilliams and Potts (1978)
Lagodon rhomboides	1.0	—	0	−15	Carrier and Evans (1976)
Dormitator maculatus	—	—	—	−36	Evans *et al.* (1974)
Platichthys flessus	0.1	0.04	1.0	−54.1 ± 4.8	Macfarlane and Maetz (1975)
Achirus lineatus	—	—	—	+5.5	Evans (1977)
Hippocampus erectus	—	—	—	−1.9	Evans (1977)

fresh waters is very low, hydrogen ions may be the dominant factor in the diffusion potential below pH 5 (at pH 3 H^+ = 1.0 mEq liter^{-1}).

Calcium ions modulate the permeability of the gill in both seawater and fresh water, but, whereas seawater always contains about 20 mEq liter^{-1}, fresh water usually contains somewhere between 0.1 and 1.0 mEq liter^{-1}, and some waters have calcium concentations that lie outside even this range. Calcium ions, by reacting with fixed anionic groups, may alter the hydration of organic structures and thereby reduce their permeability to both sodium and chloride ions. By increasing the positive charge along the pores in the membranes, calcium may differentially reduce the permeability to sodium ions, so that high concentrations of calcium are likely to decrease the P_{Na}/P_{Cl} ratio.

For a short while following transfer from seawater to fresh water, the permeability of the gill of remains unchanged and the immediate changes in sodium effluxes can be largely accounted for by the reversal of the diffusion potential (Potts and Eddy, 1973; House and Maetz, 1974). In marine teleosts the rates of efflux usually lies in the range of 10 to 25% of total body sodium every hour. Immediately following transfer the rate may decline to 5 to 10% every hour, but in euryhaline fish the rate falls in a few hours to the range of 0.1 to 1% every hour with a smaller but still negative potential (Maetz and Companini, 1966). This implies a 10-fold reduction in absolute permeability. When stenohaline marine fish are transferred to fresh water there is no large instantaneous potential change to reduce sodium loss and little or no reduction in permeability, and the fish soon die as the result of salt losses.

The fluxes across the gills of freshwater teleosts are much smaller than those across the gills of marine fish, but in both cases the active component of the fluxes must be of the same order as the passive fluxes and might therefore produce a detectable electrogenic potential. So far there is little evidence of an electrogenic potential in freshwater fish, except possibly in the goldfish, *Carassius auratus* (Maetz, 1974) and in the anomalous case of the fresh water-adapted *Fundulus* operculum, where it is apparently due to the active efflux of chloride. In sodium- and chloride-free water containing 2 m*M* imidazole sulfate, Maetz found that the goldfish produce a potential of −44 mV but that this fell to a potential −33 mV in 1 m*M* NaCl, which was independent of external chloride. Maetz interpreted this as evidence of an electrogenic sodium uptake short-circuiting a diffusion potential. This is in agreement with an observed potential of +12 mV in isotonic Ringer solution. In contrast, Eddy (1975) found that the potential in the goldfish was negligible in an isotonic solution, although he found a wide range of potentials in other solutions that were modulated by calcium. The isolated operculum of *Sarotherodon mossambicus,* previously adapted to fresh water, displays an insignificant potential of only 1 mV when bathed on both sides by Ringer solution, in marked contrast to the seawater-adapted opercular membrane in the same conditions, which generates 21 mV. The failure to detect an electrogenic ion pump in most freshwater fishes may possibly be due to the absence of well-designed experiments but is more likely a result of the requirement in freshwater fishes that both sodium and chloride ions be actively taken up and the active fluxes be of similar magnitude, whereas in some teleosts, particularly euryhaline teleosts, the active chloride fluxes are much greater than the active sodium fluxes. A freshwater teleost is more analogous to the stenohaline teleosts such as *Achirus* or *Lagodon* (Table II), where both sodium and chloride are actively extruded and the potential is very low.

Another possible contributing factor is the differing nature of the ion pumps in seawater and in fresh water. A substantial body of evidence suggests that in fresh water sodium is taken up in exchange for hydrogen ions while chloride is taken up in exchange for bicarbonate. The hydrogen ions and bicarbonate ions can be generated internally from carbon dioxide and water. Pumps of this kind might not be electrogenic. If the chloride cell has some of the characteristics of the Silva–Epstein model, then it would be electrogenic even if the driving Na^+/K^+ exchange were not electrogenic because of the asymmetry of the apical and basal membranes in the model. It is another example of how an electrically neutral pump can generate a potential when associated with ion-selective membranes.

IV. THE EFFECTS OF CALCIUM IONS ON GILL POTENTIALS

Movement from seawater to fresh water is inevitably associated with a drastic reduction in the ambient calcium concentration, which should tend to increase both P_{Na} and P_{Cl} and the P_{Na}/P_{Cl} ratio. This may partly explain why sodium efflux and the gill potentials on transfer between seawater and fresh water do not follow exactly the predictions based on Eq. (1). Ideally transfer experiments should be performed between media containing similar activities of calcium, but the activity is difficult to control because the activity coefficient of calcium in seawater is much lower than in fresh water. In *Fundulus kansae* the rate of sodium efflux is halved by the addition of 1 mg Ca^{2+} ions liter^{-1} to low-calcium fresh water, whereas the rate of efflux increases by 125% over the normal rate in calcium-free seawater (Potts and Fleming, 1971). Similarly the rate of efflux from the eel is doubled in calcium-free seawater (Bornancin *et al.*, 1972). In the rainbow trout and the brown trout, *Salmo gairdneri* and *Salmo trutta*, the potentials are calcium dependent. In 8 m*M* Ca^{2+} liter^{-1} the ratio of P_{Na}/P_{Cl} in *S. trutta* is 0.73, but this increases to 1.67 in distilled water. In solutions containing high concentrations of calcium the potential is positive, but the fish become negative in distilled water (Kerstetter *et al.*, 1970; McWilliams and Potts, 1978).

In the goldfish the rate of sodium efflux in distilled water was six times the rate in 20 mEq Ca^{2+} liter^{-1}, whereas the chloride efflux was five times as large. Correspondingly, the fish were electropositive in water containing calcium ions but become negative in calcium-free water (Eddy, 1975). The effects of calcium depletion can be enhanced by EDTA. Sodium uptake is also increased in low calcium, although the change in transepithelial potential would make uptake more difficult (Cuthbert and Maetz, 1972). Whether this is because of changes in the barrier between the sodium pump and the medium or a secondary consequence of changes in the blood concentration is not clear, however.

The addition of calcium to acid water may be beneficial, even if the quantities are insufficient to affect the pH appreciably, because they will reduce both the permeability to sodium and—by reducing the permeability to hydrogen ions—the positive potential that facilitates the sodium efflux. In freshwater teleosts and in euryhaline teleosts adapted to fresh water, the transepithelial potential is usually negative (Table V) but may become positive in the presence of a high concentration of calcium (*Salmo gairdneri*, Kerstetter *et al.*, 1970; Kerstetter and Kirschner, 1972; Goldfish, Eddy, 1975). The negative potential is probably a diffusion potential due to a greater permeability of the gills to sodium ions than to chloride ions. In the goldfish the potential disappears in isotonic sodium chloride (Eddy, 1975), whereas the effect of calcium ions can be explained by assuming that by increasing the positive charge in the pores it reduces P_{Na} more than P_{Cl}. This is confirmed by the relatively greater reduction in the sodium fluxes (Eddy, 1972).

V. THE EFFECTS OF HYDROGEN IONS ON GILL POTENTIALS

Seawater is well buffered and slightly alkaline, so the concentration of hydrogen ions is so low as to be insignificant in the generation of diffusion potentials. Fresh waters may be naturally acid, and industrially produced sulfur dioxide and oxides of nitrogen have lowered the pH of many softwater lakes and rivers in the northern hemisphere to around pH 4, 0.1 mEq H^+ liter^{-1}, at which level the hydrogen ion gradient is the major factor in the generation of the diffusion potential. Assuming that the permeabilities to sodium and chloride ions are the same at pH 4 as at pH 7, in the brown trout at pH 4, P_H/P_{Na} is 4200 in distilled water, declining to 670 in water containing 16 mEq ions liter^{-1} (McWilliams and Potts, 1978). The assumption that the permeability to sodium does not change significantly with pH is confirmed by the observation that the change in potential with pH largely accounts for the observed change in sodium loss. The remarkably high permeability of the trout gill to hydrogen ions shows that they cross the epithelium as protons, not in the hydrated form, because the hydrogen ions are 60 times as permeant as water molecules. Although P_H/P_{Na} ratio is very high in the trout gill, it is consistent with the P_H/P_{Na} ratios found in the muscle cells (Woodbury *et al.*, 1968) and algal cells (Kitasato, 1968).

Hydrogen ions have a marked effect on salt balance in fish. Hydrogen ions act as counterions to sodium in sodium uptake, which is therefore inhibited at low pH. In addition, the potential shift that occurs in acid waters increases the rate of sodium loss, which in the trout more than doubles between pH 6 and pH 4 (McWilliams and Potts, 1978).

VI. THE EFFECTS OF OTHER IONS ON TRANSEPITHELIAL POTENTIALS

Sodium and chloride are the two major ions in seawater and in blood plasma, and are the major contributors both to the diffusion component and to the active component of the potential. Hydrogen ions may be important in certain conditions and calcium ions have a marked effect on the permeability of sodium and chloride ions, but other ions also make important contributions.

Potassium has a greater effect, per unit of concentration, than has sodium. Potts and Eddy (1973) originally suggested that, as the potassium ion was smaller and more permeant than the sodium ion, the effect of external potassium on the potential was a diffusion effect; however, later experiments (e.g., Evans *et al.*, 1974; Pic, 1978) have demonstrated that external potassium specifically increases sodium effluxes to a greater extent than can be accounted for quantitatively by its effect on the potential, and it also specifically increases the chloride efflux (e.g., Pic, 1978). These effects are probably due to the enhancement of the sodium–potassium exchange following an increase in the extracellular potassium.

In addition to the effects on the active component of the potential, potassium ions probably contribute to the diffusion component as well, being smaller and more permeant than sodium ions; however, experiments with ouabain-treated gills would be required to demonstrate this.

Magnesium ions contribute to the potential in the rainbow trout (Greenwald *et al.*, 1974), presumably via a diffusion component, despite the large diameter of the hydrated ion. Hydrated calcium ions are smaller and should therefore have a greater effect per unit molal concentration differences, but the indirect effects of calcium ions on the relative and absolute permeabilities to various ions mask and confuse any diffusion component. In addition, calcium ions have an important role as intracellular messengers, and this may complicate experiments in which there are large changes in ambient calcium concentrations.

Bicarbonate ions in the external medium stimulate chloride efflux in the toadfish *Opsanus tau* but have no significant effect on the transepithelial potential (Kormanik and Evans, 1979). When added to the Ringer solution on both sides of *Fundulus* opercular preparations, they stimulate both the short-circuit current and the potential, the latter increasing from $\sim$ 10 mV in bicarbonate-free solution to >30 mV in the presence of 30 mEq bicarbonate (Degnan *et al.*, 1977). It is therefore surprising that the carbonic anhydrase inhibitor or Diamox has no effect. These experiments make it clear that a chloride pump on the Silva–Epstein model is not the only mechanism involved in chloride excretion.

VII. THE SITE OF THE SODIUM PATHWAY

The most permeable areas on the gill are probably the single-strand tight junctions between the chloride cells sharing a common apical pit (Sardet, 1980; Sardet *et al.*, 1979). Single-strand tight junctions are permeable to lanthanum, unlike multistrand junctions, and the gills of a few euryhaline marine teleosts are permeable to large molecules such as inulin and methylene blue.

The correlation between high resistance and multistrand junctions, and low resistance and single-strand junctions is well established in other tissues (Claude and Goodenough, 1973; Fromter and Diamond, 1973), but the discrimination between sodium and chloride ions must reside in the biochemical properties of the junctions.

Corroborative evidence that sodium travels mainly through the paracellular pathway in *Fundulus* is provided by the observation that triaminopyrimidine (TAP^+), which is a specific blocker of sodium movements along paracellular pathways in the gallbladder (Moreno, 1975), reduces sodium fluxes and short-circuit current by 70% in the isolated operculum (Degnan and Zadunaisky, 1980). It may be significant that it has no observable effect on *Opsanus beta*, a fish that has a low gill potential and therefore must actively transport both chloride and sodium (Howe, 1980).

VIII. THE EFFECTS OF DRUGS AND HORMONES ON THE TRANSEPITHELIAL POTENTIAL

In the analysis of the pharmacological characteristics of whole fish, or even of isolated gills, it is difficult to distinguish between the direct effects of a compound on the active transport system or on the permeability, and the indirect effects consequent on alterations in blood or saline flow through the gills. The development of the *Fundulus* opercular preparation and more recently that of *Oreochromis* [*Sarotherodon* (*Tilapia*)] (Foskett *et al.*, 1979) has clarified the situation, but it should again be borne in mind that *Fundulus* and *Sarotherodon* both possess almost pure chloride pumps and may not be representative of the generality of marine teleosts.

Chloride effluxes from the gills and the short-circuit current in the opercular preparations are under the control of α- and β-adrenergic receptors. They are stimulated by β-adrenergic activators such as isoproterenol, and they are inhibited by adrenaline and α-adrenergic activators such as arteronal. The α blocker phentolamine reverses adrenaline inhibition (Shuttleworth, 1978; Foskett and Hubbard, 1981; Degnan and Zadunaisky, 1982). Stimulation by way of the β receptors appears to operate through the agency of cyclic AMP. Theophylline and aminophylline, which stimulate the short-circuit current, also increase cyclic AMP through inhibition of phosphodiesterase (Degnan and Zadunaisky, 1982).

Furosemide is a specific inhibitor of chloride transport in the kidney, and thiocyanate ions, which mimic chloride ions, both inhibit active chloride transport in many tissues and reduce the short-circuit current in the *Fundulus* operculum. In the intact mullet, thiocyanate reduces the potential in seawater by a few millivolts, perhaps eliminating E_A and it halves the chloride efflux (Pic and Maetz, 1975). Ouabain blocks chloride efflux and the active component of the potential when applied serosally in all preparations. All these compounds may be used in investigating the active component of the potential. TAP^+ blocks the paracellular sodium pathway but unfortunately affects the chloride pump as well. Calcium probably acts mainly on the paracellular pathway. Lanthanum appears to displace calcium ions in the goldfish but does not reduce the permeability to other ions as calcium does (Eddy and Bath, 1979).

The hormone prolactin, which helps to adapt teleosts to fresh water, in the long term causes dedifferentiation of chloride cells. In the short term it reduces both the active transport of chloride and the apical membrane chloride conductance (Foskett, 1981). In both cases it converts the opercular epithelia toward the freshwater state.

ACKNOWLEDGMENTS

I am indebted to A. W. Potts and M. G. A. Potts for the calculation of Fig. 1.

REFERENCES

Bornancin, M., Cuthbert, J., and Maetz, J. (1972). The effects of calcium on branchial sodium fluxes in the sea-water adapted eel *Anguilla anguilla*. *J. Physiol. (London)* **222,** 487–496.

Carrier, J. C., and Evans, D. H. (1976). The role of environmental calcium in freshwater survival of the marine teleost *Lagodon rhomboides*. *J. Exp. Biol.* **65,** 529–538.

Claude, P., and Goodenough, D. A. (1973). Fracture faces of *zonulae occludentes* from 'tight' and 'leaky' epithelia. *J. Cell Biol.* **58,** 390–400.

Cuthbert, A. W., and Maetz, J. (1972). The effects of calcium and magnesium on sodium fluxes through gills of *Carassius auratus*. *J. Physiol. (London)* **221,** 633–643.

Degnan, K. J., and Zadunaisky, J. A. (1980). Ionic contributions to the potential and current across the opercular epithelium. *Am. J. Physiol.* **238,** R231–R239.

Degnan, K. J., and Zadunaisky, J. A. (1982). The opercular epithelium: An experimental model for teleost gill osmoregulation and chloride secretion. *In* "Chloride Transport in Biological Membranes" (J. A. Zadunaisky, ed.), pp. 295–318. Academic Press, New York.

Degnan, K. J., Karnaky, K. J., and Zadunaisky, J. A. (1977). Active chloride transport in the *in vitro* opercular skin of a teleost (*Fundulus heteroclitus*), a gill-like epithelium rich in chloride cells. *J. Physiol. (London)* **271,** 155–191.

Dharmamba, M., Bornancin, M., and Maetz, J. (1975). Environmental salinity and sodium chloride exchanges across the gill of *Tilapia mossambica*. *J. Physiol. (Paris)* **70,** 627–636.

Eddy, F. B. (1975). The effect of calcium on the gill potentials and on sodium and chloride fluxes in the goldfish *Carassius auratus*. *J. Comp. Physiol.* **96,** 131–142.

Eddy, F. B, and Bath, R. N. (1979). Effects of lanthanum on sodium and chloride fluxes in the goldfish *Carassius auratus*. *J. Comp. Physiol.* **129,** 145–150.

Evans, D. H. (1969). Sodium, chloride and water balance of the intertidal teleost, *Pholis gunnelis*. *J. Exp. Biol.* **50,** 179–190.

Evans, D. H. (1980). Kinetic studies of ion transport by fish gill epithelium. *Am. J. Physiol.* **238,** R224–R230.

Evans, D. H., and Cooper, K. (1976). The presence of Na/Na and Na/K exchange sodium extrusion by three species of fish *Nature* (*London*) **259,** 241–242.

Evans, D. H., Carrier, J. C., and Bogan, M. (1974). The effect of external potassium on the electrical potential measured across the gills of the teleost *Dormitator maculatus*. *J. Exp. Biol.* **70,** 213–220.

Fletcher, C. R. (1976). A phenomenalogical description of active transport. *In* "Perspectives in Experimental Biology" (P. Spencer Davies, ed.), pp. 55–64. Pergamon, Oxford.

Fletcher, C. R. (1977). Electrical potential differences across salt transporting membranes. *J. Theor. Biol.* **67,** 255–268.

Fletcher, C. R. (1978a). Osmotic and ionic regulation in the cod (*Gadus callarias* L.). I. Water balance. *J. Comp. Physiol.* **124,** 149–156.

Fletcher, C. R. (1978b). Osmotic and ionic regulation in the cod (*Gadus callarius* L). II. Salt balance. *J. Comp. Physiol.* **124,** 157–168.

Foskett, J. K. (1981). Chloride secretion by teleost opercular membrane: Circuit analysis of the effects of prolactin. *Ann. N.Y. Acad. Sci.* **372,** 644–645.

Foskett, J. K., and Hubbard, G. M. (1981). Hormonal control of chloride secretion by teleost opercular membrane. *Ann. N.Y. Acad. Sci.* **372,** 643.

Foskett, J. K., and Scheffey, C. (1982). The chloride cell. Definitive identification as the salt secreting cell in teleosts. *Science* **215,** 164–166.

Foskett, J. K., Turner, T., Logsdon, C., and Bern, H. (1979). Electrical correlates of chloride-cell development in subopercular membranes of the tilapia *Sarotherodon mossambicus* transferred to seawater. *Am. Zool.* **19,** 998.

Foskett, J. K., Craig, L. D., Turner, T., Machen, T. E., and Bern, H. (1981). Differentiation of the chloride extrusion mechanism during seawater adaptation of a teleost fish, the cichlid *Saratherodon mossambicus*. *J. Exp. Biol.* **93,** 209–224.

Frömter, E., and Diamond, J. (1973). Route of passive ion permeation in epithelia. *Nature* (*London*), *New Biol.* **235,** 9–13.

Girard, J. P., and Payan, P. (1980). Ion exchange through respiratory and chloride cells in freshwater- and seawater-adapted teleosteans. *Am. J. Physiol.* **238,** R260–R268.

Greenwald, L., Kirschner, L. B., and Sanders, M. (1974). Sodium efflux and potential difference across the irrigated gill of the seawater adapted rainbow trout *Salmo gairdneri*. *J. Gen. Physiol.* **64,** 135–147.

House, C. R. (1963). Osmotic regulation in the brackish water teleost *Blennius pholis*. *J. Exp. Biol.* **40,** 87–104.

House, C. R., and Maetz, J. (1974). On the electrical gradient across the gill of the sea water adapted eel. *Comp. Biochem. Physiol. A* **47A,** 917–924.

Howe, D. (1980). Osmotic and ionic regulation in the euryhaline marine teleost, *Opsanus beta*. Ph.D Dissertation, Duke University, Durham, North Carolina.

Karnaky, K. J., Degnan, K. J., and Zadunaisky, J. A. (1977). Chloride transport across isolated opercular epithelium of killifish: A membrane rich in chloride cells. *Science* **195,** 203–205.

Kerstetter, T. H., and Kirschner, L. B. (1972). Active chloride transport by the gills of rainbow trout (*Salmo gairdneri*). *J. Exp. Biol.* **56,** 263–273.

Kerstetter, T. H., Kirschner, L. B., and Rafuse, D. A. (1970). On the mechanism of sodium ion transport by the irrigated gills of rainbow trout (*Salmo gairdneri*). *J. Gen. Physiol.* **56,** 342–359.

Keys, A. B. (1931). Chloride and water secretion and absorption by the gills of the eel. *Z. Vergl. Physiol.* **15,** 364–388.

Kirschner, L. B., Greenwald, L., and Sanders, M. (1974). On the mechanism of sodium extrusion across the irrigated gill of the seawater adapted rainbow trout (*Salmo gairdneri*). *J. Gen. Physiol.* **64,** 148–165.

Kitasato, H. (1968). Influence of H^+ on the membrane potential and ion fluxes of *Nitella*. *J. Gen. Physiol.* **52,** 60–87.

Kormanik, G. A., and Evans, D. H. (1979). HCO_3-stimulated Cl efflux in the gulf toadfish acclimated to sea water. *J. Exp. Zool.* **208,** 13–16.

Macfarlane, N. A. A., and Maetz, J. (1975). Acute response to a salt load of the NaCl excretion mechanisms of the gill of *Platichthys flessus* in sea water. *J. Comp. Physiol.* **102,** 101–113.

McWilliams, P. G., and Potts, W. T. W. (1978). The effects of pH and calcium concentration on gill potentials in the brown trout, *Salmo trutta*. *J. Comp. Physiol.* **126,** 277–286.

Maetz, J., and Bornancin, M. (1975). Biochemical and biophysical aspects of salt excretion by chloride cells. *Fortschr. Zool.* **23,** 327–362.

Maetz, J., and Campanini, G. (1966). Potentials transépithéliaux de la branchie d'Anguille *in vivo* et eau douce et eau de mer. *J. Physiol. (Paris)* **58,** 248.

Maetz, J., and Pic, P. (1975). New evidence for a Na/K and Na/Na exchange carrier linked with the Cl^- pump in the gill of *Mugil capito*. *J. Comp. Physiol.* **102,** 85–100.

Marshall, W. S. (1977). Transepithelial potential and short-circuit current across the isolated skin of *Gillichthys mirabilis* (Teleostei: Gobiidae), acclimated to 5% and 100% sea water. *J. Comp. Physiol.* **114,** 157–165.

Marshall, W. S., and Bern, H. (1980). Ion transport across the isolated skin of the teleost *Gillichthys mirobilis*. *In* "Epithelial Transport in the Lower Vertebrates" (B. Lahlou, ed.), pp. 337–350. Cambridge Univ. Press, London and New York.

Moreno, J. (1975). Blockage of gallbladder tight junction cation-selective channels by 2, 4, 6-trans-aminopyrimidium (TAP). *J. Gen. Physiol.* **66,** 117–128.

Nonette, G., Colin, D. A., and Nonette, L. (1982). Na^+ and Cl^- transport and intercellular junctions in the isolated skin of a marine teleost *Blennius pholis* L. *J. Exp. Zool.* **224,** 39–44.

Pic, P. (1978). A comparative study of the mechanism of Na^+ and Cl^- excretion by the gill of *Mugil capito* and *Fundulus heteroclitus:* Effects of stress. *J. Comp. Physiol.* **123,** 155–162.

Pic, P., and Maetz, J. (1975). Differénce de potential transbranchiale et flux ioniques chez *Mugil capito* adapté à eau de mer. Importance de l'ion Ca^{++}. *C. R. Hebd. Seances Acad. Sci.* **280D,** 983–986.

Potts, W. T. W., and Eddy, F. B. (1973). Gill potentials and sodium fluxes in the flounder *Platichthys flessus*. *J. Comp. Physiol.* **87,** 29–48.

Potts, W. T. W., and Fleming, W. R. (1971). The effects of environmental calcium and ovine prolactin on sodium balance in *Fundulus kansae*. **55,** 63–76.

Potts, W. T. W., Fletcher, C. R., and Eddy, F. B. (1973). An analysis of the sodium and chloride fluxes in the flounder *Platichthys flessus*. *J. Comp. Physiol.* **87,** 21–28.

Sardet, C. (1980). Freeze fracture of the gill epithelium of euryhaline teleost fish. *Am. J. Physiol.* **238,** R207–R212.

Sardet, C., Pisam, M., and Maetz, J. (1979). The surface epithelium of teleostean fish gills. Cellular and junctional adaptation. *J. Cell Biol.* **80,** 97–117.

Shuttleworth, T. J. (1978). The effect of adrenaline on the potentials in the isolated gill of the flounder *Platichthys flessus*. *J. Comp. Physiol.* **124,** 129–136.

Shuttleworth, T. J., Potts, W. T. W., and Harris, J. N. (1974). Bioelectric potentials in the gills of the flounder *Platichthys flessus*. *J. Comp. Physiol.* **94,** 321–329.

Smith, P. G. (1969). The ionic relations of *Artemia salina* (L.). I. Measurements of electrical potential difference and resistance. *J. Exp. Biol.* **51,** 727–738.

Woodbury, J. W., White, S. H., and Weakly, J. N. (1968). High membrane permeability of frog skeletal muscle. *Abstr. Int. Congr. Physiol. Sci., 24th, 1968,* p. 472.

5

THE CHLORIDE CELL: THE ACTIVE TRANSPORT OF CHLORIDE AND THE PARACELLULAR PATHWAYS

J. A. ZADUNAISKY

Departments of Physiology and Biophysics and of Ophthalmology
New York University Medical Center
New York, New York

FISH PHYSIOLOGY, VOL. XB

ISBN 0-12-350432-5

I. INTRODUCTION

The gills have respiratory as well as secretory or osmoregulatory functions, and the latter are as important as the functions of the kidney in the regulation of the composition of the internal medium and the water balance of the fish. The anatomic element directly associated with secretion of salts is the chloride cell located on the gill filaments, mainly on its proximal end. These large columnar cells are present in greater quantities in the seawater-adapted fish and are less frequent in the freshwater-adapted fish. Despite a controversy concerning their existence, their function, and their appropriate name that was extended for some 50 years, it is evident in the light of more recent investigations that the chloride cell is in fact the site of salt secretion, and that it specifically secretes chloride ions. Thus, the original nomenclature of these cells is appropriate and should be used instead of the more doubtful terms, "mitochondria-rich cells" or "Keys–Willmer cells," which avoided interpretation of their function.

II. HISTORICAL BACKGROUND ON THE CHLORIDE CELLS

Homer Smith (1930) localized the "head region" as the source of electrolytes secreted to the outside medium in fish studied with the rubber diaphragm technique, separating the head from the rest of the body of the fish. His contention was that the gill was extremely important in replacing the kidney in the saltwater fish because of its drinking of a highly concentrated salt solution (seawater) and its need to excrete the excess salts and thus maintain a constant composition of the body fluids with respect to electrolytes.

Ancel Keys perfected a technique of gill perfusion very adequate for the times (1931) and proved that salts decreased as their chlorides disappeared from the perfusate bathing the blood side of the gills and increased in the external fluid consisting of seawater. After confirming the secretion of salts in his isolated perfused heart–gill preparations, Keys demonstrated with Wilmer in the gills of teleosts the presence of special cells with the histological characteristics of secretory cells (Keys and Willmer, 1932). Considering the available methodology of the times, he provided good experimental evidence as well as the prediction of an anatomic entity on the basis of physiological experiments.

Ancel Keys was a fellow of the National Research Council of America, who did the perfusion experiment leading to the observation of chloride secretion in the Zoophysiological Laboratory in Copenhagen, Denmark, under the direction of August Krogh. Also, he collaborated with Krogh in the development of the "syringe pipette for analytical usage" (Krogh and Keys, 1931), used in his determination of chloride by the methods of Van Slyke (1923) and Rehberg (1926). With these two tools—that is, a highly sensitive analytical method for 1932 and a good perfusion technique—Keys demonstrated to his satisfaction and that of others that chloride is indeed secreted by the gills of the seawater-adapted eel.

After performing these physiological experiments, Keys moved with his fellowship to Cambridge, England to work with Wilmer, in order to perform histological studies of the gills of eels. The anatomic knowledge of the gill epithelium indicated that it was composed only of flat cells of the respiratory type, and therefore it was at that time inconceivable that these simple epithelial cells could perform secretory functions. Their efforts were rewarded with the observation of larger columnar cells that were more abundant in seawater-adapted gill filaments of *Anguilla* and less frequent in the ones of freshwater-adapted specimens. With the knowledge acquired in the physiological experiments indicating chloride secretion against a gradient of concentration and the finding of cells that could be assimilated to secretory cells, Keys and co-workers called them chloride cells or chloride-secreting cells.

Despite the findings, if the methods that are now available for the measurement of minute amounts of sodium could have been used by Keys in his experiments in Copenhagen in 1931, he would probably have found that there was also a surprising amount of sodium accompanying the secretion of chloride. This is because of the need for electroneutrality in the secretion of salts such as NaCl, regardless of which one of the two ions provides the main driving force. Alas, however, the analytical methods available in 1932 were excellent for halogens but not for alkali metal ions. Although chloride ions are transported actively, as we know today, from blood to the seawater, sodium accompanies chloride in the gill secretion. The electrical potential difference (PD) across the gill epithelium, negative in the seawater, will retard some of the movement of chloride outward and accelerate the exit of sodium.

The finding of Keys and Willmer (1932) of the special secretory cells in the gill filaments was denied categorically by Bevelander (1935, 1936), who claimed that the cells they found were mucous cells and did not have other functions. Histochemistry was not the most accurate of methods to settle this dispute, and, as indicated later by Copeland (1948), positive reaction for mucus did not exclude the cells newly found by Keys and Willmer to be the site of secretion in the gill. In fact the argument of Bevelander is of no relevance today, and his findings in *Fundulus heteroclitus* showing specific mucous reaction of the chloride cells is not reproducible (Copeland, 1948).

However, this dismissal of Keys' finding introduced doubts concerning the presence and the function of these large cells in the gill filaments and set back the understanding of this system for many years. In fact, in 1960 Doyle referred to them as the "so-called chloride cells" (Doyle and Gorecki, 1961). It is possible that there could be doubt concerning chloride active transport, which by that time was not found in several model epithelial systems, such as the frog skin (Ussing and Zerahn, 1951), and sodium transport was the focus of attention. However, Copeland (1948) utilized histochemical methods for silver chloride precipitates and demonstrated a preferential localization in the apical cavity of the chloride cells of *F. heteroclitus*. These cavities were shown to exist or to be more frequently found in seawater-adapted than freshwater-adapted specimens. Philpott (1965) observed a similar precipitate in electron-microscopic sections and examined it by X-ray diffraction, confirming that it was a highly concentrated precipitate of chloride salts located near the margin of the apical cavity of the chloride cells of *F. heteroclitus* gill filaments, in fish adapted to seawater. Nevertheless, a series of lucid and extremely valuable articles inspired by J. Maetz in his laboratories first in Paris and then at the Station Zoologique in Villefranche sur mer (France) (for review, see Maetz and Bornancin, 1975) followed the general trend in the field of secretion and ascribed to sodium transport the main driving force across the gill. Radioactive sodium efflux in intact fish demonstrated a number of important changes during secretion that coincided with the existence of a (Na^+,K^+) pump in the gill epithelium (Maetz, 1969).

Furthermore, specific inhibitors of sodium transport such as Ouabain had an effect on radiolabeled sodium efflux from the gills. All observations at this time pointed to the existence of (Na^+,K^+)-ATPase in the chloride cells that was responsible for the sodium secretion in seawater-adapted fish. In fact, the gill is very rich in (Na^+,K^+)-ATPase (Sargent *et al.*, 1980), and by autoradiographic methods radiolabeled ouabain can be observed binding to sites on the membranes of the tubular system of the chloride cells (Karnaky *et al.*, 1976).

The need to clarify the discrepancy between the negative electrical potential difference and the claim of active sodium transport prompted experimentation with the opercular epithelium, a layer that in the case of *F. heteroclitus* (Burns and Copeland, 1950) contains chloride cells identical to those of the gill. Its geometrical characteristics permit placing this membrane, once dissected, between two well-stirred salt solutions in an Ussing chamber (Ussing and Zerahn, 1951). Electrical properties and ion fluxes can be measured under these conditions, and a distinction between active and passive movements of ions can be determined. In fact, Degnan *et al.* (1977) proved that opercular epithelia transported chloride outwards, and later that the density of chloride cells was proportional to the rate of chloride transport (Karnaky *et al.*, 1979, 1984).

It is again surprising and perhaps serendipitous that the cells were called chloride cells, because the confirming evidence has only recently been demonstrated with good biophyscial methods in vitro.

The evidence pointing to the existence of a (Na^+,K^+) pump is albeit correct, because the chloride secretion is inhibited by ouabain. However, this is related to the sodium dependence of chloride secretion in epithelia, and it will be discussed later on in the chapter.

III. CHLORIDE CELLS IN THE OPERCULAR EPITHELIUM

The intricate anatomy of the gill osmoregulatory epithelium makes it almost impossible to use a direct biophysical approach to the study of ion and water movements by the chloride cell "*in situ*," and alternative approaches to the study of its functions had to be developed. A successful approach has been the use of extrabranchial layers of epithelia containing high concentrations of chloride cells in regions other than the osmoregulatory epithelium of the gill (Degnan *et al.*, 1977). The existence of extrabranchial chloride cells was described by Burns and Copeland (1950), and a more detailed study confirming the identical anatomic features of the mitochondria-rich cells in the base of the gill filaments and in the opercular epithelium and head region of *Fundulus heteroclitus* was published by Karnaky and Kinter (1977). Since the first report on the use of the isolated opercular epithelium for the study of chloride cell function in *F. heteroclitus*, several other similar preparations have been developed, namely the skin of the teleost *Gillichthys mirabilis* (Marshall, 1977), the opercular epithelium of *Fundulus grandis* (Krasny and Zadunaisky, 1978), and the opercular epithelium of telapia, *Oreochromis* (formerly *Sarotherodon*) *mossambicus* (Foskett *et al.*, 1981).

The identical nature of the chloride cells of the isolated opercular or skin preparations to the gill chloride cell has permitted conclusions on the secretion of chloride, the origin of gill electrical potentials, the influence of salt concentrations, the adaptation to seawater and fresh water, hormonal effects, the influence of drugs, and the nature of the paracellular or shunt pathways. These results can be extrapolated to the functions of the osmoregulatory epithelium of the gill in teleosts. Although evidence has accumulated on euryhaline teleosts, extrapolation to the gill function of elasmobranchs or other chloride cell-containing osmoregulatory epithelia in other life forms might be premature, despite the solid evidence available to explain the function of chloride cells in teleosts.

A. Histology and Cell Density

The tissue lining the opercular cavity of *Fundulus heteroclitus* consists of a stratified epithelium with an underlying layer of connective tissue. A schematic illustration of a section through the epithelium and the features of the chloride cells present is shown in Fig. 1. The bottom portion shows the several types of cells that can be recognized: mucous cells, pavement cells, nondifferentiated cells, and chloride cells. The pavement cells form a continuous layer facing the cavity of the operculum and in this sense have the same aspect as pavement cells of the skin. This continuous layer is interrupted by the contact with the exterior made by the mucous cells and the piths of the chloride cells. The chloride cells extend from the basal lamina to the external environment. The density of chloride cells in the operculum of *F. heteroclitus* was estimated by Karnaky and Kinter (1977) to be from 50 to 70% of the total population of cells of the epithelium.

Figure 2 shows the density of chloride cells observed after staining with the vital dye DASPMI (dimethylaminostyrylmethylpyridiniumiodine; Bereiter-Hahn, 1976). The prominent masses of the chloride cells can be seen in great density in this picture obtained after bathing the opercular lining with a 50 m*M* solution of the fluorescent dye dissolved in teleosts Ringer solution (Zadunaisky, 1979). Vital fluorescent dyes such as DASPEI (dimethylaminostyrylethylpyridiniumiodide), a fluorophore that binds preferentially to mitochondria, were developed by Bereiter-Hahn (1976) in Britton Chance's laboratory for the study of the redox states of mitochondria in tissue cultures. The use of these dyes, which during the first several minutes of binding do not affect cell metabolism, has been very useful in the study of opercular epithelium of *F. heteroclitus*, *Fundulus grandis* (Krasny, 1981), the skin of *Gillichthys mirabilis* (Marshall and Nishioka, 1980), and the operculum of telapia (Foskett *et al.*, 1981).

Chloride cell counts in *F. heteroclitus* gill filaments represent approximately 6% of the total number of cells (Karnaky and Kinter, 1977), whereas in the opercular epithelium the density was 4×10^5 cells cm^{-2}. In *F. grandis* the cell count obtained by Krasny (1981) was 2.5×10^5 cells cm^{-2}. In the case of the skin of *Gillichthys mirabilis* adapted to 200% seawater a cell density of 6.3×10^4 cells cm^{-2} has been reported (Marshall and

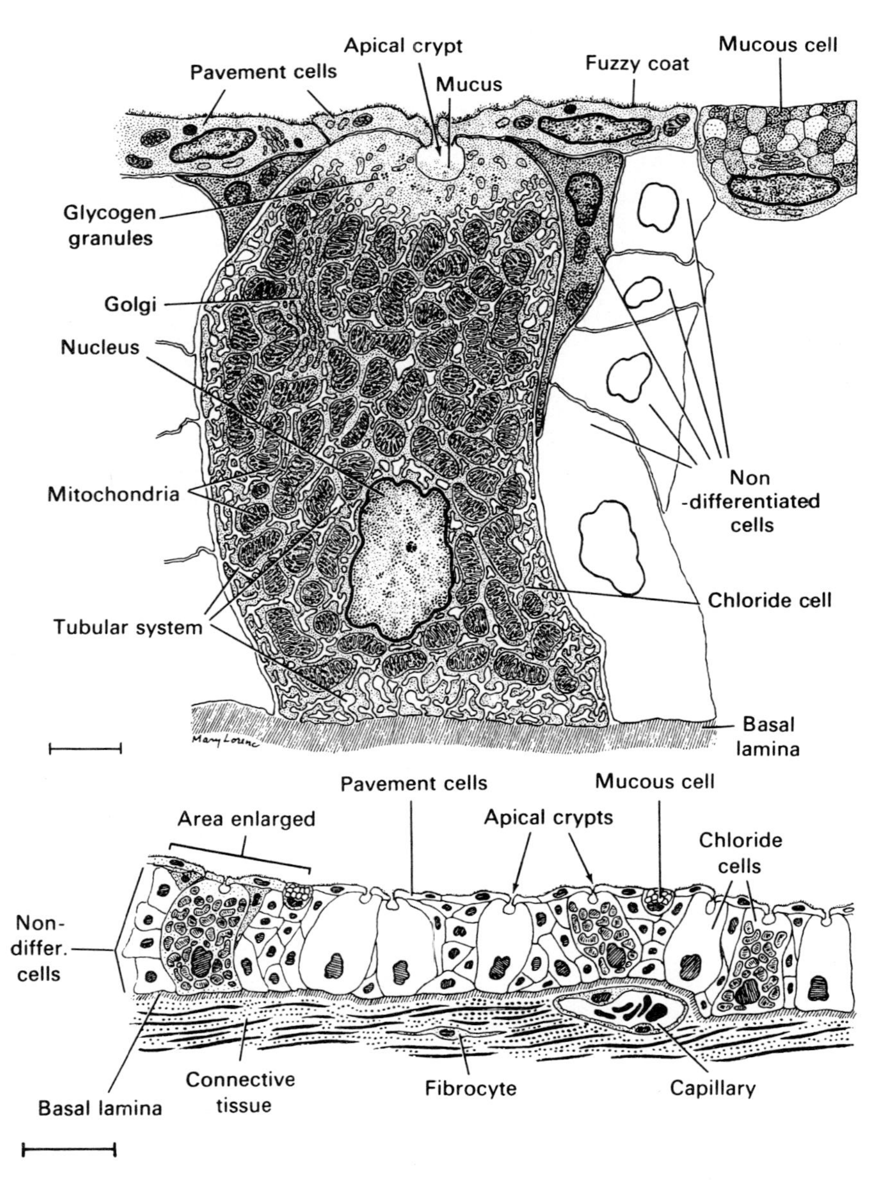

Fig. 1. Schematic representation of the ultrastructural details of the opercular epithelium of *Fundulus heteroclitus* showing the mitochondria-rich chloride cells, which are identical to the ones found in the fish gills. This epithelium transports chloride ions toward the saltwater side. (From Degnan *et al.*, 1977.)

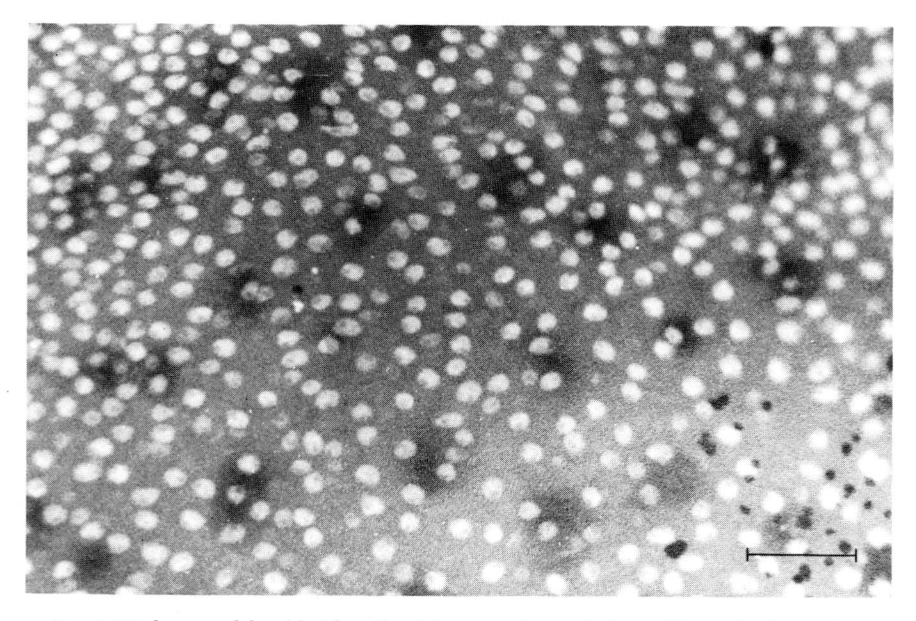

Fig. 2. Vital stains of the chloride cells of the opercular epithelium of *Fundulus heteroclitus*. Note the high density of the mitochondria-rich cells. Photograph obtained by epifluorescence microscopy of a freshly dissected opercular epithelium bathed in 50 μM DASPMI in Ringer solution. Bar = 100 μm. (From Zadunaisky, 1979.)

Nishioka, 1980). For the opercular epithelium of telapia (*Sarotherodon mossambicus*), the chloride cell density in specimens fully adapted to seawater was approximately 6×10^4 cells cm^{-2}.

Therefore, the range of cell density in teleost operculi or skin fully adapted to seawater is from 0.6 to 4×10^5 cells cm^{-2}. Chloride cell density has been correlated to chloride-secretory rates and adaptation to fresh water and seawater. These results are discussed further on in this chapter.

The high density of chloride cells identical to the gill chloride cells thus permits the use of these isolated epithelia as models for the osmoregulatory functions of the gill chloride cells.

B. Basic Principles for the Study of Isolated Layers Rich in Chloride Cells

The basic principles of electrophysiology and membrane transport applied to the isolated opercular membranes consist of the comparison between the behavior of living membranes and inert, nonliving membranes. In the case of a passive inert membrane the flux ratio (J_{in}/J_{out}) and the electrochemical gradient are linked by Ussing's equation (Ussing, 1949):

$$\frac{J_{in}}{J_{out}} = \frac{a_{i1}}{a_{i2}} e^{zF\Delta E/RT} \tag{1}$$

Equation (1) applies to membranes that do not exhibit active ion transport. The concentration gradient affects the flux ratio because of the diffusion of ionic species down the gradient. The electrical potential difference is the driving force that affects the unidirectional flux of the charged species. Experimentally the potential difference is measured in membranes bathed on both sides by similar solution, eliminating the concentration gradient. Thus, the flux ratio becomes a function only of the force of the electric field. In passive membranes the left side of Eq. (1) is numerically identical to the right side of the equation, but in membranes that actively transport ions the flux ratio is not equal to the electrochemical component of the right term of the equation. In biological membranes the flux ratio can be predicted on the basis of Eq. (1) and compared to the value obtained experimentally by measuring the ionic fluxes with a radioactive species, and the potential difference across the membrane. In the case of the isolated operculum of *Fundulus heteroclitus* the predicted flux ratio for chloride ions was 11.0 and the value observed experimentally was 1.38 (Degnan and Zadunaisky, 1979). This great difference indicates that an extra force is applied to the chloride ion to move 11 times faster in the outward direction, toward the negatively charged side of the tissue, than toward the inside or blood side, which is the positively charged side of the isolated preparation. In the case of sodium for the same operculum, the predicted flux ratio was 0.94 and the observed ratio was 1.38. The small difference between the predicted and observed ratios for sodium was not statistically significant. Therefore, sodium moves passively whereas chloride is actively transported outwards against an electrochemical gradient.

A further step in the analysis of fluxes and electrical potentials is the use of the short-circuit current technique (Ussing and Zerahn, 1951). In this case a circuit is constructed that opposes the spontaneous potential difference produced by the membrane. Thus, under conditions of zero potential difference and no concentration gradient, Eq. (1) reduces to

$$\frac{J_{in}}{J_{out}} = 1 \tag{2}$$

The flux ratio has a value of 1 if the membrane is inert or the ionic species moves passively across the membrane, but it is greater or smaller if the ion is actively transported across the membrane. In the case of the opercular epithelium of *F. heteroclitus* the flux ratio under short-circuiting conditions for chloride ion was more than 4 instead of 1 because of a large value of the efflux component. The flux ratio for sodium ion under the same conditions was near 1. Again chloride is actively transported outwards while sodium moves passively across the membrane. Short-circuiting and isotopic flux techniques permit the measurement of current carried by each ion species. The total electrical current determined as the short-circuit current (I_{sc}) is equal to the algebraic sum of the current carried by each ionic species according to its charge and direction. Thus,

$$I_{sc} = I_{n_1} \pm I_{n_2} \pm \cdots\cdots \pm I_{n\infty} \tag{3}$$

The ionic fluxes in terms of mass per unit time and area of membrane are converted into μA per unit area and compared to the experimentally determined short-circuit current. The opercular epithelium of *F. heteroclitus* maintains currents between 80 and 160 μA cm^{-2}. The net ionic flux of chloride found in a series of experiments published by Degnan *et al.* (1977) was 163 μA cm^{-2}, whereas the total current observed was 159 μA cm^{-2}. Thus, practically all the current across the operculum was carried by chloride ions leaving no room for other ionic species to contribute to the current. In fact, there was no net flux of sodium across the tissue.

C. Chloride Cell-Rich Opercular Membranes

The method utilized for the isolation and study of isolated chloride cell-rich opercular epithelia is described in detail by Degnan *et al.* (1977) and Degnan and Zadunaisky (1979). The opercular epithelium is exposed by cutting away the gills and the branchiostegal rays of the operculum. The epithelium covering the inside of the opercular flap is gently teased away under a dissecting microscope in order to free it from the underlying bone. The tissue is then floated over the aperture of one-half of a lucite chip covered with Sylgard to avoid edge damage to the preparation. The tissue can be pinned down with tips of insect microdissecting pins, and then the second half of the chip is positioned on top of the tissue thus "sandwiched" to be placed in the Ussing chamber. Initially a chamber used for small areas and thin epithelia from the eye of vertebrates (Zadunaisky and Degnan, 1976) was used, with an aperture of 0.07 cm^2. More recently areas of 0.3 cm^3

or larger have been used according to the area of chloride cell-rich opercular epithelium available for the experiment. Identical techniques have been used for the study of operculi or skin of *Fundulus grandis* (Krasny and Zadunaisky, 1978), *Gillichthyis mirabilis* (Marshall and Burns, 1981), and telapia (Foskett *et al.*, 1981).

IV. RESULTS OBTAINED IN THE ISOLATED CHLORIDE CELL MEMBRANES

A. Electrical Properties

The electrical potential difference of opercular epithelia of *Fundulus heteroclitus* oscillates between 10 and 35 mV. The mean value for a series of experiments was 18.7 ± 1.2 mV for 64 preparations. The short-circuit current was 136.5 ± 11.1 μA cm^{-2} for the same group, with an average electrical resistance of 173.7 ± 12.1 Ω cm^2 (Degnan *et al.*, 1977).

There are important seasonal variations in these parameters that have some relationship to the hormonal changes that *Fundulus* undergoes during its yearly cycle. The electrical values are highest in May, with tissues that initially start with a low potential difference (PD) when mounted but then increase steadily and stabilize at a high value. In July instead the values are significantly lower, averaging 12.2 mV for the potential difference, 75.2 μA cm^{-2} for the I_{sc}, and an increase in electrical DC resistance to 213.9 Ω cm^2. These changes of 50 to 60% decrease in PD and I_{sc} with a similar increase in resistance have to be taken into account when experiments are performed. The reason for lower current and potential values in July could be associated to the breeding period of *F. heteroclitus* and the hormonal changes associated with it. The increase in resistance certainly correlates well with effects produced by prolactin, which will be increased during the breeding period. The seasonal changes, however, do not reflect a basic change in transport functions, because the correlation between net fluxes of chloride and the short-circuit current was maintained during all periods of the year.

The orientation of the potential difference, outside negative in the isolated preparations of all types of operculi and skins so far examined, correlates well with the orientation and values that have been obtained in intact fish adapted to seawater. For instance, Shuttleworth *et al.* (1974) observed 18 mV in the flounder saltwater side negative. Potts has reviewed in this volume (see Chapter 4, this volume) the orientation of the gill potential and the values obtained by several authors in intact fish. It can be concluded from his Table II that for the isolated skins and operculum that have been examined isolated *in vitro* until the writing of this chapter, the values and orientation of the potential difference are similar to those obtained in the intact fish.

It should be pointed out that the determination of gill electrical potential in the fish *in vivo* already indicated that the negative potential in the salt-water side was an index of the secretion of an anion, the most abundant being chloride. However, because of the interest in sodium active transport in the field of membranes in the 1950s and 1960s, investigators tended to interpret the electrophysiology and flux measurements in intact fish as manifestations of active sodium extrusion (see Maetz and Bornancin, 1975). The difficulty in assessing the real driving forces stimulated some speculation; however, only a few investigators considered chloride active transport as an alternative explanation to the source of the outside negative gill potential.

The basic electrophysiological values for the several isolated preparations containing chloride cells are shown in Table I. The orientation of the potential difference (PD) is consistently serosal negative in all species. The electrophysiological values show some spread, but are all within a similar range. The electrical potential difference is less of an indication of the function of the chloride cells than is the short-circuit current. This is because of the presence of diffusion potentials in the generation of the transepithelial PD, and, without a knowledge of the actual unidirectional fluxes and partial conductances, it reflects a number of not easily interpretable variables. However, the short-circuit current is directly proportional to the function and number of chloride cells. Except for *Gillichthys*, the values for I_{sc} are high when compared to other ion-secretory epithelia. Especially in *F. heteroclitus* and telapia, the levels of current are higher or similar to the values obtained, for example, in the isolated skins of *Rana* or the bladder of *Bufo*.

Table I

Electrical Properties of Chloride Cell-Rich Opercular Epithelia[a]

Species	PD (mV)	I_{sc} ($\mu A\ cm^{-2}$)	R ($\Omega\ cm^2$)	Reference
Fundulus heteroclitus	18.0 (10–40)	138 (50–350)	173 (50–450)	Degnan *et al.* (1977)
Gillichthys mirabilis	14	23.6	540	Marshal (1977)
Fundulus grandis	8.2	52.8	194	Krasny and Zadunaisky (1978)
Sarotherodon mossambicus (telapia)	21.5	100.6	259	Foskett *et al.* (1981)

[a] PD, electrical potential difference; I_{sc}, short-circuit current; R, electrical direct-current resistance.

The electrical resistance would indicate that in general these are in the category of leaky epithelia, and therefore rather high values for the passive unidirectional fluxes of some ionic species are to be expected. As indicated, these electrophysiological parameters show seasonal variations and are very different in specimens that are euryhaline after adaptation to seawater. Aspects of electrical properties during or after full adaptation to seawater are discussed further on.

B. Ion Substitution and Ion Fluxes across the Chloride Cell

The PD and I_{sc} are extremely sensitive to the chloride concentration in the solutions bathing the isolated chloride cell-rich epithelia (Degnan *et al.*, 1977). In *Fundulus* the effect of chloride replacement with a nonpermanent anion such as sulfate or methylsulfate is shown in Fig. 3. The I_{sc} and the PD drop rapidly and in some of the experiments decrease to zero. This is an important indication of the involvement of chloride in the secretory process

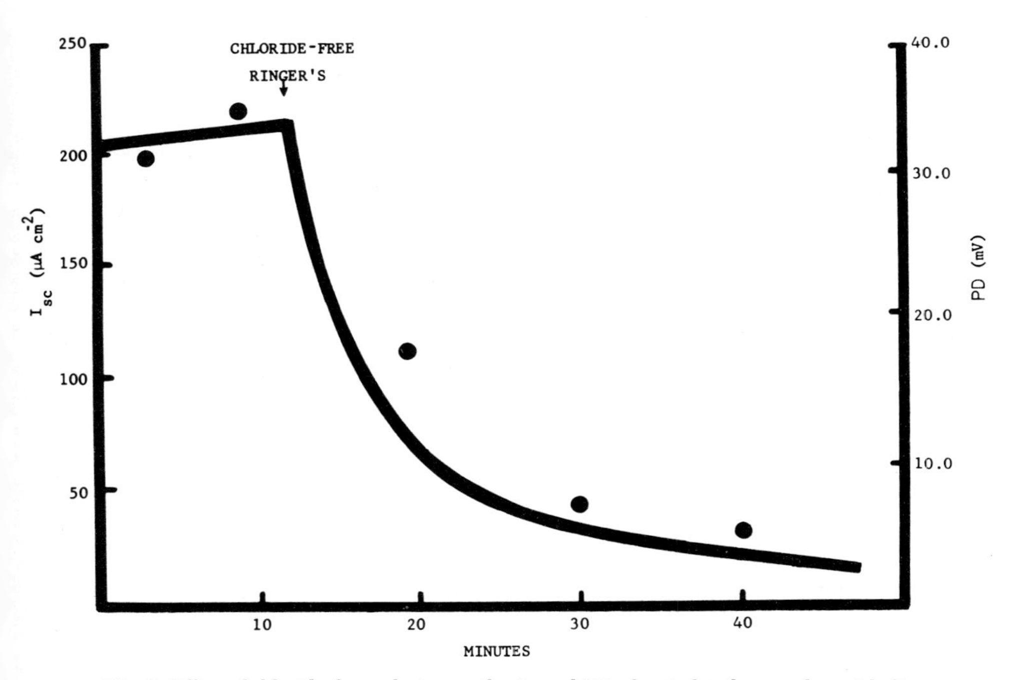

Fig. 3. Effect of chloride-free solution on the I_{sc} and PD of an isolated opercular epithelium of *Fundulus heteroclitus*. (From Degnan *et al.*, 1977.)

of the chloride cell. The observation is identical to other previously described chloride-transporting epithelia such as the frog cornea (Zadunaisky, 1966), where all of the current is carried by chloride ions.

The electrical properties are extremely sensitive also to the concentration of bicarbonate in the Ringer solutions. A more detailed presentation will be found in Section IV,G. However, here the influence on the electrical properties is discussed. The formula for the teleost Ringer solution utilized in the original experiments with *F. heteroclitus* by Karnaky *et al.* (1977) contains 16 m*M* sodium bicarbonate. This value is somewhat high, but, if a lower bicarbonate concentration is used as in the composition of Forster's teleost Ringer, it is found that the electrophysiological values are at a minimu. Increases in bicarbonate concentration produce increases in electrophysiological values, similar to those found *in vivo* across the gill. The concentration of 16 m*M* seems to be the most appropriate because it is closer to the actual bicarbonate concentration in blood of teleost fish.

C. Ion Fluxes and Short-Circuit Current (I_{sc})

Seawater-adapted *Fundulus heteroclitus* opercular membranes, short-circuited and bathed on both sides with Ringer solution, gassed with either 100% oxygen or 95% oxygen and 5% CO_2, exhibit a net chloride secretion equivalent to the I_{sc} with no net movement of sodium (Degnan *et al.*, 1977). The chloride effluxes, that is in the direction from blood to seawater, range in both conditions between 2.5 and 21 μEq cm^{-2} hr^{-1}, and the chloride influxes range from 0.36 to 14.5 μEq cm^{-2} hr^{-1}. The mean chloride efflux was between four and five times greater than the mean chloride influx, resulting in a net efflux of 6 to 7 μEq cm^{-2} hr^{-1}. This represents 160 μA cm^{-2} of current circulating through the preparation, and the average short-circuit current measured in the same experiments was 158 μA cm^{-2}. Statistically this is good evidence that there is a chloride active transport from blood to seawater. It also indicates that this is an electrogenic chloride system and that most probably sodium follows the gradient created by the driving force of chloride.

The fluxes of sodium when measured under identical conditions were found to be 2.63 μEq cm^{-2} hr^{-1} for sodium efflux and 2.95 μEq cm^{-2} hr^{-1} for sodium influx, the small difference between the two not being statistically significant and confirming that practically all the current is carried by chloride ions. As indicated previously, bicarbonate has an important effect on the current; however, this evidence of a net chloride efflux under short-circuit conditions can exclude the presence of a neutral transepithelial

Cl^-/HCO_3^- exchange, but it does not exclude the possibility of a Cl^-/HCO_3^- exchange across one of the two membranes.

D. Ion Fluxes under Open-Circuit Conditions

The fluxes of sodium and of chloride under open-circuit conditions were examined in isolated opercular epithelia of seawater-adapted *Fundulus heteroclitus* by Degnan and Zadunaisky (1979). The predicted sodium flux ratio was 0.94, and the observed ratio was 1.14; the difference between these two values was not statistically significant, and it indicated that sodium behaves as a passive ion across this membrane. The mean predicted chloride flux ratio was 11.4 on the basis of the flux equation, and the observed ratio was 1.38; the difference between these last two means was highly significant. The addition of ouabain at 10^{-6} *M* in the serosal solution produced a significant reduction in the sodium efflux, while it did not have any significant effect on the sodium influx. The agreement between the predicted and observed sodium flux ratios after ouabain treatment suggested that this effect could be completely attributed to the depolarization of the epithelium secondary to inhibition of (Na^+,K^+)-ATPase. The general results of this study indicated that sodium is near thermodynamic equilibrium and the net Cl^- flux is the natural driving force for the chloride cells.

E. Evidence That the Chloride Cell Is Responsible for the Net Chloride Transport

The flat sheet of opercular epithelium contains considerable numbers of chloride cells, and it is clear that potential difference, short-circuit current, and ion flux will be directly related to the presence of so many chloride cells. However, as is shown in Fig. 1, there are other cellular elements in the opercular epithelium. There are pavement cells, nondifferentiated cells, and mucous cells, as well as a layer of connective tissue and some muscle fibers.

In order to prove that the density of chloride cells is directly related to the transport rate, Karnaky *et al.* (1984) compared by regression analysis the chloride cell numbers versus the short-circuit current, revealing a correlation coefficient of 0.89. This is shown in Fig. 4 and Table II. Each chloride cell can generate approximately 1 nA of current, and these results provide compelling evidence for a chloride-secretory role for the chloride cells in this tissue. Furthermore, in the same analysis a comparison was made of two epithelial regions, one containing many chloride cells (the opercular epithelium of *Fundulus heteroclitus*) and the other, the epithelium lining the

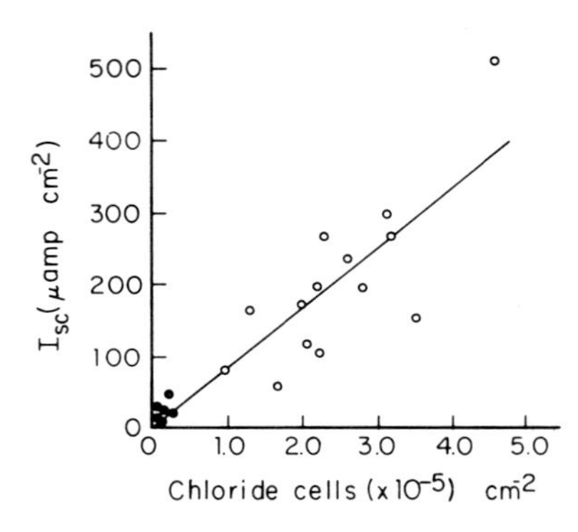

Fig. 4. Correlation between cell density and I_{sc} in isolated epithelia of *Fundulus heteroclitus*. Filled dots, cells from the epithelium of the roof of the mouth; White dots, cells from the opercular epithelium. Density and current are very low in the epithelium of the roof of the mouth as compared to the opercular epithelium. (Plotted from data of Karnaky *et al.*, 1984).

roof of the mouth of the same fish. What was found was that the active chloride secretion measured by mounting both tissues and determining chloride active transport shows values in the opercular epithelium that were several times greater than in the roof epithelium. The fluorescent dye DASPMI was used to quantitate the number of chloride cells in these two epithelia.

This type of analysis, mainly the correlation between number of cells in the epithelium and short-circuit current, was studied also in the isolated skin of telapia by Foskett *et al.* (1981). In telapia (*Sarotherodon mossambicus*), the opercular membranes of a specimen adapted to fresh water have low conductances and low currents; they do not transport chloride, and simultaneously there is a very low number of degenerated chloride cells. As the animals are placed in seawater the isolated epithelia slowly start to pump chloride, and this is coincident with the full development of chloride cells. This is shown in Fig. 5. If the short-circuit current is used as an index of the appearance of chloride cells in the isolated opercular epithelium of telapia, then the first sign that the transport is appearing is observed within 24 hr after the transfer to seawater and continues to increase until in 1 or 2 weeks following saltwater transfer, it reaches a new level of approximately 100 μA cm^{-2}. Regression analysis was also performed on data from this species by the same authors, and it provided an excellent linear regression that correlated the current with cell diameters. In the isolated opercular epithelium of *Fundulus grandis*, Krasny (1981) performed this regression analysis and observed the same correlation: the increase in the number of cells is directly related to the value of the short-circuit current. Finally, the same type of analysis was applied by Marshall and Nishioka (1980) in their study of the marine teleost *Gillichthys mirabilis*.

Another mode of analysis of the presence of low-resistance pathways at the level of secretory cells was the use of the vibrating probe technique. This

Table II

Correlation between Cellular Characteristics and the Short-Circuit Current of Chloride Cells[a,b]

Epithelium type and species	I_{sc}/chloride cell (nA)	Crypt diameter[c] (μm)	I_{sc}/chloride cell apical surface area (mA cm^{-2})	Method	Reference
Opercular					
Fundulus heteroclitus	1.0	3.8	4.5	DASPMI—Ussing chamber	Karnaky *et al.* (1984)
Sarotherodon mossambicus	2.5	3.0	17.7	Vibrating probe	Foskett *et al.* (1983)
Jaw					
Gillichthys mirabilis	0.65	3.5	3.4	DASPEI—Ussing chamber	Marshall and Nishioka (1980)

[a]Calculated from data of Karnaky *et al.* (1984).

[b]All data are from seawater-adapted animals, except *Gillichthys* data, which include animals adapted to seawater and to double-strength seawater.

[c]Crypt shape is assumed to be hemispherical.

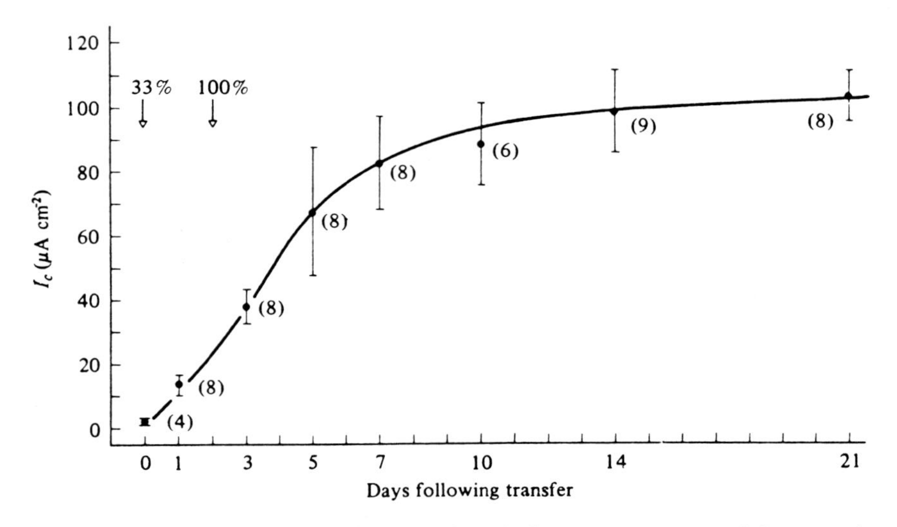

Fig. 5. Increase in I_{sc} during adaptation from fresh water to seawater of the opercular epithelium of telapia (*Sarotherodon mossambicus*). (From Foskett *et al.*, 1981.)

method was used to localize the high conductance and chloride current specifically to the chloride cells, thereby establishing that these cells definitely are the extrarenal salt-secretory cells in the fish. These experiments were performed in skins of *Sarotherodon mossambicus* (telapia) by Foskett and Scheffey (1982).

It is obvious that all of this evidence definitely indicates that the chloride cells, in a quantitative manner, are responsible for the secretion. In the rest of this chapter we will refer directly to all of the functions of the isolated epithelia as functions of the chloride cells as far as osmoregulation is concerned.

F. Effect of Anoxia and Drugs That Affect Transport on the Properties of the Chloride Cells

The results obtained in *Fundulus heteroclitus* are described here (Degnan *et al.*, 1977; Karnaky *et al.*, 1977). The effect of anoxia, when the cells are bathed on both sides with Ringer solution, and then gassed with nitrogen, produces within 30 min a decline of the chloride current and potential difference that closely resembles the effects observed in perfused gills of flounders (Shuttleworth *et al.*, 1974). These results indicate that the potential difference across the epithelia is dependent on metabolic energy and is not purely a result of diffusion potentials when bathed in both sides with Ringer solution. Ouabain, the inhibitor of the (Na^+,K^+)-ATPase that has been shown to inhibit chloride transport in a variety of epithelia (Zadunaisky *et al.*, 1963, in frog skin; Burg and Green 1973, in ascending loop of Henle; Stoff *et al.*, 1977, in perfused rectal glands), at 10^{-5} *M* applied to the isolated chloride cell preparation produces a steady and irreversible decline in the short-circuit current and potential difference across the epithelium that reaches lower steady-state levels around 5 to 10% of their control levels in 60 to 90 min.

Furosemide, the specific inhibitor of chloride transport in the kidney (Burg *et al.*, 1973) and in the red blood cell (Brazy and Gunn, 1976), causes the current and potential difference to decline to near zero values in about 30 min. The inhibitory effects of furosemide were not readily reversed by rinsing the chambers several times with fresh Ringer solution.

Another known inhibitor of chloride transport, in this case a competitor for the transport site, thiocyanate (Zadunaisky *et al.*, 1971), had no effect at 10^{-3} *M* in both bathing solutions; however, at higher concentrations thiocyanate caused a decline in the current and potential difference of about 50% of their control level within 10 min, which was always followed by a spontaneous tendency to return to the control level. In other systems the inhibition by thiocyanate (SCN^-) is not complete, only partial, and the mechanism has already been described by Durbin (1964), who demonstrated competitive kinetics between SCN^- and chloride for the Cl^- transport system in the gastric mucosa. It is very likely that thiocyanate inhibited chloride efflux across the opercular epithelium by competing for the same carrier site and that the spontaneous recovery of the current and potential difference may have resulted from active SCN^- transport by the chloride cells. Injections of thiocyanate to intact eels and flounders (Epstein *et al.*, 1973) reduced chloride efflux by 50%.

Theophylline, known to stimulate chloride transport in the cornea (Chalfie *et al.*, 1972) as well as in the isolated rectal gland of the dogfish, caused variable stimulation in the current and potential difference that reached maximum levels within 30 min. These stimulatory effects of theophylline were variable, ranging from slight to doubling these parameters, and in every instance there was lowering of the electrical resistance of the tissue.

Diamox, the well-known inhibitor of carbonic anhydrase, at 10^{-3} *M* on both sides bathing the chloride cells produces slight stimulation in the short-circuit current and potential difference, and these changes were surprisingly significant. They could be subject to question, since such high doses of this inhibitor may have produced generalized cellular effects unrelated to carbonic anhydrase activity. A carbonic anhydrase-mediated Cl^-/HCO_3^- exchange operating across the basolateral membranes of the gill chloride cell of seawater-adapted fish has been proposed by Maetz (1971). In the present opercular epithelial studies no inhibition was observed even with a few experiments at 10^{-2} *M* diamox. There was no explanation for the slight stimulatory effect of diamox, but the point is that there was no inhibition by this agent. It is possible that other inhibitors of carbonic anhydrase might have different effects on the same chloride cells.

Amiloride at a concentration of 10^{-5} *M* produced very small decreases in the short-circuit current of the order of 10 $\mu A\ cm^{-2}$ that were insignificant when compared to the 100–200 $\mu A\ cm^{-2}$ produced across these epithelia.

Amphotericin B at 10^{-5} *M* appeared to have no effect on the short-circuit current.

Several of these agents or drugs that are related to transport mechanisms or specific effects that are used to interpret membrane phenomena were also tested for their action on chloride fluxes. In all instances either stimulations or inhibitions were correlated with an increase of chloride transport by the chloride cells of the epithelia studied (Degnan *et al.*, 1977).

G. Bicarbonate

It has been mentioned before in this chapter that HCO_3^- had a stimulatory effect on the current and potential difference across the epithelium of *Fundulus heteroclitus*. The titration of sodium bicarbonate into the Ringer bathing both sides of the epithelium produced corresponding increases in the current and potential difference, and this is illustrated in Fig. 6. In the absence of exogenous HCO_3^- the current and potential difference were comparatively low. It can be observed that maximum values of current are obtained with 10 to 30 m*M* of bicarbonate. The effect seems to be more clearly attributed to the presence of bicarbonate in the serosal side of the preparation. Effects of pH could have been produced under the experimental conditions in which bicarbonate was tested on the chloride cells. Isolated preparations were then bathed in phosphate buffer and the pH changed accordingly. This is shown in Fig. 6. It is observed that the changes in current and potential difference with increases of pH occur in the opposite direction from the ones produced by bicarbonate.

Although specific effects of pH here cannot be interpreted, the matter of the bicarbonate effect at the time of writing of this chapter is still a problem of study. Actions of the agents described earlier with specific effects on transport have also been tested in other preparations rich in chloride cells under similar conditions, and the results leave no doubt that the general findings are extremely similar to those obtained in *F. heteroclitus*.

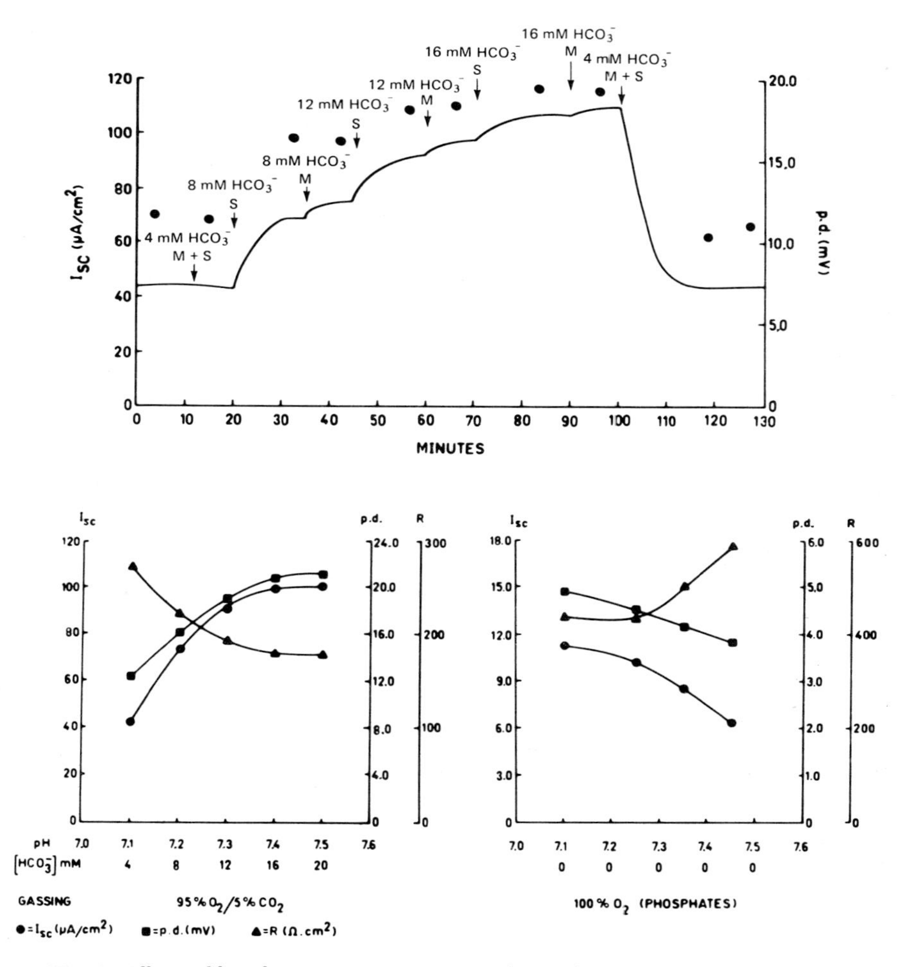

Fig. 6. Effects of bicarbonate concentration on electrical properties of isolated opercular epithelium of *Fundulus heteroclitus*. On the left, note the increase in current produced by bicarbonate. On the right, pH changes produce effect on current and potential differences that are in opposite direction to the effects of bicarbonate.

H. Adaptation to High and Low Salinity of Chloride Cells of *Fundulus heteroclitus*

In the hope of obtaining chloride cells adapted to fresh water and to seawater, isolated epithelia containing chloride cells were obtained from specimens adapted to both conditions. The surprise was that epithelia obtained from killifish (*F. heteroclitus*) adapted to fresh water when bathed in Ringer on both sides showed a potential difference and current that was not different from those adapted to seawater (Degnan *et al.*, 1977). However, Mayer-Gostan and Zadunaisky (1978) were able to adapt killifish slowly to fresh water and to mount epithelia that showed practically no potential difference. Also, treatment with prolactin resulted in epithelia that showed no potential difference or current when mounted with Ringer on both sides. The paradox of the findings in *F. heteroclitus* by Degnan *et al.* (1977) can be explained on the basis of the rapid and quick adaptation to the presence of sodium chloride in the outside bathing solution.

In the case of other species such as *Sarotherodon mossambicus* (telapia), the chloride cells disappear or degenerate when changed from seawater to fresh water. When the chloride cells are fully developed they maintain potential difference and current in saltwater telapia, as in the case of *F. heteroclitus*. It is very likely that the quick adaptation from seawater to fresh water that the killifish shows is due to local mechanisms resulting in a change in the conductance of the epithelium of the gill where the chloride cells are located. When the fish are exposed to fresh water there may be a local, immediate tightening of the membranes especially at the level of the tight junctions, and therefore the whole system behaves like a more impermeable, nonleaky epithelium. As soon as sodium chloride is detected in the apical side of the chloride cell, there is a rapid tendency to leakiness and for movement of chloride in the outward direction. This will be commented on further on, in discussing adaptation to salinity change and the mechanism of prolactin effects on the isolated chloride cells.

V. RECEPTORS AND HORMONAL EFFECTS

A. Catecholamine Receptors

1. β Receptors

The addition of isoproterenol (Degnan *et al.*, 1977; Degnan and Zadunaisky, 1979) at concentrations of 10^{-8} *M* produces stimulation of the short-circuit current of the isolated chloride cells bathed with Ringer on both sides, that reaches a maximum at concentrations of 10^{-5} *M*. The effect is extremely rapid, as shown in Fig. 7, and it is blocked by the specific inhibitor propranolol. Therefore, the chloride cell contains β receptors that are extremely sensitive to its natural stimulant. The Cl^- fluxes were increased proportionally to the increases in the current, and the average stimulation due to isoproterenol was of the order of 30%; however, there was great

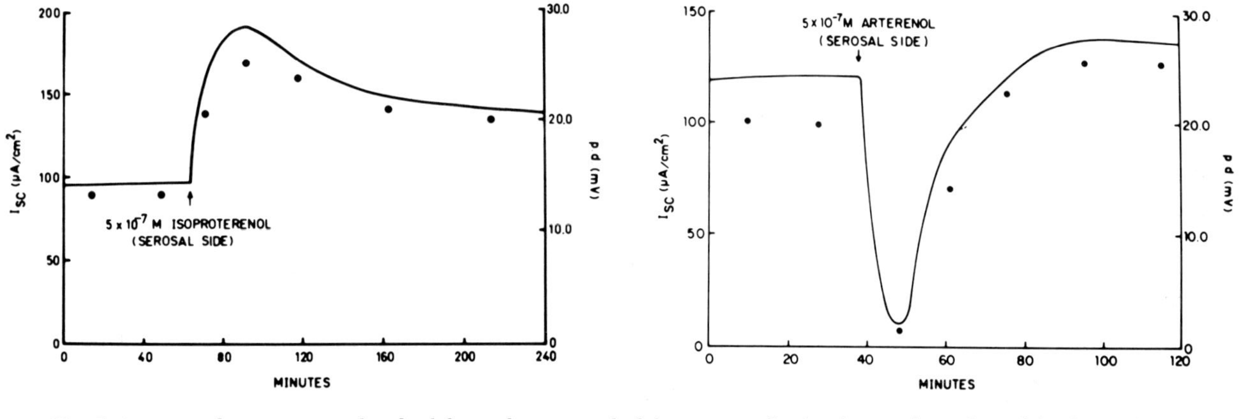

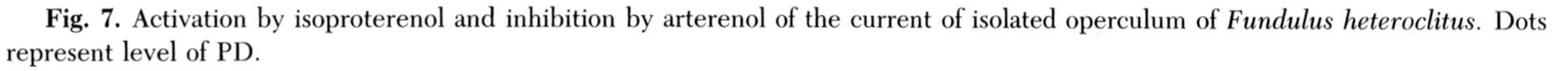

Fig. 7. Activation by isoproterenol and inhibition by arterenol of the current of isolated operculum of *Fundulus heteroclitus*. Dots represent level of PD.

variability and in some cases the current increased to 100% of its basic value (Degnan and Zadunaisky, 1979). These increases in chloride transport are accompanied by increases in the cyclic AMP content of the tissues. Mendelsohn *et al.* (1981) were able to show a significant increase in cyclic AMP during incubation of these opercular epithelia with isoproterenol.

2. α RECEPTORS

The action of arterenol on the chloride cell can be observed in Fig. 7. It is observed that at a dose of 10^{-7} *M* noradrenaline there is rapid inhibition of the current with a recovery that occurs within 20 to 30 min. α-Adrenergic activation by noradrenaline inhibited the chloride efflux by 66%. As shown by Mendelsohn *et al.* (1981), this was not accompanied by significant changes in the content of cyclic AMP in the tissue.

There is an antagonistic effect of β stimulation and α stimulation in the isolated chloride cells as studied in this manner, which can be correlated to some of the data that exist on the effect of catecholamines on gill function. In fact, Keys and Bateman (1932) were the first to demonstrate the depressive effect of adrenaline on the branchial chloride secretion. Using the gill perfusion method developed by Keys in the eel, they noticed that adrenaline not only reduced or abolished chloride secretion, but it produced considerable vasomotor activity. Pic *et al.* (1975) studied the mullet and Girard (1976) and Shuttleworth (1978) the trout. They reported similar effects of adrenaline on branchial chloride secretion. Shuttleworth (1978) also reported a depressive effect of adrenaline on the transgill potential and suggested that this chloride secretion was electrogenic. These effects are, however, complicated by the branchial hemodynamic changes, which in themselves could account for a considerable amount of the reduction in chloride secretion.

Other studies of Shuttleworth (1978) and Staggs and Shuttleworth (1984) have correlated the actions of β-stimulation and α-stimulation in perfused isolated gill preparations by determining the hemodynamics flow distribution and pressure and the electrogenic transgill potential. During constant-flow perfusion the effects that were found were (1) an increase in efferent flow due to the catecholamines with a corresponding decrease in venous flow and (2) a transitory rise in afferent pressure followed by a subsequent decline to original levels. There was concomitantly a decrease in the transepithelial electrogenic potential; therefore, epinephrine shows mostly its α-inhibitory effect and very little or none of the stimulatory effect of the β receptor. This coincides with the observations of the actions of epinephrine on the operculum published by Degnan and Zadunaisky (1979). With isoproterenol alone M. Davis and T. J. Shuttleworth (private communication, 1983) have found a small but significant increase in the transepithelial electrogenic potential. It is likely that the preparation utilized—the perfused gill of the flounder—has a very low density of β-receptors, although it responds very well to cyclic AMP and forskolin. One interpretation of the general effects found in the vessels and in the transport epithelium could be that the effect of epinephrine itself is to increase the flow to the arterial system and reduce flow to the arteriovenous system. This presumably enhances the supply of oxygenated blood to the muscles and other body tissues and at the same time inhibits transport. Regulation of ion transport in the gill depends on the interaction between this adrenergic inhibition and the peptidergic stimulation via glucagon.

Fosket *et al.* (1982b) have tested the effects of epinephrine on the isolated chloride cell-rich skin of telapia and found that, as in *Fundulus heteroclitus*, epinephrine has mostly an inhibitory effect by activation of α-receptors. The minimum effective doses were found to be very similar to the ones in *Fundulus*. Isoproterenol activation of β-receptors was not tested in these preparations. Marshall and Bern (1980) have also tested catecholamines on the isolated skins of *Gillichthys*, and they find that concentrations of 10^{-7} *M* epinephrine reduce the electrical parameters within 2 or 3 min of addition to the serosal side. Recovery from epinephrine inhibition is not spontaneous; in this case, thorough rinsing several times was necessary to accomplish full recovery of the current. Phenylephrine, an α-agonist, reduced the short-circuit current, and the effect followed a time course similar to that of epinephrine and norepinephrine. In this case the β-agonist isoproterenol was tested, and it produced a transient increase in electrical potential difference that was in effect opposite to epinephrine or epinephrine plus phenylephrine. In this case, this epithelium shows exactly the same behavior as the one of *F. heteroclitus*, with only a small increase due to β-stimulation.

B. Acetylcholine Receptors

Inhibition of the chloride-secretory mechanisms of the isolated opercular epithelium of *F. heteroclitus* was produced by acetylcholine and other cholinergic agonists and antagonists (Rowing and Zadunaisky, 1978). Acetylcholine caused a dose-dependent decrease in the short-circuit current from its resting level to near zero. It was more effective when applied serosally. The response was not associated with contraction of the underlying musculature. The muscarinic agonists carbachol methacholine and muscarine also caused similar dose-dependent decreases in the short-circuit current. The muscarinic antagonist homatropine caused a parallel shift of the acetylcholine dose–response curve to the right indicative of competitive inhibition. Nicotine produced no change in the short-circuit current, and the tissue remained responsive to carbachol, eliminating the possible involvement of a nicotinic receptor. The fluxes of chloride measured with ^{36}Cl showed that the decrease in short-circuit current was accounted for by an inhibition of chloride secretion. Pharmacological parameters for acetylcholine alone in the presence of eserine and in the presence of both eserine and homatropine and for carbachol were calculated, and the PD_2 for acetylcholine was not significantly changed by eserine. The esterase-resistant agonist carbachol had a PD_2 not significantly different from these values.

It is suggested that either low levels of acetylcholine in the tissue occur or the enzyme is distant from the receptor. Since these studies of hormones occur in the absence of innervation because the tissue is isolated, we have to assume that the muscarinic receptors are located in the chloride cells. The possibility exists that the nerves that still remain in the tissue could be releasing some factors. However, it is very unlikely that these concentration levels would be such that they would interfere with these *in vitro* studies (Rowing and Zadunaisky, 1978). The actual significance of the presence of these cholinergic inbibitors is not very clear at this moment. Other tissues have been shown to contain high levels of acetylcholine, and they are not innervated. The epithelium of the cornea is known to have high levels of acetylcholine, and its function is not clear (Howard, Zadunaisky, and Dunn, 1975).

C. Glucagon

The effect of glucagon on chloride secretion was tested using the skin of telapia (Foskett *et al.*, 1982b), and it was found to have an effect that was dose dependent with a minimum effective dose of 10^{-9} M and a maximum stimulation of 72% of the current at 10^{-5} M. This stimulating effect of glucagon was potentiated by IBMX (protein inhibitor of phosphodiesterase). The suggestion was made that IBMX and epinephrine acted on the current by changing tissue conductance, suggesting that this agent acted antagonistically on a nonconductive transport mechanism.

D. Vasoactive Intestinal Polypeptide (VIP)

Again in the skin of telapia VIP was tested, and it produced an increase of the current in epinephrine-inhibited tissues; this effect was also potentiated by IBMX.

It is not clear whether these hormones, glucagon and VIP, have any stimulatory function *in vivo*, and the available knowledge is much less than that for catecholamines.

E. Prolactin

Prolactin has a long history as a hormone that controls osmoregulation by the gills (Maetz and Burnancin, review of 1975). Mayer-Gostan and Zadunaisky (1978) observed an inhibition of chloride secretion in isolated chloride cells of the *Fundulus heteroclitus* opercular epithelium by prolactin and concluded that the chloride cell is the site of action of the regulating effect of prolactin. However, it is clear from existing data in intact fish that the mechanism of prolactin is not fast enough to account for the rapid adaptation of fish like *Fundulus* from fresh water to seawater and vice versa. In a detailed study performed by Foskett *et al.* (1982a), the effects of prolactin were tested on the transport properties of opercular membranes from the seawater-adapted telapia (*Seratherodon mossambicus*). They found that prolactin injections into the fish decreased both the current and the conductance of the epithelia in a dose-dependent manner. The conclusion of an interesting electrophysiological analysis by means of equivalent circuits for the paracellular conductances of the epithelium, was that the ratio of paracellular to active transport pathway conductances associated with chloride cells is constant in the epithelium of telapia. The differences in G_T (total conductance) and I are due to parallel changes in this conductance. Prolactin may effectively remove chloride cells from these membranes as well as inhibit, reversibly, active pathways or ionic conductance of the remaining cells. The effect of prolactin is well documented in intact fish, and the finding in the isolated preparation is of importance. Prolactin per se, however, does not have an effect when added directly to the bathing solution in vitro, but it has to be injected previously into the fish.

It is very likely that two mechanisms of adaptation for the euryhaline fish exist:

1. A very quick one occurs initially and will depend on the salt concentration in the environment. A reduction in salt concentration will tend to make the tissue tighter; that is, it will change the paracellular pathway so that the membrane will become more impermeable. Its conductance will decrease, and this will reduce the entry of sodium and chloride with a subsequent modification of the chloride-secretory system.
2. A second, more slow mechanism will be based on hormonal effects. It is not clear yet how prolactin works; however, it is obvious that prolactin is not controlling the very rapid change that is needed for *Fundulus*, for in-

stance, to be removed from seawater, washed with distilled water, and then put directly into a beaker of fresh water without damage to the specimen in question. Prolactin is a hormone that produces long-term adaptations, and in this sense it is probably very important for the permanent adaptation of the fish to seawater.

F. Cortisol

Cortisol is the predominant corticoid in the teleost fish (Henderson *et al.*, 1970) and has long been implicated in saltwater adaptation (see Maetz, 1974). Injections of cortisol into specimens of telapia with subsequent isolation and mounting of the opercular epithelium showed that this steroid produces a significant increase in the density of cells that were observed by staining with DASPMI, in comparison to animals that were injected with saline alone, without cortisol (Foskett *et al.*, 1981). The size and fine structure of the cells is similar to those observed in the cells of opercular membranes of noninjected fish. The authors have noted that the absolute increase in cell density induced by cortisone is the same as the maximal change in cell density induced by saltwater adaptation. Both treatments, saltwater adaptation and cortisol injections into freshwater fish, cause the cell density to increase approximately 2.5 times. However, there were great differences in the electrophysiological properties of the isolated tissues between the two groups. Cortisol treatment did not produce Cl^- secretion but saltwater adaptation did, despite the presence of normal abundant chloride cells during cortisol treatment. Here there is a discrepancy between the presence of cells and the presence of transport, but it has to be understood that even if the cells are present, the animals still remained in fresh water. Perhaps the rapid signal for secretion was lacking. It would be interesting to pursue these studies and, after injection of seawater-adapted euryhaline fish with cortisol, observe the response to very small doses of sodium and chloride in the apical membrane to see if a rapid adaptation to greater secretory rates occurs.

G. Urotensin I and Urotensin II

Marshall and Burns (1981) studied the neurosecretory peptine urotensin I and found that it stimulated *in vitro* the short-circuit current across the skin of the marine teleost *Gillichthys mirabilis*, which contains chloride cells. Urotensin I also reversed the previous inhibition of the current induced by epinephrine. Urotensin II stimulated untreated skins that had low currents. Urotensin II and epinephrine were able to reduce the current and net flux of chloride by decreasing chloride efflux. The phosphodiesterase inhibitor

IBMX stimulated the current and chloride net flux by increasing both in the direction of secretion. The results of these authors suggest that urotensin I can antagonize the effect of urotensin II and epinephrine on active chloride transport by chloride-secretory cells of these marine teleosts.

VI. PASSIVE SODIUM AND CHLORIDE MOVEMENTS

A. The Paracellular Shunt Pathway and Ionic Conductances for the Chloride Cell

The results discussed up to this point indicate evidence that the unidirectional sodium fluxes and the chloride influx (seawater to blood) behave passively in the isolated preparations containing a high density of chloride cells. However, strong statements have been published indicating that in the marine teleost gill epithelium sodium is actively secreted or exchanged for other ions, and that chloride is exchanged for other anions. In the review of Maetz and Bornancin of 1975, these exchanges are proposed on the basis of experiments in the intact animals or perfused gills.

A more detailed study of the sodium and chloride movements in the isolated layers containing chloride cells was needed to define the pathways through and around the cells and to permit definition of these avenues of ion movement, as well as to clarify whether this predicted information in the intact fish is confirmed or not. As will be seen in the following, the exchanges proposed by Maetz cannot be confirmed, and this is not only on the basis of the characteristics of these newly developed preparations but on the basis of a very strict analysis of the paracellular pathway.

The ideal condition to study the cellular and paracellular pathways of ions across epithelial layers consists of measuring potential differences and ionic activities within the cell by means of intracellular microelectrodes and intracellular ion-selective microelectrodes. However, in the case of the chloride cell, because of the abundance of other cells in the epithelium, attempts to do this experiment of impaling the cells have not been successful as of the writing of this chapter (S. Helman and J. A. Zadunaisky, unpublished observations, 1981). It is thus necessary to resort to somewhat more indirect evidence, but not less consequential than the measurement of intracellular potential differences. The approach of Degnan and Zadunaisky (1980a) consisted of (1) an examination of the relationship between the sodium conductance and the total ionic conductances of the tissue, (2) the response of the unidirectional sodium fluxes to voltage clamping at levels higher and lower than those spontaneously produced by the membrane, (3) the detection of the effects of drugs that affect the junctions between the cells, such as TAP (thiaminopyrimidine) on the sodium fluxes and the fluxes of a nonelectrolyte such as urea while the total tissue conductance was measured, (4) examination of the effect of unilateral sodium substitutions on the sodium and urea fluxes, and (5) examination of the nature of the chloride influx across the epithelium and comparison of it to total tissue conductance. Each approach will be discussed separately.

1. Relationships between Sodium Movements and Total Ionic Conductance

The conductance of the pathway of a passively and independently moving ion, such as sodium, was examined by Hodgkin in 1951, and the following equation was developed for conductance.

$$G_{Na^+} = \frac{z^2F^2}{RT} J^{Na^+} \tag{4}$$

In this equation the conductance G is the sodium conductance, and it is a function of F (Faraday constant), R (gas constant), and T (absolute temperature) times the flux of sodium J^{Na^+}. Under steady-state conditions and short-circuiting, the total ionic conductance is given by the ratio of $I_{sc}/\Delta E$, and the sodium permeability (P_{Na^+}) can be approximated under a variety of conditions by the ratio of the flux of sodium over the sodium concentration, expressed as $J^{Na^+}/[Na^+]$. In this case $[Na^+]$ is the activity in the compartment from which the flux originates. If one uses Eq. (4) as the sodium fluxes across the layer of chloride cells of the operculum with identical solutions of Ringer in both sides and short-circuited, that is under conditions in which the ΔE is zero, G_{Na^+} can be obtained and compared to G_T, which is the total electrical conductance of the tissue measured by detection of changes in voltage by small-current pulses. The result of these experiments showed that the ratio of G_{Na^+}/G_T for all preparations ranged from 0.41 to 0.69, giving a combined mean of 0.54. This indicates that from these analyses, 54% of the total ionic conductance was a sodium conductance. When the potential difference is zero the percentage represents the mean partial ionic conductance, and this comes out directly from the values and characteristics of this equation.

The relationship for the total conductance versus the sodium conductance is shown in Fig. 8. The figure combines data from influxes and effluxes, because it was observed that the same correlation can be obtained using influxes or effluxes. The reason for this initial caution of separating them is

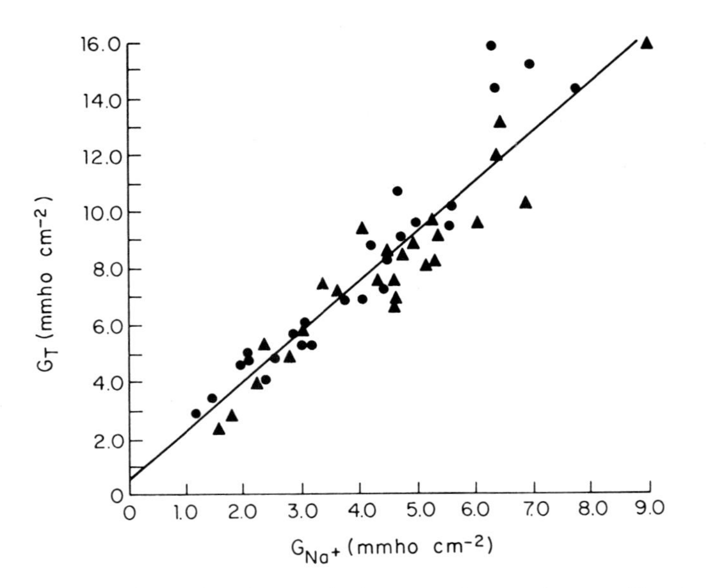

Fig. 8. Plot of total electrical conductance (G_T) versus sodium conductance (G_{Na^+}) across the isolated operculum of *Fundulus heteroclitus*. Note the excellent correlation between the two parameters indicating that sodium moves through a single-barrier paracellular conductive pathway. (From Degnan and Zadunaisky, 1980a.)

that despite the information that the fluxes of sodium are passive, one is never sure which of the two would be more amenable to this examination; it does happen, however, that in this tissue sodium really moves passively and in a high proportion, practically all of it, through the paracellular pathway.

A comparison of the urea permeability and the total ionic conductance, shown in Fig. 9, shows that there is no correlation between permeability of urea and electrical conductance.

2. The Unidirectional Sodium Fluxes under Different Levels of Voltage Clamping

It is possible to distinguish the extracellular pathway from the pathway to the cell by measuring the voltage dependence of the unidirectional fluxes (Mandel and Curran, 1972). The Nernst–Planck Equation describes the net flux of an ion under the influence of an electrochemical gradient, and if this equation is integrated under a uniform barrier by the constant field assumption of Goldman (1943), it is possible to arrive at an equation for the flux of the following form:

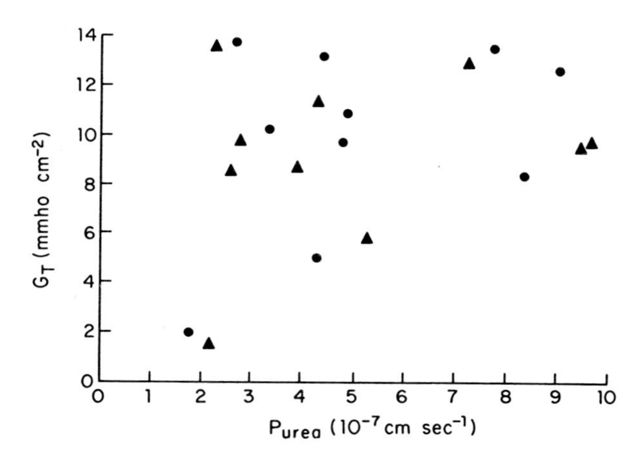

Fig. 9. Plot of total electrical conductance (G_T) versus urea fluxes. Note the lack of correlation between these two parameters. (From Degnan and Zadunaisky, 1980a.)

$$J_{\text{net}} = -\frac{D_i}{d}\frac{F\Delta E z_i}{RT}\left[\frac{C_i^o - C_i^i \exp(z_i F\Delta E/RT)}{1-\exp(z_i F\Delta E/RT)}\right] \tag{5}$$

Here D is the diffusion coefficient of the ion, d is the thickness of the barrier, C_i^o is the ion activity inside and outside, ΔE is the potential across the membrane, and the other symbols are the same as in the previous equation. If one assumes that the net flux arises from two independent unidirectional fluxes, the equation for the unidirectional flux of sodium, for example, can be expressed as

$$J^{Na^+} = \frac{P_{Na^+} zF\Delta E/RT}{1-\exp(-zF\Delta E/RT)} \times C \tag{6}$$

where P_{Na^+} is equal to D_i/d and represents the sodium permeability and C the sodium activity of the compartment from which the flux originates. If the sodium activity of the compartment from which the flux originates is kept constant and sodium permeability is assumed to be independent of voltage, then the relationship between sodium unidirectional flux when the preparation is short-circuited (the potential difference is equal to zero) and at other levels of voltage clamp (the potential is not equal to zero) is given by the following equation:

$$J^{Na^+}_{\Delta E \neq 0} = J^{Na^+}_{\Delta E = 0}\frac{zF\Delta E/RT}{1-\exp(-zF\Delta E/RT)} \tag{7}$$

The relationship expressed by this last equation was applied to the unidirectional sodium fluxes across the chloride cell layer (Degnan and Zadunaisky, 1980a), and the results were as follows. For each flux direction (J_{sm} and J_{ms}, where sm = serosal to mucosal, ms the reverse), the flux at zero voltage was measured, as well as at two-clamp voltages, ±25 mV from zero. Comparison of the flux values at the two voltage clamp levels shows that there were no significant differences in the measured and predicted parameters and therefore allows these fluxes to be described as absolutely passive and crossing only one rate-limiting barrier. This condition also supposes the assumption that P_{Na^+} was voltage independent within the applied voltage range of this experiment. This type of behavior is usually attributed to movement through the paracellular shunt pathway with a single tight junction functioning as the rate-limiting barrier (see also Bruus *et al.*, 1976).

In order to test whether this barrier was sensitive to blockage of sodium movement by amiloride, the known diuretic that affects specific sodium channels in the apical cell membrane, and amphotericin B, another compound that is known to influence transcellular sodium pathways, these drugs were tested under the described conditions. The results indicated that there was no effect of amiloride on the sodium fluxes, and amphotericin B produced a small reduction in the fluxes that was not statistically significant. The total electrical conductances in either direction were also not affected by these two compounds that are known to affect the cellular pathway.

3. Action of Thiaminopyrimidine (TAP) on Sodium and Urea Fluxes and Tissue Conductances

Thiaminopyrimidine has been reported to block paracellular sodium movements specifically and decrease the ionic conductance across the gallbladder (Moreno, 1975a,b). Subsequently, however, it has been reported that TAP has effects on cellular cation pathways (Lewis and Diamond, 1976). Still, when TAP has a very definite effect on the tissue it serves as a useful probe of paracellular permeability, particularly when other criteria demonstrate extracellular sodium movements. The pronounced effect of TAP on the short-circuit current across the epithelium is observed in Fig. 10, which shows that TAP reduced the sodium fluxes and tissue conductances over 70%. However, the ratio G_{Na^+}/G_T remained relatively constant (around 0.5), indicating that this compound reduced the conductance of all ions proportionally. Comparable reductions of the conductances were observed, while no effect on the urea fluxes were found. In 20 preparations examined, 10 mM TAP inhibited the short-circuit current by 64%. This compound is not without some deleterious effects also on this tissue; after exposure to TAP for more than 2 hr there was spontaneous recovery of the current and

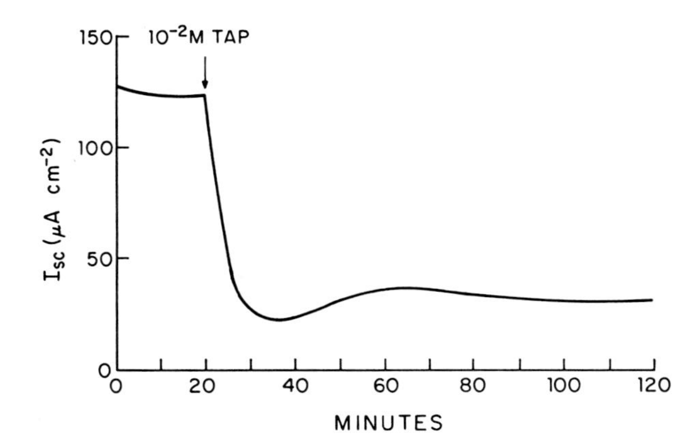

Fig. 10. Effect of TAP on the I_{sc} of an isolated opercular epithelium of *Fundulus heteroclitus*. Although this compound has other secondary effects, this pronounced, rapid action indicates blocking effect on the sodium paracellular pathways. (From Degnan and Zadunaisky, 1980a.)

reversal of the effect with an increase in the urea fluxes and total electrical conductance, reflecting a general deterioration of the tissue. Thus, results with this agent have to be taken with caution. When these results are considered together with the two previous types of analyses, as well as with the ones that follow, it is obvious that sodium moves passively through the paracellular pathway and not through the cell.

4. Sodium Substitution: Effects on Urea and Sodium Fluxes

Mucosal sodium substitution resulted in 60 to 76% reduction in the unidirectional radioisotopic sodium fluxes and total electrical conductance, respectively. However, similar substitutions had no effect on the urea fluxes, while they produced reductions of about 70% in total electrical conductance in the preparations studied for urea. In contrast, serosal sodium substitution—that is, basolateral or blood side substitution of sodium—had no significant effect on the sodium flux, while it decreased the total electrical conductance and had no effect on the urea fluxes.

The results just described show the sensitivity of the sodium pathway to mucosal sodium, that is, outside sodium. However, the response to mucosal sodium substitution produced greater effects in the reduction of G_T than in the mean partial sodium conductance of this tissue. This observation can be explained by the sodium dependence of the partial conductance of another ion than sodium. In fact, studies demonstrated that bilateral sodium substitution of sodium for a nonpermeant cation produced effects on chloride secretion across the epithelium (Degnan and Zadunaisky, 1980a). The effect of mucosal sodium substitution on total electrical conductance is explained by a reduction of sodium conductance through the shunt pathway, as well as by the reduction in chloride conductance of the mucosal membrane. The effect of the serosal sodium substitution can be attributed to the reduction in chloride conductance of the serosal membrane.

It can be observed that it is dangerous, then, to make definite conclusions when ionic interactions are occurring. These ionic interactions are very important and are the consequence of the existence of ion coupling. The coupling in this case is of Na^+ and Cl^-. It is clear here that unilateral substitutions indicate the passive nature of sodium movements through the paracellular pathway, and discard the presence of Na^+/Na^+ exchange and suggest the possibility of a separate urea pathway. At the same time, one can conclude that the sodium sensitivity of the chloride transport is quite remarkable and produces definitive effects in the measurements of total transepithelial conductance.

5. The Characteristics of the Passive Chloride Influx across the Chloride Cell Layer

The same type of experiments as those performed for sodium and urea were done for chloride influx, the passive flux. The lack of correlation between chloride flux from mucosa to serosa and total electrical conductance (G_T) is shown in Fig. 11. It can be observed that in comparison to the very strict dependence of sodium conductance on total electrical conductance, there is no correlation and the passive influx of chloride behaves practically like the movements of urea. The mean P_{Cl^-} calculated from this experiment was 1.24×10^{-6} cm sec^{-1}. This value represents 13% of the mean P_{Na^+} and was only 22% of the total electrical conductance in these experiments. In increasing and decreasing the voltage by 25 mV above and below zero, it was seen that the flux of chloride from blood to seawater was a passive diffusional flux traversing only one rate-limiting barrier. Thiaminopyrimidine had no effect on this flux while it reduced the total electrical conductance by 44% in these experiments, but it did significantly inhibit the active chloride flux in the blood to seawater direction. This inhibitory effect of TAP on the active chloride efflux across the operculum was similar to the effect of TAP on the active chloride transport across flounder intestine. The voltage studies, however, resulted in a very interesting observation. The passive chloride flux is an extracellular one, and the insensitivity of this flux to TAP and mucosal sodium substitution suggests an extracellular pathway separate from the

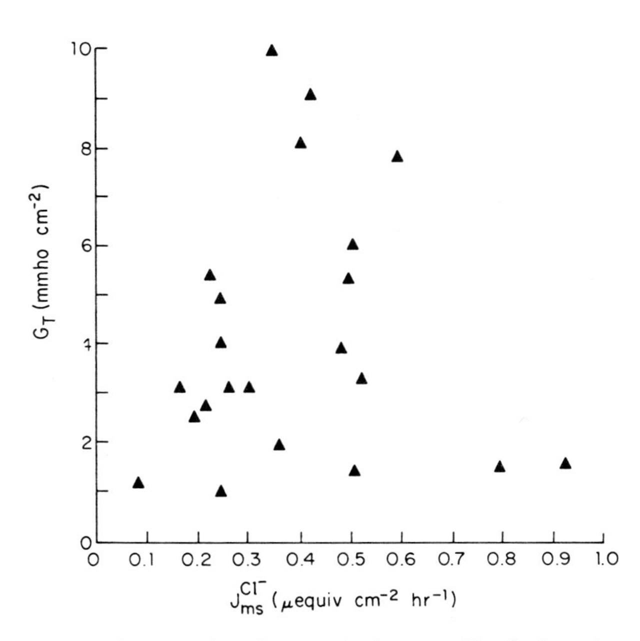

Fig. 11. Plot of total electrical conductance (G_T) versus chloride fluxes (passive). Note the lack of correlation of the passive chloride flux and the total electrical conductance. (From Degnan and Zadunaisky, 1980a.)

sodium pathway. In all respects the passive flux of chloride and the fluxes of urea were similar, suggesting that these may traverse the same pathway.

B. Conclusions from Study of Paracellular Pathway

The preceding results confirm the finding that sodium moves passively across the chloride cell-rich preparations, and in addition it indicates the absence of Na^+/Na^+ exchange diffusion. Because of the nature of the experiment performed in better biophysically controlled conditions than intact fish gill effluxes, the conclusion can be expressed that the exchanges proposed in the past are not really tenable. A thorough discussion of the results presented earlier can be found in Degnan and Zadunaisky (1980a). The main conclusions are that (1) the passive chloride flux and the sodium fluxes in both directions occur through the extracellular compartments, and (2) there are different pathways for chloride and urea and for sodium. The results, however, agree with observations that sodium fluxes across intact seawater-adapted trout gill were passive diffusional fluxes. The results do not agree with reports of Na^+/Na^+ and Cl^-/Cl^- exchanges proposed by Motais *et al.* (1966), and the Na^+/K^+ exchanges proposed by Maetz (1969). Neither agrees with the results on active sodium efflux proposed by Potts and Eddy (1973) across the seawater-adapted flounder gill. The observations that permitted the postulation of these exchanges was the "trans" effect of external Na^+ and Cl^- efflux rates resembling typical saturation kinetics. However, similar observations have been made in the isolated opercular epithelium preparation (Degnan and Zadunaisky, 1980b) where such exchanges are not operative.

The transfer of fish from seawater to fresh water is accompanied by rapid reduction in the branchial sodium and chloride efflux rates (Motais *et al.*, 1966) and are followed by a delayed secondary reduction believed to be under neurohumoral control (Pickford andPhillips, 1959; Mayer-Gostan and Hirano, 1976). Most probably these effects are due to prolactin and cortisol. In the opercular epithelium lowering the mucosal Na^+ and/or changing the chloride concentration reduces the chloride secretion rate (Degnan and Zadunaisky, 1980a), and lowering the sodium concentration shown in studies of the paracellular pathway reduced the sodium efflux rate. These changes, apparently limited by the turnover rate of the mucosal solution in these chambers, are rapid enough to account for the fast branchial efflux changes observed in intact fish. External sodium and chloride appear necessary to maintain the P_{Cl^-} of the apical membrane, whereas external sodium appears necessary to maintain the paracellular sodium shunt pathway and the overall transepithelial resistance. Such permeability changes explain the acute branchial response of the fish to changes in salinity. This refers to the rapid adaptation before the neurohumoral factor becomes operative. The reduction in the branchial permeability when fish are transferred from seawater to fresh water as well as the "trans" effect of external Na^+ and Cl^- on the branchial efflux rate can then be explained without invoking exchange diffusion.

VII. WATER MOVEMENTS ACROSS THE CHLORIDE CELL

In isolated, well-controlled thermodynamic conditions as those provided by the opercular epithelium of *Fundulus heteroclitus*, very little has been done on water movements. Preliminary experiments were performed by Brown and Zadunaisky (1982). The epithelium was mounted in a special system able to detect nanoliters of water movements across the tissue. The system utilized was the apparatus developed by Bourguet *et al.* (1964). In Fig. 12 and Table III it can be observed that there is a basic rate of water

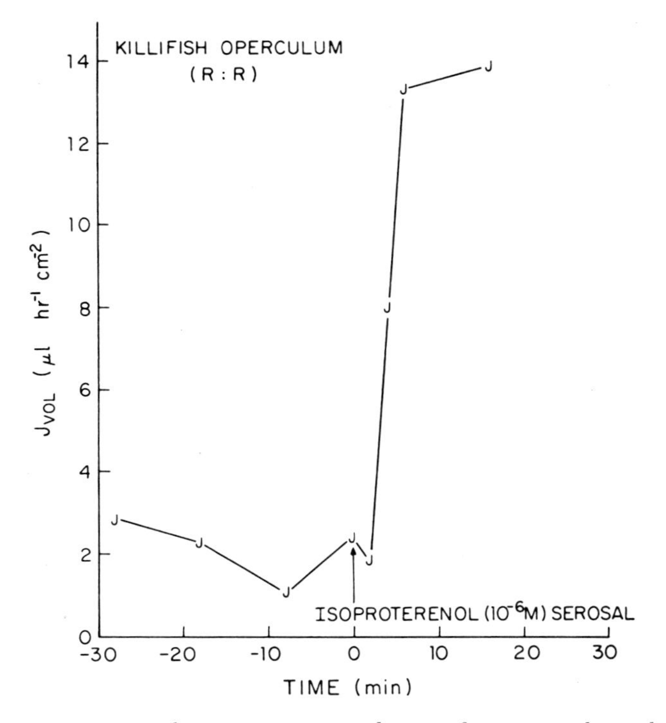

Fig. 12. Measurements of water movements and action of isoproterenol on isolated opercular epithelia. Note the basal water movement and the rapid stimulation by the β stimulant. (Data of Brown and Zadunaisky, 1982.)

movement that is very rapidly enhanced by the administration of isoproterenol. Isoproterenol (see Section V,A) in this preparation is a stimulant of the chloride secretion. It appears that the water movement follows the ion flux, and it would be rather important to study further water movement across the tissue in order to understand the normal mechanism as well as the humoral control.

VIII. A MODEL FOR THE CHLORIDE CELL

The ideal model for the chloride cell should describe parameters that will incorporate the ionic activities inside the cell, the paracellular pathways, and the location of specific pumps and receptors, and will provide better knowledge of this complicated and very exciting type of secretory cell. However, only some of the information has been obtained, and the main piece of

Table III

Water Flux from Serosa to Mucosa across Isolated Opercular Epithelia of *Fundulus heteroclitus*

	J_{vol} (μl hr^{-1} cm^{-2})	PD (mV)	T_{sc} (μA cm^{-2})	R (Ω cm^{-2})	G_T (mS cm^{-2})
High-resistance tissues[a] (n = 5)	7.2 ± 4.2[b]	8.6 ± 1.7	36.4 ± 11.5	270 ± 28.4	3.92 ± 0.43
Low-resistance tissues[a] (n = 7)	0.0	8.6 ± 1.8	70.3 ± 11.0	118 ± 14.0	9.4 ± 1.6

[a]High-resistance tissues, a group with a DC resistance greater than 170 Ω cm^{-2}. All tissues of the low-resistance group did not show any water flow at all. The net water flux is in the direction of the chloride net flux. The tissues were mounted without osmotic or pressure gradient with Ringer solutions in both sides. (Data of Brown and Zadunaisky, 1982.)

[b]Mean ± SEM.

information lacking is the determination of intracellular ionic activities. However, enough information is available to describe the behavior of the system. It can be easily assimilated to other chloride-secretory systems where intracellular ionic activities have been obtained. This does not mean that the information should not be obtained specifically for the chloride cell. In this section we are going to attempt an analysis of what information is available to put together a model such as the one shown in Fig. 13. We have the following information:

1. There is a net transepithelial movement of chloride.
2. This transepithelial movement of chloride is sodium dependent.
3. The sodium–potassium pump inhibitor ouabain stops the chloride transport.
4. The loop diuretic furosemide is able to inhibit profoundly the chloride transport.
5. Activation of β-receptors or the addition of cyclic AMP (E. J. Degnan and J. A. Zadunaisky, unpublished) can stimulate Cl^- transport, as in many of the other secretory epithelia that show the same characteristics.
6. Studies by means of histochemical techniques for the localization of (Na^+,K^+)-ATPase on the basis of electron-microscopic detection of precipitates, as well as autoradiography of titrated ouabain, show that this enzyme is located in the basolateral aspect of the chloride cell.

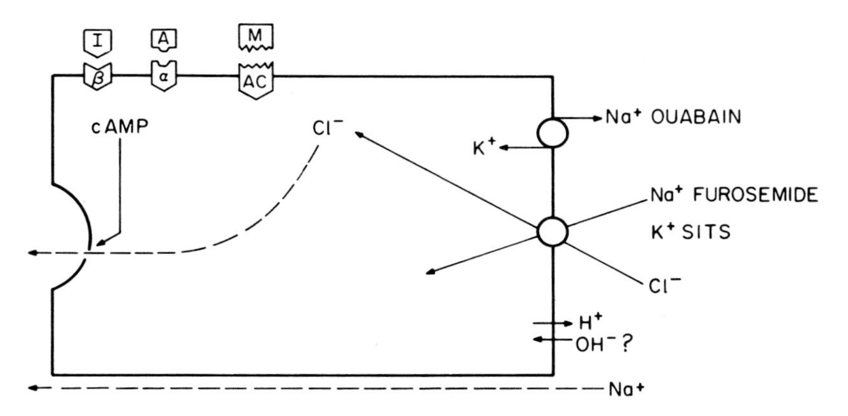

Fig. 13. Diagram indicating the main characteristics of the chloride cell that permit the construction of a working model. The (Na^+,K^+) pump on the basolateral side of the cell maintains the Na^+ low inside the cell. The Na^+,Cl^-,K^+ carrier permits Cl^- entry, and diffusion across the crypt occurs.

7. Amiloride has practically no effect on this preparation, indicating that sodium is not moving through the cellular pathway.
8. Inhibitors of cyclic AMP such as theophylline produce increases in the current that are very similar to those found in other chloride-secretory tissues.
9. The effect of other agents that increase cyclic AMP in the cells, such as glucagon and VIP, reverse the action of cyclic AMP phosphodiesterase inhibitors.
10. Thiocyanate, a typical competitor of chloride transport, has very clear effects on the chloride cells, either in opercular preparations or in ion secretion through the gills.
11. SITS has an inhibitory effect on the chloride current of the chloride cells of the opercular epithelium of the *Fundulus heteroclitus*, which points to the existence of a sodium–chloride coupler such as the one found in the basolateral membrane of many other chloride-secretory systems.
12. The reduction of potassium in the basolateral side of the opercular epithelium produces drastic inhibition of the chloride current, as shown in Fig. 14, indicating the possibility that a (Na^+,Cl^-,K^+) coupler also exists in this preparation.

All of the aforementioned reasons could permit the construction of a model as the one shown in Fig. 13. However, this has to be considered as a working model to be tested and to stimulate further experimental research to conclude exactly how the cell performs. In the model presented here, the entry of chloride into the cell is dependent on sodium concentration and potassium. This is shown as a coupler on the basolateral membrane; the (Na^+,K^+) pump also located in the same membrane keeps the sodium low inside the cell, providing the driving force for chloride accumulation in the cell. Chloride diffuses passively through the apical membrane, in this case shown at the level of the cript of the chloride cell. On the basolateral

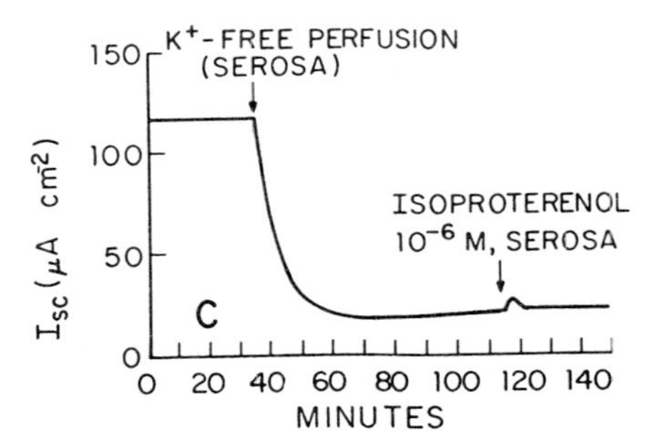

Fig. 14. Requirements of K^+ for the active chloride transport across the opercular epithelium. Note the important effect that K^+ has on the I_{sc}. (From Degnan and Zadunaisky, 1980b.)

membrane there are a number of receptors; the better known ones perform through cyclic AMP, and we have suggested here that cyclic AMP modifies the permeability of the apical membrane. This is not at all understood or known in the chloride cell; however, it is the case in other chloride-secretory epithelia, especially in the cornea (Zadunaisky *et al.*, 1979, 1982; Nagel and Reinach, 1980; Reuss *et al.*, 1983). This last mechanism of action of cyclic AMP is assumed to be mediated by a phosphorylating step on the proteins of the membranes that changes the permeability to chloride in the specific chloride channel located there.

The consequences of this model for chloride secretion are very clear for the fish exposed to seawater: it can get rid of the salts that are absorbed through the intestine and maintain salt and water balance. However, in the case of sodium reabsorption during exposure to fresh water (in both freshwater and euryhaline fish), the picture is not as clear. It is very likely that the first step is a detection of the sodium chloride concentration on the apical side, with drastic inhibition of Cl^- secretions. The next step can consist of the tightening of the paracellular pathway at the level of the occluding junctions to avoid the leak of Na^+ and Cl^- from the blood down the gradient. Under these rapid adaptative conditions, then, the chloride cell might be transformed into a reabsorptive organ. However, this is not clear at all. In the fresh water-adapted eel and telapia, most probably sodium chloride reabsorption is occurring through the gill without the presence or with the presence of very few degenerating chloride cells. It is thus not clear whether the chloride cells are there to perform both functions of secretion in seawater and reabsorption in freshwater conditions. The reduced number of cells could perform this task, but this is a question to be answered in the future, and only speculation can be exercised at this point. In *Fundulus* there is no dramatic change in number of cells in fresh water.

IX. CHLORIDE TRANSPORT IN OTHER TISSUES OR ORGANS

In the past the concept has been that chloride is a passive ion that moves down the gradient and follows the driving force of sodium or other specific cations. However, this concept has subsequently been thoroughly revised. In 1963 Zadunaisky *et al.* demonstrated chloride active transport from fresh water to blood in the isolated skin of frogs (*Leptodactylus ocellatus*). This finding was pursued, and it was found that (1) it was not affected by ADH, (2) that there were two components to the movement of chloride, one of low affinity and the other of high affinity (Fischbarg *et al*,1967), and (3) that the system had a specific dependence on sodium concentration. The active transport of the skin of frogs is very similar to the active transport found in the ascending loop of Henle by Burg *et al.* (1973). In the red cell Cabantchik and Rothstein (1976) definitely showed that a specific protein, band 3 protein, represents about 50% of the protein of the cell membrane and is involved in the movements of chloride. Kinetic studies by Gunn in 1982 and molecular studies by Knauf and Grinstein (1982) have advanced and confirmed these observations.

Concerning the secretory chloride transport, the most characteristic example has been the corneal epithelium, especially in frogs, where all the short-circuit current is due to the movement of chloride outwards in a secretory manner (see review by Zadunaisky, 1978). In this system, the sodium dependence, the action of furosemide, ouabain, and thiocyanate, the high chloride activity inside the cell (Zadunaisky *et al.*, 1979), and the activating effect by epinephrine through cyclic AMP (Chalfie *et al.*, 1972) were clear indications that the movement of chloride could be only secretory and due to active transport in conditions in which thermodynamically the preparations were impeccable. In the comparative field the rectal gland of the dogfish, an accessory sodium chloride-secretory system, performs by the active secretion of chloride into the lumen of the gland that then is secreted to the outside (Stoff *et al.*, 1977). Chloride active transport has been found also in peripheral nerves (Keynes, 1963), as well as in the central nervous system in both neurons and glial cells. For an update of the state of chloride transport in different organs the reader is referred to a recent book that has been published containing information about different systems (Zadunaisky, 1982).

Therefore, the chloride cell falls into the category of several cells that have similar secretory characteristics.

X. BRIEF CORRELATION OF FINDINGS IN THE INTACT GILL OR INTACT FISH WITH STUDIES OF ISOLATED CHLORIDE CELLS

The main conclusion is that the driving force for the well-known sodium chloride secretion through gills of seawater-adapted fish is chloride transport. In a sense this dissipates the doubts and interpretations that were held for a few decades based on the assumption that the driving force was due to a cation. The action of several hormones and pharmacological agents is easily interpreted on the basis of the presence of this phenomenon, and no contradictions are really found between the phenomenon in the gill and that in the isolated chloride cells. Therefore, no attempt will be made to stress the obvious; rather, it would be more interesting to look into the future and ask what are the problems, difficulties, or contradictions in order to advance the knowledge of how osmoregulation through the gills occurs.

The first problem that comes to mind is the dissociation between the regulatory effects of hormones, especially the catecholamines, on the circulation and on the osmoregulatory portion of the gills. A valid attempt was made by Shuttleworth (1978), who has taken into account the information on the secretion of chloride through the gill and the regulation by catecholamines, and studied the electrical potential together with the hydrodynamics of the gill. His findings seem to indicate that in fact there are effects that can be correlated. Although the measurement of the negative potential difference across the gill as an index is very valid under conditions in which it is impossible to short-circuit or to use any other method, it is clear that the potential difference is the function of many parameters, not only of the active electrogenic secretion of chloride, but also of the state of permeability including paracellular pathways. Moreover, these might not be directly related to the chloride cell or to the osmoregulatory functions of the gill.

The actual nature of change induced by NaCl on gill permeability that permits the rapid adaptation from seawater to fresh water in euryhaline fish should be pursued, looking not for electron-microscopic changes but for molecular events. The absorptive process should now be studied *in vitro*, because much of the secretory process seems to be more clearly understood, as stressed in this chapter. Cells with the aspect of other cells engaged in active salt transport have been seen to develop in the gills of *Carcinus meanas* during exposure to low concentrations of sodium chloride in the intact crab. The relationship between water movements and salt movements should be pursued, following perhaps the initial experiments presented here under that heading (Section VII). Attempts should also be made to impale the chloride cell and study intracellular potentials and intracellular activities. The separation of chloride cells from the gills and the study of the molecular events associated with transport could also provide very much needed information in this field.

REFERENCES

Bereiter-Hahn, J. (1976). Dimethylaminostyrylmethylpyridiniumiodine (DASPMI) as a fluorescent probe for mitochondria in situ. *Biochim. Biophys. Acta* **423,** 1–14.

Bevelander, G. (1935). A comparative study of the branchial epithelium in fishes, with special reference to extrarenal excretion. *J. Morphol.* **57,** 335–352.

Bevelander, G. (1936). Branchial glands in fishes. *J. Morphol.* **59,** 215–224.

Brazy, P. C., and Gunn, R. B. (1976). Furosemide inhibition of chloride transport in human red blood cells. *J. Gen. Physiol.* **68,** 583–599.

Brown, J., and Zadunaisky, J. A. (1982). Fluid movements across anion-secreting epithelia. *Fed. Proc., Fed. Am. Soc. Exp. Biol.* **41,** 1266.

Bruus, K., Kristensen, P., and Larsen, E. H. (1976). Pathways for chloride and sodium transport across toad skin. *Acta Physiol. Scand.* **97,** 31–47.

Burg, M., and Green, N. (1973). Function of the thick ascending limb of Henle's loop. *Am. J. Physiol.* **224,** 659–668.

Burg, M., Stoner, L., Cardinal, J., and Green, N. (1973). Furosemide effect on isolated perfused tubules. *Am. J. Physiol.* **225,** 119–124.

Burns, J., and Copeland, D. E. (1950). Chloride excretion in the head region of *Fundulus heteroclitus. Biol. Bull.* (*Woods Hole*) **99,** 381–385.

Cabantchik, Z. I., and Rothstein, A. (1974). Membrane proteins related to anion permeability of human red blood cells. I. Localization of disulfonic stilbene binding sites in proteins involved in permeation. *J. Membr. Biol.* **15,** 207–226.

Chalfie, M., Neufeld, A. H., and Zadunaisky, J. A. (1972). Action of epinephrine and other cycle AMP mediated agents on the chloride transport of the frog cornea. *Invest. Ophthalmol.* **11,** 644–650.

Copeland, D. E. (1948). The cytological basis of chloride transfer in the gills of *Fundulus heteroclitus. J. Morphol.* **82,** 201–227.

Copeland, D. E., and Fitzgerald, A. T. (1968). The salt absorbing cell in the gills of the blue crab (*Callinectes sapidus*) with a note on modified mitochondria. *Z. Zellforsch. Mikrosk. Anat.* **92,** 1–22.

Degnan, K. J., and Zadunaisky, J. A. (1979). Open-circuit sodium and chloride fluxes across isolated opercular epithelia from the teleost *Fundulus heteroclitus. J. Physiol.* (*London*) **294,** 484–495.

Degnan, K. J., and Zadunaisky, J. A. (1980a). Passive sodium movements across the opercular epithelium: The paracellular shunt pathway and ionic conductance. *J. Membr. Biol.* **55,** 175–185.

Degnan, K. J., and Zadunaisky, J. A. (1980b). Ionic contributions to the potential and current across the opercular epithelium. *Am. J. Physiol.* **238,** R231–R239.

Degnan, K. J., and Zadunaisky, J. A. (1981). The sodium and chloride dependence of active Cl^- secretion across the opercular epithelium. *Fed. Proc., Fed. Am. Soc. Exp. Biol.* **40,** 370.

Degnan, K. J., and Zadunaisky, J. A. (1982). The opercular epithelium experimental model for teleost gill osmoregulation and chloride secretion. *In* "Chloride Transport in Biological Membranes" (J. A. Zadunaisky, ed.), pp. 215–318. Academic Press, New York.

Degnan, K. J., Karnaky, K. J., Jr., and Zadunaisky, J. A. (1977). Active chloride transport in the *in vitro* opercular skin of a teleost (*Fundulus heteroclitus*), a gill-like epithelium rich in chloride cells. *J. Physiol.* (*London*) **271,** 155–191.

Degnan, K. J., Zadunaisky, J. A., and Mayer-Gostan, N. (1980). The bicarbonate sensitivity of chloride secretion across the opercular epithelium. *Bull. Mt. Desert Isl. Biol. Lab.* **20,** 48–50.

Doyle, W. L., and Gorecki, D. (1961). The so-called chloride cell of the fish gill. *Physiol. Zool.* **34,** 81–85.

Epstein, F. H., Maetz, J., and de Renzis, G. (1973). Active transport of chloride by the teleost gill: Inhibition by thiocyanate. *Am. J. Physiol.* **224,** 1295–1299.

Evans, D. H. (1980). Kinetic studies of ion transport by fish gill epithelium. *Am. J. Physiol.* **238,** R224–R230.

Fischbarg, J., Zadunaisky, J. A., and Fish, F. W. (1967). Dependence of sodium and chloride transport on chloride concentration in isolated frog skin. *Am. J. Physiol.* **213,** 963–969.

Foskett, J. (1981). Chloride secretion to teleost opercular membrane: Circuit analysis of the effects of prolactin. *Ann. N.Y. Acad. Sci.* **372,** 644–645.

Foskett, J. K., and Schelley, C. (1982). The chloride cell: Definitive identification as the salt-secretory cell in teleosts. *Science* **215,** 164–166.

Foskett, J. K., Hogsdon, C. D., Turner, T., Machen, T. E., and Bern, H. A. (1981). Differentiation of the chloride extrusion mechanisms during seawater adaptation of the teleost fish, the cichlid *Sarotherodon mossambicus*. *J. Exp. Biol.* **93,** 209–224.

Foskett, J. K., Machen, T. E., and Bern, H. A. (1982a). Chloride secretion and conductance of teleost opercular membrane: Effect of prolactin. *Am. J. Physiol.* **242,** R380–R389.

Foskett, J. K., Hubbard, G. M., Machen, T., and Bern, H. A. (1982b). Effect of epinephrine, glucagon, and vasoactive intestinal polypeptide on chloride secretion by teleost opercular membrane. *J. Comp. Physiol.* **146,** 27–34.

Foskett, J. K., Bern, H. A., Machen, T. E., and Conner, M. (1983). Chloride cells and the hormonal control of teleost fish osmoregulation. *J. Exp. Biol. Rev.* **5** (in press).

Girard, J. P. (1976). Salt excretion by the perfused head of trout adapted to sea water and its inhibition by adrenaline. *J. Comp. Physiol. B* **111B,** 77–91.

Goldman, D. E. (1943). Potential, impedence and rectification in membranes. *J. Gen. Physiol.* **27,** 37–60.

Gunn, R., Dalmark, M., Tosteson, D., and Wieth, J. (1973). Characteristics of chloride transport in human red blood cells. *J. Gen. Physiol.* **61,** 185–206.

Henderson, I. W., Chan, D. K. O., Sandor, T., and Chester Jones, I. (1970). The adrenal cortex and osmoregulation in teleosts. *Mem. Soc. Endocrinol.* **18,** 31–55.

Hodgkin, A. L. (1951). The ionic basis of electrical activity in nerve and muscle. *Biol. Rev. Cambridge Philos. Soc.* **26,** 339–409.

Karnaky, K. J., Jr., and Kinter, W. B. (1977). Killifish opercular skin: a flat epithelium with a high density of chloride cells. *J. Exp. Zool.* **199,** 355–364.

Karnaky, K. J., Jr., Kinter, L. B., Kinter, W. B., and Stirling, C. E. (1976). Teleost chloride cell. II. Autoradiographic localization of gill Na, K-ATPase in killifish Fundulus heteroclitus adapted to low and high saline environments. *J. Cell Biol.* **70,** 157–177.

Karnaky, K. J., Jr., Degnan, K. J., and Zadunaisky, J. A. (1977). Chloride transport across isolated opercular epithelium of killifish: A membrane rich in chloride cells. *Science* **195,** 203–295.

Karnaky, K. J., Jr., Degnan, K. J., and Zadunaisky, J. A. (1979). Correlation of chloride cell number and short-circuit current in chloride-secreting epithelia of Fundulus heteroclitus. *Bull. Mt. Desert Isl. Biol. Lab.* **19,** 109–111.

Karnaky, K. J., Jr., Degnan, K. J., Garretson, L. T., and Zadunaisky, J. A. (1984). Identification and quantification of mitochondria-rich cells in transporting epithelia. *Am. J. Physiol.* (in press).

Keynes, R. (1963). Chloride in the squid axon. *J. Physiol. (London)* **169,** 690–705.

Keys, A. B. (1931). The heart-gill preparation of the eel and its perfusion for the study of a natural membrane *in situ*. *Z. Vergl. Physiol.* **15,** 352–363.

Keys, A. B., and Bateman, J. B. (1932). Branchial responses to adrenaline and to pitressin in the eel. *Biol. Bull. (Woods Hole, Mass.)* **63,** 327–336.

Keys, A. B., and Willmer, E. N. (1932). "Chloride-secreting cells" in the gills of fishes with special reference to the common eel. *J. Physiol. (London)* **76,** 368–378.

Knauf, P. A., and Grinstein, S. (1982). Functional sites of the red cell anion exchange protein: Use of bimodal chemical probes. *In* "Chloride Transport in Biological Membranes" (J. A. Zadunaisky, ed.), pp. 61–90. Academic Press, New York.

Krasny, E. J. (1981). Ion transport properties of the opercular epithelium of *Fundulus grandis*. Doctorate Thesis, University of Miami, Coral Gables, Florida.

Krasny, E., and Zadunaisky, J. A. (1978). Ion transport properties of the isolated opercular epithelium of *Fundulus grandis*. *Bull. Mt. Desert Isl. Biol. Lab.* **18,** 117–118.

Krogh, A., and Keys, A. B. (1931). A syringe-pipette for precise analytical usage. *J. Chem. Soc.* pp. 2436–2440.

Lewis, S. A., and Diamond, J. M. (1976). Na^+ transport by rabbit urinary bladder, a tight epithelium. *J. Membr. Biol.* **28,** 1–40.
Maetz, J. (1969). Seawater teleosts: Evidence for a sodium-potassium exchange in the branchial sodium-excreting pump. *Science* **166,** 613–615.
Maetz, J. (1971). Fish gills: Mechanisms of salt transfer in fresh water and sea water. *Philos. Trans. R. Soc. London, Ser. B* **262,** 209–249.
Maetz, J., and Bornancin, M. (1975). Biochemical and biophysical aspects of salt excretion by choloride cells in teleosts. *Fortschr. Zool.* **23,** 322–362.
Mandel, L. J., and Curran, P. F. (1972). Response of the frog skin to steady-state voltage clamping. I. The shunt pathway. *J. Gen. Physiol.* **59,** 503–518.
Marshall, W. S. (1977). Transepithelial potential and short-circuit current across the isolated skin of Gillichthys mirabilis (Teleostei: Gobiidae), acclimated to 5% and 100% seawater. *J. Comp. Physiol. B* **114B,** 157–165.
Marshall, W. S. (1981). Active transport of Rb+ across skin of the marine teleost *Gillichthys mirabilis. Am. J. Physiol.* **241,** F482–F486.
Marshall, W. S., and Bern, H. A. (1979). Teleostean urophysis: Urotensin II ion transport across the isolated skin of a marine teleost. *Science* **204,** 519–521.
Marshall, W. S., and Bern, H. A. (1980). Ion transport across the isolated skin of the teleost Gillichthys mirabilis. *In* "Epithelial Transport in the Lower Vertebrates" (B. Lahlou, ed.), pp. 337–350. Cambridge Univ. Press, London and New York.
Marshall, W. S., and Nishioka, R. S. (1980). Relation of mitochondria-rich chloride cells to active chloride transport in the skin of a marine teleost. *J. Exp. Zool.* **214,** 147–156.
Mayer-Gostan, N., and Hirano, T. (1976). The effects of transecting the IXth and Xth central nerves on hydromineral balance in the eel Anguilla anguilla. *J. Exp. Biol.* **64,** 461–475.
Mayer-Gostan, N., and Zadunaisky, J. A. (1978). Inhibition of chloride secretion by prolactin in the isolated opercular epithelium of Fundulus heteroclitus. *Bull. Mt. Desert Isl. Biol. Lab.* **18,** 106–117.
Mendelsohn, S. A., Cherksey, B., and Degnan, K. J. (1981). Adrenergic regulation of chloride secretion across the opercular epithelium: The role of cyclic AMP. *J. Comp. Physiol. B* **145B,** 29–35.
Moreno, J. (1975a). Blockage of gallbladder tight junction cation-selective channels by 2,4,6-triaminopyrimidinium (TAP). *J. Gen. Physiol.* **66,** 97–115.
Moreno, J. (1975b). Routes of nonelectrolyte permeability in gallbladder: Effects of 2,4,6-triaminopyrimidinium (TAP). *J. Gen. Physiol.* **66,** 117–128.
Motais, R., Garcia-Romeu, F., and Maetz, J. (1966). Exchange diffusion effect and euryhalinity in teleosts. *J. Gen. Physiol.* **50,** 391–422.
Nagel, W., and Reinach, P. (1980). Mechanism of stimulation by epinephrine of active transepithelial Cl transport in isolated frog cornea. *J. Membr. Biol.* **56,** 73–79.
Philpott, C. W. (1965). Halide localization in the teleost chloride cell and its identification by selected area electron diffraction. *Protoplasma* **60,** 7–23.
Pic, P., Mayer-Gostan, N., and Maetz, J. (1975). Branchial effects of epinephrine in the seawater-adapted mullet. II. Na^+ and Cl^- extrusion. *Am. J. Physiol.* **228,** 441–447.
Pickford, G. E., and Phillips, J. G. (1959). Prolactin, a factor in promoting survival of hypophysectomized killifish in freshwater. *Science* **139,** 454–455.
Potts, W. T. W., and Eddy, F. B. (1973). Gill potentials and sodium fluxes in the flounder Platichthys flesus. *J. Cell. Comp. Physiol.* **87,** 29–48.
Rehberg, P. B. (1926). The determination of chlorine in blood and tissues by microtitration. *Biochem. J.* **20,** 483–485.
Reuss, L., Reinach, P., Weinman, S. A., and Grady, T. P. (1983). Intracellular ion activities

and Cl^- transport mechanisms in bullfrog corneal epithelium. *Am. J. Physiol.* **244,** C336–C347.

Rowing, G. M., and Zadunaisky, J. A. (1978). Inhibition of chloride transport by acetylcholine in the isolated opercular epithelia of *Fundulus heteroclitus*. Presence of a muscarinic receptor. *Bull. Mt. Desert Isl. Biol. Lab.* **18,** 101–104.

Sargent, J. R., Bell, M. V., and Kelly, K. F. (1980). The nature and properties of sodium ion plus potassium ion-activated adenosine triphosphatase and its role in marine salt secreting epithelia. *In* "Epithelial Transport in the Lower Vertebrates" (B. Lahlou, ed.), pp. 251–269. Cambridge Univ. Press, London and New York.

Shuttleworth, T. J. (1978). The effect of adrenaline on potentials in the isolated gills to the flounder (*Platichthys flesus* L.). *J. Comp. Physiol.* **124,** 129–136.

Shuttleworth, T. J., Potts, W. T. W., and Harris, J. N. (1974). Bioelectrical potentials in the gills of the flounder *Platichthys flesus*. *J. Cell. Comp. Physiol.* **94,** 321–323.

Smith, H. W. (1930). The absorption and excretion of water and salts by marine teleosts. *Am. J. Physiol.* **93,** 485–505.

Staggs, L., and Shuttleworth, T. J. (1984). Effects of catecholamines on transport and hemodynamics. *Am. J. Physiol.* (in press).

Stoff, J. S., Silva, P., Field, M., Forrest, J., Stevens, A., and Epstein, F. H. (1977). Cyclic AMP regulation of active chloride transport in the rectal gland of marine elasmobranchs. *J. Exp. Zool.* **199,** 443–448.

Ussing, H. H. (1949). The distinction by means of tracers between active transport and diffusion. *Acta Physiol. Scand.* **19,** 43–51.

Ussing, H. H., and Zerahn, K. (1951). Active transport of sodium as the source of electric current in the short-circuited isolated frog skin. *Acta Physiol. Scand.* **23,** 110–127.

Van Slyke, D. D. (1923). The determination of chlorides in blood and tissues. *J. Biol. Chem.* **58,** 523–529.

Zadunaisky, J. A. (1966). Active transport of chloride in frog cornea. *Am. J. Physiol.* **211,** 506–512.

Zadunaisky, J. A. (1978). Transport in eye epithelia: The cornea and crystalline lens. *In* "Membrane Transport in Biology" (G. Giebisch, D. C. Tosteson, and H. H. Ussing, eds.), Vol. 3, pp. 307–335. Springer-Verlag, Berlin and New York.

Zadunaisky, J. A. (1979). Characteristics of chloride secretion in some nonintestinal epithelia. *In* "Mechanisms of Intestinal Secretion" (H. J. Binder, ed.), pp. 53–64. Alan R. Liss, Inc., New York.

Zadunaisky, J. A., ed. (1982). "Chloride Transport in Biological Membranes." Academic Press, New York.

Zadunaisky, J. A., and Degnan, K. J. (1976). Passage of sugars and urea across the isolated retina pigment epithelium of the frog. *Exp. Eye Res.* **23,** 191–196.

Zadunaisky, J. A., and Degnan, K. J. (1980). Chloride active transport and osmoregulation. *In* "Epithelial Transport in the Lower Vertebrates" (B. Lahlou, ed.), pp. 185–197. Cambridge Univ. Press, London and New York.

Zadunaisky, J. A., Candia, O., and Chiarandini, D. J. (1963). The origin of the short-circuit current in the isolated skin of the South American Frog, *Leptodactylus ocellatus*. *J. Gen. Physiol.* **47,** 393–402.

Zadunaisky, J. A., Lande, M. A., and Hafner, J. (1971). Further studies on chloride transport in the frog cornea. *Am. J. Physiol.* **221,** 1832–1836.

Zadunaisky, J. A., Spring, K. R., and Shindo, T. (1979). Intracellular chloride activity in the frog corneal epithelium. *Fed. Proc., Fed. Am. Soc. Exp. Biol.* **38,** 1059.

Zadunaisky, J. A., Wiederholt, M., and Evans, A. (1982). Action of papaverine, Bay K 552 and interaction with cAMP in the isolated frog corneal epithelium. *Bull. Mt. Desert Isl. Biol. Lab.* **22,** 106–108.

6

HORMONAL CONTROL OF WATER MOVEMENT ACROSS THE GILLS

J. C. RANKIN
School of Animal Biology
University College of North Wales
Bangor, Wales

LIANA BOLIS
Istituto di Fisiologia Generale
Messina, Italy

I. INTRODUCTION

The great majority of fish species live in either hypoosmotic- or hyperosmotic media (i.e., fresh water or seawater) and face a constant problem of osmotic water gain or loss. Fish with body fluids of almost the same osmotic concentration as the environment, such as marine elasmobranchs and hagfish, have a very rapid turnover of body water (100% per hr or more) which is not disadvantageous, since there is no osmotic water loss (Payan and Maetz, 1971). Fish exposed to osmotic dilution or concentration of their body fluids

ISBN 0-12-350432-5

have evolved reduced permeabilities to water, and several hormones have been implicated in the regulation of water permeability. Study of this phenomenon has been severely hampered by the technical problems involved in the measurement of water fluxes across gills. These have been reviewed by Isaia (Chapter 1, this volume), but it is worth considering in some detail how they affect investigations of endocrine involvement in order to assess the significance of experimental data that are of necessity based on questionable methods.

In whole-animal studies it is generally assumed that measured water fluxes across the integument involve only the gills, the skin playing a negligible role. The evidence in support of this assumption that will stand up to detailed scrutiny is very limited, although convincing. This does raise an interesting question in view of findings that the skin plays an important role in oxygen uptake in some fish (Kirsch and Nonnotte, 1977; Nonnotte and Kirsch, 1978; Steffensen *et al.*, 1981). It is generally thought that fish cannot avoid detrimental osmotic water fluxes, because they need to have a respiratory exchanger of large surface area (the gills), which cannot be impermeable to water if it is to be permeable to oxygen. How, then, can an impermeable skin take up a significant proportion of the body's oxygen consumption in some fish?

A. Cutaneous Water Exchange

A common misconception is that the surface area of the skin is negligible in comparison with that of the gills. For example, Loretz (1979) misinterprets Parry (1966), who misinterprets Gray (1954) in assuming that the skin has only about 2% of the surface area of the gills in the goldfish he is studying, which he further assumes to have a gill area of 400 cm^2 100 g body $weight^{-1}$. The skin area of 100-g goldfish would therefore appear to be 8 cm^2 (4 cm^2 on either side!). There is really no excuse for misunderstanding Gray, although he quotes skin area in square centimeters and gill area in square millimeters, since he gives a very clear pictorial representation of gill area in relation to skin area (see his Fig. 1). It is worth quoting a few results from Gray (1954) to correct the misconceptions that may have arisen (Table I).

Only in very active fish does the gill area greatly exceed the skin area. The only ratio Gray recorded greater than 18.28 was 48.54 from a single specimen of false albacore (*Gymnosarda alleterata*), and this is the figure that has created the lasting impression. The lowest value he quotes, from a single specimen of sand flounder (*Lophosetta maculata*), gives a ratio of 0.90 and, since Gray did not include the fins in his measurement of skin area, it is

Table I

Ratios of Gill Areas to Skin (Excluding Fin) Areas[a]

Species	Gill area/skin area
Toadfish (*Opsanus tau*)	1.25
Southern flounder (*Pseudopleuronectes americanus*)	1.25
American eel (*Anguilla rostrata*)	1.93
Sea trout (*Cynoscion regalis*)	4.10
Mullet (*Mugil cephalus*)	6.54
Mackerel (*Scomber scombrus*)	8.38
Dolphin fish (*Coryphaena hippuris*)	11.69
Menhaden (*Brevoortia hippurus*)	18.28

[a]From Gray (1954).

clear that the cutaneous exchange area can easily exceed that of the gills in some inactive fish, particularly if the linings of the buccal and opercular cavities are taken into account.

If the skin is not negligible in area compared to the gills, is it really impermeable to water but permeable to oxygen? Results of in vitro experiments, with skin mounted in chambers between two solutions, cannot necessarily be used as evidence for low water permeability. Fish skin is very thick, and it is not surprising that it forms an efficient barrier to water diffusion. The outer layers are, however, well vascularized, and isolated skin preparations will tell us nothing about water exchange between blood and external medium unless they are perfused. Nevertheless even the nonperfused isolated skin of an elasmobranch, *Scyliorhinus canicula*, is relatively permeable to water, with a diffusional permeability coefficient of 10^{-4} cm sec^{-1} (Payan and Maetz, 1970), suggesting that about 15% of the water exchange measured in vivo might be across the skin (Payan and Maetz, 1971). However, the authors considered that the *in vitro* flux was exaggerated because of damage to the skin. Experiments on three eel tails perfused at physiological pressures and flows from a peristaltic pump via the cannulated dorsal aorta and suspended in a vigorously stirred freshwater bath gave very low diffusional permeability coefficients of 6.1 and 6.9×10^{-8} and 1.4×10^{-7} cm sec^{-1} (J. C. Rankin and M. B. Bennett, unpublished observations).

Another approach to distinguishing between cutaneous and branchial permeabilities has been to calculate gill blood flow by Fick principle from the clearance of tritiated water and compare it with cardiac output measured by an electromagnetic flowmeter. In the eel approximate equality was found between values obtained by the two methods (Motais *et al.*, 1969), but since the existence of prelamellar arteriovenous shunts in eels (Dunel and Laurent, 1977) means that branchial blood flow does not equal cardiac output, and in any case the flowmeter results were of limited accuracy, these data do not prove that the skin is not involved in water exchange. The best evidence that it is not comes from the results of Kirsch (1972), in which volume changes in the anterior compartments (containing the head and gills of an eel) and posterior compartments (containing the rest of the body, including the majority of the skin) were measured directly. In fresh water the volume of the posterior compartment decreased by 0.058 ± 0.049 ml hr^{-1} kg^{-1} compared with a decrease of 1.87 ± 0.19 ml hr^{-1} kg^{-1} in the anterior compartment. Osmotic uptake from the posterior compartment could conceivably have been partly balanced by fecal losses, and the loss of fluid from the anterior compartment included water drunk [measured by esophageal cannulation by Kirsch *et al.* (1975) at 2.75 ml hr^{-1} kg^{-1} in freshwater eels in later experiments], in addition to water entering the gills by osmosis, but 24 hr after adaptation to seawater the volume of the posterior compartment was stable, showing that no measurable osmotic loss across the skin was occurring. This result seems conclusive, so although the question of how eel skin can be permeable to oxygen but not to water remains an interesting one, we can safely assume that osmotic fluxes measured *in vivo* are indeed across the branchial epithelium.

B. Methodology

As explained by Isaia (Chapter 1, this volume), osmotic water fluxes can only be measured indirectly in vivo. The result can only be arrived at by difference after drinking rates, urine flow rates, fecal losses, and body weight changes are taken into account. (Even with the apparatus used by Kirsch, in which anterior and posterior changes can be separated, the anterior changes are a balance between branchial osmotic fluxes, drinking, and changes in the volume of the head.) Following are a number of simplifying assumptions that are commonly made:

1. Drinking is negligible in freshwater fish (often untrue).
2. Urine flow rate is negligible in marine fish.
3. Fecal losses are always negligible.
4. Body weight remains constant.

Although some or all of these assumptions may be true under steady-state conditions, it is by no means certain that this will be the case during perturbations in water balance brought about by hormone injections or ablation of endocrine organs. For example, adrenaline injection produces a large diuretic response in seawater-adapted eels (Babiker, 1973) that, if reproduced in mullet, would have accounted for the change in water balance interpreted by Pic *et al.* (1974) as an increase in branchial osmotic permeability. Their conclusion is probably valid because adrenaline is not diuretic in other marine fish, such as the cod (Ungell and Rankin, 1984), but without checking for renal effects in mullet (rather difficult in 5- to 41-g animals) one cannot be sure. Even if it were possible to measure all the parameters over the same period of time (no one has yet devised a method of continuous recording of weight changes of a fish in water), the possible stress to the animal, with release of catecholamines known to have profound effects on water balance, must be considered.

Because of these difficulties, *in vitro* techniques have often been preferred. The simplest technique is incubation of excised gill arches (usually with the cut ends ligated). Weight changes resulting from osmotic water fluxes can then be followed. The composition of the gill tissues will change with time, so only the initial rate of weight loss or gain will reflect the in vivo situation. Isaia and Hirano (1975) found that osmotic fluxes calculated from the weight gain during the first 10 min of incubation of freshwater European eel gills in fresh water were in good agreement with those calculated from the exponential change in weight over a period of 150 min, although they admitted that rapid changes in weight made it difficult to evaluate the initial slope of the weight against time plot directly. It is worth examining this finding, since other workers have not used the initial rate of weight change to assess hormone effects.

The rate of weight increase levels off for several reasons. Isaia and Hirano (1975) found that the sodium content of freshwater eel gills incubated in fresh water decreased by 17.5% during a 30-min period, during which time weight increased by about 12.5%. A reasonable assumption would be that the osmolarity of the gill fluid had decreased from around 300 to around 220 mOsm liter^{-1} during this period as a result of both diffusional loss of salts and gain of water. Internal osmolarity would therefore have been of the order of 150 mOsm liter^{-1} when equilibrium was achieved—clearly not as a result of complete osmotic equilibration. Did influx cease because the internal hydrostatic pressure (in gills ligated with steel wire at both ends) balanced the osmotic pressure? An internal pressure of about 2500 mm Hg would have been required. Since we have observed bleeding from the gills of eels during injections of catecholamines that have raised ventral aortic blood pressure to more than 50 mm Hg, it seems likely that damage to gill tissues will be the main factor affecting weight change kinetics.

Whereas seawater gills incubated in seawater shrink rather than swell, tissue damage could still be an important factor. Isaia and Hirano (1975) found an increase in sodium content of about 60% after 30 min incubation, during which time weight had decreased by only about 6%, whereas Kamiya (1967) found no changes in sodium content of seawater-adapted Japanese eel gills incubated in seawater for 75 min. The only apparent difference in technique was that she did not ligate the ends of the gill arches. It is clear that diffusional fluxes of salts, leading to attentuation of osmotic gradients, and tissue damage could have as great an effect on the results of incubation experiments as branchial osmotic permeability. Hormones could affect these processes without necessarily altering osmotic permeability; for example, by increasing passive permeability to salts they could produce an effect that could be interpreted as increasing osmotic permeability. Apart from these considerations, the lack of perfusion and ventilation would lead to large unstirred layers on both sides of the epithelium, and while these and changes in functional surface area are unlikely to affect water fluxes in vivo (Chapter 1, this volume) it is unlikely that this assumption could also be made for incubated gills. An alternative approach is perfusion of isolated gill arches or heads. This approach can be used for isotopic flux studies but not to measure net water movement, and since they often leak, measurement of effluxes may be problematical.

Before discussing results obtained, it is worth considering the magnitude of the errors that may result from the method of measurement. Isaia (Chapter 1, this volume) gives an excellent indication of the nature of the problem when he cites two studies giving a ratio between the osmotic and diffusional permeability coefficients (P_{os}/P_{dif}) of 0.15 but concludes that in reality P_{os} may be much greater than P_{dif}! P_{os} is greater than P_{dif} in many epithelia, including amphibian skin. Three explanations have been advanced to explain this discrepancy: (1) the effect of unstirred layers, which has been ruled out in fish gills *in vivo* by Isaia but which would be much greater in incubated gills, (2) pores of >3 Å diameter in which water moves by bulk flow as described by the Poiseuille equation, and (3) small pores (~2 Å diameter) through which water molecules pass in single file. The third explanation is currently in favor for epithelia as diverse as amphibian skin and mammalian renal tubules (Hebert and Andreoli, 1982).

Isaia (1982) has shown that virtually all of the diffusional water flux occurs through tissues exposed to the arterio-arterial blood pathway (primarily the secondary lamellae), and Isaia and Masoni (1976) have suggested that osmotic fluxes might occur mainly through the chloride cells. If this is so, studies on effects of hormones on osmotic permeability would better be performed on chloride cell-rich epithelia, such as *Fundulus* opercular epithelium. It is difficult, however, to see why osmosis should not occur through the respiratory epithelium if this is the site of diffusional fluxes.

Chloride cells are present in much smaller numbers than respiratory epithelial cells in gill filaments (Naon and Mayer-Gostan, 1983), and individually they present a much smaller surface area to the environment (see, for example, Fig. 8 of Laurent, 1982). A very rough estimate based on examination of eel gills is that the primary epithelium constitutes about 5% of the total surface area, and examination of a number of electron micrographs revealed that in all cases the surface area of the chloride cells was less than 10% of the area of primary epithelium illustrated. The exposed area of chloride cells must therefore be less than 0.5% of the total gill area. Since water flux is directly proportional to surface area as well as to permeability, if much more of the total osmotic water flow passes through chloride cells than respiratory cells (say, for the sake of argument, 10 times as much) and P_{osm} for the whole gill is of the order of 2×10^{-5} cm sec^{-1} (Motais *et al.*, 1969), then P_{osm} for the chloride cell surface must be $2 \times 10^{-5} \times 10/0.005$ or 4×10^{-2} cm sec^{-1}. If the suggestion is not that water passes through the whole surface of the chloride cell but through channels open to the exterior, then the only such channels visible on published electron micrographs are the intercellular junctions between chloride cells and accompanying cells, which are reported to be single-stranded or occasionally completely open junctions (Sardet *et al.*, 1979). The example given of an open junction appears to be about 10 nm across, suggesting that the surface area of the junctions must be of the order of 1% of the total surface area of the chloride cell. If in the case just considered the osmotic flow was through the junctions rather than the whole exposed surface of the chloride cell, they would need to have an osmotic permeability coefficient of around 4 cm sec^{-1}, which seems unreasonably high.

If water was indeed flowing at this rate between chloride cells, it would drag dissolved salts with it, and thus an osmotic gradient would not exist. It is in any case difficult to imagine how an osmotic gradient, due mainly to sodium and chloride ions, could be maintained across pores permeable to large molecules such as lanthanum ions. Bulk flow of solution may well occur between chloride cells, but it seems more reasonable to assume, in the absence of compelling evidence to the contrary, that osmotic fluxes occur at the site of the major diffusional fluxes of water, the secondary lamellae, leaving unanswered the question of how diffusional water fluxes measured with tritiated water (HTO) could possibly be greater than osmotic fluxes. There are plenty of reasons, in addition to the three given previously, for underestimation of diffusional water flux by HTO measurements. With all other radioactive tracers, isotope effects can be ignored, but HTO is about 11% heavier than H_2O and diffuses through water about 12% more slowly than H_2O^{18}, which, although it is the same molecular weight as HTO, should behave in a similar manner to H_2O^{16} because their principal moments of inertia are much lower than those of HTO (Wang *et al.*, 1953).

However, Mills (1973) calculated that HTO should diffuse through water only about 3% more slowly than H_2O, but this difference might well be exaggerated in diffusion through restricted spaces, as between membrane phospholipids or through narrow pores.

Loretz (1979) has shown (albeit by extrapolation from data obtained on much larger fish) that cardiac output could be a limiting factor in *in vivo* HTO efflux studies. It is sometimes assumed that if less than half the HTO is cleared from the blood on one passage through the gills, errors will be small (Chapter 1, this volume). To take a simplistic example, if all the lamellae are equally perfused and equally and uniformly permeable, and 20% of the HTO is lost in the first half of all the lamellae, then 20% of what is left, or 16% of the original amount, will be lost in the second half, making a loss of 36% rather than the 40% that it would have been had perfusion not been to some extent limiting. This 10% error is obviously small in relation to other errors, but if some lamellae are perfused more slowly than others (a distinct possibility in view of current ideas about recruitment), perfusion limitation might influence the result even if overall branchial blood or Ringer flow was high in relation to HTO clearance. If some lamellae are unperfused the effect on HTO flux will be much greater than an osmotic flux, because equilibration will be rapidly attained (Chapter 1, this volume).

Differential perfusion of lamellae could conceivably influence the results of kinetic studies. The finding of Isaia *et al.* (1978) that the main barrier to water diffusion is basal rather than apical is surprising (if the respiratory epithelial cells exchange water more readily with the external medium than with the blood, they might be expected to swell enormously on transfer of the fish to fresh water and shrink correspondingly on transfer to seawater—a phenomenon not reported to our knowledge), and it is contrary to the situation in other epithelia exposed to osmotic gradients, including amphibian skin, where the apical membranes are rate limiting (Andreoli and Schafer, 1976). If some lamellae are perfused more slowly than others (or if parts of lamellae are perfused more slowly, for example, the center of the lamellae in comparison with the marginal and basal channels), this will slow washout to the internal medium. In any case, perfusate will have a longer overall transit time the more distal a lamella is on a filament, and transit times may differ through different arches. Although in the experiments of Isaia *et al.* (1978), perfusate would probably only have taken about 12 to 18 sec on average to pass through the preparation, washout of tracer from the gill blood space might have contributed to the kinetics of the fast component observed in unloading experiments (half-time ~40 sec) or to the delay in appearance of counts in loading experiments, two parameters used to calculate the ratio between apical and basal permeabilities.

Methodological problems associated with in vitro preparations are thus likely to be as great as, if not greater than, those associated with in vivo measurements. If both sorts of techniques give results leading to similar conclusions, as in the case of catecholamines increasing water flux, this perhaps does not matter too much, but where they give opposite results (as in prolactin effects on water flux), one might be forgiven for concluding that the gill is too complex an organ to use even for qualitative studies with the techniques at our disposal.

The reason for dwelling at such great length on methods before considering the evidence for endocrine involvement in the regulation of water fluxes should now be apparent. By making the most pessimistic assessment of the methodological problems it is quite possible that the significance of some experiments has been underestimated; indeed it is to be hoped that this is the case, but the onus should be on experimenters to validate their techniques.

II. EFFECTS OF HORMONES ON GILL WATER FLUXES

Studies on hormonal control of water fluxes appear so far to have been confined mainly to pituitary and adrenal hormones, although in fish the adrenal consists of separate groups of chromaffin (catecholamine-producing) and interrenal (corticosteroid-producing) cells. Since secretion of the principal fish corticosteroid, cortisol, is under the control of pituitary adrenocorticotrophic hormone (ACTH), it is convenient to consider it together with the pituitary hormones.

A. Effects of Hypophysectomy

Hypophysectomized fish have a reduced rate of water turnover. This applies in freshwater (Lahlou and Sawyer, 1969; Lahlou and Giordan, 1970; Oduleye, 1975) and euryhaline (Potts and Fleming, 1970) teleosts and in marine elasmobranchs (Payan and Maetz, 1971). There have been several reports of reduced urine flow in hypophysectomized fish, but in the absence of data on weight changes and drinking rates this does not tell us anything about branchial osmotic permeability. Although urine flow is probably normally regulated to balance osmotic water influx (McVicar and Rankin, 1983; Rankin *et al.*, 1983), this regulation may be disrupted by experimental treatments (Lahlou and Giordan, 1970). Hypophysectomy reduced urine flow but not drinking rate (at least in comparison with controls; sham-hypophysectomized fish had a greatly increased drinking rate) in European eels (Gaitskell and Chester Jones, 1971); however, this does not necessarily imply a reduced osmotic permeability if the fish were increasing in weight, as might be expected in view of the increase in muscle water of hypophysectomized eels (Butler, 1966; Chan *et al.*, 1968).

B. Prolactin

Lahlou and Giordan (1970) measured weight changes and urine flow in goldfish, *Carassius auratus*, as well as drinking rate in a small sample, which appeared not to influence the results. They found that osmotic as well as diffusional permeability was decreased by hypophysectomy. Diffusional permeability could be maintained at or above control levels by injections of prolactin. Osmotic permeability of hypophysectomized fish could be maintained by prolactin, which also increased water turnover in intact goldfish.

Prolactin also restored the reduced water turnover following hypophysectomy in killifish, *Fundulus kansae*, in fresh water (Potts and Fleming, 1970), and prolactin restored, at least for the first few days after operation, the reduced water turnover rate of hypophysectomized dogfish, *Scyliorhinus canicula* (Payan and Maetz, 1971). In brown trout, *Salmo trutta*, in fresh water, prolactin restored the water turnover, which had been reduced by hypophysectomy (Oduleye, 1975). In these experiments there was no significant change in body water content and, as drinking rates and urine flows were measured, osmotic permeability could be calculated to have decreased following hypophysectomy, the ratio of P_{os} to P_{dif} remaining at approximately 1. As in other studies on teleosts the calcium concentration of the fresh water had an important influence on permeability; increasing the calcium concentration reduced both P_{os} and P_{dif} in hypophysectomized fish.

The picture that emerges thus far of the action of mammalian (usually ovine) prolactin is clear; it increases gill water fluxes. Experiments on incubated gills tell a very different story. Lam (1969) found that the maximum weight increase of threespine stickleback (*Gasterosteus aculeatus*, form *trachurus*) gills incubated in fresh water was reduced by injection of prolactin 30 min before the start of the experiment. Using a more realistic method of interpreting the results of similar experiments, in which allowance was made for the progressive reduction of internal osmolarity, gills from hypophysectomized goldfish were found to increase in weight at the same rate as those from control fish, although bovine prolactin, but not bovine growth hormone, reduced the weight increase of both groups equally (Ogawa *et al.*, 1973). Prolactin also decreased the rate of weight increase of gills of the Japanese eel, *Anguilla japonica*, and rainbow trout, *Salmo gairdneri*, incubated in fresh water (Ogawa, 1974, 1975). Hypophysectomy increased water influx in incubated Japanese eel gills; prolactin reduced the rate of weight increase in gills from control, sham-operated, and hypophysectomized fish (Ogawa, 1977).

Prolactin decreased osmotic influx or outflux in gills of the tilapia, *Oreochromis mossambicus* [formerly *Sarotherodon mossambicus* (formerly *Tilapia mossambica*)], incubated in hypoosmotic or hyperosmotic salines, although it had no effect on gills from fish kept in high-calcium water (Wendelaar Bonga and Ven der Meij, 1981). Since pituitary prolactin cell activity was greater in fish kept in high (hypo- or hyper-) osmotic gradients or low-calcium solutions, these authors suggested that the main osmoregulatory function of prolactin in fish was reducing water permeability of the gills in waters of low calcium or magnesium concentrations. However the low blood prolactin concentrations in the tilapia adapted to artificial seawater were not significantly changed by adaptation to calcium- or magnesium-free artificial seawater; only adaptation to hypo-osmotic media caused large increases, concentrations being seven times higher in fresh water than in artificial seawater (Nicoll *et al.*, 1981). There appears to be no correlation between blood prolactin concentrations and osmotic permeability of the gills, which is at its highest in fish adapted to isosmotic saline (Wendelaar Bonga and Van der Meij, 1981). The reduction of permeability of gills from hypophysectomized Japanese eels and the increase produced by prolactin only occur in media of low calcium content; no effects were observed in 1 mmol $liter^{-1}$ calcium chloride solution, which itself causes a rapid reduction in the rate of water influx (Ogasawara and Hirano, 1983).

A clear contradiction thus exists between in vivo and in vitro results. There could be several possible reasons for this. Apart from considerations such as whether injected doses produce plasma concentrations in the physiological range or what the duration of the action is (which apply equally to in vitro experiments), injection of a hormone into a whole animal could affect water fluxes indirectly by activation of other endocrine systems or by acting on some other aspect of the animal's physiology. Prolactin is essential for the survival of some freshwater fish, acting to stimulate sodium uptake and limit sodium loss, thus maintaining plasma osmolarity (Loretz and Bern, 1982), and the effects produced might conceivably affect water fluxes. Action on organs other than the gills might indirectly affect branchial water fluxes. In contrast, the way in which prolactin injections correct the decreased water fluxes produced by hypophysectomy tell us what the hormone is actually doing in the animal, whatever the mechanism. This is always assuming that mammalian prolactin—which, incidentally, is toxic in some fish (Gona, 1979), possibly by damaging the kidney (Gona, 1981)—acts in the same way as fish prolactin. It is by no means certain that this is the case, since ovine prolactin may produce effects in fish where endogenous prolactin does not

(Loretz and Bern, 1983), possibly by having, or containing impurities that have ACTH-like activity (Ball and Hawkins, 1976).

The same considerations apply to incubation experiments where prolactin was injected before removing the gills. It is possible, but unlikely, that prolactin can only exert its normal action in the presence of some other factor that is lacking in incubated gills. Since, although prolactin definitely does something to incubated gills, no one has proved that it actually does decrease osmotic permeability rather than influence one of the other factors discussed before, it would perhaps be wisest to mistrust all incubated gill results until someone can devise a preparation to measure effects on osmotic fluxes directly. It may be of interest to note that the question of whether prolactin increases or decreases water permeability in the Amphibia is also the subject of controversy (Brown and Brown, 1982).

C. ACTH and Cortisol

Although fewer data are available than for prolactin, at least results from in vivo and in vitro experiments are in agreement that cortisol increases branchial water fluxes. Cortisol secretion is controlled by ACTH released from the pituitary, so the effects of hypophysectomy outlined already will, in part, be due to reduced cortisol secretion. Adrenalectomy is a difficult and traumatic operation in the eel, the only fish in which it has been reported, involving the removal of parts of the anterior and posterior cardinal veins. It causes a reduction in urine flow and an increase in drinking rate in freshwater yellow European eels, with a greatly reduced difference between the two rates that could imply a reduced branchial osmotic entry (Gaitskell and Chester Jones, 1971). However, adrenalectomized silver European eels rapidly increase in weight in fresh water (Chan *et al.*, 1967), and on taking this into account gill water influx is probably similar to that of control fish.

Cortisol and ACTH produce similar increases in water turnover in intact goldfish, and cortisol reverses the reduced water turnover produced by hypophysectomy (Lahlou and Giordan, 1970). These workers found that cortisol, in contrast to prolactin, produced only small increases in osmotic flux in hypophysectomized goldfish. Injections of ACTH at least temporarily restored the reduced water turnover produced by hypophysectomy in the dogfish (Payan and Maetz, 1971). Cortisol increased the influx of water into Japanese eel gills incubated in fresh water (Ogawa, 1975), and into seawater mullet (*Chelon labrosus*) gills incubated in deionized water (but not those of mullet adapted for 7 to 30 days to fresh water; Gallis *et al.*, 1979).

D. Neurohypophysial Hormones

The lack of interest in neurohypophysial hormonal effects on gill water fluxes is surprising in view of their importance in the control of water balance in amphibia and higher vertebrates (Bentley, 1982), and their renal (Rankin *et al.*, 1983) and hemodynamic (Rankin and Maetz, 1971) actions in fish. Lahlou and Giordan (1970) found that arginine vasotocin (AVT) produced large reductions in water turnover in intact and hypophysectomized goldfish, isotocin being without effect in the latter group. Even in concentrations producing considerable vasoconstriction, AVT appears either to have little effect or to increase HTO influx in perfused eel gills (M. B. Bennett and J. C. Rankin, unpublished observations).

E. Catecholamines

Isaia (Chapter 1, this volume) has reviewed the actions of catecholamines on branchial water fluxes, so this section will not attempt to review all the literature but will discuss some of the findings from a different perspective and present some new data. It is clear that adrenaline and noradrenaline increase gill permeability to water, but the extent to which this might be an indirect result of their hemodynamic actions (Rankin and Maetz, 1971) has not been satisfactorily elucidated. Haywood *et al.* (1977) found that 10^{-6} mol liter^{-1} adrenaline increased HTO and urea effluxes by about 100% but perfusion flow by only 30% in a perfused trout head preparation. They also found that decreasing the perfusion pump output by 80% caused only a 45% decrease in water and a 38% decrease in urea effluxes, and this has been interpreted as showing "that epinephrine does not lead to an increase of the functional surface area by the intermediary of lamellar recruitment" (Isaia *et al.*, 1979). There is no reason to suppose that reduction in functional surface area should be directly proportional to reduction in perfusion, or that catecholamines should act either solely on surface area or solely on permeability.

It is clear that changes in fluxes of urea (used because it is a small molecule that diffuses readily across cell membranes) produced by catecholamines cannot be accounted for solely by increases in number of lamellae perfused. Urea efflux in perfused rainbow trout gills was increased by more than 300% by 10^{-6} mol liter^{-1} noradrenaline and by more than 400% by 10^{-5} mol liter^{-1} adrenaline (Bergman *et al.*, 1974). Only a small part of this increase could have been due to increased functional surface area, since Booth (1979) found that almost 60% of lamellae were perfused in resting rainbow trout and that this was increased by only 29% by adrenaline injection.

The concept of recruitment of lamellae is widely accepted but based on little direct quantitative evidence. In the anesthetized European eel under conditions of low blood pressure and flow, visual examination has shown that there are no unperfused lamellae, although, unless adrenaline is injected or the animal starts to recover from the anesthesia, erythrocyte flow is confined to the marginal and basal channels (Rankin, 1976). However, similar observations in the cod, *Gadus morhua,* have revealed unperfused proximal lamellae during anesthesia, which are recruited following adrenaline administration (S. Nilsson, personal communication).

Booth (1978, 1979) has provided the most direct measure of the number of lamellae perfused and has shown that the potential for increased oxygen or water transfer by lamellar recruitment is very small in relation to the sort of changes that have been observed (up to a 12.6-fold increase in oxygen uptake during exercise; Jones and Randall, 1978), even if the technique is 100% accurate. The accuracy depends on the complete exchange of dye between the injected and endogenous erythrocytes; otherwise, even if the marked erythrocytes were randomly mixed some lamellae would be expected to lack dyed cells. Booth relied on observations of the complete exchange of dye between cells *in vitro* and on the fact that all cells in the afferent branchial and filament arteries were stained. He gives no indication of the time taken for complete exchange of dye, and since the arteries were observed after thawing the tissues, the possibility that equilibration was completed after blood flow had ceased cannot be excluded. It seems at least a possibility that injection of labeled cells into what would possibly be laminar flow streams in the ventral aorta might not result in 100% efficient mixing in the short time before the lamellae are reached. The possibility that the greater pulse pressure following adrenaline injection could lead to more efficient mixing, although unlikely, cannot be ruled out on the evidence published. It seems safer to regard Booth's figures as minimum values for the number of lamellae perfused.

Sorenson and Fromm (1976) studied heat transfer across isolated perfused rainbow trout gills and found that 10^{-5} mol liter^{-1} adrenaline (a concentration that caused a 440% increase in urea efflux) caused only an 11% increase in heat transfer ($p > 0.05$), a parameter that might be expected to be influenced by functional surface area (although heat can be transmitted through all parts of the gill whether perfused or not).

Even if recruitment of additional lamellae could increase functional surface area (and recruitment of water channels must also be considered; Hughes, 1972), it does not necessarily follow that this would lead to an increase in osmotic water fluxes. Osmotic fluxes would presumably continue in unperfused unventilated lamellae until the osmotic gradients ran down, with consequent swelling or shrinkage of, and in all probability damage to, the cells. Indeed, it is perfusion with blood that carries away excess water entering freshwater gills (to be eliminated by the kidney) or excess salts entering seawater gills (to be eliminated by the chloride cells), which ensures the functional integrity of the respiratory epithelium. Davie and Daxboeck (1982) have suggested that lack of pulsatility in nonperfused lamellae may reduce water exchange across cell membranes, since the washout time for ethanol was greater in nonpulsatile gill perfusions than in pulsatile perfusions. However, in a series of experiments measuring HTO influx in isolated perfused eel gills (with vigorous external stirring), in which flow was changed from pulsatile to nonpulsatile and back, influx during the nonpulsatile periods was 1.006 ± 0.0053 (mean $\pm$ SEM, n = eight gills) of the mean influx before and after (J. C. Rankin and M. B. Bennett, unpublished observations).

It seems unlikely that changes in the number of lamellae perfused would have large effects on osmotic water flux, but they would have large effects on measured fluxes of HTO or other tracers. In efflux experiments HTO in the blood or perfusate would not reach unperfused lamellae, which would nevertheless contain water that would continue to exchange osmotically with the external medium, even if at a slower rate because of the lack of pulsatility. Similar considerations would apply to influx measurements. Equilibration of HTO would be rapid, initial fluxes being proportional to the roughly 55 mol liter^{-1} activity gradient for labeled water, compared to the 1 mol liter^{-1} or less activity gradient for unlabeled water, which is the driving force for osmotic flux. Lamellae presumably cannot remain unperfused indefinitely, so eventually the fluid contained in them would be mixed with the general circulation, causing a disproportionately large addition of volume compared to counts. Reduced perfusion of some lamellae would also greatly affect HTO fluxes because, as discussed previously, a large proportion of the HTO in the perfusate is exchanged in a transit of average duration. The ratio of P_{os} to P_{dif} would therefore be much greater than unity if significant nonperfusion of lamellae occurred. This is not observed.

In perfused eel gills HTO influx is comletely unaffected by large changes in perfusate flow rate (Table II), whereas catecholamines, acting through β adrenoceptors, produce large increases in HTO influx irrespective of their effects on pressure and flow (M. B. Bennett and J. C. Rankin, unpublished observations). The conclusion of Haywood *et al.* (1977) that catecholamines increase membrane permeability, not functional surface area, would thus appear to be correct. However, Jackson and Fromm (1981) found that a fourfold change in perfusion flow in the physiological range (about 0.1 to 0.4 ml sec^{-1} g^{-1}) through isolated rainbow trout gills caused HTO flux to

Table II

Effect of Changes in Perfusate Flow Rate on HTO Influx[a]

Perfusate flow rate (ml min^{-1})			Ratio of HTO influxes
P1	P2	P3	($2F2/F1 + F3$)
1.32	0.84	1.32	1.005
0.84	1.32	0.84	1.075
1.32	0.75	1.32	1.049
0.95	0.17	0.95	1.073
1.32	0.25	1.32	1.004

[a]Isolated eel gills were perfused with Ringer solution at constant flow from a peristaltic pump and bathed in fresh water containing HTO. Arterial outflow was collected and counted and the influx of HTO (dpm per unit time) was measured. In five gills a period of altered flow rate (P2) was interposed between two control periods (P1 and P3). Influx of HTO during the experimental period was expressed as a ratio of the mean (before and after) control influx.

increase by just under 40%. They explained this in terms of lamellar recruitment and redistribution of flow. Adrenaline (10^{-5} mol $liter^{-1}$) increased HTO fluxes by 225% in their preparation, but it is likely to decrease, rather than increase, functional surface area in a constant-flow perfusion, since in a constant flow-perfused catfish (*Ictalurus punctatus*) gill preparation, adrenaline lowers input pressure and increases red blood cell flow at the basal and midfilamental lamellae at the expense of the more distal lamellae (Holbert *et al.*, 1979).

F. Acetylcholine

Acetylcholine is not a hormone but it does affect gill water fluxes, and, although there is no proof that it acts directly on membrane permeability, Oduleye and Evans (1982) found that carbachol, acting through muscarinic receptors, increased HTO efflux from perfused toadfish (*Opsanus beta*) heads, as did adrenaline acting through β adrenoceptors. Since acetylcholine decreases the number of lamellae perfused in trout (Booth, 1979), it is difficult to see how it could increase functional surface area, raising the interesting possibility of nervous involvement in the control of branchial permeability.

III. DISCUSSION

Despite the technical problems that make quantification difficult, it is clear that hormones can modify branchial permeability to water. What is not so clear is the function of this regulation. It might be thought that gills that were permeable to oxygen but impermeable to water, as the skin of some fish appears to be, would have evolved. Fish that do not have to cope with large osmotic imbalances between the body fluids and the environment have rather permeable gills and skin (Payan and Maetz, 1970, 1971); less permeable integuments have evolved in fish that are exposed to osmotic water fluxes, but there seems to have been some limit to the extent to which this was possible.

A. Relationship between Oxygen Uptake and Water Fluxes

Probably there is a balance between the conflicting requirements for maximum permeability to oxygen and minimum permeability to water. A "trade-off" between oxygen uptake and the necessity to limit detrimental salt and water fluxes has often been assumed, but this has usually been thought of in terms of changes in functional surface area (the "lamellar recruitment" theory). In rainbow trout in fresh water, exercise is followed by increased urine production to eliminate the increased branchial osmotic entry of water (Wood and Randall, 1973), and urine flow increases with increasing oxygen consumption (Hofmann and Butler, 1979). Such changes could possibly be due to increased permeability following adrenaline release, maybe as an indirect effect of an action to increase permeability to oxygen (Chapter 1, this volume) in that, for the reasons outlined previously, changes in functional surface area may be of lesser importance than has generally been supposed. It is more difficult, however, to explain the apparent correlation between urine flow and spontaneous variations in respiratory rate observed in river lampreys, *Lampetra fluviatilis* (McVicar and Rankin, 1983), in terms other than of increased functional surface area, but the situation could well be quite different in cyclostomes, which are more distantly related to the jawed fishes than the latter are to the mammals. In fact one problem with research on fish physiology is that only a few species have been investigated using a variety of techniques. Even within the teleosts there is an enormous diversity, and attempts to draw general conclusions from a few observations on a single species are usually of doubtful validity. There are occasions when it would be very useful just to apply the same technique to several different species.

Under normoxic and resting conditions, oxygen transfer across rainbow trout gills has been shown to be primarily perfusion limited, but during heavy exercise or hypoxia it has been suggested that increase in diffusing capacity may be of great importance (Daxboeck *et al.*, 1982). This is where a catecholamine-mediated permeability increase may be involved, but this effect has only been observed in the presence of at least 10^{-6} mol liter^{-1} adrenaline or noradrenaline, and because this is the highest concentration ever reported in fish blood (Nakano and Tomlinson, 1967), the question of the physiological significance of such effects arises. Improvements in assay techniques have led to lower estimates for circulating catecholamine levels, and effects produced solely by high concentrations may only be of significance where there is sympathetic nervous involvement (see Chapter 3, Volume 10A, this series).

Confirmation of an effect of adrenaline on branchial permeability to oxygen (Chapter 1, this volume) is essential to a proper appreciation of its role in exercise in fish. Since catecholamines inhibit salt extrusion in marine fish (Chapter 5, this volume) and constrict the arteriovenous pathway that bathes the bases of the chloride cells (Girard and Payan, 1977), it is quite conceivable that for short periods of maximal activity osmoregulatory control is subordinated to the more immediate problem of oxygen uptake and supply to the tissues. It may be significant that osmotic influx in freshwater rainbow trout declined, following its initial increase, during prolonged activity (Wood and Randall, 1973); HTO influx in isolated perfused eel gills declines, following its initial increase, during prolonged exposure to catecholamines (M. B. Bennett and J. C. Rankin, unpublished observations).

B. Adaptation to Different Salinities

Another aspect of regulation of water permeability that has been studied is adaptation to either seawater or fresh water. Branchial water permeability is greater in freshwater fish, and it has been suggested that the metabolic costs of producing a dilute urine to balance branchial osmotic influx are small (Shuttleworth and Freeman, 1974). Marine fish may incur greater energy costs in replacing osmotic water loss, in that salts absorbed with the ingested seawater have to be eliminated by the chloride cells; thus, it is perhaps not surprising that they have evolved gills with lower water permeability than those of freshwater fish. It would be interesting to see if they were also less permeable to oxygen, which might not be too disadvantageous in view of the unlikelihood of encountering low oxygen partial pressures in the sea.

It is rather paradoxical that cortisol, often considered to be a "seawater hormone" (Hirano and Mayer-Gostan, 1978), increases water permeability, whereas prolactin, the "freshwater hormone" reduces it, at least in incubated gills. If in vivo studies are to be believed, prolactin increases water fluxes and thus may be involved in the increased water permeability of gills in freshwater fish. The advantages of the increased permeability, which appears to be due to some pituitary hormone (either prolactin or ACTH), are not evident, but it is probably of greater importance to the fish to reduce permeability to ions as much as possible, and this may be the main role of prolactin (for references, see Loretz and Bern, 1982).

C. Some Special Features of Cell Membranes in Relation to Fish Life

Isaia (Chapter 1, this volume) has demonstrated that cell membrane permeability determines the magnitude of the fluxes of water and other small molecules. Permeability varies with changes in environmental salinity and is influenced by several hormones, so it is worth considering the role of cell membranes in general and the special features of fish cell membranes.

The highly dynamic properties of biological membranes rely on the continuous interactions in time and space of their molecular components; lipids, proteins, and carbohydrates (Singer, 1974). The most important function of cellular membranes is to maintain constant the large difference in molecular composition between the interior of the cell and its environment.

The metabolic activities of cells control the interactions among membrane components and control the inflow and outflow of ions, metabolites, and water; when metabolic activities cease, the structure disappears, and all intracellular and extracellular components go to equilibrium without any discrimination, which is controlled only by structured membranes. The metabolic activities make possible the existence of phases adjacent to, but far from equilibrium with, each other, and dissipative metabolic structures make possible the controlled transfer of ions, metabolites, and water. The classical membrane model, which included a lipid bilayer interposed between two asymmetrical protein layers (Robertson, 198b), now takes into consideration the molecular composition of the membranes, in particular in relation to membrane fluidity in the fluid mosaic model (Singer and Nicholson, 1972). The idea of membrane fluidity is related to the fact that all the membrane constituents move freely within the structure; the molecular motion reflects the fluidity state (Frye and Edidin, 1970).

Lipids are particularly important in the determination of membrane fluidity. Indeed, increases in degree of fatty acid unsaturation and in the length of the phospholipid hydrocarbon chains, and decreases in the protein:lipid ratio and in the sphingomyelin:lectin ratio, all cause greatly increased membrane fluidity. The role of cholesterol is also relevant to the fluidity conditions of hydrocarbon chains, ensuring the fluidity of the lipid moieties at temperatures where lipids would usually be in the gel micellar form; cholesterol acts as a stabilizing agent (Chapman, 1973).

There is now good evidence that membrane activities, like operation of membrane-bound enzymes and permeability phenomena as well as antigenic responses (Kimelberg and Papahadjopoulos, 1972), are related to membrane lipid fluidity, which is dependent on molecular composition and physical factors like temperature (Kimelberg, 1977). The fact of temperature dependence of fluidity is of particular importance for fish life; poikilotherms change the saturation of their membrane lipids in relation to the environmental temperature in order to ensure successful adaptation, increasing the degree of unsaturation with decreasing temperature (Hazel and Prosser, 1974). Many organisms possess the ability to modulate their membrane fluidity; the phenomenon is termed homeoviscous adaptation (Cossins and Prosser, 1978). The temperature dependence pattern of membrane-bound enzymes, and consequently the Arrhenius plot, generally appears straight in poikilotherms (Kimelberg, 1977; Cossins and Prosser, 1978; Hazel and Prosser, 1974; Prosser and Nelson, 1981) in contrast to homeotherms, where a break appears, normally shifted in relation to temperature following thermal adaptation.

Further studies illustrate the relation between phospholipid metabolism and membrane fluidity in the transmission of biological signals (Axelrod and Hirata, 1980). It appears that the stimulation of methyltransferase activity inducing phospholipid methylation gives rise to an increase in membrane fluidity, which is particularly important for the interactions of ligands like catecholamines with plasma membranes (Loh and Law, 1980).

Red blood cell membranes of marine and freshwater fish present a very different lipid composition from those of terrestrial vertebrates in the sense that their fatty acids are highly (50%) unsaturated (Bolis and Luly, 1972) and also as far as their phospholipid distribution is concerned (Bolis and Fänge, 1979). This of course gives rise to a different molecular organization of the membranes in space and time, and as a consequence some of the transport phenomena, like transport of nonelectrolytes in erythrocytes [e.g. D-(+)-glucose], is different in fish in comparison with terrestrial vertebrates (Bolis, 1973; Bolis and Luly, 1972), and the action of some drugs on transport of nonelectrolytes is also different (Bolis *et al.*, 1982).

IV. CONCLUSIONS

The hormonal control of branchial permeability, although difficult to investigate, is of great interest both from the point of view of understanding how fish are able to live in a variety of environments and on the more general level of studying the evolution of osmoregulatory control mechanisms in the vertebrates. Fish gill membranes, because of their adaptability and the involvement of several different hormonal control mechanisms, would make excellent model systems for the study of the relationships between permeability and phospholipid fluidity, were it not for the practical problems involved.

Progress in understanding physiological mechanisms has often been the result of the development of a particularly convenient experimental preparation, and the problem with gills—whether studied *in vivo*, perfused *in situ* in head preparations, or perfused or incubated in isolation—is that it is very difficult to be sure that membrane permeability is being studied uninfluenced by hemodynamic or other factors. The development of the *Fundulus* opercular skin preparation (Chapter 5, this volume) has simplified studies on the mechanisms of ion transport in chloride cells by eliminating the need to work with a complex organ such as the gill. It is difficult to see how such simplification will be possible in the study of the permeability of the respiratory epithelium. Possibly study of gill cells grown in tissue culture (Naito and Ishikawa, 1980) could make a contribution.

REFERENCES

Andreoli, T. E., and Schafer, J. A. (1976). Mass transport across cell membranes: The effects of antidiuretic hormone on water and solute flows in epithelia. *Annu. Rev. Physiol.* **38**, 451–500.

Axelrod, J., and Hirata, F. (1980). Lipids and the transduction of biological signals through membranes. *Prog. Psychoneuroendocrinol.* **8**, 1–12.

Babiker, M. M. (1973). A comparative study of kidney function in fish with special reference to hormone action. Ph.D. Thesis, University of Wales.

Ball, J. N., and Hawkins, E. F. (1976). Adrenocortical (interrenal) responses to hypophysectomy and adenohypophysial hormones in the teleost *Poecilia latipinna*. *Gen. Comp. Endocrinol.* **28**, 59–70.

Bentley, P. J. (1982). "Comparative Vertebrate Endocrinology," 2nd ed. Cambridge Univ. Press, London and New York.

Bergman, H. L., Olson, K. R., and Fromm, P. O. (1974). The effects of vasoactive agents on the functional surface area of isolated-perfused gills of rainbow trout. *J. Comp. Physiol.* **94**, 267–286.

Bolis, L. (1973). Comparative transport of sugars across red blood cells. *In* "Comparative Physiology" (L. Bolis, K. Schmidt-Nielsen, and S. H. P. Maddrell, eds.), pp. 583–590. North-Holland Publ., Amsterdam.

Bolis, L., and Fänge, R. (1979). Lipid composition of the erythrocyte membrane of some marine fish. *Comp. Biochem. Physiol. B* **62B**, 345–348.

Bolis, L., and Luly, P. (1972). Membrane lipid pattern and nonelectrolyte permeability in *Salmo trutta* red blood cells. *Biomembranes* **3**, 357–362.

Bolis, L., Canciglia, P., and Trischita, F. (1982). The effect of chlorpromazine on monosaccharide transport in fish erythrocytes. *Pharmacol. Res. Commun.* **14**, 321–326.

Booth, J. H. (1978). The distribution of blood flow in the gills of fish: Application of a new technique to rainbow trout (Salmo gairdneri). *J. Exp. Biol.* **73**, 119–129.

Booth, J. H. (1979). The effects of oxygen supply, epinephrine and acetylcholine on the distribution of blood flow in trout gills. *J. Exp. Biol.* **83**, 31–39.

Brown, P. S., and Brown, C. (1982). Effects of hypophysectomy and prolactin on the water balance response of the newt, *Taricha torosa*. *Gen. Comp. Endocrinol.* **46,** 7–12.

Butler, D. G. (1966). Effect of hypophysectomy on osmoregulation in the European eel (*Anguilla anguilla* L.). *Comp. Biochem. Physiol.* **18,** 773–781.

Chan, D. K. O., Chester Jones, I., Henderson, I. W., and Rankin, J. C. (1967). Studies on the experimental alteration of water and electrolyte composition in the eel (*Anguilla anguilla* L.). *J. Endocrinol.* **37,** 297–317.

Chan, D. K. O., Chester Jones, I., and Mosley, W. (1968). Pituitary and adrenocortical factors in the control of the water and electrolyte composition of the freshwater European eel (*Anguilla anguilla* L.). *J. Endocrinol.* **42,** 91–98.

Chapman, D. (1973). Some recent studies of lipid, lipid-cholesterol and membrane systems. *Biol. Membr.* **2,** 91.

Cossins, A. R., and Prosser, C. L. (1978). Evolutionary adaptations of membranes to temperature. *Proc. Natl. Acad. Sci. U.S.A.* **75,** 2040–2043.

Davie, P. S., and Daxboeck, C. (1982). Effect of pulse pressure on fluid exchange between blood and tissues in trout gills. *Can. J. Zool.* **60,** 1000–1006.

Daxboeck, C., Davie, P. S., Perry, S. F., and Randall, D. J. (1982). Oxygen uptake in a spontaneously ventilating, blood-perfused trout preparation. *J. Exp. Biol.* **101,** 35–45.

Dunel, S., and Laurent, P. (1977). La vascularisation branchiale chez l'Anguille: action de l'acétylcholine et de l'adrénaline sur la répartition d'une résine polymerisable dans les differentes compartiments vasculaires. *C. R. Hebd. Seances Acad. Sci. Ser. D* **284,** 2011–2014.

Frye, L. D., and Edidin, M. (1970). The rapid intermixing of cell surface antigens after formation of mouse-human heterokaryons. *J. Cell Sci.* **7,** 319–335.

Gaitskell, R., and Chester Jones, I. (1971). Drinking rate and urine production in the European eel (*Anguilla anguilla* L.). *Gen. Comp. Endocrinol.* **16,** 478–483.

Gallis, J. L., Belloc, F., Lasserre, P., and Boisseau, J. (1979). Freshwater adaptation in the euryhaline teleost, *Chelon labrosus*. II. Effects of continuance of adaptation, cortisol treatment and environmental calcium on water influx in the isolated gill. *Gen. Comp. Endocrinol.* **38,** 11–20.

Girard, J.-P., and Payan, P. (1977). Kinetic analysis and partitioning of sodium and chloride influxes across the gills of sea water adapted trout. *J. Physiol.* (*London*) **267,** 519–536.

Gona, O. (1979). Toxic effects of mammalian prolactin on *Colisa lalia* and two other related teleostean fish. *Gen. Comp. Endocrinol.* **37,** 468–473.

Gona, O. (1981). Effects of prolactin on the kidney of a teleostean fish: Transmission electron microscopic observations. *Gen. Comp. Endocrinol.* **43,** 346–351.

Gray, I. E. (1954). Comparative study of the gill area of marine fishes. *Biol. Bull.* (*Woods Hole, Mass.*) **107,** 219–225.

Haywood, G. P., Isaia, J., and Maetz, J. (1977). Epinephrine effects on branchial water and urea flux in rainbow trout. *Am. J. Physiol.* **232,** R110–R115.

Hazel, J. R., and Prosser, C. L. (1974). Molecular mechanisms of temperature compensation in poikilotherms. *Physiol. Rev.* **54,** 620–677.

Hebert, S. C., and Andreoli, T. E. (1982). Water permeability of biological membranes. Lessons from antidiuretic hormone-responsive epithelia. *Biochim. Biophys. Acta* **650,** 267–280.

Hirano, T., and Mayer-Gostan, N. (1978). Endocrine control of osmoregulation in fish. *In* "Comparative Endocrinology" (P. J. Gaillard and H. H. Boer, eds.), pp. 209–212. Elsevier/North-Holland Biomedical Press, Amsterdam.

Hofmann, E. L., and Butler, D. G. (1979). The effect of increased metabolic rate on renal function in the rainbow trout, *Salmo gairdneri*. *J. Exp. Biol.* **82,** 11–23.

Holbert, P. W., Boland, E. J., and Olson, K. R. (1979). The effect of epinephrine and acetylcholine on the distribution of red cells within the gills of the channel catfish, *Ictalurus punctatus*. *J. Exp. Biol.* **79,** 135–146.

Hughes, G. M. (1972). Morphometrics of fish gills. *Respir. Physiol.* **14,** 1–25.

Isaia, J. (1982). Effects of environmental salinity on branchial permeability of rainbow trout, *Salmo gairdneri*. *J. Physiol.* (*London*) **326,** 297–307.

Isaia, J., and Hirano, T. (1975). Effect of environmental salinity change on osmotic permeability of the isolated gill of the eel, *Anguilla anguilla* L. *J. Physiol.* (*Paris*) **70,** 737–747.

Isaia, J., and Masoni, A. (1976). The effects of calcium and magnesium on water and ionic permeabilities in the sea water adapted eel, *Anguilla anguilla* L. *J. Comp. Physiol.* **109,** 221–233.

Isaia, J., Girard, J.-P., and Payan, P. (1978). Kinetic study of gill epithelium permeability to water diffusion in the fresh water trout *Salmo gairdneri*. Effect of adrenaline. *J. Membr. Biol.* **41,** 337–348.

Isaia, J., Payan, P., and Girard, J.-P. (1979). A study of the water permeability of the gills of freshwater- and seawater-adapted trout (*Salmo gairdneri*): Mode of action of epinephrine. *Physiol. Zool.* **52,** 269–279.

Jackson, W. F., and Fromm, P. O. (1981). Factors affecting ^3H_2O transfer capacity of isolated trout gills. *Am. J. Physiol.* **240,** R235–R245.

Jones, D. R., and Randall, D. J. (1978). The respiratory and circulatory systems during exercise. *In* "Fish Physiology" (W. S. Hoar and D. J. Randall, eds.), Vol. 7, pp. 425–501. Academic Press, New York.

Kamiya, M. (1967). Changes in ion and water transport in isolated gills of the cultured eel during the course of salt adaptation. *Annot. Zool. Jpn.* **40,** 123–129.

Kimelberg, H. K. (1977). The influence of membrane fluidity on the activity of membrane-bound enzymes. *In* "Dynamic Aspects of Cell Surface Organization" (G. Poste and G. L. Nicholson, eds.), pp. 205–239. North-Holland Publ., Amsterdam.

Kimelberg, H. K., and Papahadjopoulos, D. (1972). Phospholipid requirements for (Na^+ + K^+)-ATPase activity: Head-group specificity and fatty acid fluidity. *Biochim. Biophys. Acta* **282,** 277–292.

Kirsch, R. (1972). The kinetics of peripheral exchanges of water and electrolytes in the silver eel (*Anguilla anguilla* L.) in fresh water and in sea water. *J. Exp. Biol.* **57,** 489–512.

Kirsch, R., and Nonnotte, G. (1977). Cutaneous respiration in three freshwater teleosts. *Respir. Physiol.* **29,** 339–354.

Kirsch, R., Guinier, D., and Meens, R. (1975). L'équilibre hydrique de l'anguille européenne (*Anguilla anguilla* L.). Étude du rôle de l'oesophage dans l'utilisation de l'eau de boisson et étude de la permeabilité osmotique branchiale. *J. Physiol.* (*Paris*) **70,** 605–626.

Lahlou, B., and Giordan, A. (1970). Le controle hormonal des échanges et de la balance de l'eau chez le téléostéen d'eau douce *Carassius auratus*, intact et hypophysectomisé. *Gen. Comp. Endocrinol.* **14,** 491–509.

Lahlou, B., and Sawyer, W. H. (1969). Influence de l'hypophysectomie sur le rénouvellement de l'eau interne (étudié à l'aide de l'eau tritiée) chez le poisson rouge, *Carassius auratus*, L. *C. R. Hebd. Seances Acad. Sci., Ser. D* **268,** 725–728.

Lam, T. J. (1969). The effects of prolactin on osmotic influx of water in isolated gills of the marine threespine stickleback *Gasterosteus aculeatus* L. form *trachurus*. *Comp. Biochem. Physiol.* **31,** 909–913.

Laurent, P. (1982). Structure of vertebrate gills. *In* "Gills" (D. F. Houlihan, J. C. Rankin, and T. J. Shuttleworth, eds.), pp. 25–43. Cambridge Univ. Press, Cambridge and New York.

Loh, H. H., and Law, P. Y. (1980). The role of membrane lipids in receptor mechanisms. *Annu. Rev. Pharmacol. Toxicol.* **20,** 201–234.

Loretz, C. A. (1979). Water exchange across fish gills: The significance of tritiated-water flux measurements. *J. Exp. Biol.* **79,** 147–162.

Loretz, C. A., and Bern, H. A. (1982). Prolactin and osmoregulation in vertebrates. *Neuroendocrinology* **35,** 292–304.

Loretz, C. A., and Bern, H. A. (1983). Control of ion transport by *Gillichthys mirabilis* urinary bladder. *Am. J. Physiol.* **245,** R45–R52.

McVicar, A. J., and Rankin, J. C. (1983). Renal function in unanaesthetised river lampreys (*Lampetra fluviatilis* L.): Effects of anaesthesia, temperature and environmental salinity. *J. Exp. Biol.* **105,** 351–362.

Mills, R. (1973). Self-diffusion in normal and heavy water in the range 1–45°. *J. Phys. Chem.* **77,** 685–688.

Motais, R., Isaia, J., Rankin, J. C., and Maetz, J. (1969). Adaptive changes of the water permeability of the teleostean gill epithelium in relation to external salinity. *J. Exp. Biol.* **51,** 529–546.

Naito, N., and Ishikawa, H. (1980). Reconstruction of the gill from single-cell suspensions of the eel, *Anguilla japonica. Am. J. Physiol.* **238,** R165–R170.

Nakano, T., and Tomlinson, N. (1967). Catecholamine and carbohydrate concentrations in rainbow trout (*Salmo gairdneri*) in relation to physical disturbance. *J. Fish. Res. Board Can.* **24,** 1701–1715.

Naon, R., and Mayer-Gostan, N. (1983). Separation by velocity sedimentation of the gill epithelial cells and their ATPase activities in the seawater adapted eel *Anguilla anguilla* L. *Comp. Biochem. Biophys. A* **75A,** 541–547.

Nicoll, C. S., Wilson, S. W., Nishioka, R., and Bern, H. A. (1981). Blood and pituitary prolactin levels in tilapia (*Sarotherodon mossambicus;* Teleostei) from different salinities as measured by a homologous radioimmunoassay. *Gen. Comp. Endocrinol.* **44,** 365–373.

Nonnotte, G., and Kirsch, R. (1978). Cutaneous respiration in seven sea-water teleosts. *Respir. Physiol.* **35,** 111–118.

Oduleye, S. O. (1975). The effect of hypophysectomy and prolactin therapy on water balance of the brown trout, *Salmo trutta. J. Exp. Biol.* **63,** 357–366.

Oduleye, S. O., and Evans, D. H. (1982). The isolated perfused head of the toadfish, *Opsanus beta.* II. Effects of vasoactive drugs on unidirectional water flux. *J. Comp. Physiol.* **149,** 115–120.

Ogasawara, T., and Hirano, T. (1984). Effects of prolactin and environmental calcium on osmotic water permeability of the gills in the eel, *Anguilla japonica. Gen. Comp. Endocrinol.* **53,** 315–324.

Ogawa, M. (1974). The effects of bovine prolactin, sea water and environmental calcium on water influx in isolated gills of the euryhaline teleosts, *Anguilla japonica* and *Salmo gairdnerii. Comp. Biochem. Physiol. A* **49A,** 545–553.

Ogawa, M. (1975). The effects of prolactin, cortisol and calcium-free environment on water influx in isolated gills of Japanese eel, *Anguilla japonica. Comp. Biochem. Physiol. A* **52A,** 539–543.

Ogawa, M. (1977). The effect of hypophysectomy and prolactin treatment on the osmotic water influx into the isolated gills of the Japanese eel (*Anguilla japonica*). *Can. J. Zool.* **55,** 872–876.

Ogawa, M., Yagasaki, M., and Yamazaki, F. (1973). The effect of prolactin on water influx in isolated gills of the goldfish, *Carassius auratus* L. *Comp. Biochem. Physiol. A* **44A,** 1177–1183.

Parry, G. (1966). Osmotic adaptation of fishes. *Biol. Rev. Cambridge Philos. Soc.* **41,** 392–344.

Payan, P., and Maetz, J. (1970). Balance hydrique et minérale chez les élasmobranchs: Arguments en faveur d'un contrôle endocrinien. *Bull. Inf. Sci. Tech., Comm. Energ. At. (Fr.)* **146,** 77–96.

Payan, P., and Maetz, J. (1971). Balance hydrique chez les élasmobranches: Arguments en faveur d'un contrôle endocrinien. *Gen. Comp. Endocrinol.* **16,** 535–554.

Pic, P., Mayer-Gostan, N., and Maetz, J. (1974). Branchial effects of epinephrine in the seawater-adapted mullet. 1. Water permeability. *Am. J. Physiol.* **226,** 698–702.

Potts, W. T. W., and Fleming, W. R. (1970). The effects of prolactin and divalent ions on the permeability to water of *Fundulus kansae. J. Exp. Biol.* **53,** 317–327.

Prosser, C. L., and Nelson, D. O. (1981). The role of the nervous system in temperature adaptation of poikilotherms. *Annu. Rev. Physiol.* **43,** 281–300.

Rankin, J. C. (1976). Factors controlling pressure and permeability in gills. *Physiologist* **19,** 426.

Rankin, J. C., and Maetz, J. (1971). A perfused teleostean gill preparation: Vascular actions of neurohypophysial hormones and catecholamines. *J. Endocrinol.* **51,** 621–635.

Rankin, J. C., Henderson, I. W., and Brown, J. A. (1983). Osmoregulation and the control of kidney function. *In* "Control Processes in Fish Physiology" (J. C. Rankin, T. J. Pitcher, and R. Duggan, eds.), pp. 68–88. Croom Helm, London.

Robertson, J. D. (1981). Membrane structure. *J. Cell Biol.* **91,** 189s–204s.

Sardet, C., Pisam, M., and Maetz, J. (1979). The surface epithelium of teleostean fish gills. Cellular and junctional adaptations of the chloride cell in relation to salt adaptation. *J. Cell Biol.* **80,** 96–117.

Shuttleworth, T. J., and Freeman, R. F. H. (1974). Net fluxes of water in the isolated gills of *Anguilla dieffenbachii. J. Exp. Biol.* **60,** 769–781.

Singer, S. J. (1974). The molecular organization of membranes. *Annu. Rev. Biochem.* **43,** 805–833.

Singer, S. J., and Nicholson, G. C. (1972). The fluid mosaic model of the structure of cell membranes. *Science* **175,** 720–731.

Sorenson, P. R., and Fromm, P. O. (1976). Heat transfer characteristics of isolated-perfused gills of rainbow trout. *J. Comp. Physiol.* **112,** 345–357.

Steffensen, J. F., Lomholt, J. P., and Johansen, K. (1981). The relative importance of skin oxygen uptake in the naturally buried plaice, *Pleuronectes platessa,* exposed to graded hypoxia. *Respir. Physiol.* **44,** 269–276.

Ungell, A.-L., and Rankin, J. C. (1984). In preparation.

Wang, J. H., Robinson, C. V., and Edelman, I. S. (1953). Self-diffusion and structure of liquid water III Measurement of the self-diffusion of liquid water with H^2, H^3 and O^{18} as tracers. *J. Am. Chem. Soc.* **75,** 466–470.

Wendelaar Bonga, S. E., and Van der Meij, J. C. A. (1981). Effect of ambient osmolarity and calcium on prolactin cell activity and osmotic water permeability of the gills in the teleost *Sarotherodon mossambicus. Gen. Comp. Endocrinol.* **43,** 432–442.

Wood, C. M., and Randall, D. J. (1973). The influence of swimming activity on water balance in the rainbow trout (*Salmo gairdneri*). *J. Comp. Physiol.* **82,** 257–276.

7

METABOLISM OF THE FISH GILL

THOMAS P. MOMMSEN
Department of Biology
Dalhousie University
Halifax, Nova Scotia

I. INTRODUCTION

The branchial epithelium of fish possesses a wide variety of different physiological functions; the gills constitute the major organ for respiratory gas exchange, play important roles for ionic as well as osmotic balance, and are the main location for excretion of nitrogenous substances. All of these different processes are facilitated through the anatomic arrangement of the tissue, which includes its large surface area, its short diffusion distances, and its countercurrent arrangement of water and blood flow. In addition, these factors also result in the diffusional exchange of water across the gill.

FISH PHYSIOLOGY, VOL. XB

ISBN 0-12-350432-5

With the exception of the exchange of water, oxygen, and carbon dioxide, which seems to be due to diffusion (Isaia *et al.*, 1978; Loretz, 1979; Jackson and Fromm, 1981), all other processes just mentioned require a metabolic activity that—on a gram basis and in a resting animal—surpasses all remaining fish tissues. This high metabolic requirement is manifested in the substantial "internal" oxygen uptake by the gill filaments—uptake that is not in the realm of the respiratory function of the tissue (Johansen and Pettersson, 1981). The high oxidative throughput is also reflected in the abundance of mitochondria in specialized gill cells with defined functions, such as chloride cells (Shirai and Utida, 1970) and mucous cells (Morgan and Tovell, 1973). Obviously, the musculature of the gill filaments will consume oxygen, as will the pillar cells with their implied contractile properties (Pasztor and Kleerekoper, 1962; Vogel *et al.*, 1976; Bettex-Galland and Hughes, 1972).

Measured with two different techniques, internal oxygen uptake of gill tissue is estimated at around 100 μl O_2 (g gill weight)$^{-1}$ hr^{-1} at 15°C, which amounts to almost twice the oxygen uptake of the intact fish, on a gram basis (Johansen and Pettersson, 1981; Wood *et al.*, 1978). This means that gill tissue alone may sequester some 7% of the fish's total oxygen consumption for its own metabolism.

Although an extensive body of knowledge about the diverse physiological functions of the gill has been amassed, there is an obvious dearth of information about the biochemical performance of gill tissue. It is trivial to note that ultimately the driving force for most physiological performances is the hydrolysis of ATP, of some other form of high-energy phosphate, or a redox couple. The question remains by which metabolic pathway gill tissue furnishes the needed high-energy phosphates or the redox potential. The first part of this chapter will summarize the state of knolwedge of the metabolic support system for the gill. Later sections will cover some specific metabolic pathways that are analyzed in more detail.

II. EXOGENOUS SUBSTRATES

In apparent absence of an ample supply of endogenous oxidative substrate (see later), gill tissue will have to rely on blood-borne substrates to maintain its high metabolic throughput. Although electron micrographs of gill tissue always reveal the presence, or the alleged presence, of glycogen stores (Philpott and Copeland, 1963), biochemical analysis shows that these stores are rather limited at 5 μmol glycosyl Units (g tissue)$^{-1}$ (freshwater-adapted rainbow trout, T. P. Mommsen unpublished), a number that is dwarfed by the values commonly found in liver (170 μmol g^{-1}, rainbow trout, French *et al.*, 1981) or fast-twitch muscle (21 μmol g^{-1}, rainbow trout, Black *et al.*, 1962). Also during metabolic stress like hypoxia, change in ionic milieu (freshwater → marine), or saline perfusion of an isolated head preparation, the tissue holds on to its limited supply of glycogen (T. P. Mommsen, unpublished). Since the electron micrographs further provide evidence that there are no important deposits of oxidizable fat localized within the tissue, one is left with the conclusion that only exogenous substrates are available to maintain oxidative metabolism of the gill.

The four main substrates usually considered are glucose, lactate, amino acids, and fatty acids, all of which will be dealt with in the following paragraphs.

In general, the measurement of uptake of a substance by an organ through determining the concentration differences on either side of the organ is a well-established technique in physiological research. Problems with this technique arise, however, in organs when the blood flow is high (the gill receives the total cardiac output) and where the substrates under consideration occur in relatively high concentrations. Potential substrate utilization may therefore be concealed *in vivo* through a combination of these two effects or a general low metabolic turnover with respect to the particular substrate in question. The interpretation will be further complicated if the organ utilizes a mixture of several substrates, as is often the case.

A. Glucose

Although no comprehensive study on gill tissue enzymes or glucose requirement has been done yet, it is always presumed that the tissue cannot function without this particular substrate, since glucose is religiously added to all media utilized in the perfusion of isolated fish heads or isolated gill arches. Indeed this bias can be substantiated by analysis of gill enzyme activities. The presence of many enzymes of glycolysis has been verified for gill, including the activities of phosphofructokinase (PFK) and pyruvate kinase (PK), thought to be among the flux-limiting enzymes of glycolysis (Newsholme and Start, 1973). However, surprisingly little attention has been devoted to the activity, kinetics, and regulatory characteristics of hexokinase or glucokinase, the two ATP-dependent enzymes that catalyze the first step in the utilization of exogenous glucose. Glucokinase, a hexokinase with a high Michaelis–Menten constant (K_m) for D-glucose, seems to be absent from rainbow trout (*Salmo gairdneri*) or bowfin (*Amia calva*) gill (T. P. Mommsen, unpublished), while hexokinase occurs in relatively low activities when compared with other oxidative tissues like slow-twitch muscle or heart (Knox *et al.*, 1980; see Table II). Yet it is apparently active enough to pace a total glucose utilization of more than 0.10 μmol glucose (g gill wet weight)$^{-1}$ hr^{-1} (15°C) in tissue slices of rainbow trout gill (Table I). When analyzing glucose concentrations in blood or plasma across the gill, by sam-

Table I

Substrate Utilization in Tissue Slices from Rainbow Trout (*Salmo gairdneri*)[a,b]

Substrate	Utilization (μmol g^{-1} hr^{-1})
α-D-Glucose	0.10
L-Lactate	0.14
L-Alanine	0.07
Palmitate	0.39×10^{-3}

[a]From Mommsen (1983).

[b]Values are given in μmol substrate oxidized into CO_2 per gram of tissue wet weight per hour at 15°C and pH 7.4. Calculated from [^{14}C]CO_2 evolution data obtained from tissue slices incubated with the respective uniformly labeled substrate. Concentrations used were 2 m*M* for glucose, lactate, and alanine, and 0.25 m*M* for palmitate, respectively.

pling in ventral and dorsal aorta, no significant decrease can be detected (Soivio *et al.*, 1981). This observation is not surprising, since glucose concentrations in the ventral aorta of rainbow trout range from 4 to 5 m*M*, a concentration that would mask the rate of glucose utilization indicated earlier.

B. Lactate

In contrast to glucose, lactate has been identified unequivocally as an important oxidative substrate for gill by several authors; in each case a different aspect of lactate utilization was the focus of attention. Bilinski and Jonas (1972) used a radioisotope technique to verify gill as an organ that liberates [^{14}C]CO_2 from [3-^{14}C]lactate at higher rates than any other tissue of rainbow trout. Working on the same species, Driedzic and Kiceniuk (1976) noted a decrease in lactate across the gill after strenuous exercise in 8 of 11 fish analyzed. Soivio *et al.* (1981) confirmed this observation in resting, prehypoxic rainbow trout, although the latter authors also noticed that the small difference between ventral and dorsal aortic blood disappears after the onset of mildly hypoxic conditions.

These findings indicate that under stress conditions, in this case mild hypoxia, gill tissue replaces lactate with some other substrate, presumably one that does not pose the redox problems that can potentially arise from the utilization of lactate. In our own studies on the same subject, we were able to reproduce the results of Bilinski and Jonas (1972), yet we failed to detect any arteriovenous difference in lactate concentrations across the rainbow trout gill (T. P. Mommsen and S. F. Perry, unpublished), using a simultaneous sampling technique on noncannulated fish. Hochachka *et al.* (1979) reported on lactate utilization in cell-free gill preparations from two Amazon fish, but actual rates were not given.

On a molar basis, the oxidative utilization of lactate in tissue slices of rainbow trout gill surpasses the utilization of glucose and any other substrate analyzed (Table I). Apart from the unequivocal results of Bilinski and Jonas (1972) and the data given in Table I, the data on lactate *utilization* by the gill are ambiguous. The fact that the tissue takes up lactate from the venous blood in a few instances does not, taken on its own, necessarily only imply its position as an oxidative substrate. Lactate carbon, once accumulated in gill, could just as well be funneled into gluconeogenesis or other anabolic pathways.

Fish gill can also be identified as a lactate-consuming organ from the biochemical characteristics of the tissue lactate dehydrogenase (LDH). First, the gill isozymes show a close electrophoretic resemblance to the isozymes of fish heart (a lactate consumer), and second, they are biochemically designed to catalyze the reverse direction of lactate into pyruvate more efficiently than lactate dehydrogenases from lactate-producing tissues, such as fast-twitch muscle.

It can be concluded that lactate is actively taken up by gill, where it serves as an important oxidative substrate, at least under normoxic conditions, but the overall utilization of this substrate and its partitioning between different anabolic and catabolic pathways remains to be analyzed. The failure to detect lactate gradients across the gill in some cases is probably due to different experimental approaches and different background lactate concentrations. Since the gill receives the total cardiac output, it will obviously be difficult to detect a decrease in blood lactate if the lactate concentrations are high or if the lactate consumption under the specific experimental conditions is low.

C. Amino Acids

In the search for the elusive oxidative substrates for gill tissue, amino acids have unfortunately been largely neglected. A bulk of literature exists on the function of amino acids as potential substrates for the delivery of ammonia (i.e., the sum of ammonium ions and ammonia, the gas), but the ultimate fate of their carbon backbone remains unnoticed.

Payan and Pic (1977) showed for an isolated trout head that if the perfusing Ringer solution is supplemented with ammonia, no amino acid-derived ammonia is excreted across the rainbow trout gill. The authors also noted that only if ammonia-free perfusion medium is used does endogenous production of ammonia become significant. Although these experiments are conclusive, it is nevertheless difficult to extrapolate to "real life" for a fish and to give an in vivo estimate of the contribution of gill to ammonia excretion from endogenous sources (see later).

In the first case, where no endogenous ammonia is produced and excreted by gill tissue, transamination remains one of the means of funneling amino acid-derived carbon into ATP-yielding reactions; in the second case, both direct deamination through a dehydrogenase or transdeamination and transamination alone constitute possibilities to utilize amino acid carbon in energy production (cf. Fig. 1).

The enzymes required for the two pathways are present in gill in sufficient amounts (Walton and Cowey, 1977; McCorkle *et al.*, 1979; Table II), and our own studies reveal that at least alanine can serve as a reasonably good carbon source for rainbow trout, as judged from the appearance of $[^{14}C]CO_2$ from $[U\text{-}^{14}C]$alanine (Table I); no analysis was done to monitor the short- or long-term fate of the amino group of alanine in the tissue slice experiments. Under identical experimental conditions, the rate of CO_2 production from L-lactate was twice as high as the rate from L-alanine.

Although alanine is a poorer oxidative substrate for gill tissue than lactate or glucose, it cannot be ignored as a potentially important substrate. In contrast to lactate, blood alanine is not subjected to large fluctuations with the exercise status of the animal. Blood alanine concentrations show very

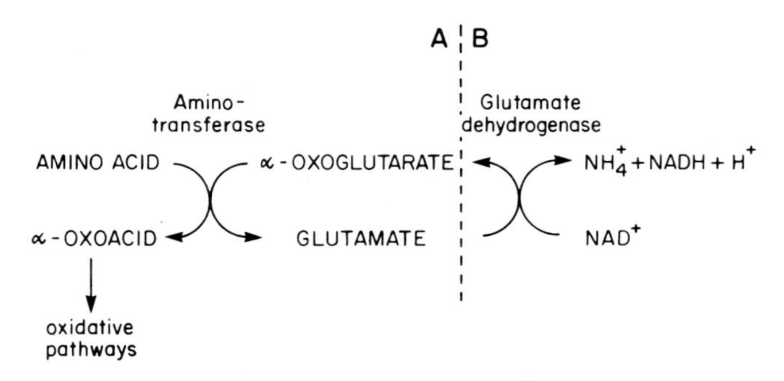

Fig. 1. Metabolism of amino acids. (A) Utilization of amino acids via transamination. Amino group acceptor is α-oxoglutarate, probably originating in the Krebs cycle. No ammonium/ammonia is generated, glutamate accumulates, and the resulting α-oxoacid feeds into oxidative pathways. (B) Glutamate dehydrogenase (GDH) reaction. A and B. Utilization of amino acids via the transdeamination pathway. α-Oxoglutarate is recycled from glutamate via the GDH reaction, and ammonium is liberated.

little variation, even during severe exercise or after starvation, whereas blood glucose or lactate levels or both may be affected. In fact, it has been suggested for dogfish (*Squalus acanthias*, Leech *et al.*, 1979) and sockeye salmon (*Oncorhynchus nerka*, Mommsen *et al.*, 1980) that alanine, which is exported from fast-twitch skeletal muscle during starvation, constitutes the major blood-borne carbon source for other tissues.

Unfortunately, other amino acids have not been tested with respect to their ability to provide carbon for oxidative metabolism. Obvious choices would be those nonessential amino acids that occur in fish blood in prominent concentrations: glycine and serine. As detailed later, in a different context, gill tissue is equipped with the enzymatic machinery to catalyze the transamination of a variety of amino acids (Fig. 1). Because in isolated fish heads supplied with ammonia in the perfusate no ammonia is generated from these amino acids, the interesting question arises of why gill tissue is fitted with all this enzyme machinery to interconvert amino acids. Evidently, there is an open field of study for a metabolic biochemist.

Of course, aspartate and alanine aminotransferases, the aminotransferases that attracted the bulk of the attention (McCorkle *et al.*, 1979; Hulbert *et al.*, 1978a,b), also possess important functions in other pathways. Fish gill is a very active tissue with high rates of protein synthesis (Haschemeyer and Smith, 1979; Haschemeyer *et al.*, 1979; Somero and Doyle, 1973), RNA and DNA turnover (Conte and Lin, 1967; Conte, 1977), and DNA synthesis, protein breakdown (Somero and Doyle, 1973), and, consequently, protein turnover (Tondeur and Sargent, 1979). This fact may help to explain the abundance of transaminases apart from their usually implied function in amino acid catabolism and their role in the production of ammonia through transdeamination. High protein turnover rates in this tissue are supported by the swiftness with which gill adapts to changes in the external environment, for example, the adjustment in number, orientation, and mitochondrial content of the chloride cells after a change from fresh water to seawater and vice versa (Shirai and Utida, 1970; Utida *et al.*, 1971). Yet another function of the transaminases is the integral part they perform in the malate–aspartate shuttle (see later, Fig. 3C).

D. Fatty Acids

With respect to the utilization of free fatty acids by the gill tissue, the information available is even more limited than in the case of glucose. The interesting article by Bilinski and Jonas (1972) on fatty acid oxidation in salmon tissues (*Oncorhynchus nerka*), however, did not include gill tissue. Our own results (Table I) reveal that the capacity of gill to oxidize palmitate is rather limited. For experimental reasons, the palmitate concentration chosen in the tissue slice experiments (0.25 mM) is substantially lower than the total amount of free fatty acids in fish blood (1–2 mM). Even if one assumes a linear increase in palmitate oxidation up to 1 mM, which approaches the value for K_m in mammals (Oram *et al.*, 1973), the utilization of palmitate in the rainbow trout lies well below the utilization rates for the other substrates listed in Table I. Overall, this conclusion is a little surprising for a tissue with free access to oxygen and a fair abundance of mitochondria. However, this observation is backed up by enzyme data that show no activity of β-ketoreductase in the crucian carp (Ekberg, 1962) and very low activity of hydroxyacyl-CoA dehydrogenase (EC 1.1.1.36) in bowfin (Table II) and rainbow trout gills (T. P. Mommsen, unpublished).

III. GILL LIPIDS

Turning the attention to lipid metabolism or lipid composition in the fish gill, one is once again faced with a rather limited supply of available information. Although electron micrographs reveal very moderate amounts of lipid deposits in fish gill (Morgan and Tovell, 1973), the tissue may possess a surprising capacity to synthesize lipids (Patton *et al.*, 1978). This preliminary suggestion by Patton *et al.* (1978) needs to be substantiated very carefully before general statements about lipid synthesis can be made, though. It is very possible that the relatively high short-term incorporation of label from [1-^{14}C]acetate measured in gill lipids of a catfish from the Amazon (*Phractophalus hemiliopterus*) merely indicates a high turnover of membrane lipids, together with the generally rapid turnover rates of gill proteins, RNA, and DNA. Suffice it to mention at this point that key dehydrogenases of the pentose phosphate pathway, which are NADP linked and therefore thought to be connected with fatty acid synthesis, occur in fish gill in 3 to 5 Units g^{-1} at 15°C (Hulbert *et al.*, 1978a,b; cf. Table II; T. P. Mommsen, unpublished) and thus surpass the rate of PFK—often thought to be the rate-limiting step in glycolysis (Newsholme and Start, 1973)—by 5- to 10-fold.

With respect to lipid composition, the teleost gill is not different from any other fish tissue: sphingomyelin, phosphatidylcholine, and phosphatidylethanolamine together amount to over 80% of the total phospholipids (Thomas and Patton, 1972; Phleger, 1978). A greater percentage of phosphatidylserine and sphingomyelin occurs in the presumed osmoregulatory organs of the European eel (*Anguilla anguilla*): gill, gut, and kidney. Generally, compounds in these organs were also labeled with ^{32}P more rapidly than in any other tissue of the eel, possibly indicating a rapid turnover of gill phospholipids rather than high synthesis rates for export of phospholipids. In addition, gill and gut of the eel showed greater turnover rates in phosphatidylethanolamine than of phosphatidylcholine (Zwingelstein *et al.*, 1975; Zwingelstein, 1979–1980).

The gill tissue of the eel contains appreciable levels of sulfolipids amounting to about 0.3 mol% of phospholipids. Biosynthesis of these *sulfolipids*—together with ^{35}S-binding proteins—is enhanced during seawater adaptation of the eels, in contrast to the *phospholipid* synthesis rates, which are unaltered during this adaptation. Sulfolipid is not, however, correlated with increased (Na^+,K^+)-ATPase activity in seawater-adapted eels (Zwingelstein *et al.*, 1980).

In parallel with other tissues in lower vertebrates and invertebrates, free fatty acid and lipid composition of fish gill adjusts during acclimation to different environmental temperatures (Anderson, 1970; Caldwell and Vernberg, 1970) and to changing external salinity (Daikoku *et al.*, 1982). The same applies to tissue enzymes (Caldwell, 1969). As mentioned elsewhere, the capacity of fish gill to utilize fatty acids for catabolic processes is rather limited, but again, further work is required to verify such a global statement.

IV. GILL ENZYMES

It is important to preface any discussion on enzyme activities in the fish gill with a recollection of the wide variety of different cell types and different cell compositions that have been described for this tissue. Therefore, the measurements of enzyme activities will merely reflect a relative potential of the gill as a whole to perform certain metabolic requirements. Necessarily, this approach will overestimate or underestimate the *real* potential for a specific cell type within the gill tissue.

Researchers interested in fish genetics look at enzymes in a completely different context, and they have contributed a wealth of information on gel electrophoretic patterns of tissue isozymes and allozymes. Some of these publications also deal with isozyme separations of gill tissue, for example for malate dehydrogenases (Starzyk and Merritt, 1980) or 6-phosphogluconate dehydrogenases (Bender and Ohno, 1968), to name but a few. In our context, however, these publications do not contribute to the understanding of gill biochemistry per se.

A. Glycolysis

In most gill studies to date, the activities of several enzymes of glycolysis have been assessed. Enzymes commonly measured include hexokinase (HK), phosphofructokinase (PFK), pyruvate kinase (PK), and lactate dehydrogenase

(LDH). In a few instances, PFK activities were measured together with fructose 1,6-bisphosphatase (F1,6BPase), and a wide range of ratios in their respective occurrences were determined (Hulbert *et al.*, 1978a,b; Knox *et al.*, 1980). The simultaneous functioning of two enzymes poised in opposite directions with the net result of ATP hydrolysis and the release of heat, has been termed "futile cycling." The enzyme pair PFK and F1,6BPase in glycolysis and gluconeogenesis, respectively, is a prime example of such a futile cycle (Katz and Rognstad, 1978), another example being the pair HK and glucose 6-phosphatase (Fig. 2). The results of the enzyme measurements for PFK and F1,6BPase in teleost gills are indicative of a certain potential for futile cycling between fructose 6-phosphate (F6P) and fructose 1,6-bisphosphate (F1,6BP), but this is already the point where this particular discussion stands: unresolved. For instance, no analysis was done on the effects of AMP on these enzymes, which generally have AMP as modulator: positive in the case of PFK and negative in the case of F1,6BPase. An attempt to put these activities into a physiological framework has yet to be supplied.

Renewed attention has been focused on futile cycling in glycolysis through the discovery of different naturally occurring most powerful effectors of PFK and F1,6BPase. The first is the modulation of enzymes in this particular futile cycle through the thiol:disulfide ratio (Pontremoli and Horecker, 1970; Gilbert, 1982), and it is even considered that this ratio may serve as a "third messenger" in response to cyclic adenosine monophosphate (cAMP) for key enzymes in metabolism *in vivo* (Gilbert, 1982).

The second effector is the compound fructose 2,6-bisphosphate (F2,6BP), which is reported to occur naturally in animal tissues, higher plants, and fungi (Pilkis *et al.*, 1982; Hers *et al.*, 1982), although the proof for its presence in fish tissues is still lacking. In rat tissues, F2,6BP is snythesized enzymatically from F6P and ATP (Fig. 2), where its synthesis is under stringent control by hormones and metabolites. In liver, its concentration lies in the micromolar range, and since turnover is not very fast, its synthesis does not cause any noticeable drain from the pool of glycolytic intermediates. The compound constitutes the most powerful inhibitor known for F1,6BPase and also functions as a potent stimulator for PFK; in addition, it may serve as a direct antagonist of cAMP. In the light of the sum of this added information, the potential for futile cycling of the F1,6BP–F6P cycle in gill tissue, the direction and rate of net flux through the system, and its regulation should be reexamined very carefully. The antagonistic function of F2,6BP with cAMP is one of the more startling observations for the gill-oriented physiologist because of the important physiological role that cAMP plays in the integration of gill function (Payan and Pic, 1977; Pic and Djabali, 1982; Djabali and Pic, 1982).

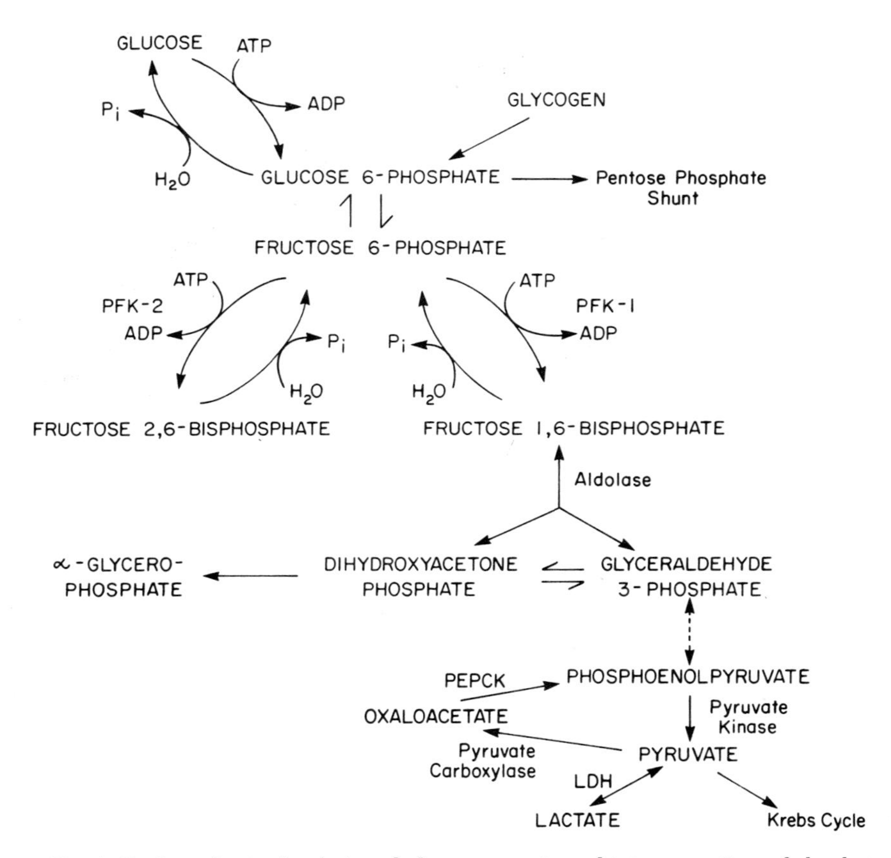

Fig. 2. Futile cycles in glycolysis and gluconeogenesis, and interconnections of glycolysis with other pathways. *Futile cycles:* (1) Substrate pair, glucose and glucose 6-phosphate. Enzymes involved are hexokinase, which cataylzes the glucose 6-phosphate-forming reaction, and glucose 6-phosphatase, which catalyzes the glucose-liberating reaction. (2) Substrate pair, fructose 6-phosphate and fructose 1,6-bisphosphate. Enzymes involved are phosphofructokinase-1 (PFK-1) for the formation of fructose 1,6-bisphosphate, and fructose 1,6-bisphosphatase to liberate fructose 6-phosphate. (3) Branching off from mainstream glycolysis and gluconeogenesis is the substrate pair fructose 6-phosphate and fructose 2,6-bisphosphate. Phospho-fructokinase-2 (PFK-2) is responsible for the enzymatic formation of fructose 2,6-bisphosphate, whereas fructose 2,6-bisphosphatase hydrolyzes it. *Connecting pathways:* (1) Glucose 6-phosphate may lead into the pentose phosphate shunt, which generates NADPH and pentoses. (2) From dihydroxyacetone phosphate (DHAP) comes the α-glycerophosphate dehydrogenase reaction for the formation of α-glycerophosphate and oxidation of cytoplasmic NADH (see Fig. 3A). (3) Pyruvate may be funneled into the Krebs cycle via the pyruvate dehydrogenase complex. (4). From pyruvate comes the bypass of reverse reaction of pyruvate kinase during gluconeogenesis. Phosphoenolpyruvate (PEP) is formed via oxaloacetate through reactions catalyzed by pyruvate carboxylase and phosphoenolpyruvate carboxykinase (PEPCK), respectively.

To give an overview of enzyme activities commonly encountered in gill tissue of fish, Table II presents the activities of a selection of enzymes from bluefin tuna and bowfin gill.

Phosphofructokinase is considered to be catalyzing one of the rate-determining steps in glycolysis, since its activity *in vitro* is similar to the maximum glycolytic rate—a statement that is valid at least for different types of muscles (Read *et al.*, 1977). The activity of PFK in fish gill tissue is generally quite low, being in the range of 0.3 to 0.5 Units g^{-1} (at 15°C) in rainbow trout, codfish (*Gadus morhua*), and flounder (*Pleuronectes platessa*) (Knox *et al.*, 1980), and even lower in the bowfin (Table II). Activities of PK in the same four species of fish are moderate at 7.5 to 24.4 Units g^{-1} (Knox *et al.*, 1980; Table II), whereas the value determined for the bluefin tuna (Table II) may point to an outstanding flux—among the fishes— through glycolysis in this warm-blooded species. The same conclusion can be drawn from the

Table II

Activities of Selected Enzymes in Bluefin (*Thunnus thynnus thynnus*) and Bowfin (*Amia calva*) Gill Tissue[a]

Enzyme[b]	Bluefin (Units/g fresh weight)[c]	Bowfin
Alanine AT (EC 2.6.1.2)	3.7	4.1
Aspartate AT (EC 2.6.1.1)	32.0	8.9
Hexokinase (EC 2.7.1.1)	1.06	0.27
Phosphofructokinase (EC 2.7.1.11)	n.d.[g]	0.19
Pyruvate kinase (EC 2.7.1.40)	103.9	24.4
Lactate DH (EC 1.1.1.27)[d]	26.0	13.2
α-Glycerophosphate DH (EC 1.1.1.8)[e]	n.d.	0.16
Malate DH (EC 1.1.1.37)[d]	46.0	23.0
Citrate synthase (EC 4.1.3.7)[f]	5.55	2.4
Glucose 6-phosphate DH (EC 1.1.1.49)	8.73	4.2
6-Phosphogluconate DH (EC 1.1.1.44)	4.36	n.d.
Hydroxyacyl-coenzyme A DH (EC 1.1.1.36)	n.d.	0.10

[a]From T. P. Mommsen and J. C. Coghlan (unpublished).

[b]All enzymes were determined spectrophotometrically under saturating conditions. AT, aminotransferase; DH, dehydrogenase.

[c]Values given in μmol substrate utilized per minute and per gram fresh tissue at 25°C (bluefin tuna, $n = 4$) and 15°C (bowfin, $n = 3$), respectively, and pH 7.0. Maximum SD, 32%.

[d]Determined in the respective NADH-consuming directions. Buffers used were 50 mM imidazole and 50 mM Tris-HCl (pH 8.1), respectively.

[e]Measured at pH 7.80.

[f]Measured at pH 8.10.

[g]n.d., not determined.

analysis of hexokinase activities, since the tuna enzyme surpasses the activities of the three cold-bodied fishes analyzed by Knox *et al.* (1980) and the bowfin by more than threefold. Generally it seems that compared with other fish tissues such as heart, slow- and fast-twitch muscle, or even kidney or brain, glycolytic capacity in gill is rather limited, with the notable exception of the warm-blooded bluefin tuna. At this point it should be noted again that enzymes catalyzing rate-limiting reactions are subject to stringent control and that activities measured under optimum conditions do not necessarily reflect activities *in vivo* (cf. Read *et al.*, 1977).

Lactate dehydrogenase is usually considered to be the redox-regulating dehydrogenase during hypoxic or anoxic periods, and generally under conditions where glycolytic flux surpasses the cell's ability to balance cytosolic redox through reoxidation of NADH within the mitochondrion. In highly glycolytic tissues such as fast-twitch fish muscle, lactate dehydrogenases are characterized, without exception, by high ratios of maximum rates in the lactate-producing over the lactate-consuming direction. Furthermore, these enzymes reveal only minor substrate inhibition through pyruvate. Fish gill LDH, however, possesses a differing biochemical setup. Judging from its kinetic and electrophoretic behavior, it can be concluded that the fish gill enzyme, which is abundant but not overly active (Table II), is very similar to the heart-type isozyme. This is consistent with the contention that lactate serves as a major carbon and energy source for gill, and it also helps to explain the observation that utilization of lactate is influenced by the redox state of the animal (Soivio *et al.*, 1981). The electrophoretic studies by Lim *et al.* (1975) and Williscroft and Tsuyuki (1970) clearly identify gill LDH as identical or very closely related to the heart type. The kinetic analyses by Gaudet *et al.* (1975), Hulbert *et al.* (1978a,b), and T. P. Mommsen (unpublished) on different fish species can be summarized as follows:

1. Gill LDHs are strongly inhibited by low (less than 1 m*M*) pyruvate concentrations.
2. They exhibit low Michaelis–Menten constants for L-lactate and pyruvate.
3. They are characterized by a low ratio of $V_{max\ forward}$ (lactate producing) over $V_{max\ reverse}$ (lactate consuming).

As would be expected for a lactate-consuming organ, the catalytic rate of bluefin and bowfin gill LDHs in the lactate-forming direction is rather low (Table II); this observation is especially striking when contrasted to over 2000 Units g^{-1} commonly found in fast-twitch muscles of skipjack tuna (*Katsuwonus pelamis*, Guppy *et al.*, 1979) or sockeye salmon (*Oncorhynchus nerka*, Mommsen *et al.*, 1980).

From the conclusion that fish gill denotes a lactate-consuming tissue, another complication arises. Unless cytosolic lipid synthesis balances the aerobic metabolism of lactate and all other cytosolic H_2 production—which is most unlikely—oxidative substrate utilization requires mechanisms of transferring reducing equilvalents bound to NAD from the cytosol into the mitochondrion—that is, hydrogen shuttle.

B. Hydrogen Shuttles

The three most common transport systems for the coordination of mitochondrial and cytoplasmic redox balance are the α-glycerophosphate (α-GP) shuttle (Fig. 3A), the malate–citrate shuttle (Fig. 3B), and the malate–aspartate shuttle (Fig. 3C).

Mainstream glycolysis proceeds from F1,6BP through the aldolase reaction into glyceraldehyde 3-phosphate (GAP) and dihydroxyacetone phosphate (DHAP; Fig. 2), where the former is subsequently oxidized into phosphoglyceric acid. However, many tissues maintain appreciable activities of α-glycerophosphate dehydrogenase (α-GPDH), which catalyzes the reduction of dihydroxyacetone phosphate into α-glycerophosphate. This enzyme plays an important role in the α-glycerophosphate shuttle (Fig. 3A) or may function to supply α-glycerophosphate for triglyceride synthesis. Although the second function poses an interesting alternative, only the first has been analyzed for gill tissue. The results of studies on several fish (Hulbert *et al.*, 1978a,b; Hochachka *et al.*, 1979; Table II) can be summarized as follows:

1. α-Glycerophosphate dehydrogenase is present in all gills analyzed.
2. Its activity, at less than 1 Unit g^{-1}, is low compared with other fish tissues.

Fig. 3. Hydrogen shuttles for the mitochondrial oxidation of cytosolic NADH. (A) α-Glycerophosphate shuttle. Cytosolic dihydroxyacetone phosphate (DHAP)—originating, for instance, from glycolysis (glyceraldehyde 3-phosphate, GAP)—is reduced with NADH as coenzyme by a cytosolic α-glycerophosphate dehydrogenase into α-glycerophosphate (α-GP). Mitochondrial α-GP is reoxidized with FAD as coenzyme into DHAP by mitochondrial α-glycerophosphate dehydrogenase. The mitochondrial dehydrogenase is located in the inner membrane but oriented toward the outside so that the substrate does not have to enter the matrix. (B) Malate–citrate shuttle. Cytosolic citrate lyase catalyzes the reaction from citrate into oxaloacetate and acetyl-coenzyme A (acetyl-CoA). Oxaloacetate is reduced with NADH to malate in the malate dehydrogenase (MDH) reaction. Mitochondrial malate is reoxidized to oxaloacetate through mitochondrial MDH; citrate is synthesized from acetyl-CoA and oxaloacetate through citrate synthase catalysis. This hydrogen shuttle employes a citrate–malate antiport and also serves to generate cytosolic acetyl-CoA. (C) Malate–aspartate shuttle. The shuttle requires the presence of cytosolic and mitochondrial forms of malate dehydrogenase and aspartate aminotransferase as well as two antiports for malate–oxoglutarate and glutamate–aspartate, respectively.

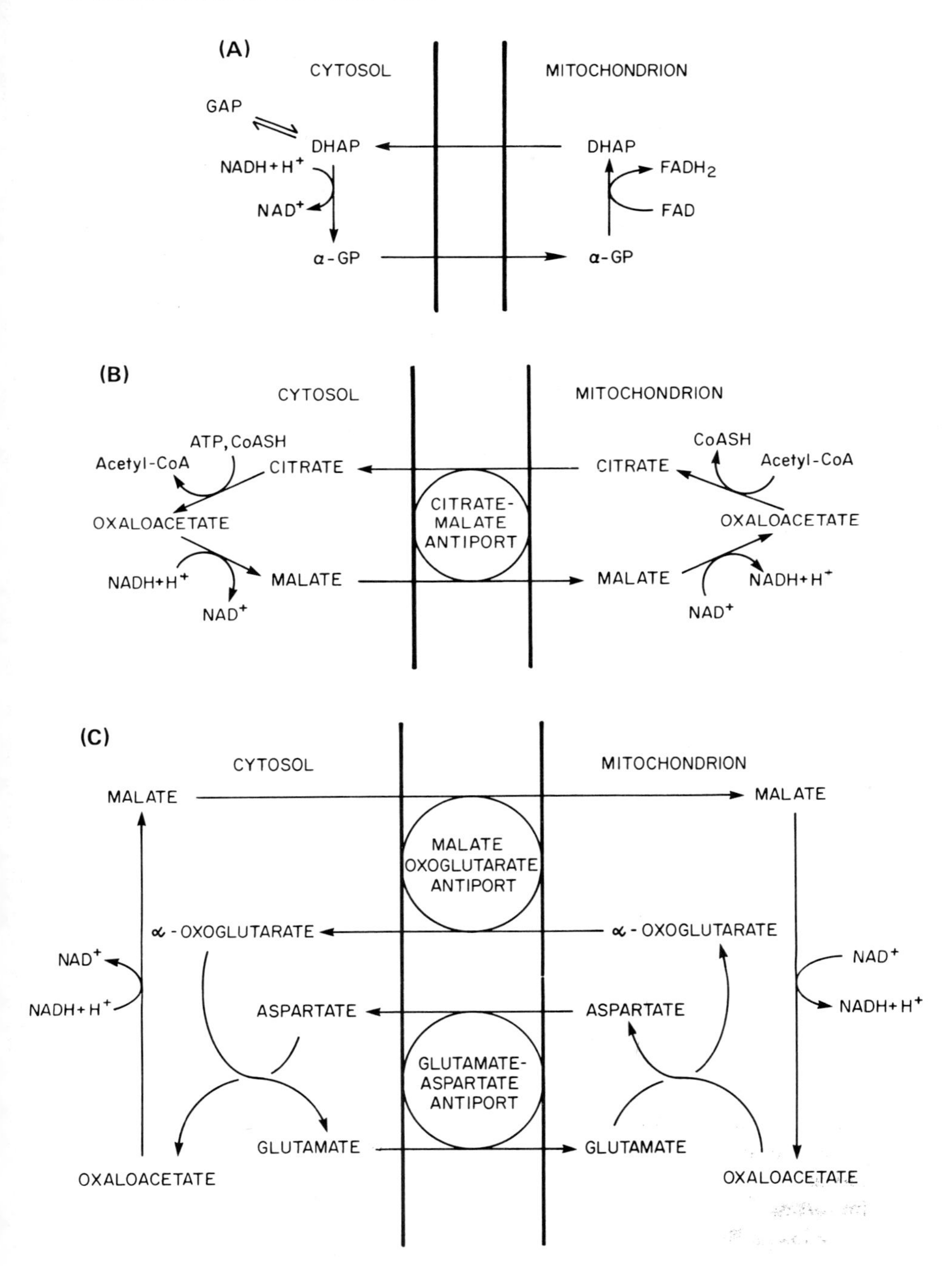
(A)
CYTOSOL
MITOCHONDRION
GAP
DHAP
NADH+H+
NAD+
α-GP
DHAP
FADH2
FAD
α-GP
(B)
CYTOSOL
MITOCHONDRION
ATP, CoASH
Acetyl-CoA
CITRATE
OXALOACETATE
NADH+H+
NAD+
MALATE
CITRATE-
MALATE
ANTIPORT
CoASH
Acetyl-CoA
CITRATE
OXALOACETATE
MALATE
NADH+H+
NAD+
(C)
CYTOSOL
MITOCHONDRION
MALATE
MALATE
MALATE
OXOGLUTARATE
ANTIPORT
α-OXOGLUTARATE
α-OXOGLUTARATE
NAD+
NADH+H+
ASPARTATE
ASPARTATE
GLUTAMATE-
ASPARTATE
ANTIPORT
GLUTAMATE
GLUTAMATE
OXALOACETATE
OXALOACETATE
NAD+
NADH+H+

3. The α-glycerophosphate cycle plays a role secondary to the malate–aspartate shuttle.

The notable exception to this generalization is the gill tissue of *Osteoglossum bicirrhosum*, where a functional α-GP cycle could be demonstrated (Hochachka *et al.*, 1979). In view of the low activity of α-GPDH in that particular fish, this is a rather puzzling observation, possibly indicating that the common "optimum" assay for the enzyme tends to underestimate its in vivo activity.

The well-understood malate–aspartate shuttle transfers reducing equivalents between the cytosol and the mitochondrion by employing two different antiport systems for malate–2-oxoglutarate and glutamate–aspartate, as well as cytosolic and mitochondrial forms of aspartate aminotransferase and malate dehydrogenase (Fig. 3C). In contrast to the α-GP cycle, the malate–aspartate shuttle is energetically more efficient, since it furnishes NADH inside the mitochondrion instead of $FADH_2$. The shuttle is active in subcellular preparations from gills of *Arapaima gigas*, and substantial activities of aspartate aminotransferase and malate dehydrogenase are readily demonstrable in gill tissue from a variety of fish, including bluefin tuna, bowfin (Table II), and Amazon fish (Hulbert *et al.*, 1978a,b). As long as no analysis with respect to the cell compartment distribution of these enzymes is available, the mere presence of required enzymes—albeit in high activities—supplies only circumstantial evidence for the operation of the shuttle.

With respect to the importance or presence of the malate–citrate shuttle in the fish gill, no conclusions can be drawn, as long as crucial information on specific enzymatic reactions is still lacking. Enzymes involved in this shuttling system are the ubiquitous cytosolic and mitochondrial forms of malate dehydrogenase, and the abundant citrate synthase (see Table II), as well as an ATP-dependent cytosolic citrate lyase (Fig. 3B). In addition to transferring reducing equivalents, this shuttle provides intramitochondrially generated acetyl-coenzyme A (without actual transfer of coenzyme A) to the cytoplasm. Therefore, the alleged presence of this shuttle system ties in nicely with the observed rapid synthesis of fatty acids (Patton *et al.*, 1978) and phospholipids in the gill (Zwingelstein, 1979–1980).

It has to be concluded that a lot more work is needed to allow generalizations regarding which shuttling systems prevail in the fish gill and regarding the importance of shuttling systems to gill oxidative metabolism per se. The malate–citrate shuttle definitely deserves more attention because of its double function of supplying acetyl-coenzyme A to the cytosol, which may be important for gill, as well as supplying reducing equivalents to the mitochondrion.

C. Citric Acid Cycle

In all oxidative animal tissues, the Krebs cycle plays a pivotal role in supplying intramitochondrially generated NADH and $FADH_2$ to the electron transfer system. Although a certain percentage of "reducing power" is supplied from the cytosol through hydrogen shuttle systems, the Krebs cycle furnishes the bulk of reducing power through its cyclic and catalytic breakdown of acetyl-coenzyme A. In the catabolism of carbohydrates and some amino acids, the pyruvate dehydrogenase complex (Fig. 2) delivers pyruvate-derived carbons (acetyl ~) to the citrate synthase (CS) in the mitochondrion. Since no information is available on the activity or metabolic regulation of this important regulatory enzyme in fish gill, the following discussion will be restricted to the Krebs cycle per se.

Regulation of flux through the Krebs cycle is achieved through kinetic control of those enzymes that *in vitro* are associated with large negative changes in free energy: citrate synthase, isocitrate dehydrogenase (ICDH), and 2-oxoglutarate dehydrogenase. With respect to the "spinning rate" of the cycle in vivo, a great deal of information can be gained from *in vitro* measurements of the activity of the 2-oxoglutarate dehydrogenase complex, which, of all the enzymes just mentioned, seems to be most closely correlated with actual fluxes through the cycle (Cooney *et al.*, 1981). To date, this enzyme has not been discussed in the literature on fish gill enzymes. Therefore, the ensuing paragraphs will deal only with citrate synthase, isocitrate dehydrogenase, and two nonregulatory enzymes of the cycle that catalyze equilibrium reactions, namely succinate dehydrogenase and malate dehydrogenase.

Surprisingly for a tissue rich in mitochondria, the maximum *in vitro* activity of gill citrate synthase is rather low, being in the range of 0.4 to 2.2 Units g^{-1} (at 25°C) for two erythrinid and two osteoglossid fish species from the Amazon (Hulbert *et al.*, 1978a,b). Gill CS activity in the warm-bodied bluefin tuna (5.55 Units g^{-1}, Table II) is not very impressive either; in fact, its activity is even lower than in skipjack (*Katsuwonus pelamis*) white muscle (7.0 Units g^{-1}), a tissue that is known neither for its abundance of mitochondria nor for its high aerobic capacity, but rather for its outstanding anaerobic burst potential, characterized by more than 2000 Units g^{-1} of lactate dehydrogenase (Guppy *et al.*, 1979).

The next enzyme in the Krebs cycle, aconitase, an equilibrium enzyme, has not been measured for the gills yet, but it is worth mentioning that aconitase activity could not be detected in liver mitochondria of the American eel, *Anguilla rostrata* (Moon and Ouellet, 1979). This observation already points to the possibility that the Krebs cycle, at least in that particular fish tissue, possesses vital cytoplasmic components. Subsequent to aconitase, which catalyzes the conversion of citrate into isocitrate, isocitrate dehydrogenase oxidizes the latter to produce 2-oxoglutarate. In the case of ICDH, interpretation is normally complicated through the simultaneous occurrence of NAD^+- and $NADP^+$-linked enzymes in the same cell, but in different compartments. It is considered that NAD^+-linked ICDH is operative in the Krebs cycle in the mitochondrion, whereas the (cytoplasmic) $NADP^+$-linked form has often been implicated as cytoplasmic source of NADPH for fatty acid and sterol synthesis. However, this scheme does not seem to apply to fish tissues without some major revisions; for comparison, one has to draw heavily from enzyme data on liver.

In isolated mitochondria from eel (*Anguilla rostrata*) liver, NAD^+-linked ICDH is not detectable (Moon and Ouellet, 1977), a finding supported through the unusual presence of mitochondrial $NADP^+$-linked ICDH; in fact, the activity of the mitochondrial enzyme exceeds the activity in the cytosol by threefold. The apparent lack of NAD^+-linked ICDH, the failure to detect mitochondrial aconitase, and the presence of an active mitochondrial NADPH/NAD^+ transhydrogenase strongly suggest that $NADP^+$-linked ICDH takes the place of the NAD^+-linked enzyme in the eel's hepatic Krebs cycle. This conclusion can be extended to rainbow trout, where NAD^+-linked ICDH is also absent from liver (Moon and Hochachka, 1971) and hardly detectable in slow-twitch muscle (Alp *et al.*, 1976), and to dogfish where heart and slow-twitch muscle show very low activities of this enzyme (Alp *et al.*, 1976) Obviously, further work is needed to allow extrapolation to gill or any generalization such as the novel metabolic scheme involving isocitrate as proposed by Moon and Ouellet (1979). As long as this highly interesting topic is unsettled, suffice it to mention that the gill tissue activities of $NADP^+$-linked ICDH are moderate at 2 to 4 Units g^{-1} (at 30°C) for perch, cod, and plaice (Skorkkowski *et al.*, 1980).

Malate dehydrogenase, in contrast, is definitely not a rate-determining enzyme in the Krebs cycle. It is generally very abundant and active in all fish tissues, including gill (Hulbert *et al.*, 1978a,b; Table II). The enzyme occurs in cytoplasmic and mitochondrial forms, which can be distinguished through kinetic and inhibitor studies or by analysis of isolated mitochondria; the relative distribution between the compartments of gill, however, remains to be analyzed. Both forms play crucial roles in the malate–aspartate and malate–citrate shuttles (cf. Fig. 3B and C) and in gluconeogenesis.

Not surprisingly for mitochondria-rich cell types, the occurrence of succinate dehydrogenase—an equilibrium enzyme—is associated with the chloride cells (Conte and Tripp, 1970). In fact, its activity is positively correlated with the increase in number of chloride cells (Shirai and Utida, 1970; Utida *et al.*, 1971) and (Na^+,K^+)-activated ATPase during saltwater adaptation in the eel (Sargent *et al.*, 1975), although earlier reports on the same subject failed to detect such a correlation (Epstein *et al.*, 1967; Conte, 1969). Succinate dehydrogenase activity in rainbow trout gill epithelium is associated with the microsomal fraction, which also reveals chloride- and bicarbonate-dependent ATPase activity (Bornancin *et al.*, 1980).

While the rate-determining step in fish gill Krebs cycle still awaits identification and a few surprises as to the actual compartmentation of aconitase and ICDH are likely, it can be concluded that the oxidative capacity of gill tissue as a whole is moderate at best. The work of Sargent *et al.* (1975), Kamiya (1972), and Naito and Ishikawa (1980) on enriched chloride cell preparations denotes the first attempts to overcome the obvious drawbacks of the "whole-gill tissue" approach, which necessarily averages over a variety of cells and, in vitro, waters down the realization of the full oxidative potential of the mitochondria-rich cells.

D. Gluconeogenesis

The biosynthetic capacity for gluconeogenesis of rainbow trout gill seems to be rather limited if present at all. This notion is supported through a variety of independent observations:

1. Bilinski and Jonas (1972) measured a low ratio between CO_2 formation from [1-^{14}C]lactate over [3-^{14}C]lactate; as a matter of fact, gill had the lowest ratio of all six rainbow trout tissues examined.
2. Knox *et al.* (1980) failed to detect phosphoenolpyruvate carboxykinase in gill, although fructose 1,6-bisphosphatase was present in low amounts. Furthermore, the activity of pyruvate carboxylase (Fig. 2) was hardly detectable in this tissue (Cowey *et al.*, 1977).
3. Lactate or alanine carbons could of course be shunted back into glucose by bypassing phosphoenolpyruvate carboxykinase and "malic enzyme" or pyruvate carboxylase through the reversal of flux through pyruvate kinase, a possibility that seems to be realized in rabbit muscle (Dyson *et al.*, 1975). Still, the activities of all these key enzymes taken together are probably too low to cause any appreciable accumulation of glucose or glycogen through gluconeogenesis.
4. This prediction was confirmed using ^{14}C-radiotracer studies on rainbow trout gill slices. After slices were incubated with [U-^{14}C]alanine or [U-^{14}C]lactate, no label could subsequently be detected in either gill glycogen or free glucose (Mommsen, 1983).

Since gill tissue is not capable of significant rates of gluconeogenesis, the endogenous glycogen stores of gill—albeit small compared with other fish tissues—must be replenished through exogenous glucose, setting up a critical competition situation. Potentially, glycolysis, glycogen synthesis, and the pentose phosphate shunt would compete for glucose 6-phosphate, which poses a quite intriguing regulatory problem for the gill cell. Another interesting aspect of nonexistent gluconeogenesis is the presence of F1,6BPase in easily detectable amounts. The possible involvement of this enzyme in the futile cycle between fructose 6-phosphate and fructose 1,6-bisphosphate has already been emphasized elsewhere (Fig. 2). A further function of this enzyme, if not reversal of glycolysis, is its allegedly important role in the pentose phosphate pathway.

E. Pentose Phosphate Pathway

A common observation in gill enzyme studies is the rather high activity of glucose 6-phosphate dehydrogenase (G6PDH), supposedly one of the rate-limiting steps of the pentose phosphate shunt (Stryer, 1981). In all four different species of Amazon fish assayed by Hulbert *et al.* (1978a,b), G6PDH was 5–10 times more active than PFK, one of the presumed rate-determining enzymes of glycolysis (Newsholme and Start, 1973). The same is true for the bowfin (Table II) and the rainbow trout (T. P. Mommsen, unpublished). Surprisingly, Hulbert *et al.* (1978a,b) do not further comment on this rather perplexing finding. In addition to G6PDH, fish gill also reveals substantial activities (at about 50% of the levels of G6PDH of 6-phosphogluconate dehydrogenase (6PGDH; EC 1.1.1.44) (Ekberg, 1962), the enzyme that catalyzes the next step in the pentose phosphate pathway.

Although the accepted route of the pentose phosphate shunt has become the center of some controversy (Williams, 1980), the biological functions of the pathway have been established for a long time: (1) the delivery of pentose phosphates for RNA and DNA synthesis, a high rate of which would be desirable for a tissue with high turnover of proteins and RNA, and (2) the supply of reducing equivalents in the form of NADPH from G6PDH and 6PGDH reactions for lipid synthesis, which may also be applicable to gill.

A third alternative might be the alleged involvement of the pentose phosphate shunt in the transport of protons. For the turtle bladder, another tissue with relatively high activities of G6PDH, Norby and Schwartz (1976) demonstrated a close coupling of proton transport to [1-^{14}C]glucose oxidation rather than to [6-^{14}C]glucose oxidation. The authors calculate that stimulation of proton transport leads to an increase in flux through the pentose phosphate shunt with a simultaneous reduction in glycolytic flux. Since the fish gill is involved in acid–base balance to a large extent, the findings of Norby and Schwartz (1976) certainly deserve some closer examination by fish-oriented transport biochemists.

As noted earlier, experimental evidence suggests that the gluconeogenic pathway is not operative in rainbow trout gill, since several key enzymes, like pyruvate carboxylase or phosphoenolpyruvate carboxykinase are absent (Knox *et al.*, 1980; Cowey *et al.*, 1977; T. P. Mommsen, unpublished). This turns the presence of fructose 1,6-bisphosphatase into a rather puzzling observation. The occurrence of an active pentose phosphate shunt in gill might help to explain this phenomenon, because F1,6BPase takes a prominent position in refunneling glyceraldehyde 3-phosphate into glucose 6-phosphate, especially in situations when the shunt is operating to supply reducing power through the complete oxidation of glucose 6-phosphate. Under these idealized conditions, the synthesis of ribulose 5-phosphate and thus RNA and DNA ceases. In this context it should be recalled that in rat liver and some other mammalian tissues, NADPH and $NADP^+$ compete in their binding to glucose 6-phosphate dehydrogenase, presumably the enzyme that catalyzes the rate-determining step for the *oxidative* part of the pentose phosphate shunt. Furthermore, not only the redox balance of the cytosol but also the adenylate energy status wield an intricate influence on the direction and the rate of flux through the shunt. Both glucose 6-phosphate and ATP are direct competitors in their binding to glucose 6-phosphate dehydrogenase.

V. NITROGEN METABOLISM

A. Ammonia Disposal

The principal nitrogenous waste product of metabolism in fishes is ammonia (i.e., the sum of ammonium ions and ammonia gas), although minor amounts of urea are excreted as well (Smith, 1929). Normally, the bulk of the ammonia is disposed of through the gills, while smaller amounts can also leave the fish via the kidneys (Forster and Goldstein, 1969; Hickman and Trump, 1969; Cameron and Wood, 1978) or the skin (Morii *et al.*, 1978).

As mentioned earlier, some controversy seems to exist about the capacity of gill tissue to contribute to the excretion of ammonia through the action of L-amino acid oxidases, transaminases, and ultimately deamination of amino acids or nucleotides. On the one hand, Goldstein *et al.* (1964) and Goldstein and Forster (1961) estimated that about 60% of the ammonia excreted through the gills of a sculpin (*Myoxocephalus scorpius*) can be accounted for by blood clearance, the remainder being contributed by the gill tissue. In support of this view Walton and Cowey (1977) measured a significant decrease in blood glutamate on passage through the gills of rainbow trout.

From their experiments on isolated trout heads, Payan and Matty (1975) conclude that gill tissue possesses only the capability for marginal "basal" ammonia excretion from the tissue itself, which amounts to about 20% of the total ammonia disposed. After improvements on their perfusion technique and a thorough reinvestigation of the same problem, Payan and Pic (1977) come to the conclusion that gill tissue itself does *not* participate in ammonia production but merely constitutes the tissue of excretion. Yet, when the Ringer perfusion medium utilized was devoid of ammonium, endogenous production was observed and ammonium was excreted into the perfusion medium as well as into the water irrigating the gill. A review of this controversy has been given by Kormanik and Cameron (1981).

B. Enzymes

Most enzymes involved in nitrogen metabolism have been described as occurring in fish gill in one form or another. From the available data, it is unfortunately not possible to put these different enzyme activities together into a comprehensive picture.

Glutamine synthetase, an enzyme that is thought to be involved in ammonia detoxification and that occurs in highest activities in the piscine brain (Webb and Brown, 1976), has been described for gill tissue of several fish, including *Ictalurus punctatus* (Wilson and Fowlkes, 1976), *Carassius auratus* (vanWaarde and Kesbeke, 1982), and *Salmo gairdneri* (C. J. French, personal communication). However, the effect of the presence of this enzyme activity cannot be detected in vivo, since there is no *net* removal of glutamate from or excretion of glutamine into blood on passage through the gill. Yet it should be recalled in this context that Walton and Cowey (1977) detected some glutamate uptake by the rainbow trout gill—a result that could not be substantiated in our own laboratory (T. P. Mommsen and C. J. French, unpublished)—and thus glutamate could serve as endogenous substrate for glutamine synthesis *inside* the gill tissue. Furthermore, small rates of synthesis and consequent glutamine excretion into the arterial blood might easily be masked through the high cardiac output and the relatively high blood concentration of this amino acid [5.9 μM (100 ml plasma)$^{-1}$, Walton and Cowey, 1977].

A metabolically interesting situation occurs in the gill of the common goldfish (*Carassius auratus*)—and most likely in other fish—where glutamine and asparagine synthetases are present as well as enzymes hydrolyzing these two amino acids (vanWaarde and Kesbeke, 1982). Opposing activities of the enzyme pairs glutaminase–glutamine synthetase and asparaginase–asparagine synthetase are almost equally high in each case. Significant levels of glutaminase are also reported for *Salmo gairdneri* (Walton and Cowey, 1977) and for *Myoxocephalus scorpius* (Goldstein and Forster, 1961).

As shown for rat kidney, the enzyme pair glutaminase–glutamine synthetase can operate as a futile cycle (Damian and Pitts, 1970), where the net flux will consequently be determined by the difference between the rates of the enzyme pair involved rather than through their absolute rates. Through this kind of arrangement a very sensitive control—possibly through the energy charge—of the direction of the *net flux* is achieved. In the rat liver, this particular futile cycle operates at a rate of about 100 to 200 nmol min^{-1} g^{-1}, resulting in a net synthesis of glutamine (Häussinger and Siess, 1979).

Gill tissue reveals considerable capacity to transaminate amino acids, where, in agreement with other fish tissues, aspartate aminotransferase seems to be the dominating enzyme (McCorkle *et al.*, 1979; Hulbert *et al.*, 1978a,b; Wilson, 1973; Table II). The enzyme plays an important part in the malate–aspartate shuttle and functions to funnel amino acid carbon into oxidative pathways (Fig. 1). The latter function is shared with alanine aminotransferase; both enzymes are ubiquitous in fish tissues, but as long as further information is missing with respect to their compartmentation, or their interplay with other tissue transaminases (e.g., branched-chain amino acid transaminases) and glutamate dehydrogenase (cf. Fig. 1), interpretation of the activity data would be premature.

The occurrence of glutamate dehydrogenase (GDH) has been reported for piscine gills numerous times (Goldstein and Forster, 1961; McBean *et al.*, 1966; Walton and Cowey, 1977; Hulbert *et al.*, 1978a,b). Biochemically, the enzyme is very similar to glutamate dehydrogenases from other fish tissues, being activated by AMP and ADP and inhibited by GDP, GTP, ATP, and the reduced and oxidized forms of the nucleotide NAD^+ (Fields *et al.*, 1978). In general, the activities of gill GDHs are lower than in other tissues of fish (Walton and Cowey, 1977; Fields *et al.*, 1978; Hulbert *et al.*, 1978a,b; Storey *et al.*, 1978) From their kinetic data of glutamate dehydrogenase from two Amazon fish with different life strategies, Fields *et al.* (1978) conclude that the fish gill GDH has no significant aminating function, since the kinetic parameters include high (i.e., well above physiological concentrations) Michaelis–Menten constants for ammonia and 2-oxoglutaric acid. The inhibition pattern suggests that the enzyme is under stringent control, through both the energy charge (Atkinson, 1977, see later) and the redox state of the appropriate cell type of the gill.

VI. ADENYLATE METABOLISM

Since the alleged ultimate force driving ion transport in gill is the hydrolysis of ATP (Silva *et al.*, 1977), and cAMP and adenosine apparently possess important hemodynamic functions (Colin and Leray, 1977, 1979; Pic, 1981; Pic and Djabali, 1982), some publications have dealt with adenylate metabolism in gill.

The concept of the energy charge,

$$EC = \frac{ATP + 0.5ADP}{ATP + ADP + AMP}$$

as regulator of cellular processes was introduced by Atkinson (1977), and it has been mentioned previously that numerous enzymes involved in gill amino acid metabolism—in addition to many others—are probably subject to control through the adenylate energy status of the cell. Included were glutamate dehydrogenase and the substrate cycle around glutamate–glutamine.

It has been shown that the energy status of the gill tissue as a functional unit strongly depends on external factors, such as environmental pH (MacFarlane, 1981) and salinity (Leray *et al.*, 1981). Although MacFarlane (1981) did not analyze different cell types in the gill epithelium, he was able to show for gill as a whole that during exposure of the killifish, *Fundulus grandis,* to pH 4.0 water, the energy charge drops from 0.89 to 0.53. This is mainly a result of a precipitous decline in the ATP concentration; in addition, the total adenylate pool decreases by more than 65% on the adaptation of the fish to pH 4.0 water and by more than 60% in the case of pH 5.0 water. Under the osmotic stress that occurs on transfer of rainbow trout from fresh water to seawater (Leray *et al.*, 1981), gill energy charge declines during the acute-stress phase but returns to the "control" range after an adaptation period of about 6 days. In the case of transfer of seawater-adapted rainbow trout to fresh water, no significant change in ATP concentrations or energy charge occurred (Leray *et al.*, 1981). The values of Leray *et al.* (1981) for the energy charge are higher than those of MacFarlane (1981), being 0.91 to 0.95 for the rainbow trout versus 0.89 in the killifish; the actual drop in the adenylate energy charge—reversible in the case of osmotic stress in the trout—is considerably lower in the trout, where the lowest values lie in the range of 0.91 to 0.88, whereas in the killifish the values are as low as 0.53, without recovery.

The dominant nucleotides in each cell are ATP and its derivatives, and in this respect the fish gill is no exception. Although its total pool of adenylates is somewhat modest at 3.5 μmol g^{-1} (T. P. Mommsen, unpublished), the results of Leray *et al.* (1981) and MacFarlane (1981) point to a significant potential and rate to interconvert adenylates in this tissue.

A central role in these interconversions has been assigned to the so-called purine nucleotide cycle, a pathway that has been thoroughly investigated in muscular tissue (Lowenstein, 1972). After the initial description of a very active AMP deaminase in fish gill, in addition to its abundance in fish muscular tissues (Makarewicz and Zydowo, 1962; Makarewicz, 1969; Walton and Cowey, 1977), subsequent research was concerned with the biochemical characterization of gill AMP deaminase (Raffin and Leray, 1980; Raffin, 1981) and with the activity of the purine nucleotide cycle in general (Leray *et al.*, 1979). A general scheme of the cycle and a few connecting pathways are presented in Fig. 4. One of the interesting aspects of the findings of Leray *et al.* (1981) is the fact that apart from the purine nucleotide cycle a whole array of other enzymes involved in adenylate metabolism are present in gill tissue.

The nonrelease of ammonia in the isolated head preparation (Payan and

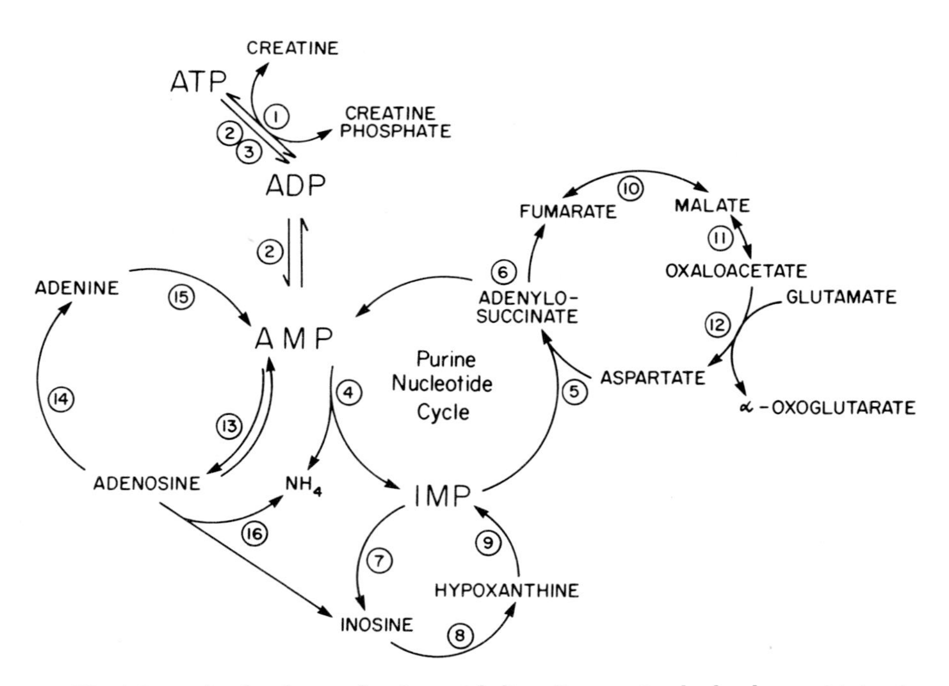

Fig. 4. Interrelated pathways of purine metabolism. Enzymes involved and present in trout gill tissue are identified by circled numbers (Leray *et al.*, 1979; T. P. Mommsen, unpublished): (1) creatine phosphokinase (EC 2.7.3.2); (2) adenylate kinase (EC 2.7.4.3); (3) adenosine triphosphatase (EC 3.6.1.4); (4) adenosine monophosphate deaminase (EC 3.5.4.6); (5) adenylosuccinate synthase (EC 6.3.4.4); (6) adenylosuccinase (EC 4.3.2.2); (7) 5′-nucleotidase (EC 3.1.3.5); (8) purine nucleoside phosphorylase (EC 2.4.2.1); (9) hypoxanthine phosphoribosyltransferase (EC 2.4.2.8); (10) fumarase (EC 4.2.1.2); (11) malate dehydrogenase (EC 1.1.1.37); (12) aspartate aminotransferase (EC 2.6.1.1); (13) 5′-nucleotidase (EC 3.1.3.5); (14) purine nucleoside phosphorylase (EC 2.4.2.1); (15) adenine phosphoribosyltransferase (EC 2.4.2.7); (16) adenosine deaminase (EC 3.5.4.4); (17) adenosine kinase (EC 2.7.1.20).

Pic, 1977) indicates that, at least in vivo, deaminating and reaminating reactions are in balance, or they are in an imbalance that lies below the detection limit for ammonia release and thus eludes the experimenter.

Creatine phosphokinase, adenylate kinase, and other kinases work in a concerted fashion to supply ATP, wherever and whenever it is needed. In the short term, AMP deaminase presence may serve as a regulator of the energy charge (Chapman and Atkinson, 1973), but in the long term, it will lead to a *net* decrease in adenylates available in the tissue. In the rainbow trout gill, two different pathways are operative that potentially deplete the adenylate pool. Mainstream degradation leads from AMP to IMP through the action of AMP deaminase, an enzyme that is strongly activated through an endogenous protease (Raffin, 1981). An alternative pathway involves 5′-nucleotidase; in both cases, inosine and/or hypoxanthine are the important final products. Although a multitude of different adenylate salvage pathways—including the purine nucleotide cycle—are present in the gill (Fig. 4), in vitro they by no means counterbalance the degradation pathways.

Somehow it seems very difficult to put all these isolated results and the obvious discrepancies between *in vivo* and *in vitro* experimentation into some sort of framework. Nevertheless, a number of interesting questions come to mind, for example concerning the feedback control of energy charge regulation (cf. van den Berghe *et al.*, 1977), the influence of *exogenous* adenylates on the cell's balance, or the compartmentation of adenylate degradative and salvage pathways. Undoubtedly, the whole topic of adenylate control and metabolism deserves much more attention, especially in the light of the impressive ATP turnover of the gill tissue.

VII. PHARMACOKINETIC FUNCTION

The gill lies between the venous and arterial circulation, and because of this strategic positioning, paired with its large endothelial surface area, it would be ideally situated to adjust concentrations of substances before they reach arterial receptors. Furthermore, gill constitutes the only capillary bed through which *all* the blood has to pass on *each* circuit through the body. The latter statement holds for most fish, with the obvious exception of the lungfish and other air-breathing fish with rearrangements of their circulatory system (Randall *et al.*, 1981). In addition, some controversy exists about the existence and physiological significance of shunting pathways within the gill (Laurent and Dunel, 1976, 1980; Bond, 1979; Holbert *et al.*, 1979).

Although a comparison with the mammalian lung seems rather far-fetched at first, it is a very informative approach with respect to future gill research, because lung occupies the same pivotal position as gill within the animal's circulatory circuit. Through their positioning as respiratory organs, susceptibility to minor changes in oxygen supply will be minimized. However, gill and lung will also be the first organs with large surface areas to be subjected to environmental influences, which makes these organs unique sites for rapid and efficient control of the hormonal status for the organism. Examples of the vast effects that environmental factors exert on the gill are the rapid and substantial changes in energy charge and adenylate pool, detectable immediately after transfer of fish from fresh water to seawater (Leray *et al.*, 1981) or after subjecting fish to low environmental pH (MacFarlane, 1981).

On the one hand, the influences of hormones on function, on hemodynamics, and as secondary messengers of the gill have been described and analyzed in some detail (Wood, 1975; Payan and Girard, 1977; Haywood *et al.*, 1977; Pic and Djabali, 1982); on the other, no work has been performed on the pharmacokinetic functions of the tissue, that is, on the effects that gill exerts on physical state and metabolism of hormones or other blood-borne substances.

In many experiments on gill hemodynamics, catecholamines are being utilized to elicit specific physiological responses. Injection of both adrenaline and noradrenaline, for instance, induces increase of branchial cAMP concentration (Pic and Djabali, 1982), but no information is available with regard to the *fate* of the hormones themselves during the gill passage after release from their respective receptors. It is at this point that a comparison with lung metabolism may show the way for some future gill research. Interestingly, some 20–40% of venous noradrenaline is taken up and metabolized on passage through the mammalian lung (Bahkle and Vane, 1974, 1977); although the mechanism of noradrenaline *removal* from the blood is still uncertain, the *metabolism* of this catecholamine is due to the action of an intramitochondrially located monoamine oxidase (Bakhle and Youdhim, 1979) and/or a cytosolic catechol-*O*-methyltransferase in the endothelial cell (Ferreira *et al.*, 1980). An even more drastic example of the pharmacokinetic activity of the lung is the almost complete (90–98%) removal of 5-hydroxytryptamine from the lung circulation (Gaddum *et al.*, 1953). In contrast, adrenaline, histamine, and peptide hormones such as antidiuretic hormone, oxytocin, and substance P are not taken up and/or not metabolized by the lung tissue of mammals; this apparent metabolic neutrality is probably due to the absence of uptake and transport mechanisms for these substances (Alabaster, 1977).

In addition to noradrenaline and 5-hydroxytryptamine, many other bioactive substances are partially cleared during their passage through the lung; these include insulin, bradykinin, acetylcholine, and some prostaglandins (Vane, 1980). The added fact that ATP too is almost completely removed in the lung circulation (Simionescu, 1980) could signify an important regulatory mechanism. To date, dephosphorylation of ATP in gill cells has always been automatically associated with active ion transport mechanisms or other ATP-driven processes. Maybe this concept deserves some revision, especially in the light of the close relationship that circulating ATP reveals with (1) 2,3-diphosphoglycerate (2,3-DPG) and (2) adenosine. In the blood of mammals, 2,3-DPG is a potent modulator of hemoglobin–oxygen affinity; interestingly, in fish this role is taken by ATP itself or by guanosine triphosphate (GTP) (Isaacks, *et al.*, 1978). Adenosine functions as a vasodilating agent in both mammals (Haddy and Scott, 1968) and fish (Colin and Leray, 1977).

Another group of substances is converted biochemically within the lung tissue and subsequently released; in this group belong (1) cortisone, which is converted to cortisol (Smith *et al.*, 1973), and (2) angiotensin I, which is hydrolyzed to angiotensin II (Ng and Vane, 1967).

The analysis and interpretation of the diverse pharmacokinetic functions of lung tissue in mammals seems to be an active and rapidly developing direction of research (Vane, 1980); unfortunately, the same cannot be said about the analysis of similar metabolic activities in the fish gill.

VIII. EPILOGUE

In the preceding discussion on metabolism, gill tissue has been treated rather globally as a more or less uniform tissue, paying little more than lip service to the tissue's composition of a variety of different cell types. A priori, this type of approach, although somewhat justified through the existing body of literature, sets major limitations to interpretation or further biochemical experimentation. A further simplification of the system is implicated by the lack of attention that has been given in the literature to the actual positioning of individual cell types in the tissue.

In correspondence with many other epithelial cells, gill cells show pronounced polarity in their orientation, a feature that is paramount to their function and that is exemplified, for instance, by the chloride cells, which apparently change their orientation (and numbers) when catadromous fish are moved from fresh water into seawater (Shirai and Utida, 1970; Utida *et al.*, 1971). Such polarity, paired with the heterogeneity of the tissue, presents obvious experimental disadvantages, independent of gill preparation *in vitro* or *in situ*. Some of these problems may be overcome by the use of membranes that are presumably similar to gill in their transport mechanisms, such as opercular epithelia (Degnan *et al.*, 1977; Foskett and Scheffey, 1982), by employing isolated cell preparations (Naon and Mayer-Gostan, 1983), or by the use of tissue culture techniques for epithelial cells (cf.

Wright, 1981), a technique that—with respect to fish in general—is only in its infancy.

Gill tissue subserves its varied functions in a wide diversity of aquatic environments, since fish occupy a broad spectrum of different ecological niches. Of all fish tissues, gill possesses the largest surface area facing the external milieu, and thus its cells will also be the first to be subjected to environmental changes, be it salinity, ionic composition, or heavy metal loading. Other factors exerting immediate effects on outward-facing gill surfaces will be temperature, oxygen tension, and proton loading, to name but a few.

One parameter of this immediate and far-reaching impact on the gill as a whole functional unit, touched on in this chapter, is the energy status of the tissue as influenced by a change in the outer ionic environment or an increase in proton concentration.

In contrast to the exterior milieu, the interior milieu of the fish generally undergoes changes that are less drastic than potential alterations in the environment. Apart from more or less predictable changes in water, oxygen, CO_2, bicarbonate, and ammonium and ammonia concentrations, relatively few other factors fluctuate. Both in seawater and in fresh water, a steep gradient in osmotic concentration exists across the thin gill epithelium between the blood of the respective fish and its surrounding medium. Gill tissue has to maintain this gradient while actively transporting selected ions in either direction and simultaneously holding on to an intracellular environment that is conducive to its diverse functions.

One scenario that shows the different emphasis that may be put on gill membrane function is exemplified by the retention of urea in some marine fish. Sharks, rays, chimaeras, and the coelacanth *Latimeria chalumnae* accumulate large amounts of urea and trimethylaminoxide (TMAO) in their blood to achieve a blood solute concentration that is equal to or above that of the surrounding seawater. In distinction from teleost fishes, which tend to lose urea across the gill (Fromm, 1963), these fish therefore possess mechanisms to retain urea and TMAO in their blood while maintaining other membrane characteristics unaltered, so that diffusional exchange of water across the gill is unaffected. This startling difference in gill membrane properties is reflected in the water:urea permeability (cm sec^{-1}) ratio, which ranges from 5 to 8 in teleosts (Haywood *et al.*, 1977; Steen and Stray-Pederson, 1975) but reaches more than 1500 in the dogfish *Scyliorhinus canicula* (Payan and Maetz, 1970), mainly because of decreased permeability for urea.

At this point it seems rather premature to attempt to develop an integrated picture with respect to probable metabolic patterns or gill-to-gill variability arising from the diverse functional demands on the gill tissue.

Metabolic fine analysis has to date been hampered mainly by the anatomic arrangement and the cell diversity of the tissue, and less by the lack of sophistication and state-of-the-art experimental procedures on the part of the researchers. However, no comparative efforts were made to relate metabolic patterns to gill function in varying or varied environments. Obviously, the fish gill presents a wide-open field for innovative research that will contribute some urgently needed information toward the better understanding of metabolic integration and regulation in this multifaceted organ.

REFERENCES

Alabaster, V. A. (1977). Inactivation of endogenous amines in the lungs. *Lung Biol. Health Dis.* **4**, 3–25.

Alp, P. R., Newsholme, E. A., and Zammit, V. A. (1976). Activities of citrate synthase and NAD^+-linked and $NADP^+$-linked isocitrate dehydrogenase in muscle from vertebrates and invertebrates. *Biochem. J.* **154**, 689–700.

Anderson, T. R. (1970). Temperature adaptation and the phospholipids of membranes in goldfish (*Carassius auratus*). *Comp. Biochem. Physiol.* **33**, 663–687.

Atkinson, D. E. (1977). "Cellular Energy Metabolism and Its Regulation." Academic Press, New York.

Bakhle, Y. S., and Vane, J. R. (1974). Pharmacokinetic function of the pulmonary circulation. *Physiol. Rev.* **54**, 1007–1045.

Bakhle, Y. S., and Vane, J. R., eds. (1977). "Metabolic Functions of the Lung," Lung Biology in Health and Disease, Vol. 4. Dekker, New York.

Bakhle, Y. S., and Youdhim, M. B. H. (1979). The metabolism of 5-hydroxytryptamine and β-phenylethylamine in perfused rat lung and *in vivo*. *Br. J. Pharmacol.* **65**, 147–154.

Bender, K., and Ohno, S. (1968). Duplication of the autosomally inherited 6-phosphogluconate dehydrogenase gene locus in tetraploid species of cyprinid fish. *Biochem. Genet.* **2**, 101–107.

Bettex-Galland, M., and Hughes, G. M. (1972). Contractile filamentous material in pillar cells of fish gills. *J. Cell Sci.* **13**, 359–370.

Bilinski, E., and Jonas, R. E. E. (1972). Oxidation of lactate to carbon dioxide by rainbow trout (*Salmo gairdneri*) tissues. *J. Fish Res. Board Can.* **29**, 1467–1471.

Black, E. C., Robertson Conner, A., Lam, K. C., and Chiu, W. G. (1962). Changes in glycogen, pyruvate, and lactate in rainbow trout *Salmo gairdneri* during and following musclular activity. *J. Fish. Res. Board Can.* **19**, 409–436.

Bond, C. E. (1979). "Biology of Fishes." Saunders, Philadelphia, Pennsylvania.

Bornancin, M., de Renzis, G., and Naon, R. (1980). Cl^-/HCO_3^--ATPase in gills of the rainbow trout: Evidence for its microsomal localization. *Am. J. Physiol.* **238**, R251–R259.

Caldwell, R. S. (1969). Thermal compensation of respiratory enzymes in tissues of the goldfish (*Carassius auratus* L.). *Comp. Biochem. Physiol.* **31**, 79–93.

Caldwell, R. S., and Vernberg, F. J. (1970). The influence of acclimation temperature on the lipid composition of fish gill mitochondria. *Comp. Biochem. Physiol.* **34**, 179–191.

Cameron, J. N., and Wood, C. M. (1978). Renal function and acid-base regulation in two Amazonian erythrinid fishes: *Hoplias malabaricus*, a water breather, and *Hoplerythrinus unitaeniatus*, a facultative air breather. *Can. J. Zool.* **56**, 917–930.

Chapman, A. G., and Atkinson, D. E. (1973). Stabilization of adenylate energy charge by the adenylate deaminase reaction. *J. Biol. Chem.* **248,** 8309–8312.

Colin, D., and Leray, C. (1977). Résponses hémodynamiques de la branchie à l'adenosine chez la truite (*Salmo gairdneri*). Etude sur tête isolée perfusée. *C. R. Hebd. Séances Acad. Sci.* **284,** 1191–1194.

Colin, D., and Leray, C. (1979). Interaction of adenosine and its phosphorylated derivatives with putative purinergic receptors in the gill vascular bed of rainbow trout. *Pflueger's Arch.* **383,** 35–40.

Conte, F. P. (1969). The biochemical aspects of salt secretion. *In* "Fish in Research" (O. W. Neuhaus and J. E. Halver, eds.), pp. 105–120. Academic Press, New York.

Conte, F. P. (1977). Molecular aspects of chloride cell formation in *Oncorhynchus*. II. Isolation and characterization of gill transfer RNAs during active salt secretion. *J. Exp. Zool.* **199,** 395–402.

Conte, F. P., and Lin, D. (1967). Kinetics of cellular morphogenesis in gill epithelium during saltwater adaptation of *Oncorhynchus* (Walbaum). *Comp. Biochem. Physiol.* **23,** 945–957.

Conte, F. P., and Tripp, M. J. (1970). Succinic dehydrogenase activity in the gill epithelia of euryhaline fishes. *Int. J. Biochem.* **1,** 129–138.

Cooney, G. J., Taegtmeyer, H., and Newsholme, E. A. (1981). Tricarboxylic acid cycle flux and enzyme activities in the isolated working rat heart. *Biochem. J.* **200,** 701–703.

Cowey, C. B., de la Higuera, M., and Adron, J. W. (1977). The effect of dietary composition and of insulin on gluconeogenesis in rainbow trout *Salmo gairdneri. Br. J. Natr.* **38,** 385–395.

Daikoku, T., Yano, I., and Masui, M. (1982). Lipid and fatty acid compositions and their changes in the different organs and tissues of guppy, *Poecilia reticulata* on sea water adaptation. *Comp. Biochem. Physiol. A* **73A,** 167–174.

Damian, A. C., and Pitts, R. F. (1970). Rates of glutaminase I and glutamine synthetase reactions in rat kidney *in vivo. Am. J. Physiol.* **218,** 1249–1255.

Degnan, K. J., Karnaky, K. J., and Zadunaiski, J. A. (1977). Active chloride transport in the *in vitro* opercular skin of a teleost (*Fundulus heteroclitus*), a gill-like epithelium rich in chloride cell. *J. Physiol. (London)* **271,** 155–191.

Djabali, M., and Pic, P. (1982). Effects of and adrenoreceptors on branchial cAMP level in seawater adapted mullet, *Mugil capito. Gen. Comp. Endocrinol.* **46,** 193–199.

Driedzic, W. R., and Kiceniuk, J. W. (1976). Blood lactate levels in free swimming trout (*Salmo gairdneri*) before and after exercise resulting in fatigue. *J. Fish. Res. Board Can.* **33,** 173–176.

Dyson, R. D., Cardenas, J. M., and Barsotti, R. J. (1975). The reversibility of skeletal muscle pyruvate kinase and an assessment of its capacity to support gluconeogenesis. *J. Biol. Chem.* **250,** 3316–3324.

Ekberg, D. R. (1962). Anaerobic and aerobic metabolism in gills of the Crucian carp adapted to high and low temperatures. *Comp. Biochem. Physiol.* **5,** 123–128.

Epstein, F. H., Katz, A. I., and Pickford, G. E. (1967). Sodium- and potassium- activated adenosine triphosphatase of gills: Role on adaptation of teleosts to salt water. *Science* **156,** 1245–1247.

Ferreira, S. H., Greene, L. J., Saldago, M. C. O., and Krieger, E. M. (1980). The fate of circulating biologically active peptides in the lungs. *Ciba Found. Symp.* [N.S.] **78,** 129–146.

Fields, J. H. A., Driedzic, W. R., French, C. J., and Hochachka, P. W. (1978). Kinetic properties of glutamate dehydrogenase from the gills of *Arapaima gigas* and *Osteoglossum bicirrhosum. Can. J. Zool.* **56,** 809–813.

Forster, R. P., and Goldstein, L. (1969). Formation of excretory products. *In* "Fish Physiology" (W. S. Hoar and D. J. Randall, eds.), Vol. 1, pp. 313–350. Academic Press, New York.

Foskett, J. K., and Scheffey, C. (1982). The chloride cell: Definitive identification as the salt-secretory cell in teleosts. *Science* **215,** 164–165.

French, C. J., Mommsen, T. P., and Hochachka, P. W. (1981). Amino acid utilization in isolated hepatocytes from rainbow trout. *Eur. J. Biochem.* **113,** 311–317.

Fromm, P. O. (1963). Studies on renal and extra-renal excretion in a freshwater teleost, *Salmo gairdneri. Comp. Biochem. Physiol.* **10,** 121–128.

Gaddum, J. H., Hebb, C. O., Silver, A., and Swan, A. A. B. (1953). 5-Hydroxytryptamine. Pharmacological action and destruction in perfused lungs. *Q. J. Exp. Physiol. Cogn. Med. Sci.* **38,** 255–262.

Gaudet, M., Racicot, J. G., and Leray, C. (1975). Enzyme activities of plasma and selected tissues in rainbow trout *Salmo gairdneri* Richardson. *J. Fish Biol.* **7,** 505–512.

Gilbert, H. F. (1982). Biological disulfides: The third messenger? Modulation of phosphofructokinase activity by thiol/disulfide exchange. *J. Biol. Chem.* **257,** 12086–12091.

Goldstein, L., and Forster, R. P. (1961). Source of ammonia excreted by the gills of the marine teleost, *Myoxocephalus scorpius. Am. J. Physiol.* **200,** 116–118.

Goldstein, L., Forster, R. P., and Fanelli, G. M. (1964). Gill blood flow and ammonia excretion in the marine teleost, *Myoxocephalus scorpius. Comp. Biochem. Physiol.* **12,** 489–499.

Guppy, M., Hulbert, W. C., and Hochachka, P. W. (1979). Metabolic sources of heat and power in tuna muscles. II. Enzyme and metabolite profiles. *J. Exp. Biol.* **82,** 303–320.

Haddy, F. J., and Scott, J. B. (1968). Metabolically linked vasoactive chemicals in local regulation of blood flow. *Physiol. Rev.* **48,** 688–707.

Haschemeyer, A. E. V., and Smith, M. A. K. (1979). Protein synthesis in liver, muscle and gill of mullet (*Mugil cephalus* L.) *in vivo. Biol. Bull.* (*Woods Hole, Mass.*) **156,** 93–102.

Haschemeyer, A. E. V., Persell, R., and Smith, M. A. K. (1979). Effect of temperature on protein synthesis in fish of the Galapagos and Perlas Islands. *Comp. Biochem. Physiol. B* **64B,** 91–95.

Häussinger, D., and Siess, H. (1979). Hepatic glutamine metabolism under the influence of the portal ammonia concentration in the perfused rat liver. *Eur. J. Biochem.* **101,** 179–184.

Haywood, G. P., Isaia, J., and Maetz, J. (1977). Epinephrine effects on branchial water and urea flux in rainbow trout. *Am. J. Physiol.* **232,** R110–R115.

Hers, H.-G., Hue, L., and Van Schaftingen, E. (1982). Fructose 2,6-bisphosphate. *Trends Biochem. Sci.* **7,** 329–331.

Hickman, C. P., and Trump, B. F. (1969). The kidney. *In* "Fish Physiology" (W. S. Hoar and D. J. Randall, eds.), Vol. 1, pp. 91–239. Academic Press, New York.

Hochachka, P. W., Schneider, D. E., and Storey, K. B. (1979). Hydrogen shuttles in gills of water versus air breathing osteoglossids. *Comp. Biochem. Physiol. B* **63B,** 57–61.

Holbert, P. W., Boland, E. J., and Olson, K. R. (1979). The effect of epinephrine and acetylcholine on the distribution of red cells within the gills of the channel catfish (*Ictalurus punctatus*). *J. Exp. Biol.* **79,** 135–146.

Hulbert, W. C., Moon, T. W., and Hochachka, P. W. (1978a). The osteoglossid gill: Correlation of structure, function and metabolism with transition to air breathing. *Can. J. Zool.* **56,** 801–808.

Hulbert, W. C., Moon, T. W., and Hochachka, P. W. (1978b). The erythrinid gill: Correlations of structure, function and metabolism. *Can. J. Zool.* **56,** 814–819.

Isaacks, R. E., Kim, H. D., and Harkness, D. R. (1978). Relationship between phosphorylated metabolic intermediates and whole blood oxygen affinity in some air-breathing and water-breathing teleosts. *Can. J. Zool.* **56,** 887–890.

Isaia, J., Maetz, J., and Haywood, G. P. (1978). Effects of epinephrine on branchial non-electrolyte permeability in rainbow trout. *J. Exp. Biol.* **74,** 227–237.

Jackson, W. F., and Fromm, P. O. (1981). Factors affecting ^3H_2O transfer capacity of isolated perfused trout gills. *Am. J. Physiol.* **240,** R235–R245.

Johansen, K., and Pettersson, K. (1981). Gill O_2 consumption in a teleost fish, *Gadus morhua. Respir. Physiol.* **44,** 277–284.

Kamiya, M. (1972). Sodium-potassium-activated adenosinetriphosphatase in isolated chloride cells from eel gills. *Comp. Biochem. Physiol. B* **43B,** 611–617.

Katz, J., and Rognstad, R. (1978). Futile cycling in glucose metabolism. *Trends Biochem. Sci.* **3,** 171–174.

Knox, D., Walton, M. J., and Cowey, C. B. (1980). Distribution of enzymes of glycolysis and gluconeogenesis in fish tissues. *Mar. Biol.* **56,** 7–10.

Kormanik, G. A., and Cameron, J. N. (1981). Ammonia excretion in animals that breathe water: A review. *Mar. Biol. Lett.* **2,** 11–23.

Laurent, P., and Dunel, S. (1976). Functional organization of the teleost gill. I. Blood pathways. *Acta Zool. (Stockholm)* **57,** 189–209.

Laurent, P., and Dunel, S. (1980). Morphology of gill epithelia in fish. *Am. J. Physiol.* **238,** R147–R159.

Leech, A. R., Goldstein, L., Chan, C. J., and Goldstein, J. M. (1979). Alanine biosynthesis during starvation in skeletal muscle of spiny dogfish. *J. Exp. Zool.* **207,** 73–80.

Leray, C., Raffin, J. P., and Winninger, C. (1979). Aspects of purine metabolism in the gill epithelium of rainbow trout, *Salmo gairdneri. Comp. Biochem. Physiol. B* **62B,** 31–40.

Leray, C., Colin, D. A., and Florentz, A. (1981). Time course of osmotic adaptation and gill energetics of rainbow trout *Salmo gairdneri* R. following abrupt changes in external salinity. *J. Comp. Physiol.* **144,** 175–181.

Lim, S. T., Kay, R. M., and Bailey, G. S. (1975). Lactate dehydrogenase isozymes of salmonid fish. Evidence for unique and rapid functional divergence of duplicated H4 lactate dehydrogenases. *J. Biol. Chem.* **250,** 1790–1800.

Loretz, C. A. (1979). Water exchange across fish gills: The significance of tritiated-water flux measurements. *J. Exp. Biol.* **79,** 147–162.

Lowenstein, J. M. (1972). Ammonia production in muscle and other tissues: The purine nucleotide cycle. *Physiol. Rev.* **52,** 382–414.

McBean, R. L., Neppel, M. J., and Goldstein, L. (1966). Glutamate dehydrogenase and ammonia excretion in the eel (*Anguilla rostrata*). *Comp. Biochem. Physiol.* **18,** 909–920.

McCorkle, F. M., Chambers, J. E., and Yarbrough, J. D. (1979). Seasonal effects on selected tissue enzymes in channel catfish, *Ictalurus punctatus. Comp. Biochem. Physiol. B* **62B,** 151–153.

MacFarlane, R. B. (1981). Alterations in adenine nucleotide metabolism in the Gulf killifish (*Fundulus grandis*) induced by low pH water. *Comp. Biochem. Physiol. B* **68B,** 193–202.

Makarewicz, W. (1963). AMP aminohydrolase and glutaminase activities in the kidneys and the gills of some freshwater vertebrates. *Acta Biochim. Pol.* **10,** 363–369.

Makarewicz, W., and Żydowo, M. (1962). Comparative studies on some ammonia-producing enzymes in the excretory organs of vertebrates. *Comp. Biochem. Physiol.* **6,** 269–275.

Mommsen, T. P. (1983). Biochemical characterization of the rainbow trout gill. *J. Comp. Physiol.* **154,** 191–198.

Mommsen, T. P., French, C. J., and Hochachka, P. W. (1980). Sites and patterns of protein and amino acid utilization during the spawning migration of salmon. *Can. J. Zool.* **58,** 1785–1799.

Moon, T. W., and Hochachka, P. W. (1971). Temperature and enzyme activity in poikilotherms. Isocitrate dehydrogenase in rainbow trout liver. *Biochem. J.* **123,** 695–705.

Moon, T. W., and Ouellet, G. (1979). The oxidation of tricarboxylic acid cycle intermediates, with particular reference to isocitrate, by intact mitochondria isolated from the liver of the American eel, Anguilla rostrate LeSueur. *Arch. Biochem. Biophys.* **195,** 438–452.

Morgan, M., and Tovell, P. W. A. (1973). The structure of the gill of the trout, *Salmo gairdneri* (Richardson). *Z. Zellforsch. Mikrosk. Anat.* **142,** 147–162.

Morii, H., Nishikata, K., and Tamura, O. (1978). Nitrogen excretion of mudskipper fish *Periophthalmus cantonensis* and Boleophthalmus pectinirostris in water and on land. *Comp. Biochem. Physiol. A* **60A,** 189–193.

Naito, N., and Ishikawa, H. (1980). Reconstruction of the gill from single cell suspensions of the eel, *Anguilla japonica. Am. J. Physiol.* **238,** R165–R170.

Naon, R., and Mayer-Gostan, N. (1983). Separation by velocity sedimentation of the gill epithelial cells and their ATPases in the seawater adapted eel *Anguilla anguilla* L. *Comp. Biochem. Physiol. A.* **75A,** 541–547.

Newsholme, E. A., and Start, C. (1973). "Regulation in Metabolism." Wiley, New York.

Ng, K. K. F., and Vane, J. R. (1967). The conversion of angiotensin I to angiotensin II. *Nature (London)* **216,** 762–766.

Norby, L. H., and Schwartz, J. H. (1976). Energetics of urinary acidification—role of pentose shunt. *Fed. Proc., Fed. Am. Soc. Exp. Biol.* **35,** 703.

Oram, J. F., Bennetch, S. L., and Neely, J. R. (1973). Regulation of fatty acid utilization in isolated perfused rat hearts. *J. Biol. Chem.* **248,** 5299–5309.

Pasztor, V. M., and Kleerekoper, H. (1962). The role of gill filament musculature in teleosts. *Can. J. Zool.* **40,** 785–802.

Patton, J. S., Haswell, M. S., and Moon, T. W. (1978). Aspects of lipid synthesis, hydrolysis, and transport studied in selected Amazon fish. *Can. J. Zool.* **56,** 787–792.

Payan, P., and Girard, J.-P. (1977). Adrenergic receptors regulating patterns of blood flow through the gills of trout. *Am. J. Physiol.* **232,** H18–H23.

Payan, P., and Maetz, J. (1970). Balance hydrique et minérale chez les élasmobranches. Arguments en faveur d'un contrôle endocrinïen. *Bull. Inf. Sci. Tech., Comm. Energ. At. (Fr.)* **146,** 77–96.

Payan, P., and Matty, A. J. (1975). The characteristics of ammonia excretion by a perfused isolated head of trout (*Salmo gairdneri*): Effect of temperature and CO_2-free Ringer. *J. Comp. Physiol.* **96,** 167–184.

Payan, P., and Pic, P. (1977). Origine de l'ammonium excreté par les branchies chez la Truite (*Salmo gairdneri*). *C. R. Hebd. Séances Acad. Sci.* **284,** 2519–2522.

Philpott, C. W., and Copeland, D. F. (1963). Fine structure of chloride cells from three species of *Fundulus. J. Cell Biol.* **18,** 389–404.

Phleger, C. F. (1978). Gill phospholipids of Amazon fishes. *Can. J. Zool.* **56,** 793–794.

Pic, P. (1981). Control adrénergique de la concentration branchiale en AMPc chez *Mugil capito* adapté à l'eau douce. *C. R. Hebd. Séances Acad. Sci.* **292,** 863–866.

Pic, P., and Djabali, M. (1982). Effects of catecholamines on cyclic AMP in the gill of seawater-adapted mullet, *Mugil capito. Gen. Comp. Endocrinol.* **46,** 184–192.

Pilkis, S. J., El-Maghrabi, M. R., McGrane, M., Pilkis, J., Fox, E., and Claus, T. H. (1982). Fructose 2,6-bisphosphate—a mediator of hormone action at the fructose 6-phosphate/fructose 1,6-bisphosphate substrate cycle. *Mol. Cell. Endocrinol.* **25,** 245–266.

Pontremoli, S., and Horecker, B. L. (1970). Fructose 1,6-diphosphatase from rabbit liver. *Curr. Top. Cell. Regu.* **2,** 173–199.

Raffin, J. P. (1981). AMP deaminase from trout gill. Localization of the activating proteinase. *Mol. Physiol.* **1,** 223–234.

Raffin, J. P., and Leray, C. (1980). Comparative study on AMP deaminase in gill, muscle and blood of fish. *Comp. Biochem. Physiol. B.* **67B,** 533–540.

Randall, D. J., Burggren, W. W., Farrell, A. P., and Haswell, S. M. (1981) "The Evolution of Airbreathing in Vertebrates." Cambridge University Press.

Read, G., Crabtree, B., and Smith, G. H. (1977). Activities of 2-oxoglutarate dehydrogenase and pyruvate dehydrogenase in hearts and mammary glands of some ruminants and non-ruminants. *Biochem. J.* **164,** 349–355.

Sargent, J. R., Thomson, A. J., and Bornancin, M. (1975). Activities and localization of succinic dehydrogenase and Na^+/K^+-activated adenosine triphosphatase in the gills of eels (*Anguilla anguilla*). *Comp. Biochem. Physiol. B* **51B,** 75–79.

Shirai, N., and Utida, S. (1970). Development and degeneration of the chloride cell during seawater and freshwater adaptation of the Japanese eel, *Anguilla japonica. Z. Zellforsch. Mikrosk. Anat.* **103,** 247–264.

Silva, P., Solomon, R., Spokes, K., and Epstein, F. H. (1977). Ouabain inhibition of gill Na-K-ATPase: Relationship to active chloride transport. *J. Exp. Zool.* **199,** 419–426.

Simionescu, M. (1980). Ultrastructural organization of the alveolar-capillary unit. In "Metabolic Activities of the Lung" *Ciba Found. Symp.* [N.S.] **78,** 11–36.

Skorkkowski, E. F., Biegniewska, A., Aleksandrowics, Z., and Swierczynski, J. (1980). Comparative studies on NADP-linked dehydrogenases in some tissues of fish and crustacenas. *Comp. Biochem. Physiol. B* **65B,** 559–562.

Smith, B. T., Torday, J. S., and Giroud, C. J. P. (1973). The growth promoting effect of cortisol on human fetal lungs. *Steroids* **22,** 515–524.

Smith, H. W. (1929). The excretion of ammonia and urea by the gills of fish. *J. Biol. Chem.* **81,** 729–742.

Soivio, A., Nikinmaa, M., Nyholm, K., and Westman, K. (1981). The role of gills in the responses of *Salmo gairdneri* during moderate hypoxia. *Comp. Biochem. Physiol. A* **70A,** 133–139.

Somero, G. N., and Doyle, D. (1973). Temperature and rates of protein degradation in the fish *Gillichthys mirabilis. Comp. Biochem. Physiol. B* **46B,** 463–474.

Starzyk, R. M., and Merritt, R. B. (1980). Malate dehydrogenase isozymes in the longnose dace, *Rhinichthys cataractae. Biochem. Genet.* **18,** 755–764.

Steen, J. B., and Stray-Pederson, S. (1975). The permeability of fish gills with comments on the osmotic behaviour of cellular membranes. *Acta Physiol. Scand.* **95,** 6–20.

Storey, K. B., Guderley, H. E., Guppy, M., and Hochachka, P. W. (1978). Control of ammoniagenesis in the kidney of water- and air-breathing osteoglossids: Characterization of glutamate dehydrogenase. *Can. J. Zool.* **56,** 845–851.

Stryer, L. (1981). "Biochemistry." Freeman, San Francisco, California.

Thomas, A. J., and Patton, S. (1972). Phospholipids of fish gills. *Lipids* **7,** 76–78.

Tondeur, F., and Sargent, J. R. (1979). Biosynthesis of macromolecules in chloride cells in the gills of the common eel, *Anguilla anguilla*, adapting to sea water. *Comp. Biochem. Physiol. B.* **62B,** 13–16.

Utida, S., Kamiya, M., and Shirai, N. (1971). Relationship between the activity of Na^+-K^+-activated adenosine triphosphatase and the number of chloride cells in eel gills with special reference to seawater adaptation. *Comp. Biochem. Physiol. A.* **38A,** 443–447.

van den Berghe, G., Bronfman, M., Vanneste, R., and Hers, H.-G. (1977). The mechanism of adenosine triphosphate depletion in the liver after load of fructose. A kinetic study of liver adenylate deaminase. *Biochem. J.* **162,** 601–609.

Vane, J. R. (1980). "Metabolic Activities of the Lung," Ciba Found. Symp. No. 78. Excerpta Medica, Amsterdam.

vanWaarde, A., and Kesbeke, F. (1982). Nitrogen metabolism in goldfish, *Carassius auratus* (L.). Activities of amidases and amide synthetases in goldfish tissues. *Comp. Biochem. Physiol. B.* **71B,** 599–603.

Vogel, W., Vogel, V., and Pfautsch, M. (1976). Arterio-venous anastomoses in rainbow trout gill filaments. *Cell Tissue Res.* **167,** 373–385.

Walton, M. J., and Cowey, C. B. (1977). Aspects of ammoniogenesis in rainbow trout, *Salmo gairdneri. Comp. Biochem. Physiol. B* **57B,** 143–149.

Webb, J. T., and Brown, G. W. (1976). Some properties and occurrence of glutamine synthetase in fish. *Comp. Biochem. Physiol. B* **54B,** 171–175.

Williams, J. F. (1980). A critical examination of the evidence for the reactions of the pentose pathway in animal tissues. *Trends Biochem. Sci.* **5,** 315–321.

Williscroft, S. N., and Tsuyuki, H. (1970). Lactate dehydrogenase systems of rainbow trout—evidence for polymorphism in liver and additional subunits in gills. *J. Fish Res. Board Can.* **27,** 1563–1567.

Wilson, R. P. (1973). Nitrogen metabolism in channel catfish, *Ictalurus punctatus.* I. Tissue distribution of aspartate and alanine aminotransferases and glutamic dehydrogenases. *Comp. Biochem. Physiol. B* **46B,** 617–624.

Wilson, R. P., and Fowlkes, P. L. (1976). Activity of glutamine synthetase in channel catfish tissues determined by an improved tissue assay method. *Comp. Biochem. Physiol. B* **54B,** 365–368.

Wood, C. M. (1975). A pharmacological analysis of the adrenergic and cholinergic mechanisms regulating branchial vascular resistance in the rainbow trout (*Salmo gairdneri*). *Can. J. Zool.* **53,** 1569–1577.

Wood, C. M., McMahon, B. R., and McDonald, D. G. (1978). Oxygen exchange and vascular resistance in the totally perfused rainbow trout. *Am. J. Physiol.* **234,** R201–R208.

Wright, E. M. (1981). Transepithelial transport in cell culture. Introduction. *Am. J. Physiol.* **240,** C91.

Zwingelstein, G. (1979–1980). Les effects de l'adaptation à l'eau de mer sur le métabolisme lipidique du poisson. *Oceanis* **5,** 117–130.

Zwingelstein, G., Meister, R., and Brichon, G. (1975). Métabolisme comparé des phospholipides des organes effecteurs de l'osmorégulation chez l'anguille européenne (*Anguilla anguilla*). *Biochemie* **57,** 609–622.

Zwingelstein, G., Portoukalian, J., Rebel, G., and Brichon, G. (1980). Gill sulfolipid synthesis and seawater adaptation in euryhaline fish. *Anguilla anguilla. Comp. Biochem. Physiol. B* **65B,** 555–558.

8

THE ROLES OF GILL PERMEABILITY AND TRANSPORT MECHANISMS IN EURYHALINITY*

DAVID H. EVANS

Department of Zoology
University of Florida
Gainesville, Florida
and
Mt. Desert Island Biological Laboratory
Salsbury Cove, Maine

I. INTRODUCTION

The vast majority of fish species can only tolerate salinities similar to those in which they reside, that is, either fresh water or seawater. However, there are significant numbers of species that possess physiological mecha-

*The research in our laboratories in Miami and Gainesville, Florida, and in Salsbury Cove, Maine (MDIBL), has been supported by various grants from the National Science Foundation, most recently PCM 81-04046.

ISBN 0-12-350432-5

nisms that are as yet ill-defined but that enable them to survive over a wide range of salinities, either in the environment or in the laboratory (Table I). The best-studied examples of these euryhaline species are members of the families Salmonidae, Cyprinodontidae, Anguillidae, Pleuronectidae, and Mugilidae (respectively, salmon and trout, killifish, eels, flounders, and mullet). These and various other species are able to maintain quite consistent blood osmolarities and ionic concentrations over a wide range of salinities [see Holmes and Donaldson (1969) or Evans (1979) for representative data on blood ionic determinations], even though, by definition, the vectors of ionic and osmotic regulatory processes must be reversed as the animal crosses the line of isosmolarity. It is the aim of this chapter to outline the present state of our knowledge of the mechanisms of euryhalinity, particularly in reference to gill function. The branchial mechanisms involved in salt and water regulation by freshwater or marine fish have been reviewed by other authors in this volume, as well as in other volumes (Evans, 1979, 1980a,b, 1982b; Evans *et al.*, 1982; Kirschner, 1977, 1979, 1980; Maetz, 1971, 1974; Maetz and Bornancin, 1975; Maetz *et al.*, 1976; Potts, 1976, 1977) and will therefore not be outlined in detail in this chapter. We will deal mainly with ionic transport mechanisms, but also will describe what little is known about branchial water movement.

II. THE SEAWATER TO FRESH (BRACKISH) WATER TRANSITION

The decision to deal with the movement into lowered salinities first is basically arbitrary but is predicated on the assumption that fish made this transition very early in their evolution, some 510 million years ago. The earliest heterostracan fossils found to date were apparently marine (Repetski, 1978), but it is generally accepted that quite early fish were inhabiting waters whose salinities were far below that of seawater (Moyle and Cech, 1982). In addition to greater evolutionary interest in this transition, it will become apparent that we know much more about the physiology involved in the movement into lowered salinities than we do about the movement in the opposite direction.

Movement into salinities that are hypo-osmotic to the body fluids dictates that the mechanisms of saltwater osmoregulation are inhibited, whereas those allowing maintenance of body fluids more concentrated than the surrounding medium are activated, or "invented" in the case of the entry of the first vertebrates into lowered salinities. In terms of branchial salt transport, this means that extrusion of Na^+, Cl^-, and possibly other ions must

Table I

Euryhaline Fish Families[a]

Petromyzontidae (4)	Anablepidae (4)	Toxotidae (3)
Carcharhinidae (1)	Poeciliidae (4)	Coracinidae (2)
Pristidae (1)	Melanotaeniidae (5)	Scatophagidae (3)
Dasyatidae (1)	Atherinidae (2)	Rhinoprenidae (2)
Ascipenseridae (4)	Neostethidae (4)	Nandidae (5)
Lepisosteidae (5)	Phallostethidae (5)	Embiotocidae (1)
Notopteridae (5)	Syngnathidae (1)	Cichlidae (4)
Clupeidae (2)	Gasterosteidae (3)	Pomacentridae (1)
Engraulidae (2)	Synbranchidae (5)	Mugilidae (2)
Elopidae—Megalopidae (2)	Amphipnoidae (5)	Polynemidae (2)
Anguillidae (3)	Synanceiidae (1)	Cheimarrhichthyidae (5)
Salmonidae (3)	Cottidae (2)	Dactyloscopidae (1)
Retropinnidae (5)	Centropomidae (2)	Bovichthyidae (3)
Aplochitonidae (5)	Percichthyidae (3)	Congrogadidae (1)
Galaxiidae (5)	Serranidae (1)	Notograptidae (2)
Osmeridae (3)	Theraponidae (1)	Blenniidae (1)
Plecoglossidae (3)	Kuhliidae (3)	Eleotridae (1)
Salangidae (4)	Apogonidae (1)	Gobiidae (2)
Chanidae (2)	Sillaginidae (2)	Kraemeriidae (1)
Cyprinidae—*Tribolodon* (5) only	Carangidae (1)	Gobioididae (3)
Ariidae (1)	Leiognathidae (2)	Trypauchenidae (2)
Plotosidae (3)	Lutjanidae (1)	Microdesmidae (1)
Aspredinidae—Aspredininae (5)	Lobotidae (2)	Kurtidae (4)
Gadidae (1)	Gerreidae (1)	Anabantidae (5)
Ophidiidae (1)	Haemulidae (1)	Gobiesocidae (2)
Batrachoididae (1)	Sparidae (1)	Bothidae (1)
Exocoetidae (1)	Sciaenidae (1)	Pleuronectidae (1)
Belonidae (1)	Monodactylidae (2)	Soleidae (2)
Oryziatidae (5)	Pempheridae (1)	Cynoglossidae (1)
Cyprinodontidae (4)	Leptobramidae (2)	Tetraodontidae (2)

[a]The numbers 1–5 indicate salinity in which most members of a given family may be found, as well as the relative number of entrants into other salinities: 1, Mostly in seawater, rarely in fresh water; 2, frequently brackish water, occasionally fresh water; 3, mostly brackish water with members in both fresh water and seawater—catadromous or anadromous; 4, frequently brackish water, occasionally seawater; 5, mostly fresh water, occasionally brackish. On this scale, 0 would be totally seawater and 6, totally fresh water.

cease, while extraction of these now-needed ions from the medium must commence. Moreover, the rate of ionic uptake must balance the net diffusional plus urinary loss of those ions in order for blood NaCl concentrations to be maintained relatively stable. Thus, not only vectors, but magnitudes of ionic movements across the gills must be modulated. Of course, we must assume that nonbranchial osmoregulatory parameters such as oral ingestion and urinary ionic excretion are also modulated appropriately for euryhalinity to be achieved; however, these regulatory processes are not the subject of this chapter and in fact have been rarely investigated.

This interplay between salt loss and extraction in hypo-osmotic salinities has been described by various investigators (Shaw, 1961; Evans, 1975a; Kirschner, 1979) and is diagrammed in Fig. 1. If we assume that the ionic extraction mechanisms are carrier mediated, the rate of uptake can be described by a saturation curve analogous to that defined by Michaelis and Menten for an enzymatic reaction (Michaelis and Menten, 1913). If we further assume that ionic loss is purely dependent on "Fickian" kinetics (i.e., proportional to permeability, surface area, and ionic gradient) and therefore linear,* by definition, ionic regulation for a given species is only possible at external salinities (designated critical salinity minimum, CSM, in Fig. 1) equal to or above those in which uptake is equal to efflux. This relationship is defined by Eq. (4).

Since

$$J_{in} = \frac{J_{max}\,(\mathrm{CSM})}{K_m + (\mathrm{CSM})} \tag{1}$$

and

$$J_{out} = (D/l)(\mathrm{SA})C_i - C_o \tag{2}$$

and

$$J_{in} = J_{out} \tag{3}$$

$$\frac{J_{max}\,(\mathrm{CSM})}{K_m + (\mathrm{CSM})} = (D/l)(\mathrm{SA})C_i - C_o \tag{4}$$

where J_{in} = ionic influx; J_{out} = ionic efflux; J_{max} = maximum, carrier-mediated rate of ionic uptake; CSM = critical salinity minimum; K_m = the external ionic concentration at which 50% of the carrier molecules are bound to the ion species (substrate) involved, inversely related to the affinity of the carrier for the substrate; D = the diffusion coefficient of the membrane for the ion species involved; l = the length of the diffusion pathway (D/l is equivalent to the epithelial permeability to the ion species); SA = the sur-

*For graphic and calculation simplicity we have dictated that the net efflux remains constant as the external salinity falls. This is not strictly true [see Eq. (2)] but avoids the use of computers when calculating the CSM (see later and Fig. 2) without changing the conclusions drawn.

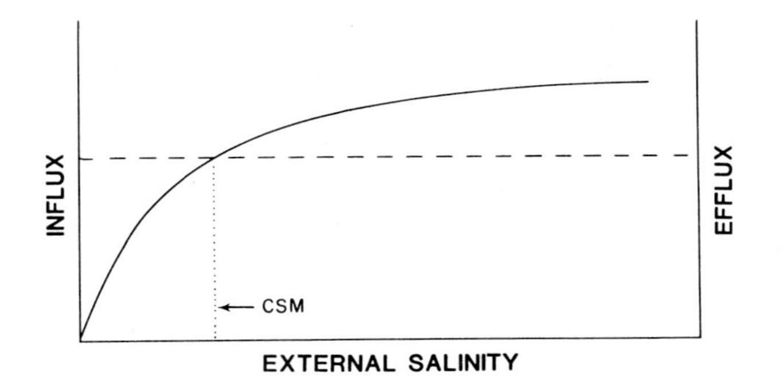

Fig. 1. Relationship between ionic efflux (---; assumed to be constant over a range of salinities), ionic uptake (——), and external salinity. For ease of calculation and graphic clarity, ionic efflux is set to be constant over the range of salinities used. This, of course, will not be strictly true, since Eq. (2) dictates that the net ionic efflux will increase as the external salinity falls. However, despite this oversimplification, the relationships between ionic uptake and loss hold true. The critical salinity minimum (CSM) is that external salinity where uptake is equal to or exceeds loss.

face area of the epithelium; C_i = ionic concentration in the blood (or, more properly, epithelial cell); and C_o = ionic concentration in the external medium ($C_i - C_o$ = ionic gradient favoring a net ionic efflux).

If Eq. (4) defines ionic balance in lowered salinities, then it should be obvious that the CSM can be lowered by either altering the rate of ionic loss or altering the kinetics of ionic uptake. Thus, the CSM is lowered by any or all of the following: (1) decrease in the ionic gradient, $C_i - C_o$, (2) decrease in the diffusion coefficient of the membrane, D, (3) increase in the diffusion path, l, (4) decrease in the surface area, SA, (5) increase in J_{max} by increasing the number of carrier molecules, or their rate of loading and unloading of the ionic substrate, and (6) decrease in K_m, which is actually an increase in the affinity of the carrier for the ionic substrate.

A. Factors Controlling Ionic Losses

Equation (2) notes that ionic losses can be regulated by variations in ionic gradient, in branchial permeability (D/l), and/or surface area. Reduction in any of these components of the efflux could reduce the CSM—Eq. (4)—and, theoretically, allow entry into reduced salinities.

While there is little doubt that the fall in blood NaCl levels associated with entry into brackish/fresh waters (see representative data in Holmes and Donaldson, 1969; Evans, 1979) can reduce the ionic gradient between the blood and the external medium, and hence the diffusional efflux—see Eq. (1)—this reduction is not sufficient to account for the fact that the Na^+ and Cl^- efflux from freshwater fish is only some 1–10% of that found from marine teleosts (see Table II and Evans, 1979, for representative flux rates). One must conclude, therefore, that reduction in the permeability and/or the surface area are the dominant parameters controlling ionic losses in lowered salinities. Unfortunately, we do not have sufficient gill morphometric data to be able to delineate differences in either diffusional path length (l) or surface area (SA) between marine and freshwater fish. Preliminary data indicate that the lamellar surface areas ($cm^2\ g^{-1}$) are actually similar in marine and freshwater species (Hughes and Morgan, 1973). Moreover, these parameters are probably limited by the constraints of gas exchange, rather than ion regula-

Table II

Na^+ and Cl^- Effluxes from Selected Marine, Freshwater, and Euryhaline Teleosts[a]

Species	Ion	Efflux (μmol 100 $g^{-1}\ hr^{-1}$)
Seawater		
Mugil capito	Na^+	3927
	Cl^-	2094
Cottus scorpius	Na^+	1077
Gobius cruentatus	Na^+	1160
Chromis chromis	Na^+	3500
Anguilla anguilla	Na^+	1450
	Cl^-	1200
Platichthys flesus	Na^+	2600
	Cl^-	1240
Tilapia mossambica	Na^+	2937
Fundulus heteroclitus	Na^+	2050
Salmo gairdneri	Na^+	254
Fresh water		
Carassius auratus	Na^+	9
	Cl^-	13–21
Anguilla anguilla	Na^+	48
	Cl^-	3–5
Platichthys flesus	Na^+	22
	Cl^-	1–2
Tilapia mossambica	Na^+	224
Fundulus heteroclitus	Na^+	60
Salmo gairdneri	Na^+	23

[a] Data from Evans (1979).

tion, so we might propose that they remain basically unaltered as the fish enters a lowered salinity. Nevertheless, an extensive examination of the morphometrics of gills from various marine and freshwater fish species (or, even better, comparison between a series of euryhaline species in high and low salinities) is necessary before we can definitely conclude that variations in l and SA play only a limited role in euryhalinity.

Nevertheless, it appears that alteration in the diffusion coefficient (D, often confused with permeability, D/l) plays an important role in euryhalinity. Parenthetically, it is interesting to note that this limitation of D may not have been important in the initial entry of the vertebrates into lowered salinities. This proposition is supported by our preliminary data indicating an extremely low rate of both Na^+ and Cl^- efflux from the hagfish, *Myxine glutinosa* (D. H. Evans and K. More, unpublished). In fact these effluxes are similar to those described for either marine elasmobranchs or freshwater teleosts (Evans, 1979). This is especially surprising in view of the fact that hagfish are nearly isonatric and isochloric to seawater and therefore have little "need" to limit ionic permeability. If indeed relative ionic impermeability is the primitive vertebrate condition, one must ask why modern marine teleost fishes have secondarily adopted a relatively great ionic permeability, which may preclude their reentry into lowered salinities. The question remains unanswered. Of course, one should add the proviso that this apparently low ionic permeability of hagfish could merely be secondary to a much reduced branchial surface area. Unfortunately, we have no morphometric comparisons of gill area for this interesting group of fish.

Additional data indicate that the passive branchial efflux of at least Na^+ (presumably dependent on D, but variations of l or SA cannot be ruled out) can be controlled by both intrinsic and extrinsic factors. Fish prolactin seems to be the dominant intrinsic factor controlling salt loss (see reviews by Hirano, 1977; Lahlou, 1980; Loretz and Bern, 1982). Initial studies by Burden (1956) demonstrated that the pituitary was necessary for the survival of killifish (*Fundulus heteroclitus*) in fresh water. Subsequent replacement therapy experiments (Pickford and Phillips, 1959) showed that the critical hormone was prolactin. Many studies followed immediately showing that prolactin was necessary for the freshwater survival of many species of fish (see review by Ensor and Ball, 1972). Prolactin's central role in low-salinity adaptation by fish has been substantiated by subsequent studies demonstrating that prolactin-secreting cells are activated in low salinities (Dharmamba and Nishioka, 1968; Wendelaar Bonga and Van der Meij, 1981) and that the blood levels of the hormone increase substantially when fish are acclimated to a reduced salinity (Nicol *et al.*, 1981). Flux studies by Potts and Evans (1966) and Maetz *et al.* (1967a,b) showed clearly that prolactin injection into

Table III

Effect of Hypophysectomy and Prolactin Injection on the Na^+ Efflux and Influx across *Fundulus heteroclitus* Transferred from Seawater to Fresh Water[a]

Fluxes (μmol 100 g^{-1} hr^{-1})	Control	Hypophysectomy	Prolactin injection
Efflux	23	50	26
Influx	34	21	27

[a] Data from Maetz *et al.* (1967).

hypophysectomized fish prompted a significant reduction in the efflux of Na^+, with little stimulation of Na^+ influx (Table III). It was initially hypothesized that the sodium-retaining ability of prolactin was associated with stimulation of mucous secretion, since many studies showed that prolactin treatment was followed by a proliferation of mucous cells (e.g., Ogawa, 1970; Olivereau and Lemoine, 1971; Marshall, 1976). However, Marshall (1978) has shown that dilute mucous solutions "could not impede the passive movement of ions sufficient to explain the sodium-retaining effect of prolactin." Additional data (Table IV) indicate that prolactin can significantly decrease the electrical conductance of the isolated opercular membrane of seawater-adapted tilapia, *Sarotherodon mossambicus* (Foskett *et al.*, 1982; see Chapter 5, this volume, for a general discussion of the opercular membrane preparation and its use as a model system for fish branchial ionic extrusion mechanisms), which is probably associated with a rearrangement of the epithelium, including reduction in the number of chloride cell to chloride cell junctions, thought to have relatively high electrical conductances (Sardet *et al.*, 1979). Some data indicate that prolactin may also function to "turn off" the branchial ionic extrusion mechanisms, which would be a liability in lower salinities (see later).

It appears clear that extrinsic factors such as calcium concentrations may also play a significant role in determining euryhalinity by alteration of passive ionic permeability. The ability of calcium to promote survival of stenohaline marine fishes in fresh waters was first systematically investigated by Breeder (1934), and other studies have demonstrated a central role for calcium as an external control of fish epithelial ionic permeability. Cuthbert and Maetz (1972) found that calcium-chelating agents in the external medium increased Na^+ efflux from goldfish, *Carassius auratus*, and Eddy (1975) found that addition of calcium to the external medium reduced both the Na^+ and Cl^- efflux from the same species. More relevant to a discussion of euryhalinity, Carrier and Evans (1976) demonstrated that the pinfish, *Lagodon rhomboides*, is able to tolerate hypo-osmotic salinities only if relatively substantial concentrations (5 mM) of calcium are present (see later). Unfortunately, whereas there is little doubt that relatively high ambient calcium concentrations can promote the survival of marine fish in lowered salinities (Breeder, 1934; Hulet *et al.*, 1967), the usually low concentrations of calcium found in most fresh waters preclude the entry of marine fish into these waters, presumably, unless they have intrinsic permeability effectors such as relatively high concentrations of prolactin. Nevertheless, in regions of the world that have hard fresh waters (e.g., the Bahamas and Florida), one often finds entry of normally stenohaline marine fish into the freshwater fauna (Neill, 1957).

B. Factors Controlling Cessation of NaCl Efflux

Of course, entry into lowered salinities dictates a cessation of the active extrusion of salt, which is a requisite for survival in seawater. A large body of evidence indicates that (Na^+,K^+)-activated ATPase is associated with salt extrusion by at least marine teleost fishes and that branchial activity of the enzyme is inversely related to salinity (see Karnaky, 1980, for a review, and other chapters in this volume). Maetz *et al.* (1968) found that injection of prolactin into intact, seawater-acclimated *Fundulus heteroclitus* inhibited the Na^+ efflux rate by some 50%, and Pickford *et al.* (1970) found that prolactin inhibited (Na^+,K^+)-activated ATPase in the same species. This certainly is consistent with the proposition that prolactin may play a central role in adaptation to reduced salinities. Unfortunately, this very interesting finding has not been extended by investigations of other species. Foskett *et al.* (1982) have shown that prolactin inhibited the short-circuit current and net Cl^- extrusion across the isolated opercular epithelium of the tilapia, *Sarotherodon mossambicus* (Table IV), probably both by direct effects on the active transport pathway (the effect is reversible by inhibition of the

Table IV

Effect of Prolactin on Conductance, Short-Circuit Current, and Net Cl^- Efflux across the Isolated Opercular Skin from Seawater-Adapted *Sarotherodon mossambicus*[a]

Experimental group	Conductance (msec cm^{-2})	Short-circuit current ($\mu A\ cm^{-2}$)	Net Cl^- efflux ($\mu mol\ cm^{-2}\ hr^{-1}$)
Saline controls	4.94	120.0	3.97
Prolactin injected[b]	1.75	17.7	0.88

[a]Data from Foskett *et al.* (1982).

[b]Prolactin injections were 10 $\mu g\ g^{-1}\ day^{-1}$ for 5 days.

breakdown of intracellular cyclic AMP) and by dedifferentiation of some of the "chloride cells." The result would be both a fall in the level of active extrusion and a reduction in passive conductance (see earlier), both requisites for entry into lowered salinities.

Other evidence indicates that epinephrine may also play a critical role in inhibiting the active extrusion of ions as the euryhaline fish enters reduced salinities. Pic *et al.* (1975) found that injection of epinephrine into intact mullet (*Mugil cephalus*) inhibited both Na^+ and Cl^- efflux, via α-adrenergic receptors. Subsequently, Girard (1976) used the perfused trout (*Salmo gairdneri*) head to demonstrate that this α-mediated epinephrine inhibition of Na^+ extrusion could be separated frm the hemodynamic effects of the drug. In addition, Shuttleworth (1978) found that the transepithelial electrical potential (TEP) across isolated flounder (*Platichthyes flesus*) gills perfused on both mucosal and serosal surfaces with teleost Ringers solution was significantly inhibited by the addition of 10^{-6} *M* epinephrine. This inhibition was reversed by the addition of the α-adrenergic blocker phentolamine. Since this TEP was presumably secondary to electrogenic active transport, Shuttleworth proposed that the epinephrine inhibited the active extrusion pump itself, via α receptors. More direct evidence for this proposition comes from recent work (Table V) with the isolated opercular epithelium of *F. heteroclitus*. Under short-circuited conditions, epinephrine (via α-adrenergic receptors) inhibits both the short-circuit current and the net efflux of Cl^- (Degnan *et al.*, 1977; Degnan and Zadunaisky, 1979). It is interesting to note that this α-mediated inhibition of active Cl^- extrusion is not mediated via changes in the intracellular concentration of cyclic AMP (Mendelsohn *et al.*, 1981). Nevertheless, it appears quite clear that increased epinephrine secretion could play a central role in adaptation to lowered salinities. Unfortunately, we have no published data on relative activities of the chromaffin cells or blood epinephrine levels during acclimation to reduced salinities.

Table V

Effect of Epinephrine on Short-Circuit Current and Net Cl^- Efflux across the Isolated Opercular Skin from Seawater-Adapted *Fundulus heteroclitus*[a]

Experimental group	Short-circuit current ($\mu A\ cm^{-2}$)	Net Cl^- efflux ($\mu mol\ cm^{-2}\ hr^{-1}$)
Control	122.0	4.49
Epinephrine (10^{-6} *M*)	67.6	2.60

[a] Data from Degnan *et al.* (1977).

C. The Role of Changes in the Transepithelial Electrical Potential (TEP)

The TEPs across marine fish range from approximately −5 mV (blood relative to medium) to +35 mV, with the majority falling above +10 mV (Evans, 1980b). These TEPs are usually the results of a greater cationic than anionic permeability, and an electrogenic Cl^- extrusion mechanism (see other chapters in this volume). When marine teleosts are transferred to fresh water or sodium-free solutions in the laboratory, the TEP rapidly falls by 5 to 75 mV (e.g., Potts and Eddy, 1973; Kirschner *et al.*, 1974; Evans and Cooper, 1976), sufficient to account theoretically for significant reductions in at least sodium loss. These reversals of the TEP could account for a reduction in the net loss of Na^+ during entry into lowered salinities, but one must remember that there would be a concomitant increase in the passive loss of Cl^-. Moreover, it is unlikely that many species experience such rapid entries into dilute solutions in nature. It is more likely that dilutions are sufficiently gradual that changes in the TEP play rather insignificant roles in controlling ionic losses.

D. Alterations of Branchial Permeability to Water

The fish gill not only faces a reversal of ionic gradients as the fish enters hypo-osmotic salinities, it also faces a reversed osmotic gradient, producing a net uptake of water. Initial studies of the diffusional fluxes of water across various species of teleosts demonstrated that freshwater species or euryhaline forms adapted to fresh water had a slightly higher diffusional water permeability (measured as the flux of tritiated water) than marine or seawater-adapted individuals (Evans, 1969). However, it is likely that these differences in apparent diffusional water permeability were secondary to the relatively high saltwater calcium concentrations, since other data have shown that calcium has pronounced effects on diffusional water permeability of the fish branchial epithelium (Potts and Fleming, 1970; Isaia and Masoni, 1976). Studies of the net osmotic flux of water across isolated gill arches have shown that osmotic flow is apparently rectified across the fish branchial epithelium. This is demonstrated by the finding that arches from the eel, *Anguilla anguilla* (Table VI), and *Salmo gairdneri* adapted to either fresh water or seawater gain water from a freshwater incubation medium much faster than they lose water into saltwater incubation medium, even though the osmotic gradient across the gill in seawater is approximately three times that across the gill in fresh water (Isaia and Hirano, 1975; Isaia *et al.*, 1979).

Table VI

Net Water Fluxes across Isolated Gills from *Anguilla anguilla*[a]

Adaptation medium	Incubation medium	
	Seawater	Fresh water
Fresh water		
Net flux[b]	−75.4	+178
Osmotic permeability[c]	+90.7	+673
Seawater		
Net flux	−19.1	+327
Osmotic permeability	+24.1	+1079

[a]Data from Isaia and Hirano (1975).

[b]Net fluxes are in mg hr^{-1} (100 mg dry $wt)^{-1}$; negative flux denotes net efflux, positive flux denotes net influx.

[c]Osmotic permeabilities are in mg hr^{-1} (100 mg dry $wt)^{-1}$ Osm^{-1}.

For instance, in the case of *A. anguilla*, the calculated osmotic permeability in seawater-acclimated eels is only 2–13% of that calculated for freshwater-acclimated eels (Table VI). These data confirmed intact animal studies that demonstrated that the apparent osmotic permeability was greater for freshwater fish (as measured by the urine flow) than for marine species (as measured by rates of oral ingestion of the medium) (Motais *et al.*, 1969).

Isaia *et al.* (1979) suggested that this apparent rectification of osmotic water flow could be secondary to (1) an increase in the ionic permeability in seawater with a concomitant fall in the reflection coefficient and hence real osmotic gradient across the gill, (2) reduced calcium concentrations in fresh water, or (3) alteration in the dimensions of the intercellular channels, which may affect the resistance to osmotic water flow. Unfortunately, we have no published data allowing us to determine which of these changes may play a major role in the discrepancies between the osmotic permeabilities of freshwater versus marine gills. Nevertheless, it appears that the gills may possess regulatory mechanisms that can limit osmotic water fluxes in either salinity. Isaia and Hirano (1975) found that gills isolated from freshwater-adapted eels gained weight much more slowly in fresh water (and hence had a lower osmotic permeability) than those isolated from seawater-adapted eels. Conversely, gills from seawater-adapted individuals lost weight in seawater much more slowly than gills from freshwater-adapted individuals (Table VI). Interestingly, gills isolated from freshwater-adapted *S. gairdneri* also gained weight in freshwater solutions more slowly than those isolated from seawater-adapted individuals, but gills isolated from seawater-adapted individuals lose water in saltwater solutions faster than those isolated from freshwater-adapted fish (Isaia *et al.*, 1979). Correlated with this apparent lack of osmotic control in the seawater-adapted trout is the finding that this species cannot survive acute transfer to full-strength seawater (Lahlou *et al.*, 1975). Nevertheless these data indicate that one adaptation allowing entry into lowered salinities may be the ability to lower the osmotic permeability of the branchial epithelium.

Interestingly, prolactin and epinephrine, each known to play important roles in ion balance in lowered salinities, both affect water permeability of the fish gill. Prolactin appears to decrease osmotic uptake of water, because it has been shown that isolated gills from prolactin-injected sticklebacks, *Gasterosteus aculeatus* (Lam, 1969), goldfish, *Carassius auratus* (Ogawa *et al.*, 1973), and eels, *Anguilla japonica* (Ogawa, 1974, 1975) displayed a decreased uptake of water from freshwater solutions. However, in other studies, prolactin appears to increase the diffusional flux of water across both *Fundulus kansae* (Potts and Fleming, 1970) and goldfish (Lahlou and Giordan, 1970). One can only speculate that differences in experimental technique can explain these differing results. Pic *et al.* (1974) found that the efflux of tritiated water from both seawater- and freshwater-acclimated mullet, *Mugil capito*, increased after epinephrine injection. Moreover, the effect was inhibited by concomitant injection of the β-adrenergic blocker propranolol and was therefore probably mediated via a β-adrenergic receptor. Since epinephrine-injected animals also lost weight more rapidly in seawater, these authors proposed that both diffusional and osmotic water permeabilities were stimulated by the catecholamine. Subsequent studies, using isolated heads and gills of *S. gairdneri* (Haywood *et al.*, 1977; Isaia *et al.*, 1979; Jackson and Fromm, 1981) and gulf toadfish, *Opsanus beta* (Oduleye and Evans, 1982), have corroborated the stimulatory action of epinephrine on the transbranchial flux of water (e.g. Table VII). These studies have corroborated earlier data on intact animals that showed that stress or swimming increased the water flux across goldfish (Lahlou and Giordan, 1970), trout (Wood and Randall, 1973), and eels (Isaia and Masoni, 1976). However, one could propose (e.g., Bergmann *et al.*, 1974) that this epinephrine effect was secondary to an increased branchial surface area, produced by an increase in the number of perfused lamellae (lamellar recruitment), which has been demonstrated in the epinephrine-treated trout (Booth, 1979) and catfish, *Ictalurus punctatus* (Holbert *et al.*, 1979).

Haywood *et al.* (1977) found, using the perfused trout head, that mechanical reduction of the afferent perfusion flow by 80% reduced the diffusional water efflux by only 45%, whereas stimulation with epinephrine doubled the water efflux and only increased the perfusate flow by 33%. Haywood and colleagues proposed that these results indicated that epinephrine also had direct effects on water permeability, but one must be quite cautious

Table VII

Effect of Epinephrine on the Net Water Fluxes across the Gills of *Salmo gairdneri*[a,b]

Adaptation/incubation medium	Diffusion water flux (ml hr^{-1} 100 g^{-1})		Osmotic net flux[c] [mg hr^{-1} (100 mg dry wt)$^{-1}$]	
	Control	Epinephrine (10^{-6} *M*)	Control	Epinephrine (10^{-6} *M*)
Seawater	+6.4	+10.5	−112	−156
Fresh water	+11.0	+32.4	+41	+57

[a]Data from Isaia *et al.* (1979).
[b]Diffusional fluxes were measured on perfused heads, osmotic fluxes on ligated gill arches.
[c]Negative flux denotes net efflux, positive flux denotes net influx.

in equating percentage changes in flow/pressure with percentage changes in functional gill surface area. We simply have no data on this relationship, and it may very well be that only slight changes in the measured flows/pressures result in quite substantial changes in the number of lamellae perfused and hence (theoretically) functional surface area for diffusional and osmotic water movements. It therefore appears that the actual direct effects of epinephrine on gill water permeability remain to be determined. At this point it is probably best to propose that the effects are a combination of changes in surface area, diffusional distance, and the diffusion coefficient to water of the membrane, that is, an increased (D/l)(SA) (Jackson and Fromm, 1981; Oduleye and Evans, 1982).

Thus, it appears that prolactin is necessary in lowered salinities for the reduction in both ionic and osmotic permeabilities, while epinephrine stimulates active uptake of at least Na^+, and osmotic uptake of water. With the exception of this latter effect, these responses to both hormones are adaptive for osmoregulation in reduced salinities.

E. The Origin of Ionic Extraction Systems

It had been commonly thought that the dominant factor limiting entry of marine species into brackish or fresh water was their inability to extract ions from the external medium. After all, cessation of drinking, alteration of renal water and solute losses, and modulation of the J_{out} across the branchial epithelium are quantitative rather than qualitative changes. However, ionic extraction from hypo-osmotic salinities is via active transport systems that are relatively well defined for freshwater species, but are an ionic liability in marine species and, hence, are probably not present (Maetz and Garcia Romeu, 1964).

Relatively early ecological data (Breeder, 1934) could have been (but never were) used to question this dogma, because it was shown that some normally stenohaline marine species (e.g., the horse-eye jack, *Caranx latus*) were living in freshwater lakes on Andros Island, the Bahamas. Analysis of the water showed that it contained relatively high (1.0–1.5 m*M*) concentrations of calcium. Although the ability of Ca^{2+} to lower passive permeability (and presumably, in this case, J_{out}) is well documented (e.g., Lowenstein, 1980), it should have been noted that, unless the resistance to salt loss is infinite (i.e., J_{out} reduced to zero), survival would have been possible only if some means of ionic uptake were also present in these fish.

Data have been accumulating that the mechanisms for NaCl uptake are actually resident in the branchial epithelium of marine fish. In early studies we found that at least in the euryhaline molly, *Poecilia latipinna*, influx of Na^+ into seawater-adapted individuals was saturable, rather than linear as expected with a simple diffusion system (Evans, 1973). Moreover, Na^+ influx was inhibited by external ammonium ions and protons, as well as potassium (Evans, 1975b). We proposed that these preliminary data indicated that at least Na^+/NH_4^+ ionic exchange was resident even in the saltwater branchial epithelium and that permeability, rather than the lack of the appropriate ionic extraction system, limited the entry of most species into lowered salinities (Evans, 1975a). More recent studies (Evans, 1977, 1982a; Evans *et al.*, 1979; Bentley *et al.*, 1976) have strengthened this view and have indicated that both Na^+/NH_4^+ and Na^+/H^+ ionic exchange systems are present in at least nine species of marine teleosts and elasmobranchs (Table VIII). We first proposed that the presence of at least the ionic extraction systems for Na^+ in marine teleosts and elasmobranchs was a remnant from earlier evolution in fresh water for both groups (Evans, 1975a), but could say nothing about the origin of the system in vertebrate evolution. However, additional data (Evans, 1980d, 1984a) indicate that at least Na^+/H^+ and Cl^-/HCO_3^- exchanges are present in the branchial epithelium of the marine hagfish, *Myxine glutinosa*. Since this group has apparently never entered fresh water during its evolution (Hardisty, 1979) it is apparent that Na^+ and Cl^- extraction systems predated the entry of the early vertebrates into brackish and fresh water.

Two relevant questions come to mind: (1) Why were these ionic exchange systems present even before the vertebrates entered fresh water, and (2) do they represent an ionic liability to the usually hyporegulating marine teleost or elasmobranch? It appears that Na^+/NH_4^+, Na^+/H^+, and Cl^-/HCO_3^- ionic exchange systems arose, not for the influx of NaCl, but

Table VIII

Evidence for the Presence of Na^+/NH_4^+ and Na^+/H^+ Exchange Systems in Marine Fish

1. Na^+ influx in low salinities is saturable, rather than linear (Evans, 1973; Bentley *et al.*, 1976).
2. Na^+ influx is reduced by the addition of NH_4^+, H^+, or amiloride to the external medium (Evans, 1975b; Bentley *et al.*, 1976; Carrier and Evans, 1976; Evans *et al.*, 1979).
3. Na^+ efflux can be stimulated by high external NH_4^+ concentrations (Evans, 1977).
4. NH_4^+ and/or H^+ efflux is dependent on external Na^+ concentration (Evans, 1977, 1982a; Evans *et al.*, 1979).
5. Injection of NH_4^+ stimulates Na^+ influx (Payan and Maetz, 1973; Evans, 1977).

for the efflux of unwanted nitrogen and the control of pH. It is relatively clear now that these branchial ionic exchange mechanisms play an important role in marine and freshwater fish acid–base regulation (see reviews by Cameron, 1978; Heisler, 1980; Evans, 1984b). Thus, influx of NaCl, so important in lowered salinities, was coupled to the extrusion of unwanted metabolites, presumably (as first proposed by Krogh, 1938) to maintain some semblance of electroneutrality across the branchial epithelium. Its use in ion regulation in lowered salinities was therefore a secondary acquisition.

The role of these ionic exchange systems in ionic balance (or imbalance) in marine fish remains to be determined. A single study (Evans, 1980c) demonstrated that, in the marine gulf toadfish (*Opsanus beta*), the Na^+ load presented by Na^+/NH_4^+ exchange was small but significant (approximately 10% of the net influx, twice as large as oral ingestion of sodium). However, this species apparently faces a significant diffusional gain of Na^+ (Evans and Cooper, 1976), contrary to some other teleosts (Evans, 1980b; Kirschner, 1980), so it would be of great interest to determine the significance of Na^+ influx via Na^+/NH_4^+ or Na^+/H^+ exchange in species that have little or no net diffusional influx of Na^+ in seawater. All marine teleosts apparently face a significant net diffusional influx of Cl^- (Evans, 1980b; Kirschner, 1980), so it is apparent that Cl^-/HCO_3^- exchange presents a rather insignificant Cl^- load to marine teleost species. However, marine elasmobranchs apparently have quite small diffusional gains of both Na^+ and Cl^- (Evans, 1979, 1980a), so even the relatively small influx of NaCl generated by excretion of NH_4^+, H^+, and/or HCO_3^- may be significant. Myxinid hagfish are isotonic to seawater but maintain blood Na^+ and Cl^- concentrations slightly but significantly above those in seawater (Hardisty, 1979). Diffusional losses are kept to a minimum by a relatively low permeability to both ions (D. H. Evans and K. More, unpublished), and one might assume that Na^+/NH_4^+, Na^+/H^+, and Cl^-/HCO_3^- exchanges are involved in the maintenance of these ionic differentials. However, we have no direct evidence for this proposition.

Thus, the evidence allows us to propose that the acquisition of ionic extraction systems was not limiting to the entry of the first vertebrates into brackish or fresh water and presumably is not now limiting the entry of modern fish species into these reduced salinities. It is therefore obvious that euryhalinity must be limited (in addition to the limitations of ionic efflux outlined previously) by the efficiency of the ionic exchange systems providing for the entry of needed Na^+ and Cl^-.

F. Factors Controlling Ionic Uptake Rates

Equation (1) illustrates that modulation of either the number of carrier molecules (increase in J_{max}) or their efficiency at transporting the substrate (decline in K_m) will increase the rate of uptake at a given external salinity. It should be obvious that, as with the other factors in Eq. (1), these two kinetic parameters are not mutually exclusive, that is, it is possible that both will be modulated in order to prompt entry into lowered salinities. Do euryhaline or freshwater fishes have a reduced K_m or an increased J_{max} relative to marine species? Unfortunately, so few data have been gathered on the kinetics of ionic uptake by any fish species that we are unable to provide an answer to this very important question. Nevertheless, discussion of some preliminary data is illuminating. Table IX lists the kinetic analyses for uptake of Na^+ by

Table IX

Kinetics of Na^+ Uptake by Fish in Fresh Water and Seawater

Species and medium	K_m (mM liter^{-1})	J_{max} (μmol 100g^{-1} hr^{-1})	Reference
Fresh water			
Fundulus heteroclitus	2.0	—	Potts and Evans (1966)
Platichthys flesus	0.8	35	Maetz (1971)
Salmo gairdneri	0.5	40	Kerstetter *et al.* (1970)
Salmo gairdneri[a]	0.02	35	Wood and Randall (1973)
Carassius auratus	0.3	65	Maetz (1973)
Poecilia latipinna	8.0	1200	Evans (1973)
Lagodon rhomboides	5.0	700	Carrier and Evans (1976)
Seawater			
Poecilia latipinna	17	3300	Evans (1973)
Lagodon rhomboides	22	3200	Carrier and Evans (1976)

[a]Adapted to very low external Na^+ concentrations (~40 μM).

the very few fish species that have been investigated. It is clear from these sparse data that firm conclusions are unwarranted, but nevertheless it appears that Na^+ uptake in fresh water vis-à-vis that in seawater is associated with an increase in the efficiency of the extraction system, but an apparent fall in J_{max}. However, one must be reminded that the Na^+ fluxes across seawater-acclimated teleosts may display apparent saturation kinetics because of either Na^+/Na^+ exchange diffusion or changes in the transepithelial electrical potential, especially if the influxes are measured at relatively high external sodium concentrations (see Evans *et al.*, 1973, for a discussion of this point). The presence of either, or both, phenomena may produce an overestimate of both K_m and J_{max}. The kinetic analyses of Na^+ influx into both *Poecilia latipinna* and *Lagodon rhomboides* were performed at external concentrations below 120 m*M* sodium, so exchange diffusion and TEP effects were minimized, but it is still quite possible that the differences between the kinetic parameters in the two salinities were overestimated. More data are certainly needed.

Other data suggest that the rate of Na^+ uptake may be modulated by fish hormones. Payan *et al.* (1975) found that both Na^+ influx and NH_4^+ efflux were stimulated by addition of epinephrine (10^{-5} *M*) to the Ringer solution perfusing the isolated, perfused trout head, presumably via stimulation of Na^+/NH_4^+ exchange (Table X). Experiments with adrenergic receptor-blocking agents (β vs α = propranolol vs phentolamine) indicated that this stimulation was mediated via β-adrenergic receptors, exclusive of any hemodynamic effects. Hormones from the pituitary may also be involved in stimulation of Na^+ uptake, since Dharmamba and Maetz (1972) and Maetz *et al.* (1967a) found that hypophysectomy of *Sarotherodon mossambicus* and *Fundulus heteroclitus*, respectively, is followed by a slight reduction in Na^+ influx. Maetz *et al.* (1964b) found that injection of either isotocin or arginine vasotocin into intact goldfish produced a significant stimulation of Na^+ in-

Table X

Effect of Epinephrine on the Influx of Na^+ and Efflux of Ammonia across the Isolated, Perfused Head of *Salmo gairdneri*[a]

Epinephrine concentration (*M*)	Fluxes (μmol $100g^{-1}$ hr^{-1})	
	Na^+ influx	Ammonia efflux
10^{-7}	21.5	19.8
10^{-5}	39.6	37.5

[a]Data from Payan *et al.* (1975).

flux, so the putative pituitary hormone may be from the neurohypophysis. Interestingly, inhibitory pituitary hormones may also play a role in the control of Na^+ uptake. Maetz *et al.* (1967b) found that hypophysectomy of the eel (*Anguilla anguilla*) was followed by stimulation rather than inhibition of Na^+ influx. Nonpituitary hormones also may modulate rates of Na^+ uptake. Maetz *et al.* (1964a) showed that extracts from the urophysis stimulated Na^+ influx into the goldfish, and interrenalectomy of the eel was followed by a decline in Na^+ uptake (Chester Jones *et al.*, 1967; Henderson and Chester Jones, 1967; Maetz, 1969a), which could be restimulated by injection of either aldosterone (Henderson and Chester Jones, 1967) or cortisol (Chester Jones *et al.*, 1967).

However, all these data on noncatecholamine hormones have been gathered on intact experimental animals. Since modulation of the active uptake of Na^+ is only one of many ways that Na^+ influx could be changed (e.g., hemodynamic, TEP, and permeability changes could have been important), these data must be considered preliminary and should certainly be extended using more modern techniques. Moreover, in no case has an alteration in J_{max} been differentiated from an alteration in K_m. It is also important to note that the time course of stimulation of Na^+ influx by these hormones has generally been found to be relatively short (i.e., in minutes). Since alteration of either the affinity of the Na^+ transport carrier or its concentration would presumably involve activation of nuclear events, one might propose that the time course should be much longer. This problem should be investigated. No studies have been published on the hormonal control of Cl^- uptake. S. F. Perry, P. Payan, and J.-P. Girard (personal communication) have reported that epinephrine stimulates (via α-adrenergic receptors) Cl^- uptake by the isolated, perfused trout head, but these are preliminary data.

G. Relative Roles of Efflux versus Influx Modulation in Acclimation to Lowered Salinities

The foregoing demonstrates that a reduction in ionic efflux and/or a stimulation of ionic uptake (via a reduction of K_m and/or an increase in J_{max}) is probably under hormonal control in fish, but the relative roles played by these two controlled variables remain to be determined. For comparative purposes, one can calculate, using Eq. (1) the theoretical reduction in the CSM produced by, for instance, an alteration of either K_m, J_{max}, or J_e by 50%. This approach was recently utilized by Kirschner (1979) and can be illustrated by Fig. 2.

These simple calculations indicate quite clearly, given the oversimplification of equivalent theoretical changes in the three variables, that alterations

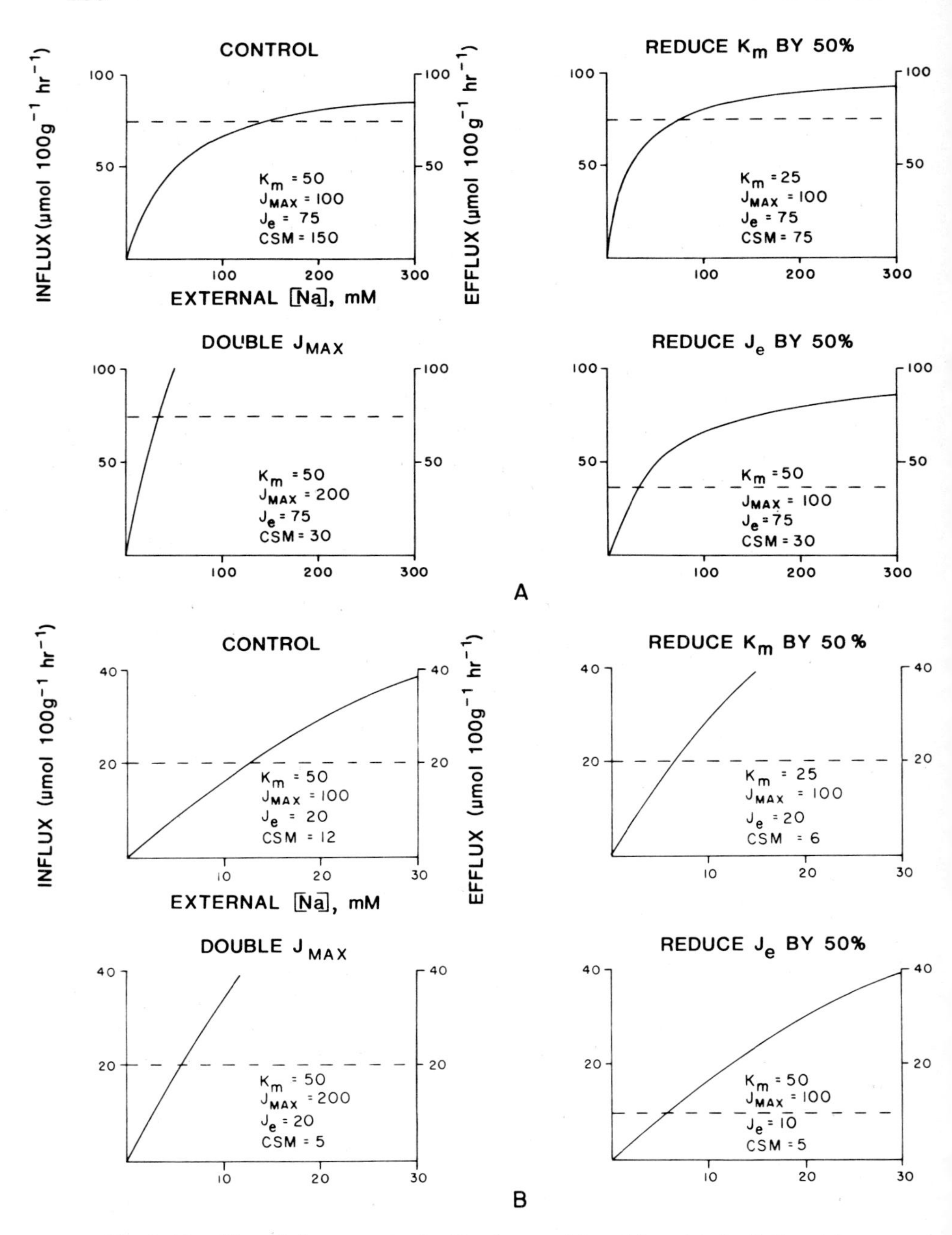

Fig. 2. The effect of alterations in the K_m, J_{max}, and J_e on the critical salinity minimum (CSM), given (A) relatively high K_m and high J_e, (B) relatively high K_m and low J_e, (C) relatively

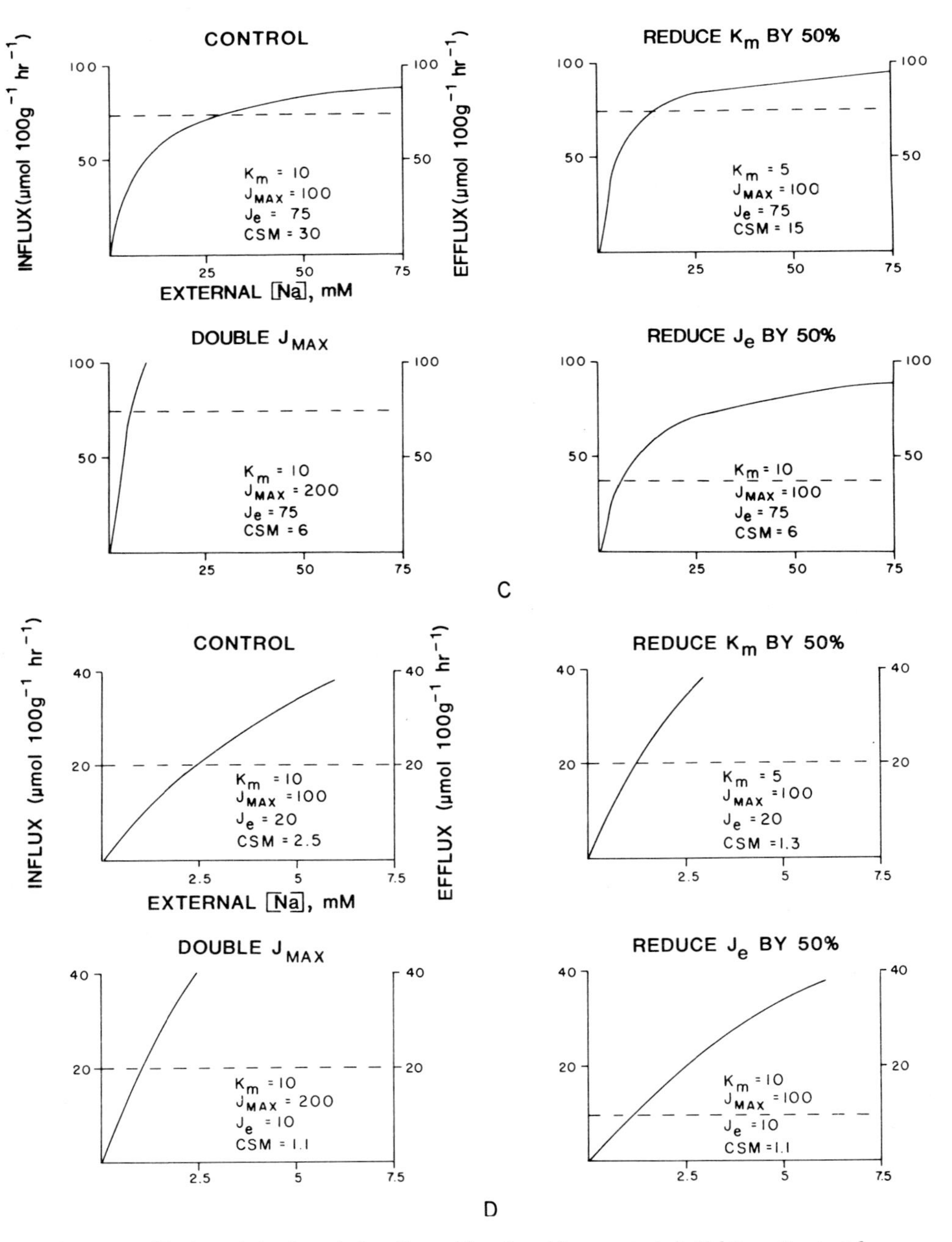

low K_m and high J_e, (D) relatively low K_m and low J_e, with a constant, initial J_{max}. See text for details. As in Fig. 1, J_e is assumed to remain constant with salinity, influx (——), and efflux (---).

in either the number of carriers (i.e., increase in J_{max}) or the efflux of ions (potentially mediated via permeability and/or surface area reductions) could play a more substantial role in prompting entry into lowered salinities than reduction of K_m. Unfortunately, we have very few experimental data that can be compared with these theoretical calculations. The data for *Poecilia latipinna* and *Lagodon rhomboides* in Table IX indicate that these euryhaline species lower both their K_m and J_{max} during acclimation to fresh water, so it appears that modulation of affinity rather than number of carriers is important in these species. Unfortunately, we have no data on the magnitude of the reduction of Na^+ efflux during acclimation to fresh water in *P. latipinna.* However, a recent study of the mechanisms of euryhalinity in *L. rhomboides* may give us a clearer picture of the interaction of intake versus loss vis-à-vis euryhalinity (Carrier, 1974; Carrier and Evans, 1976). This study found that *Lagodon* can tolerate direct transfer to salinities as low as 35% seawater (170 m*M* sodium) but can only survive in 5 m*M* NaCl artificial fresh water if the solution also contains 10 m*M* calcium. Under these conditions, survival is relatively indefinite. However, if animals that have been acclimated to this solution are transferred to 5 m*M* NaCl solutions that are calcium free, they die within 2.5 hr and display a 50% reduction in total body sodium, concomitant with a trebled efflux of Na^+. Other experiments have shown that this species, even when acclimated to seawater, has a component of the Na^+ influx that is inhibited by the addition of ammonia, acid, or amiloride to the external solution. Thus, presumably Na^+/NH_4^+ and Na^+/H^+ exchange mechanisms are present and allow entry into the lowered salinities, as long as the epithelial permeability is maintained relatively low by external Ca^{2+}. Moreover, acclimation to the lowered salinity is associated with a decline in both the K_m and J_{max} (Table IX).

However, it is also clear that modulation of Na^+ efflux, exclusive of Ca^{2+} effects, also plays a very important role. The efflux of Na^+ from pinfish acclimated to 5 m*M* NaCl–10 m*M* Ca^{2+} was some 5% of the total body sodium per hour, only 6% of that measured in individuals acclimated to seawater (which also contained 10 m*M* Ca^{2+}). Thus, although this rate of Na^+ efflux is clearly dependent on the presence of Ca^{2+} in the low salinity (see earlier), it is still significantly below that displayed by the seawater-acclimated individuals, indicating that acclimation to the lowered salinity has been associated with a decline in passive epithelial permeability. This interrelationship between the measured effluxes and uptake in *L. rhomboides* acclimated to seawater (S. W. ACC'L) and high-calcium fresh water [F. W. ACC'L (+Ca)] is diagrammed in Fig. 3. It is clear that, given the measured changes in the kinetics of Na^+ uptake, if the passive loss of Na^+ had not been reduced during acclimation, the fish would not have been able to balance loss and uptake. In fact, *Lagodon* is unable to tolerate direct transfer

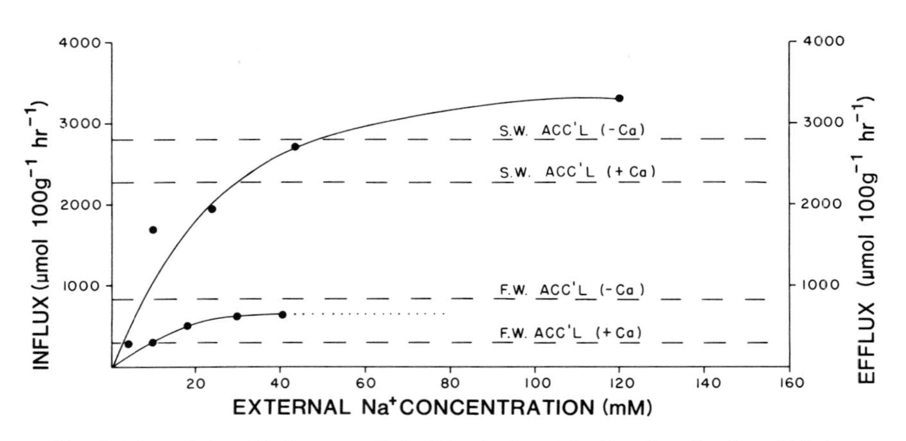

Fig. 3. The relationship between Na^+ efflux (---), uptake (——), and external Na^+ concentration in seawater- and fresh water (5m*M* NaCl)-acclimated *Lagodon rhomboides*. See text for details. (Data from Carrier, 1974.)

to even high-calcium fresh water; it must first be acclimated to approximately 35% seawater for some days (Carrier, 1974). It therefore appears that intrinsic control of permeability (via prolactin?) plays a substantial role in acclimation to reduced salinities, even in those species that maintain epithelial permeabilities that are sensitive to external Ca^{2+} concentrations. Moreover, it is important to note that even the presence of Ca^{2+} and preadaptation to brackish water did not reduce the Na^+ efflux (presumably permeability) of *Lagodon* to that of freshwater species. Its rate of efflux (5% total body sodium per hour) is still some five times that displayed by a number of freshwater fish species (Evans, 1979). Thus, one might propose that its CSM is still rather high compared to freshwater species, because of a combined low-affinity Na^+ uptake system and relatively high Na^+ permeability. Unfortunately, we know nothing about the mechanisms of Cl^- balance in this (or any other) species of euryhaline fish.

The foregoing indicates the importance of low epithelial permeability in entry into lowered salinities. However, the relative importance may be overestimated because it has only been studied directly in a species (*L. rhomboides*) with a relatively low-affinity Na^+ uptake system (see Table IX). It would obviously be of great importance to examine the kinetics of both efflux and uptake in a more euryhaline species, one that is not dependent on extrinsic factors such as calcium for entry into lowered salinities. Nevertheless, these scanty data indicate that the interplay between uptake and efflux probably defines the degree of euryhalinity of a given species. Whether modulation of the kinetics of uptake or loss is the determining factor probably depends on the species. We certainly need more data on this interesting subject, especially dealing with the interplay of Cl^- loss and uptake.

H. Elasmobranch Euryhalinity

Although there are numerous reports of the entry of a few species of sharks into fresh water (e.g., Smith, 1931; Thorson, 1972; Thomerson *et al.*, 1977), the fact remains that only members of the ray family Potamotrygonidae are actually resident in fresh water. Thus, it appears that euryhalinity in elasmobranchs is rather limited. Some evidence indicates that both sharks and rays possess mechanisms for at least Na^+/NH_4^+ and Na^+/H^+ exchange (Payan and Maetz, 1973; Bentley *et al.*, 1976; Evans *et al.*, 1979; Evans, 1982a), although the kinetics of these uptake systems have not been determined. Moreover, it is well documented that elasmobranchs display an extremely low ionic permeability, in the range of that described for freshwater teleosts (see Evans, 1979, for a review of elasmobranch osmoregulation). However, elasmobranchs also display a remarkably high permeability to water, which apparently can be reduced somewhat during acclimation to reduced salinities (Payan *et al.*, 1973), but which theoretically could result in a substantial influx of water. However, preliminary evidence indicates that the limitations to elasmobranch euryhalinity may be more complex than a simple lack of control of water permeability. Carrier and Evans (1973) measured Na^+, Cl^-, and water fluxes across *Potamotrygon* and found that it has retained the low ionic and high water permeability of its marine ancestors. In addition, our calculations of the theoretical net osmotic influx of water indicates that it could be balanced by urine flows similar to those described by Smith (1931) for the sawfish (*Pristis microdon*) in fresh water. Thus, modulation of water permeability may not be a requisite for entry of the elasmobranchs into reduced salinities. One might argue that the high urinary excretion necessary to balance the osmotic inflow of water might result in substantial ionic losses; however, Smith (1931) measured a urinary Cl^- loss from *P. microdon* of some 7 μM 100 g^{-1} hr^{-1}, only 25% of the total Cl^- efflux measured from *Potamotrygon* (Carrier and Evans, 1973). Although comparisons between species are fraught with inaccuracies, it is still clear that the actual factors limiting euryhalinity in elasmobranchs are not known and should be investigated.

III. THE FRESH WATER TO SEAWATER TRANSITION

Movement into salinities that are hyperosmotic to the body fluids dictates that the mechanisms of freshwater osmoregulation be inhibited while those of seawater osmoregulation are activated. In terms of branchial salt transport this means, in theory, that uptake of Na^+ and Cl ceases, while NaCl extrusion mechanisms begin to balance the net influx of salt produced by the diffusive uptake of NaCl across the gills and the gut salt uptake subsequent to the necessary (for osmotic balance) oral ingestion of seawater. The reader is referred to other chapters in this volume, and those cited in Section I, for discussions of the general mechanisms of saltwater osmoregulation.

Our recent findings that the mechanisms for Na^+ uptake are active in marine fishes, secondary to acid–base regulation and nitrogen excretion (see review by Evans, 1984b), indicate that the ionic extraction systems important in freshwater osmoregulation probably are not inhibited in seawater. The similarity between the ammonia excretion rates in marine and freshwater fishes (e.g., Evans, 1977, 1982a; Cameron and Kormanik, 1982; McDonald *et al.*, 1982; Smatresk and Cameron, 1982) supports this conclusion. However, one also might propose that the stoichiometry of the transporter may also be altered to avoid substantial loading while maintaining necessary ammonia efflux. This interesting proposition should be examined. Nevertheless, stimulation of the ionic extrusion mechanisms seems to be the major branchial event that is limiting with respect to entry into seawater for most freshwater fishes. One should not forget, however, that entry into hyperosmotic environments is only possible if the species in question initiates (or increases) oral ingestion of the medium and reduces urine water loss in order to maintain water balance. Smith (1930) had originally demonstrated that freshwater teleosts do not drink, presumably because they are already facing a net osmotic influx of water. However, it has now been shown that some species of teleosts do indeed ingest the medium when in fresh water (Potts and Evans, 1966; Potts *et al.*, 1967; Eddy and Bath, 1979), but at a rather low rate, which is presumably balanced by an increase in renal free-water clearance. Therefore, initiation of oral ingestion as the fish enters hypertonic environments may be a quantitative rather than qualitative change.

Reduction of urinary water output as a euryhaline species enters seawater is presumably due to a reduced glomerular filtration rate (secondary to the shutdown of some glomeruli?) and increased reabsorption of water by the renal–urinary bladder epithelia (Evans, 1979). Contrary to the situation during entry into fresh or brackish water, alteration of the passive branchial permeability to salts does not appear to play a major role in prompting euryhalinity of freshwater fish. Indeed, marine teleosts maintain an extremely high branchial salt permeability, relative to that seen in freshwater species (Evans, 1979). The question why marine teleosts maintain relatively high Na^+ and Cl^- permeabilities, especially since the primitive conditions appear to be relative salt impermeability (see earlier), remains unanswered. As indicated earlier, modulation of branchial water permeability is also apparently not a prerequisite for entry into seawater, since the published data indicate that the gill water permeability in marine teleosts is only slightly lower (probably resulting from environmental Ca^{2+} concentrations) than that found in freshwater teleosts (Evans, 1969; Motais *et al.*, 1969). Nevertheless, it does appear that modulation of the osmotic permeability is possible, since gills isolated from *Anguilla anguilla* acclimated to seawater lose water in that salinity more slowly than gills isolated from freshwater-acclimated individuals (Isaia and Hirano, 1975) (Table VI).

The remainder of this section will therefore be concerned solely with what we know about the activation of the ionic extrusion mechanisms in the branchial epithelium. With the notable exceptions of members of the families Anguillidae (catadromous eels), Salmonidae (anadromous trout and salmon), and the order Cyprinodontiformes (killifish, pup fish, and mosquito fish), the number of fish species resident in fresh water that can and do enter the marine environment is rather limited. Moreover, they are usually in taxa thought to be secondary invaders of the freshwater environment, whose closest relatives are marine groups. This infrequent entry into seawater by freshwater species is especially interesting considering the fact that this transition was a major step in early vertebrate evolution, which led to the very diverse marine teleost and elasmobranch fauna.

The best-studied group remains the salmonids (see recent reviews by Hoar, 1976; Folmar and Dickhoff, 1980; Eddy, 1982). The anadromous salmonids (predominantly of the genus *Oncorhynchus*) enter fresh water as adults, after maturation in seawater, in order to breed. The newly hatched salmon remain in the parent stream for from 2 months to 2 years (depending on the species and various environmental factors) and then go through behavioral, morphological, and physiological changes (termed smoltification) that are correlated with their migration into and tolerance of seawater (Folmar and Dickhoff, 1980). An interesting study demonstrates how critical the timing is to this salinity transition. Chum salmon (*O. keta*), contrary to the "usual" *Oncorhynchus*, are spawned quite close to the sea, and return to the high salinities within a few days or weeks, after the yolk sac has been absorbed. Iwata *et al.* (1982) have found that the ability to osmoregulate in high salinities (monitored as the time it takes to attain blood Na^{+} levels characteristic of the fully seawater-adapted state) is inversely proportional to size (and presumably, therefore, time spent in fresh water). Other species of *Oncorhynchus* (e.g. *kisutch*, the coho salmon) have to attain a much bigger size before tolerance of seawater is possible (Conte *et al.*, 1966).

Bath and Eddy (1977a,b) have presented the most complete study of the effects of acute transfer of a salmonid (in this case the rainbow trout, *Salmo gairdneri*) to a distinctly hyperosmotic salinity (67% seawater). The only other group that has received similar interest is the freshwater eels (Kirsch,

1972; Kirsch and Mayer-Gostan, 1973). Both the trout and eel studies demonstrate that oral ingestion of the medium increases significantly within a few hours after transfer into a hyperosmotic salinity; however, the activation of the ionic extrusion mechanisms is a process that takes days (Fig. 4). Indeed, one might propose that euryhalinity is limited in most freshwater species because these ionic extrusion pathways are simply not part of their repertoire of ionic transport mechanisms. This is to be contrasted with the situation in marine species, which already have the ionic transport systems necessary for salt uptake in dilute salinities secondary to the needs of acid–base balance and nitrogen excretion (see earlier). Thus, euryhalinity may be limited by the lack of the proper transport mechanisms when moving from fresh water to seawater, but by the kinetics of uptake vis-à-vis ionic permeability when moving from seawater to fresh water (see earlier).

A. Activation of the Salt Extrusion Mechanisms

These whole-animal studies have shown that the time course of activation of salt extrusion mechanisms can be measured in days. Foskett *et al.* (1981) have examined the kinetics of activation of the Cl^- extrusion mechanism

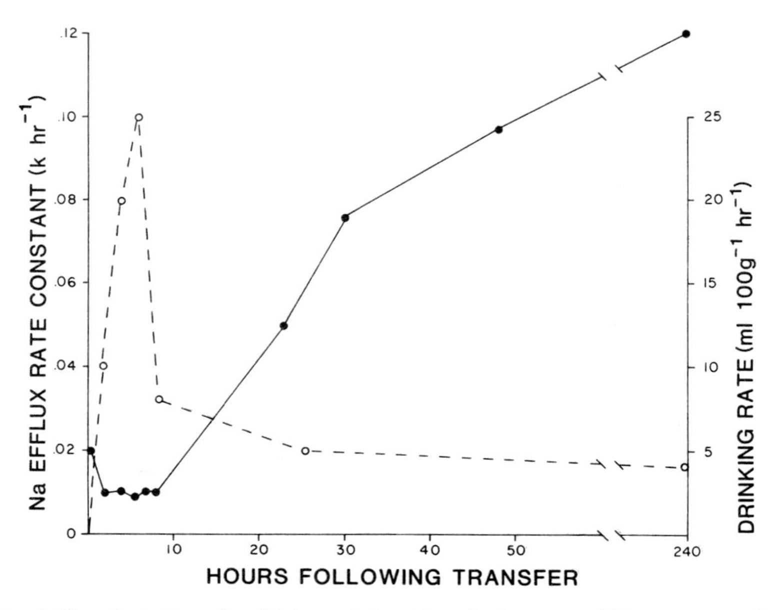

Fig. 4. The effect of transfer of *Salmo gairdneri* from fresh water to 67% seawater on the rate of Na^+ efflux (●----●) and oral ingestion of the medium (○---○). See text for details. (Redrawn from Bath and Eddy, 1979b.)

exhibited by the opercular skin isolated from tilapia. They found that it took approximately 5 days in 100% seawater (after an initial 2 days in 33% seawater) before the skins displayed the short-circuit current characteristic of the seawater-acclimated epithelium. In addition, they found that the density of "chloride cells" increased rapidly during the 2 days in 33% seawater and the first day in seawater, but then declined over the next 2 days and remained relatively stable for the next 2 weeks. The continued activation of the short-circuit current after the first day in seawater was associated with a significant increase in the diameter of the chloride cells (Fig 5).

These cytological studies provide the most recent corroboration of relatively early studies that demonstrated clearly that cellular differentiation may play a central role in the adaptation of at least two species of salmon to increased salinities. Conte (1965) found that inhibition of cell division by X irradiation of coho salmon (*Oncorhynchus kisutch*) resulted in loss of the ability to hyporegulate in seawater, and Conte and Lin (1967) found (by measuring the rate of incorporation of tritiated thymidine) that the turnover rate for interlamellar cells was 16 days in freshwater-adapted coho and chinook (*Oncorhynchus tshawytscha*), but only 6 days in seawater-adapted individuals. The actual interlamellar cell involved in this cytodifferentiation was undetermined in these studies, but one might suppose that the chloride cell was the site of action. We now have direct evidence that this is the cell that is directly involved in extrusion of Cl^- by the marine teleost gill (e.g., Foskett and Scheffey, 1982, and other chapters in this volume), and early cytological work by Copeland (1950) and Philpott and Copeland (1963)

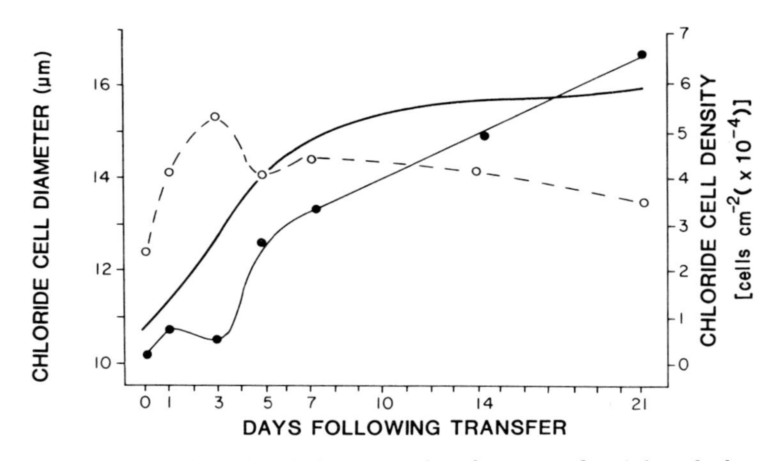

Fig. 5. The effect of transfer of tilapia (*Sarotherodon mossambicus*) from fresh water to seawater on the chloride cell diameter (●----●) and density (○---○), and short-circuit current (——) of the isolated opercular skin. See text for details. (Redrawn from Foskett *et al.*, 1981.)

showed distinctive changes in this cell (most notably the formation of an apical crypt) after adaptation to seawater. Moreover, subsequent studies have shown that the acclimation of teleosts to hyperosmotic salinities is associated with elaboration of the basolateral infoldings of the chloride cell (Karnaky *et al.*, 1976; Pisam, 1981) and an increase in the number of mitochondria associated with these basolateral infoldings (Karnaky *et al.*, 1976). Importantly, the number of interconnecting strands in the "tight junctions" between adjacent chloride cells and between chloride and respiratory cells is significantly reduced after acclimation to seawater, and the junctions are permeable to lanthanum (Sardet *et al.*, 1979). The apical crypt appears to be associated with pits on the surface of the interlamellar regions of the branchial filament, whose diameter decreases markedly after adaptation to increased salinities (Hossler *et al.*, 1979).

These cytological alterations with salinity are associated with biochemical changes. Maetz *et al.* (1969) showed that injection of actinomycin D into seawater-adapted *Anguilla anguilla* decreased the efflux of Na^+ within 4 to 5 days, with the treated fish dying on the sixth day with abnormally high levels of blood Na^+ and Cl^-. In addition, injection of freshwater-adapted eels inhibited the stimulation of Na^+ efflux normally seen after transfer to seawater (Fig. 6). Since actinomycin D inhibits the synthesis of messenger RNA (transcription), it appears that adaptation to seawater involves the production of some sort of protein. Indeed, Evans and Mallery (1975) found a distinct increase in gill protein content during acclimation of the fat sleeper (*Dormitator maculatus*) to seawater, and Stagg and Shuttleworth (1982) have recently shown that acclimation of the euryhaline flounder, *Platichthys flesus*, to seawater is associated with a significant increase in the protein content of gill homogenates. It is apparent that at least one protein whose concentrations may be enhanced during high-salinity acclimation is (Na^+,K^+)-activated ATPase. Epstein *et al.* (1967) found that acclimation of *Fundulus heteroclitus* to seawater was associated with a profound increase in branchial (Na^+,K^+)-activated ATPase activities, and a subsequent study (Jampol and Epstein, 1970) showed that the gill epithelium from marine teleosts in general had higher activities of this transport enzyme than freshwater species. Since these initial studies, various species have been described that either do not display altered enzyme activities in seawater (Kirschner, 1969) or actually have higher activities in fresh water (Lasserre, 1971; Gallis and Bourdichon, 1979; Doneen, 1981), but the general pattern still remains that increased salinity is usually associated with increased activity of (Na^+,K^+)-activated ATPase, measured as either specific activity (μmol P_i (mg protein)$^{-1}$ hr^{-1} (e.g., Karnaky *et al.*, 1976) or tritiated ouabain binding (e.g., Hossler *et al.*, 1979). The proposition that the enzyme is associated with ionic extrusion in seawater was supported by the finding in

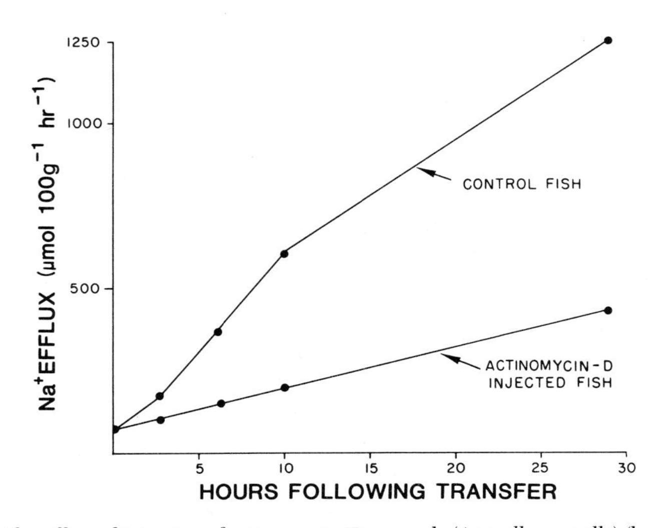

Fig. 6. The effect of injection of actinomycin D into eels (*Anguilla anguilla*) (before transfer from fresh water to seawater) on the rate of Na^+ efflux. See text for details. (Redrawn from Maetz *et al.*, 1969.)

at least three species that the time course of the stimulation of Na^+ efflux was similar to the time course of activation of the enzyme (Forrest *et al.*, 1973a; Bornancin and de Renzis, 1973; Evans and Mallery, 1975). In addition, various studies (Maetz, 1969b; Evans *et al.*, 1973) had shown that Na^+ efflux was at least partially dependent on saltwater K^+ concentrations, presumably secondary to the extrusion of Na^+ via Na^+/K^+ ionic exchange.

However pleasing this model for Na^+ extrusion by the marine teleost branchial epithelium was, we know now that Na^+ is in electrochemical equilibrium across the gills of many species of marine teleosts (see Evans, 1980b, for a more complete discussion of this subject), and that the (Na^+,-K^+)-activated ATPase is actually localized on the basolateral membrane of the chloride cell, in the wrong position for the active extrusion of Na^+ via Na^+/K^+ exchange (Karnaky, 1980). In fact, this means that the early data demonstrating similar time courses for activation of Na^+ efflux and activation of (Na^+,K^+)-activated ATPase were actually the result of two parallel and relatively independent processes: (1) increase in cation permeability of the branchial epithelium and origin of the saltwater TEP and (2) the initiation of volume regulation of the branchial cells and/or Cl^- extrusion dependent on (Na^+,K^+)-activated ATPase (Silva *et al.*, 1977). The proposed role of (Na^+,K^+)-activated ATPase in Cl^- extrusion across the marine teleost gill need not be reviewed here (see other chapters in this volume); suffice it to say that the enzyme apparently provides the electrochemical driving force necessary for the coupled basolateral uptake of Na^+ and Cl^-, which has been described in many epithelial tissues (Frizzell *et al.*, 1979). The role of (Na^+,K^+)-activated ATPase in volume regulation by branchial cells has never been examined, especially during salinity transfer, but the enzyme presumably plays a role in this volume regulation, as it does in other cells (MacKnight and Leaf, 1977). This recent revision of our understanding of the role of (Na^+,K^+)-activated ATPase in the extrusion of unwanted salt across the marine teleost branchial epithelium points to the problem of equating enzyme activity changes with alterations in the active extrusion (or uptake) of various cations or anions across epithelia. Until the actual role of the enzyme in any transport process is known, one cannot simply equate activity levels with presumed transport levels.

Whatever the enzymatic basis for activation of salt extrusion in saltwater teleosts, it appears that, at least in *Salmo gairdneri*, increased energy expenditure, as measured by a rapid fall in the gill ATP concentrations, ATP:ADP ratio, and energy charge (ATP + 0.5ADP/ATP + ADP + AMP), characterizes transfer into seawater (Leray *et al.*, 1981). This utilization of the ATP pool during the initial (approximately 4-hr) period after transfer apparently is not associated with normal oxidative pathways, because the authors did not find any significant increase in the oxygen consumption of the gill tissue, despite the use of a very sensitive assay (Leray *et al.*, 1981). Nordlie (1978) reviewed the literature pertaining to variations in oxygen consumption by either tissues or whole animals during salinity variation, and found that the responses of the 13 species of euryhaline fishes that had been examined could be divided into four categories depending on whether oxygen consumption changed with salinity and at what salinity oxygen consumption maxima or minima were displayed. It is clear that definitive statements about the energetic costs of euryhalinity cannot be made at present.

In summary, it is obvious that acclimation to increased salinities is associated with activation of morphological, physiological, and biochemical changes in the fish branchial epithelium. However, our knowledge of the structural and biochemical bases of salt extrusion and volume regulation by this tissue is still too imprecise to allow us to make definite statements about changes in the gills that can be correlated with entry into seawater. Our modeling is hindered by the fact that many of the data were gathered before the recent appreciation of the role of basolateral (Na^+,K^+)-activated ATPase and Cl^- transport in the extrusion of salt. Moreover, the opercular skin of *Fundulus* and *Sarotherodon* and chin skin of *Gillichthys* (see Chapter 5, this volume) provide model systems that can be biophysically controlled but may not be representative of all marine teleosts. A significant number of seawater-acclimated teleosts (some of which are euryhaline) maintain transepithelial electrical potentials relatively far removed from the Na^+ equilibrium potential and therefore presumably actively extrude Na^+ (Evans, 1980b; Chapter 10). Unfortunately, with the exception of the fat sleeper (*Dormitator maculatus;* Evans *et al.*, 1973; Evans and Mallery, 1975), none of the euryhaline species has been examined during the transfer from brackish/fresh water to seawater. It is clear that more data are needed on the mechanisms of Na^+ extrusion by these species, and the onset of Na^+ extrusion after transfer to hyperosmotic salinities.

B. Hormonal Control of the Activation of Salt Extrusion

Unfortunately, the same data gap is apparent when reviewing the literature on the hormonal control of branchial transport in marine teleosts. With a few recent exceptions, studies of the factors involved in stimulating salt extrusion come from the middle 1970s when it was assumed that Na^+ extrusion was active and directly driven through branchial (Na^+,K^+)-activated ATPase (see Lahlou, 1980, for a review). Thus, while the data form a basis for further discussion and investigation, the conclusions regarding mechanisms are often outdated.

Pioneering work by Mayer *et al.* (1967) demonstrated that eels (*Anguilla anguilla*) that had had their interrenal tissue removed (unfortunately, the steroid-producing tissue that is the antecedent of the tetrapod adrenal cortex is so diffuse in fishes that interrenalectomy is possible only in the anguillids) were unable to tolerate transfer to seawater and died within 48 hr. The interrenalectomized animals displayed a significantly increased blood Na^+ concentration and a reduced rate of isotopically measured Na^+ efflux (Fig. 7). Importantly, they found that injection of cortisol (the dominant corticosteroid of fishes) into the ablated fishes prolonged survival in seawater, decreased the blood Na^+ concentrations, and increased the Na^+ efflux to control levels. The authors therefore proposed that Na^+ extrusion by marine teleosts was under the control of the interrenal, via the messenger cortisol. In subsequent studies, Forrest *et al.* (1973a) demonstrated that injection of cortisol into *Anguilla rostrata* stimulated both the increase in Na^+ efflux and activity of (Na^+,K^+)-activated ATPase seen after transfer to seawater. However, the effects on the flux and enzyme were separable by alteration of the dose of hormone used. Nevertheless they proposed that "cortisol appears to be a hormonal mediator of saltwater adaptation, both facilitating increased Na transport and inducing an increased activity of (Na^+,K^+)-activated ATPase in the gill," but that other factors also may be involved. A concurrent study showed even more clearly that, while cortisol is probably important in stimulating Na^+ efflux, its effects are not direct.

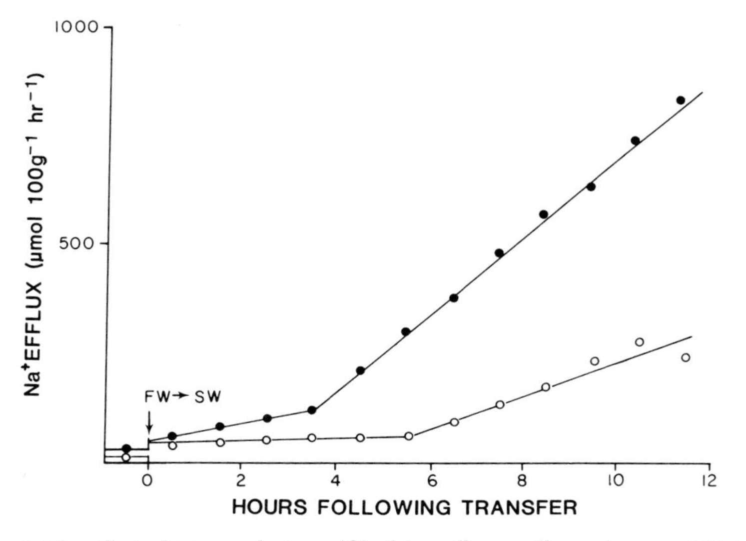

Fig. 7. The effect of interrenalectomy (○) of *Anguilla anguilla* on the rate of Na^+ efflux measured after transfer from fresh water (FW) to seawater (SW) (●) Sham-operated *A. Anguilla*. See text for details. (Redrawn from Mayer *et al.*, 1967.)

Forrest *et al.* (1973b) found that the peak of blood cortisol concentrations occurred at day 2 after transfer of *A. rostrata* into seawater, while both Na^+ efflux and (Na^+,K^+)-activated ATPase activity increased only after day 2 and reached an initial peak at day 5 (Fig. 8). It therefore appeared that a "burst of cortisol secretion . . . plays an important role in conditioning the animal to withstand the osmotic stress of full immersion in the sea" (Forrest *et al.*, 1973b).

Some evidence indicates that cortisol's role may be primarily via stimulation of morphological rather than biophysical events. Injection of cortisol into freshwater-acclimated *A. rostrata* stimulated the production and differentiation of chloride cells. However, the cells were subepithelial, not exposed to the external medium (Doyle and Epstein, 1972). Recent experiments with the tilapia opercular skin epithelium support these data. Injection of cortisol into freshwater-acclimated tilapia stimulated an increase in chloride cell density, but had no effect on the size of individual cells. In addition, the opercular cells. In addition, the opercular membranes from freshwater-acclimated, cortisol-injected tilapia did not display any of the electrophysiological characteristics of opercular membranes isolated from seawater-acclimated fish (Foskett *et al.*, 1981) These authors conclude that cortisol may be involved in the differentiation of chloride cells during acclimation to seawater but that some other, as yet undefined, factor is neces-

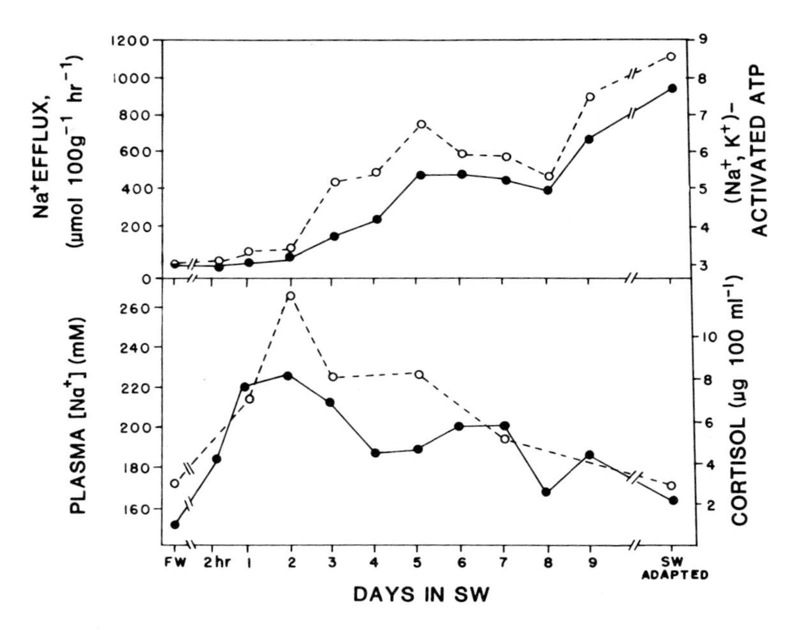

Fig. 8. The effect of transfer of *Anguilla rostrata* from fresh water (FW) to seawater (SW) on the Na^+ efflux ●——●) branchial (Na^+,K^+)-activated ATPase activity (○---○), and plasma Na^+ (●——●) and cortisol (○---○) concentrations. See text for details. (Redrawn from Forrest *et al.*, 1973a,b.)

sary to activate the Cl^- extrusion system and the high cationic permeability characteristic of the marine opercular skin and branchial epithelium. Thus, early evidence seemed to implicate cortisol in the direct control of Na^+ extrusion; however, subsequent physiological and morphological data indicate that it is more likely involved in the morphological differentiation that is associated with acclimation to increased salinities.

The other major hormone implicated in playing a role in seawater acclimation is epinephrine. The dominant action of this catecholamine seems to be inhibition of salt extrusion, via α-adrenergic receptors (see earlier; Pic *et al.*, 1975; Girard, 1976; Degnan *et al.*, 1977). However, it has been found (Table XI) that stimulation of β-adrenergic receptors in the opercular skin of *Fundulus* with isoproterenol produces a rapid increase in the Cl^- excretion rate (Degnan *et al.*, 1977; Degnan and Zadunaisky, 1979) via activation of intracellular cyclic AMP (Mendelsohn *et al.*, 1981). This β-mediated stimulation by catecholamines is especially interesting, because other studies have shown that Na^+ influx via Na^+/NH_4^+ exchange is stimulated by the same system in the perfused trout head (Payan, 1978). The use of a single

Table XI

Effects of β-Adrenergic Stimulation on the Short-Circuit Current and Net Cl^- Efflux across the Isolated Opercular Skin of *Fundulus heteroclitus*[a]

Experimental group	Short-circuit current ($\mu A\ cm^{-2}$)	Net Cl^- efflux ($\mu mol\ cm^{-2}\ hr^{-1}$)
Control	93.2	3.55
10^{-5} *M* Isoproterenol	137.0	4.87

[a]Data from Degnan and Zadunaisky (1979).

hormone–receptor system to stimulate both Cl^- (and secondarily Na^+) efflux and Na^+ uptake seems maladaptive and therefore unlikely, especially for euryhaline fish, which probably must modulate the two transport systems independently. It is clear that epinephrine has the potential to play an essential role in controlling euryhalinity. The physiology of the action of this hormone on the fish branchial epithelium is certainly a fertile area for future research.

IV. SUMMARY

It is difficult to summarize a long chapter, but it is clear that some main points should be stressed. It is relatively clear that the major limitation to entry of marine species into reduced salinities is the balance between salt loss and uptake, rather than the presence or absence of a particular transport system. It appears that loss can be controlled by some typical vertebrate hormones (e.g., prolactin and epinephrine), and extrinsic factors such as calcium, but much remains to be learned about the actual modes of actions of these effector agents. Ionic uptake apparently can be stimulated by various factors (epinephrine is the best studied), but it is not clear at present what role is played by hemodynamic changes versus direct stimulation of a transport step. Entry of freshwater fishes into the brackish water or the marine environment is apparently actually limited by the presence or absence of the requisite ionic extrusion systems. Hormones such as cortisol and epinephrine can stimulate the morphological and biochemical changes that promote salt extrusion by euryhaline species, but it appears that most groups of freshwater fishes simply do not have the transport system in their branchial epithelium. Modifications of the diffusional and osmotic permeabilities to water apparently do take place, so that euryhaline forms are able to minimize the osmotic water movements that must be balanced by oral ingestion of seawater and production of large volumes of urine in fresh water. However, the reader should be left with the distinct impression that our data base for, and models of, euryhalinity have broadened in the past few years, but that many, many questions remain unanswered, or only partially answered. Some examples will suffice to summarize what I perceive are major lacunae in our data or thinking:

1. What is the role of alteration of gill surface area—SA in Eq. (4)—and/or diffusional distance (l) vis-à-vis changes in diffusion coefficients (D) in the obvious reduction in salt fluxes measured across freshwater versus marine species of fishes?
2. Is a relatively small D a primitive vertebrate condition? If it is, why do modern marine teleosts apparently maintain a relatively high D?
3. Is there a correlation between a relatively high ionic permeability in some species of euryhaline fish and a sensitivity to external Ca^{2+} concentrations? Do true euryhaline species maintain a lower ionic permeability than stenohaline, or Ca^{2+}-sensitive species?
4. Can alterations in the transepithelial electrical potential play a role in euryhalinity?
5. What is the morphological cause of and physiological importance of the apparent rectification of water flow producing a lower calculated osmotic water permeability in marine fish than freshwater fish?
6. What are the morphological correlates of the finding that *Anguilla anguilla* is able to modify its osmotic permeability so that net water movements are minimized in either fresh water or seawater? Are other euryhaline species capable of doing this?
7. Do prolactin and epinephrine exert their effects on water permeability directly (i.e., via D) or via alterations in either SA or l?
8. Is Cl^-/HCO_3^- exchange present in the marine teleost and elasmobranch epithelium?
9. How significant is the net salt influx into marine fishes (teleosts, elasmobranchs, and hagfish) produced by the presence of Na^+/NH_4^+, Na^+/H^+, and possibly Cl^-/HCO_3^- exchange?
10. Do euryhaline species modulate K_m and J_{max} as well as efflux? What is the relative importance of the modulation of each of these parameters on salt balance of a given euryhaline species as it acclimates to a lowered salinity?
11. What is the hormonal control of salt uptake? Is it via alterations of hemodynamics, K_m, or J_{max}?
12. What is the relative importance of external modulation (via Ca^{2+}) versus internal modulation (via hormones) of permeability for entry of calcium-sensitive species of marine teleosts into brackish or fresh waters?

13. What does limit elasmobranch euryhalinity?
14. What is the morphological and biochemical basis for the presence or absence of the salt extrusion system in various groups of freshwater fish? Do some stenohaline, freshwater species have "chloride cells"?
15. Is (Na^+,K^+)-activated ATPase functioning predominantly in salt extrusion or branchial cell volume regulation? What is its role in ionic uptake in fresh water?
16. Do marine species, which apparently gain some Na^+ by passive uptake down electrochemical gradients and by oral ingestion of the medium (followed by active uptake from the gut), actually extrude Na^+ actively, contrary to the prevailing model for marine teleost salt extrusion (Silva *et al.*, 1977)?
17. What are the energetics of ionic and osmotic regulation? Can one separate the two? Does it take more energy to ionoregulate and osmoregulate in fresh water or in seawater?
18. If cortisol mainly effects the morphological differentiation of the "chloride cell," what hormones actually stimulate salt extrusion in seawater?
19. What is the actual role of epinephrine in ionic regulation in any salinity, exclusive of its hemodynamic effects? How can it stimulate Cl^- extrusion and Na^+ uptake, both via β-adrenergic receptors?

It is hoped that at least some of these questions can be answered in the next few years by interested readers of this chapter.

ACKNOWLEDGMENTS

Carol Binello, Grace Russell, and Alicia Fowler typed various sections of the manuscript, and Bill Adams drew the figures. Their excellent assistance is always appreciated. The data for Table I were gathered by Dr. Carter Gilbert and by George Burgess and J. B. Miller of the Florida State Museum, Gainesville, Florida.

REFERENCES

Bath, R. N., and Eddy, F. B. (1979a). Ionic and respiratory regulation in rainbow trout during rapid transfer to sea water. *J. Comp. Physiol.* **134**, 351–357.

Bath, R. N., and Eddy, F. B. (1979b). Salt and water balance in rainbow trout (*Salmo gairdneri*) rapidly transferred from freshwater to seawater. *J. Exp. Biol.* **83**, 193–202.

Bentley, P. J., Maetz, J., and Payan, P. (1976). A study of the unidirectional fluxes of Na and Cl across the gills of the dogfish, *Scyliorhinus canicula* (Chondrichthyes). *J. Exp. Biol.* **64**, 629–637.

Bergmann, H. L., Olson, K. R., and Fromm, P. O. (1974). The effects of vasoactive agents on the functional surface area of isolated-perfused gills of rainbow trout. *J. Comp. Physiol.* **94,** 267–286.

Booth, J. H. (1979). The effects of oxygen supply, epinephrine and acetylcholine on the distribution of blood flow in trout gills. *J. Exp. Biol.* **83,** 31–39.

Bornancin, M., and de Renzis, G. (1973). Evolution of the branchial sodium outflux and its components, especially the Na/K exchange and the Na-K dependent ATPase activity during adaptation to sea water in *Anguilla anguilla*. *Comp. Biochem. Physiol.* **43A,** 577–591.

Breeder, C. M. (1934). Ecology of an oceanic freshwater lake, Andros Island, Bahamas, with special reference to its fishes. *Zoologica (N.Y.)* **18,** 57–88.

Burden, C. E. (1956). The failure of hypophysectomized *Fundulus heteroclitus* to survive in fresh-water. *Biol. Bull. (Woods Hole, Mass.)* **110,** 8–28.

Cameron, J. N. (1978). Regulation of blood pH in teleost fish. *Respir. Physiol.* **33,** 129–144.

Cameron, J. N., and Kormanik, G. A. (1982). The acid-base responses of gills and kidneys to infused acid and base loads in the channel catfish, *Ictalurus punctatus*. *J. Exp. Biol.* **99,** 143–160.

Carrier, J. C. (1974). Physiological effects of environmental calcium on freshwater survival and sodium kinetics in the stenohaline marine teleost, *Lagodon rhomboides*. Ph.D. Dissertation, University of Miami, Coral Gables, Florida.

Carrier, J. C., and Evans, D. H. (1973). Ion and water turnover in the freshwater elasmobranch *Potamotrygon* sp. *Comp. Biochem. Physiol. A* **45A,** 667–670.

Carrier, J. C., and Evans, D. H. (1976). The role of environmental calcium in freshwater survival of the marine teleost, *Lagodon rhomboides, J. Exp. Biol.* **65,** 529–538.

Chester Jones, I.,Henderson, I. W., Chan, D. K. O., and Rankin, J. C. (1967). Steroids and pressor substances in bony fish with special reference to the adrenal cortex and corpuscles of Stannius of the eel (*Anguilla anguilla* L.). *Int. Cong. Ser.—Excerpta Med.* **132,** 135–145.

Conte, F. P. (1965). Effects of ionizing radiation on osmoregulation in fish *Oncorhynchus kisutch*. *Comp. Biochem. Physiol.* **15,** 293–302.

Conte, F. P., and Lin, D. (1967). Kinetics of cellular morphogenesis in gill epithelium during sea water adaptation of *Oncorhynchus* (Walb.). *Comp. Biochem. Physiol.* **23,** 945–957.

Conte, F. P., Wagner, H. H., Fessler, J., and Gnose, C. (1966). Development of osmotic and ionic regulation in juvenile coho salmon *Oncorhynchus kisutch*. *Comp. Biochem. Physiol.* **18,** 1–15.

Copeland, D. E. (1950). Adaptive behavior of the chloride cell in the gill of *Fundulus heteroclitus*. *J. Morphol.* **87,** 369–379.

Cuthbert, A. W., and Maetz, J. (1972). The effects of calcium and magnesium on sodium fluxes through gills of *Carassius auratus*. *J. Physiol. (London)* **221,** 633–643.

Degnan, K. J., and Zadunaisky, J. (1979). Open-circuit sodium and chloride fluxes across isolated opercular epithelia from the teleost *Fundulus heteroclitus*. *J. Physiol. (London)* **294,** 483–495.

Degnan, K. J. Karnaky, K. J., Jr., and Zadunaisky, J. A. (1977). Active chloride transport in the *in vitro* opercular skin of a teleost (*Fundulus heteroclitus*), a gill-like epithelium rich in chloride cells. *J. Physiol. (London)* **271,** 155–191.

Dharmamba, M., and Maetz, J. (1972). Effects of hypophysectomy and prolactin on sodium balance of *Tilapia mossambica* in fresh water. *Gen. Comp. Endocrinol.* **19,** 175–183.

Dharmamba, M., and Nishioka, R. S. (1968). Response of prolactin-secreting cells of *Tilapia mossambica* to environmental salinity. *Gen. Comp. Endocrinol.* **19,** 175–183.

Doneen, B. A. (1981). Effects of adaptation to sea water, 170% sea water and to fresh water on activities and subcellular distribution of branchial Na-K-ATPase, low- and high affinity Ca-

ATPase, and ouabain-insensitive ATPase in *Gillichythys mirabilis*. *J. Comp. Physiol.* **145,** 51–61.

Doyle, W. L., and Epstein, F. H. (1972). Effects of cortisol treatment and osmotic adaptation on the chloride cells of the eel *Anguilla rostrata*. *Cytobiologie* **6,** 58–73.

Eddy, F. B. (1975). The effect of calcium on gill potentials and on sodium and chloride fluxes in the goldfish, *Carassius auratus*. *J. Comp. Physiol.* **96,** 131–142.

Eddy, F. B. (1982). Osmotic and ionic regulation in captive fish with particular reference to salmonids. *Comp. Biochem. Physiol. B* **73B,** 125–141.

Eddy, F. B., and Bath, R. N. (1979). Ionic regulation in rainbow trout (*Salmo gairdneri*) adapted to fresh water and dilute sea water. *J. Exp. Biol.* **83,** 181–192.

Ensor, D. M., and Ball, J. N. (1972). Prolactin and osmoregulation in fishes. *Fed. Proc., Fed. Am. Soc. Exp. Biol.* **31,** 1615–1623.

Epstein, F., Katz, A. I., and Pickford, G. E. (1967). Sodium- and potassium-activated adenosine triphosphatase of gills: Role in adaption of teleosts to saltwater. *Science* **156,** 1245–1247.

Evans, D. H. (1969). Studies on the permability to water of selected marine, freshwater and euryhaline teleosts. *J. Exp. Biol.* **50,** 689–703.

Evans, D. H. (1973). Sodium uptake by the sailfin molly, *Poecilia latipinna:* Kinetic analysis of a carrier system present in both fresh-water-acclimated and sea-water acclimated individuals. *Comp. Biochem. Physiol. A* **45A,** 843–850.

Evans, D. H. (1975a). Ionic exchange mechanisms in fish gills. *Comp. Biochem. Physiol. A* **51A,** 491–495.

Evans, D. H. (1975b). The effects of various external cations and sodium transport inhibitors on sodium uptake by the sailfin molly, *Poecilia latipinna*, acclimated to sea water. *J. Comp. Physiol.* **96,** 111–115.

Evans, D. H. (1977). Further evidence for Na^+/NH_4^+ exchange in marine teleost fish. *J. Exp. Biol.* **70,** 213–220.

Evans, D. H. (1979). Fish. *In* "Osmotic and Ionic Regulation in Animals" (G. M. O. Maloiy, ed.), Vol. 1, pp. 305–390. Academic Press, New York.

Evans, D. H. (1980a). Osmotic and ionic regulation by freshwater and marine fishes. *In* "Environmental Physiology of Fishes" (M. A. Ali, ed.), pp. 93–122. Plenum, New York.

Evans, D. H. (1980b). Kinetic studies of ion transport by fish gill epithelium. *Am. J. Physiol.* **238,** R224–R230.

Evans, D. H. (1980c). Na^+/NH_4^+ exchange in the marine teleost, *Opsanus beta:* Stoichiometry and role in Na^+ balance. *In* "Epithelial Transport in the Lower Vertebrates" (B. Lahlou, ed.), pp. 197–205. Cambridge Univ. Press, London and New York.

Evans, D. H. (1980d). Acid and ammonia excretion by *Squalus acanthias* and *Myxine glutinosa:* Effect of hypercapnia, acid injection and Na-free sea water. *Bull. Mt. Desert Isl. Biol. Lab.* **20,** 60–63.

Evans, D. H. (1982a). Mechanisms of acid extrusion by two marine fishes: The teleost, *Opsanus beta*, and the elasmobranch, *Squalus acanthias*. *J. Exp. Biol.* **97,** 289–299.

Evans, D. H. (1982b). Salt and water exchange across vertebrate gills. *In* "Gills" (D. F. Houlihan, J. C. Rankin, and T. J. Shuttleworth, eds.), pp. 148–171. Cambridge Univ. Press, London and New York.

Evans, D. H. (1984a). Gill Na^+/H^+ and Cl^-/HCO_3^- exchange mechanisms evolved before the vertebrates entered fresh water. *J. Exp. Biol.* (in press).

Evans, D. H. (1984b). The role of branchial and dermal epithelia in acid-base regulation in aquatic vertebrates. *In* "Acid-base Regulation in Animals" (N. Heisler, ed.). Elsevier, Amsterdam (in press).

Evans, D. H., and Cooper, K. (1976). The presence of Na^+/Na^+ and Na^+/K^+ exchange in sodium extrusion by three species of fish. *Nature (London)* **259,** 241–242.

Evans, D. H., and Mallery, C. H. (1975). Time course of sea water acclimation by the euryhaline teleost, *Dormitator maculatus:* Correlation between potassium stimulation of sodium efflux and Na^+/K^+ activated ATPase activity. *J. Comp. Physiol.* **96,** 117–122.
Evans, D. H., Mallery, C. H., and Kravitz, L. (1973). Sodium extrusion by a fish acclimated to sea water: Physiological and biochemical description of a Na^+ - for - K^+ exchange system. *J. Exp. Biol.* **58,** 627–636.
Evans, D. H., Kormanik, G. A., and Krasny, E. J., Jr. (1979). Mechanisms of ammonia and acid extrusion by the little skate, *Raja erinacea. J. Exp. Zool.* **208,** 431–437.
Evans, D. H., Claiborne, J. B., Farmer, L., Mallery, C. H., and Krasny, E. J., Jr. (1982). Fish gill ionic transport: Methods and models. *Biol. Bull.* (*Woods Hole, Mass.*) **163,** 108–130.
Folmar, L. C., and Dickhoff, W. W. (1980). The parr-smolt transformation (smoltification) and seawater adaptation in salmonids. *Aquaculture* **21,** 1–38.
Forrest, J. N., Jr., Cohen, A. D., Schon, D. A., and Epstein, F. H. (1973a). Na transport and Na-K-ATPase in gills during adaptation to sea water: Effects of cortisol. *Am. J. Physiol.* **224,** 709–713.
Forrest, J. N., Jr., MacKay, W., Gallagher, B., and Epstein, F. H. (1973b). Plasma cortisol response to saltwater adaptation in the American eel. *Am. J. Physiol.* **224,** 714–717.
Foskett, J. K., and Scheffey, C. (1982). The chloride cell: Definitive identification as the salt-secretory cell in teleosts. *Science* **215,** 164–166.
Foskett, J. K., Logsdon, C. D., Turner, T., Machen, T. E., and Bern, H. A. (1981). Differentiation of the chloride extrusion mechanisms during seawater adaptation of a teleost fish, the Cichlid *Sarotherodon mossambicus. J. Exp. Biol.* **93,** 209–224.
Foskett, J. K., Machen, T. E., and Bern, H. A. (1982). Chloride secretion and conductance of teleost opercular membrane: Effects of prolactin. *Am. J. Physiol.* **242,** R380–R389.
Frizzell, R. A., Field, M., and Schultz, S. G. (1979). Sodium-coupled chloride transport by epithelial tissues. *Am. J. Physiol.* **236,** F1–F8.
Gallis, J. L., and Bourdichon, M. (1976). Changes of (Na-K) dependent ATPase activity in gills and kidneys of two mullets, *Chelon labrosus* (Risso) and *Liza ramade* (Risso) during fresh water adaptation. *Biochimie* **58,** 625–627.
Girard, J. P. (1976). Salt excretion by the perfused head of trout adapted to sea water and its inhibition by adrenaline. *J. Comp. Physiol.* **111,** 77–91.
Hardisty, M. W. (1979). "Biology of Cyclostomes." Chapman & Hall, London.
Haywood, G. P., Isaia, J., and Maetz, J. (1977). Epinephrine effects on branchial water and urea flux in rainbow trout. *Am. J. Physiol.* **232,** R110–R115.
Heisler, N. (1980). Regulation of the acid-base status in fish. *In* "Environmental Physiology of Fishes" (M. A. Ali, ed.), pp. 123–162. Plenum, New York.
Henderson, I. W., and Chester Jones, I. (1967). Endocrine influences on the net extrarenal fluxes of sodium and potassium in the European eel (*Anguilla anguilla* L.). *J. Endocrinol.* **37,** 319–325.
Hirano, T. (1977). Prolactin and hydromineral metabolism in vertebrates. *Gunma Symp. Endocrinol.* **14,** 45–59.
Hoar, W. S. (1976). Smolt transformation: Evolution, behavior and physiology. *J. Fish. Res. Board Can.* **33,** 1234–1252.
Holbert, P. W., Boland, E. J., and Olson, K. R. (1979). The effect of epinephrine and acetylcholine on the distribution of red cells within the gills of the channel catfish (*Ictalurus punctatus*). *J. Exp. Biol.* **79,** 135–146.
Holmes, W. N., and Donaldson, E. M. (1969). The body compartments and the distribution of electrolytes. *In* "Fish Physiology" (W. S. Hoar and D. J. Randall, eds.), Vol. 1, pp. 1–89. Academic Press, New York.
Hossler, F. E., Ruby, J. R., and McIlwain, T. D. (1979). The gill arch of the mullet, *Mugil*

cephalus. II. Modification in surface ultrastructure and Na, K-ATPase content during adaptation to various salinities. *J. Exp. Zool.* **208,** 403–420.

Hughes, G. M., and Morgan, M. (1973). The structure of fish gills in relation to their respiratory function. *Biol. Rev. Cambridge Philos. Soc.* **48,** 419–475.

Hulet, W. H., Masel, S. J., Jodrey, L. H., and Wehy, R. G. (1967). The role of calcium in the survival of marine teleosts in dilute seawater. *Bull. Mar. Sci.* **17,** 677–688.

Isaia, J., and Hirano, T. (1975). Effect of environmental salinity change on osmotic permeability of the isolated gill of the eel, *Anguilla anguilla* L. J. Physiol. (*Paris*) **70,** 737–747.

Isaia, J., and Masoni, A. (1976). The effects of calcium and magnesium on water and ionic permeabilities in the sea water adapted eel. *Anguilla anguilla* L. *J. Comp. Physiol.* **109,** 221–233.

Isaia, J., Payan, J. P., and Girard, J. P. (1979). A study of the water permeability of the gills of freshwater- and seawater-adapted trout (*Salmo gairdneri*): Mode of action of epinephrine. *Physiol. Zool.* **52,** 269–279.

Iwata, M., Hasegawa, S., and Hirano, T. (1982). Decreased seawater adaptability of chum salmon (*Oncorhynchus keto*) fry following prolonged rearing in freshwater. *Can. J. Fish. Aquat. Sci.* **39,** 509–514.

Jackson, W. A., and Fromm, P. O. (1981). Factors affecting ^3H_2O transfer capacity of isolated perfused trout gill. *Am. J. Physiol.* **240,** R235–R245.

Jampol, L. M., and Epstein, F. H. (1970). Sodium-potassium-activated adenosine triphosphatase and osmotic regulation by fishes. *Am. J. Physiol.* **218,** 607–611.

Karnaky, K. J., Jr. (1980). Ion-secreting epithelia: Chloride cells in the head region of *Fundulus heteroclitus*. *Am. J. Physiol.* **238,** R185–R198.

Karnaky, K. J., Jr., Ernst, S. A., and Philpott, C. W. (1976). Teleost chloride cell. I. Response to gill Na, K-ATPase and chloride cell fine structure to various environments. *J. Cell Biol.* **70,** 144–156.

Kerstetter, T. H., Kirschner, L. B., and Rafuse, D. D. (1970). On the mechanisms of sodium ion transport by the irrigated gills of rainbow trout (*Salmo gairdneri*). *J. Exp. Biol.* **56,** 342–359.

Kirsch, R. (1972). The kinetics of peripheral exchanges of water and electrolytes in the silver eel (*Anguilla anguilla* L.) in fresh water and in sea water. *J. Exp. Biol.* **57,** 489–512.

Kirsch, R., and Mayer-Gostan, N. (1973). Kinetics of water and chloride exchanges during adaptation of the European eel to sea water. *J. Exp. Biol.* **58,** 105–121.

Kirschner, L. B. (1969). ATPase activity in gills of euryhaline fish. *Comp. Biochem. Physiol.* **29,** 871–874.

Kirschner, L. B. (1977). The sodium chloride excreting cells in marine vertebrates. *In* "Transport of Ions and Water in Animals" (B. L. Gupta, R. B. Moreton, J. L. Oschman, and B. J. Wall, eds.), pp. 427–452. Academic Press, New York.

Kirschner, L. B. (1979). Control mechanisms in crustaceans and fishes. *In* "Mechanisms of Osmoregulation in Animals. Maintenance of Cell Volume" (R. Gilles, ed.), pp. 157–222. Wiley, New York.

Kirschner, L. B. (1980). Comparison of vertebrate salt-excreting organs. *Am. J. Physiol.* **238,** R219–R223.

Kirschner, L. B., Greenwald, L., and Sanders, M. (1974). On the mechanism of sodium extrusion across the irrigated gill of sea water-adapted rainbow trout (*Salmo gairdneri*). *J. Gen. Physiol.* **64,** 148–165.

Krogh, A. (1938). The active absorption of ions in some freshwater animals. *Z. Vergl. Physiol.* **25,** 335–350.

Lahlou, B. (1980). Les hormones dans l'osmorégulation des poissons. *In* "Environmental Physiology of Fishes" (M. A. Ali, ed.), pp. 201–240. Plenum, New York.

Lahlou, B., and Giordan, A. (1970). Le contrôle hormonal des échanges et de la balance de l'eau chez le Téléostéen d'eau douce *Carassius auratus,* intact et hypophysectomise. *Gen. Comp. Endocrinol.* **14,** 491–509.

Lahlou, B., Crenesse, D., Bensahla-Talet, A., and Porthe-Nibelle, J. (1975). Adaptation de la truite d'élevage à l'eau de mer. Effets sur les concentrations plasmatiques, les échanges branchiaux et le tranport intestinal du sodium. *J. Physiol. (Paris)* **70,** 593–603.

Lam, T. J. (1969). The effect of prolactin on osmotic influx of water in isolated gills of the marine threespine stickleback *Gasterosteus aculeatus* L. form trachurus. *Comp. Biochem. Physiol.* **31,** 909–913.

Lasserre, P. (1971). Increase of (Na+K)-dependent ATPase activity in gills and kidneys of two euryhaline marine teleosts, *Crenimugil labrosus* (Risso, 1826) and *Dicentrarchus labrax* (Linnaeus 1758), during adaptation to fresh water. *Life Sci.* **10,** Part II, 113–119.

Leray, C., Colin, D. A., and Florentz, A. (1981). Time course of osmotic adaptation and gill energetics of rainbow trout (*Salmo gairdneri* R.) following abrupt changes in external salinity. *J. Comp. Physiol.* **144,** 175–181.

Loretz, C. A., and Bern, H. A. (1982). Prolactin and osmoregulation in vertebrates. *Neuroendocrinology* **35,** 292–304.

Lowenstein, W. R. (1980). Cell-to-cell communication. permeability, formation, genetics and functions of the cell-cell membrane channel. *In* "Membrane Physiology" (T. E. Andreoli, J. R. Hoffman, and D. D. Fanestil, eds.), pp. 335–356. Plenum, New York.

McDonald, D. G., Walker, R. L., Wilkes, P. R. H., and Wood, C. M. (1982). H excretion in the marine teleost *Parophrys vetulus. J. Exp. Biol.* **98,** 403–414.

MacKnight, A. D. C., and Leaf, A. (1977). Regulation of cellular volume. *Physiol. Rev.* **57,** 510–573.

Maetz, J. (1969a). Observations on the role of the pituitary-interrenal axis in the ion regulation of the eel and other teleosts. *Gen. Comp. Endocrinol., Suppl.* **2,** 299–316.

Maetz, J. (1969b). Sea water teleosts: Evidence for a sodium-potassium exchange in the branchial sodium-excreting pump. *Science* **166,** 613–615.

Maetz, J. (1971). Fish gills: Mechanisms of salt transfer in fresh water and sea water. *Philos. Trans. R. Soc. London, Ser. B* **262,** 209–251.

Maetz, J. (1973). Na^+/NH_4^+, Na^+/H^+ exchanges and NH_3 movement across the gill of *Carassius auratus. J. Exp. Biol.* **58,** 255–275.

Maetz, J. (1974). Aspects of adaptation to hypo-osmotic and hyper-osmotic environments. *Biochem. Biophys. Perspect. Mar. Biol.* **1,** 1–167.

Maetz, J., and Bornancin, M. (1975). Biochemical and biophysical aspects of salt secretion by chloride cells in teleosts. International Symposium on Excretion. *Fortschr. Zool.* **23,** 322–362.

Maetz, J., and Garcia Romeu, F. (1964). The mechanism of sodium and chloride uptake by the gills of a fresh water fish. *Carassius auratus.* II. Evidence for Na^+/Na^+ and HCO_3^- exchanges. *J. Gen. Physiol.* **47,** 1209–1227.

Maetz, J., Bourquet, J., and Lahlou, B. (1964a). Urophyse et osmoregulation chez *Carassius auratus. Gen. Comp. Endocrinol.* **4,** 401–414.

Maetz, J., Bourquet, J., Lahlou, B., and Hourdry, J. (1964b) Peptides neurohypophysaires et osmoregulation chez *Carassius auratus. Gen. Comp. Endocrinol.* **4,** 508–522.

Maetz, J., Sawyer, W. H., Pickford, G. E., and Mayer, N. (1967a). Evolution de la balance minerale du sodium chex *Fundulus heteroclitus* au cours du transfert d'eau de mer en eau douce: Effets de l'hypophysectomie et de la prolactine. *Gen. Comp. Endocrinol.* **8,** 163–176.

Maetz, J., Mayer, N., and Chartier-Baraduc, M. M. (1967b). La balance minérale du sodium chez *Anguilla anguilla* en eau de mer, en eau douce et au cours du transfert d'un mileiu a l'autre: Effets de l'hypophysectomie et de la prolactine. *Gen. Comp. Endocrinol.* **8,** 177–188.

Maetz, J., Motais, R., and Mayer, N. (1968). Isotopic kinetic studies on the endocrine control of teleostean ionoregulation. *Int. Congr. Ser.—Excerpta Med.* **184,** 225–232.

Maetz, J., Nibelle, J., Bornancin, M., and Motais, R. (1969). Action sur l'osmorégulation de l'anguille de divers antibiotiques inhibiteurs de la synthèse des protéines ou du renouvellement cellulaire. *Comp. Biochem. Physiol.* **30,** 1125–1151.

Maetz, J., Payan, P., and De Renzis, G. (1976). Controversial aspects of ionic uptake in freshwater animals. *Perspect. Exp. Biol., Proc. Anniv. Meet. Soc. Exp. Biol., 50th, 1974* Vol. 1, pp. 77–92.

Marshall, W. S. (1976). Effects of hypophysectomy and ovine prolactin on the epithelial mucus-secreting cells of the Pacific staghorn sculpin, *Leptocottus armatus* (Teleostei: Cottidae). *Can. J. Zool.* **54,** 1604–1609.

Marshall, W. S. (1978). On the involvement of mucous secretion in teleost osmoregulation. *Can. J. Zool.* **56,** 1088–1091.

Mayer, N., Maetz, J., Chan, D. K. O., Forster, M., and Chester Jones, I. (1967). Cortisol, a sodium excreting factor in the eel (*Anguilla anguilla* L.) adapted to seawater. *Nature (London)* **214,** 1118–1120.

Mendelsohn, S. A., Cherksey, B. D., and Degnan, K. J. (1981). Adrenergic regulation of chloride secretion across the opercular epithelium: The role of cyclic AMP. *J. Comp. Physiol.* **145,** 29–35.

Michaelis, L., and Menten, M. L. (1913). Die Kinetik der Invertinwirkung. *Biochem. Z.* **49,** 486–501.

Motais, R., Isaia, J., Rankin, J. C., and Maetz, J. (1969). Adaptive changes of the water permeability of the teleostean gill epithelium in relation to external salinity. *J. Exp. Biol.* **51,** 529–546.

Moyle, P. B., and Cech, J. J., Jr. (1982). "Fishes. An introduction to Ichthyology." Prentice-Hall, Englewood Cliffs, New Jersey.

Neill, W. T. (1957). Historical biogeography of present day Florida. *Bull., Fla. State Mus.* **2,** 175–220.

Nicol, C. S., Farmer, S. W., Nishioka, R. S., and Bern, H. A. (1981), Blood and pituitary prolactin levels in tilapia (*Sarotherodon mossambicus:* Teleostei) from different salinities as measured by homologous radioimmunoassay. *Gen. Comp. Endocrinol.* **44,** 365–373.

Nordlie, F. G. (1978). The enfluence of environmental salinity on respiratory oxygen demands on the euryhaline teleost, *Ambassis interrupta* Bleeker. *Comp. Biochem. Physiol. A* **59A,** 271–274.

Oduleye, S. O., and Evans, D. H. (1982). The isolated perfused head of the toadfish, *Opsanus beta.* II. Effects of vasoactive drugs on unidirectional water flux. *J. Comp. Physiol.* **149,** 115–120.

Ogawa, M. (1970). Effects of prolactin on the epidermal mucous cells of the goldfish, *Carassius auratus* L. *Can. J. Zool.* **48,** 501–503.

Ogawa, M. (1974). The effects of bovine prolactin, sea water and environmental calcium on water influx in isolated gills of the euryhaline teleosts, *Anguilla japonica* and *Salmo gairdnerii. Comp. Biochem. Physiol. A* **49A,** 545–553.

Ogawa, M. (1975). The effects of prolactin, cortisol and calcium-free environment on water influx in isolated gills of Japanese eel, *Anguilla japonica. Comp. Biochem. Physiol. A* **52A,** 539–543.

Ogawa, M., Yagasaki, M., and Yamazaki, F. (1973). The effect of prolactin on water influx in the isolated gills of the goldfish, *Carassius auratus* L. *Comp. Biochem. Physiol. A* **44A,** 1177–1183.

Olivereau, M., and Lemoine, A. M. (1971). Action de la prolactine chez l'anguille intacte et hypophysectomisée. VII. Effet sur la teneur en acide sialique (N-acetyl-neuroaminique) de la peau. *Z. Vergl. Physiol.* **73,** 34–43.

Payan, P. (1978). A study of the Na^+/NH_4^+ exchange across the gill of the perfused head of the trout (*Salmo gairdneri*). *J. Comp. Physiol.* **124,** 181–188.

Payan, P., and Maetz, J. (1973). Branchial sodium transport mechanisms in *Scyliorhinus canicula:* Evidence for Na/NH_4 and N/H exchanges and for a role of carbonic anhydrase. *J. Exp. Biol.* **58,** 487–502.

Payan, P., Goldstein, L., and Forster, R. P. (1973). Gills and kidneys in ureosmotic regulation in euryhaline skates. *Am. J. Physiol.* **224,** 367–372.

Payan, P., Matty, A. J., and Maetz, J. (1975). A study of the sodium pump in the perfused head preparation of the trout *Salmo gairdneri* in freshwater. *J. Comp. Physiol.* **104,** 33–48.

Philpott, C. W., and Copeland, D. E. (1963). Fine structure of chloride cells of three species of Fundulus. *J. Cell Biol.* **18,** 389–404.

Pic, P., Mayer-Gostan, N., and Maetz, J. (1974). Branchial effects of epinephrine in the seawater-adapted mullet. I. Water permeability. *Am. J. Physiol.* **226,** 698–702.

Pic, P., Mayer-Gostan, N., and Maetz, J. (1975). Branchial effects of epinephrine in the seawater adapted mullet. II. Na and Cl extrusion. *Am. J. Physiol.* **228,** 441–447.

Pickford, G. E., and Phillips, J. B. (1959). Prolactin, a factor in promoting the survival of hypophysectomized killifish in fresh water. *Science* **130,** 454–455.

Pickford, G. E., Griffith, R. W., Torretti, J., Hendler, E., and Epstein, F. H. (1970). Branchial reduction and renal stimulation of (Na, K)-ATPase by prolactin in hypophysectomized killifish in freshwater. *Nature (London)* **228,** 378–379.

Pisam, M. (1981). Membranous systems in the "chloride cell" of teleostean fish gill: their modifications in response to the salinity of the environment. *Anat. Rec.* **200,** 401–414.

Potts, W. T. W. (1976). Ion transport and osmoregulation in marine fish. *Perspec. Exp. Biol., Proc. Annu. Meet. Soc. Exp. Biol., 50th, 1974* Vol. 1, pp. 65–75.

Potts, W. T. W. (1977). Fish gills. *In* "Transport of Ions and Water in Animals" (B. L. Gupta, R. B. Moreton, J. L. Oschman, and B. J. Wall, eds.), pp. 453–480. Academic Press, New York.

Potts, W. T. W., and Eddy, F. B. (1973). Gill potentials and sodium fluxes in the flounder *Platichthys flesus*. *J. Comp. Physiol.* **87,** 29–48.

Potts, W. T. W., and Evans, D. H. (1966). The effects of hypophysectomy and bovine prolactin on salt fluxes in fresh-water-adapted *Fundulus heteroclitus*. *Biol. Bull. (Woods Hole, Mass.)* **131,** 362–368.

Potts, W. T. W., and Fleming, W. R. (1970). The effects of prolactin and divalent ions on the permeability to water of *Fundulus kansae*. *J. Exp. Biol.* **53,** 317–327.

Potts, W. T. W., Foster, M. A., Rudy, P. P., and Parry Howells, G. (1967). Sodium and water balance in the cichlid teleost *Tiplapia mossambica*. *J. Exp. Biol.* **47,** 461–470.

Repetski, J. E. (1978). A fish from the upper Cambrian of North America. *Science* **200,** 529–531.

Sardet, C., Pisam, M., and Maetz, J. (1979). The surface epithelial of teleostean fish gills. Cellular and tight junctional adaptations of the chloride cell in relation to salt adaptations. *J. Cell Biol.* **80,** 96–117.

Shaw, J. (1961). Studies on ionic regulation in *Carcinus maenas* (L.). *J. Exp. Biol.* **38,** 135–152.

Shuttleworth, T. J. (1978). The effect of adrenaline on potentials in the isolated gills of the flounder (*Platichthys flesus* L.). *J. Comp. Physiol.* **124,** 129–136.

Silva, P., Solomon, R., Spokes, K., and Epstein, F. H. (1977). Ouabain inhibition of gill Na^+-K^+-ATPase: Relationship to active chloride transport. *J. Exp. Zool.* **199,** 419–427.

Smatresk, N. J., and Cameron, J. N. (1982). Respiration and acid-base physiology of the spotted gar, a bimodal breather. *J. Exp. Biol.* **96,** 281–293.

Smith, H. W. (1930). The absorption and excretion of salts by marine teleosts. *Am. J. Physiol.* **93,** 480–505.

Smith, H. W. (1931). The absorption and excretion of water and salts by the elasmobranch fishes. I. Freshwater elasmobranchs. *Am. J. Physiol.* **98,** 279–295.

Stagg, R. M., and Shuttleworth, T. J. (1982). Na, K-ATPase, ouabain binding and ouabain-sensitive oxygen consumption in gills from *Platichthys flesus* adapted to seawater and freshwater. *J. Comp. Physiol.* **147,** 93–99.

Thomerson, J. E., Thorson, T. B., and Hempel, R. L. (1977). The bull shark, *Carcharhinus leucas,* from the upper Mississippi River near Alton, Illinois. *Copeia* pp. 166–168.

Thorson, T. B. (1972). The status of the bull shark, *Carcharhinus leucas,* in the Amazon River. *Copeia* pp. 601–605.

Wendelaar Bonga, S. E., and Van der Meij, J. C. A. (1981). Effect of ambient osmolarity and calcium on prolactin cell activity and osmotic water permeability of the gills in the teleost *Sarotherodon mossambicus. Gen. Comp. Endocrinol.* **43,** 432–442.

Wood, C. M., and Randall, D. J. (1973). The influence of swimming activity on water balance in the rainbow trout (*Salmo gairdneri*). *J. Comp. Physiol.* **82,** 257–276.

9

THE PSEUDOBRANCH: MORPHOLOGY AND FUNCTION

PIERRE LAURENT and SUZANNE DUNEL-ERB
Laboratoire de Morphologie Fonctionnelle
et Ultrastructurale des Adaptations
Centre National de la Recherche Scientifique
Strasbourg, France

I. INTRODUCTION

In June 1785, A. Broussonet read the first description of the pseudobranch in front of the Académie Royale des Sciences:

> Cette partie qui n'a été décrite par aucun auteur peut être regardée comme une petite ouie. . . ; elle est distincte des ouies et située dans leur cavité de chaque côté, vers la base des opercules, immédiatement après l'élévation que forment les orbites. Le plus souvent elle décrit un arc. . . ; elle est ainsi que les ouies, composées de lames rangées en file, mais qui vont en décroissant vers les extrémités. Ces lames ne sont point, comme les ouies, placées deux à deux mais simples. . . ; elles ne sont jamais fixées sur un arc osseux, elles forment à leur base une espèce de bourrelet, et la membrane qui tapisse l'intérieur de la cavité les recouvre en partie . . . elle est surtout très apparente dans les poissons dont Artedi a formé une classe particulière sous la dénomination d'Acanthoptérygiens . . . J'en ai fait la mention sous le nom de pseudobranchia. . . . (A. Broussonet: Mémoire pour servir à l'histoire de la Respiration des Poissons, in Mémoires de l'Académie Royale, 1785, 174 et suiv. partly quoted).*

*This part, which has not been described by any author, can be regarded as a small gill. . . ; it is different from a gill and is located in the gill cavity, on each side, near the base of the

FISH PHYSIOLOGY, VOL. XB

ISBN 0-12-350432-5

This introduction well emphasizes the main characteristics of the teleostean pseudobranch, including that it is a hemibranch and that it is maximally developed in the more highly evolved fish.

Broussonet considered the pseudobranch a small gill whose function is related to respiration. This hypothesis was challenged later when Hyrtl (1838) demonstrated that the organ is supplied with arterialized blood provided by the efferent artery of the first branchial arch. The respiratory function was forsaken for the concept of an association of the pseudobranch with vision (Müller, 1839). This new hypothesis gained popularity until it was challenged in its turn (Laurent, 1967; Laurent and Dunel, 1964). Pseudobranch participation in osmoregulation was disputed until it became clear that chloride cells are present in certain but not all species' pseudobranchs.

In elasmobranchs the spiracular hemibranch described by Müller (1839) has been considered the equivalent of a pseudobranch shifted forward from the mandibular arch into the spiracle. It is only recently that its function has been questioned on the basis of its first ultrastructural study (see Laurent, 1976).

II. MORPHOLOGY OF THE PSEUDOBRANCH

A. The Teleost Pseudobranch

In the teleosts the pseudobranch is located in the cranial part of the subopercular cavity to which it is attached by one of its sides. The pseudobranch is basically a hemibranch, in other words, a gill arch supporting only one row of filaments. Each of them bears a series of lamellae as in the gill (cf. Chapter 2, Volume XA, this series).

Direct observation of the trout operculum reveals the existence of two vessels: one comes into the pseudobranch by its inner side (the afferent pseudobranchial artery); the other comes out from the outer side (the efferent pseudobranchial artery).

1. Origin of the Blood Supply

After Hyrtl (1838), who stated that pseudobranch is supplied with arterialized blood, Müller (1839) described different patterns of vascularization. In *Salmo* (Fig. 1a) the pseudobranchial artery is a ventral tributary of

---

operculum, immediately after the protrusion formed by the orbits. Most often, it is arc shaped. Like the gills it is composed of lamellae that are aligned and decreasing in size toward the edges. These lamellae do not occur in pairs like in the gills but individually. . . ; they are never attached to a bony arch, they form a kind of flap at their base, and they are partly covered by the membrane lining the cavity. . . . It (this part) is particularly visible in those fish classified by Artedi as acanthopterygian. . . . I have mentioned it under the name of pseudobranch.

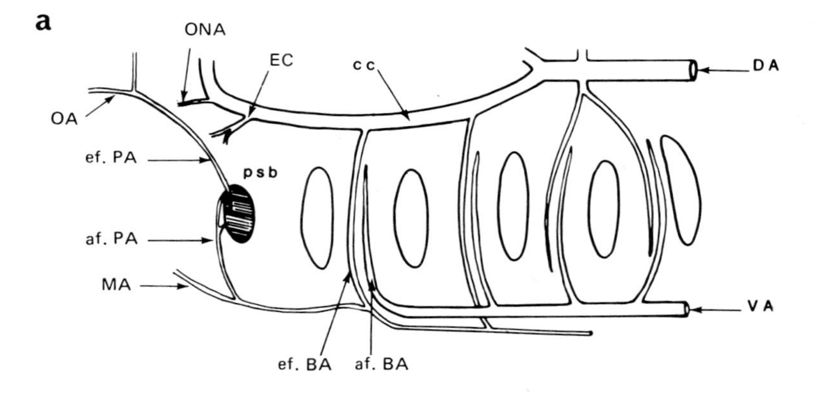

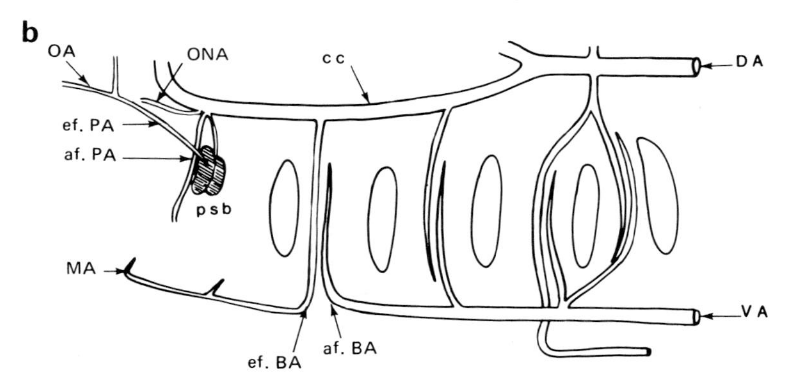

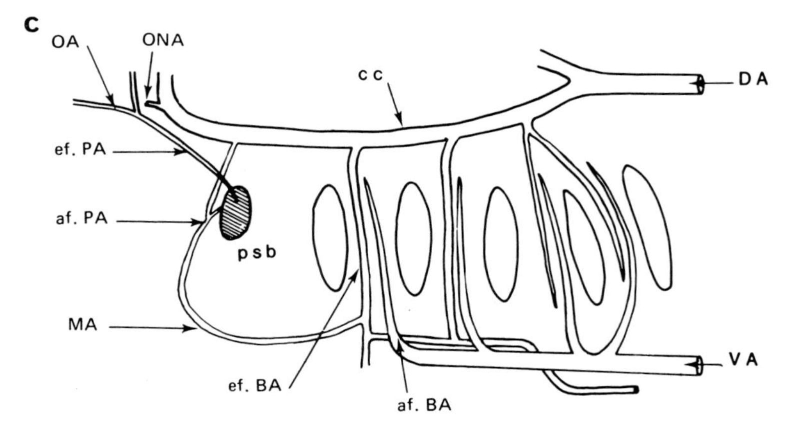

Fig. 1. The three patterns of pseudobranch vascularization. (a) In *Salmo* the afferent pseudobranchial artery (af. PA) comes from the efferent branchial artery (ef. BA) by a ventral route. (b) In *Esox* the afferent pseudobranchial artery emanates from the cephalic circle (cc). (c) In *Perca* the afferent pseudobranchial artery has a mixed origin: the cephalic circle and the efferent branchial artery. In all three cases the efferent pseudobranchial artery (ef. PA) gives rise to the ophthalmic artery (OA). af. BA, afferent branchial artery; DA, dorsal aorta; VA, ventral aorta; MA, mandibular artery; EC, external carotid; ONA, orbitonasal artery; psb, pseudobranch.

the efferent branchial artery of the first arch. In contrast, in *Esox* the pseudobranchial artery emanates from the cephalic circle (Fig. 1b). A mixed pattern of vascularization partly resembling both schemas is found in *Perca* (Fig. 1c). Since then, several studies have been devoted to this question (Allis, 1912; Stork, 1932; Kryzanovsky, 1934), emphasizing the evolution and development of the pseudobranchial arteries in relation to the aortic arches. In addition, certain studies aimed at establishing a functional viewpoint that teleost larvae depend on the pseudobranch for respiration during the "gap" between the disappearance of the vitelline vesicle used for gas exchanges and the appearance of the gills as gas exchangers (Kryzanovsky, 1934). During the development of the trout, the pseudobranch is first supplied by two tributaries of the ventral aorta: the aortic arteries of the mandibular and the hyoidean arches, which join before reaching the pseudobranch. Eight days after hatching, the efferent branchial artery of the third arch ventrally connects to these pseudobranchial arteries, the anastomosis that still connects them with the ventral aorta disappears, and by this time all blood entering the pseudobranch is oxygenated (Dunel, 1975).

2. Distribution of the Vasculature

Histological serial sections and cast preparations demonstrate that the pseudobranch has an arterio-arterial vasculature similar to that of gill. The afferent pseudobranchial artery gives rise to successive arteries supplying the filaments. These arteries in their turn branch out in a great number of lamellar arteries. These short vascular segments supply the pillar capillaries of two or three lamellae each. An identical system of blood vessels collects the blood from the lamellae and sends it into the efferent pseudobranchial artery. Microfil preparation shows the vascular organization of the pseudobranch of the flounder (Fig. 2a) and the trout (Fig. 2b).

The lack of central venous sinuses in the pseudobranch or at least their extreme reduction is one of the characteristic differences of the gill filament. This reduction is related presumably to the coalescing of the lamellae, since a central venous sinus occurs within the pseudobranch bearing free lamellae (*Gobius*). In addition, a nutrient vasculature is present within the pseudobranch. It develops from the afferent pseudobranchial artery and runs mostly on the efferent side of the pseudobranch, that is, on its external side.

Fig. 2. Vascular organization of the pseuobranch. Microfil preparation. Bar = 1 mm. (a) A marine teleost (flounder). The filaments (F) are free. Note the density of the arteriovenous vascular network at the base of the organ. (b) a freshwater teleost (trout). The filaments (F) and lamellae (L) are fused. The afferent pseudobranchial artery (af.PA) runs on the internal side (i.e., attached to the opercular wall) and the efferent pseudobranchial artery (ef.PA) on the external side. ef.FA, efferent filament artery.

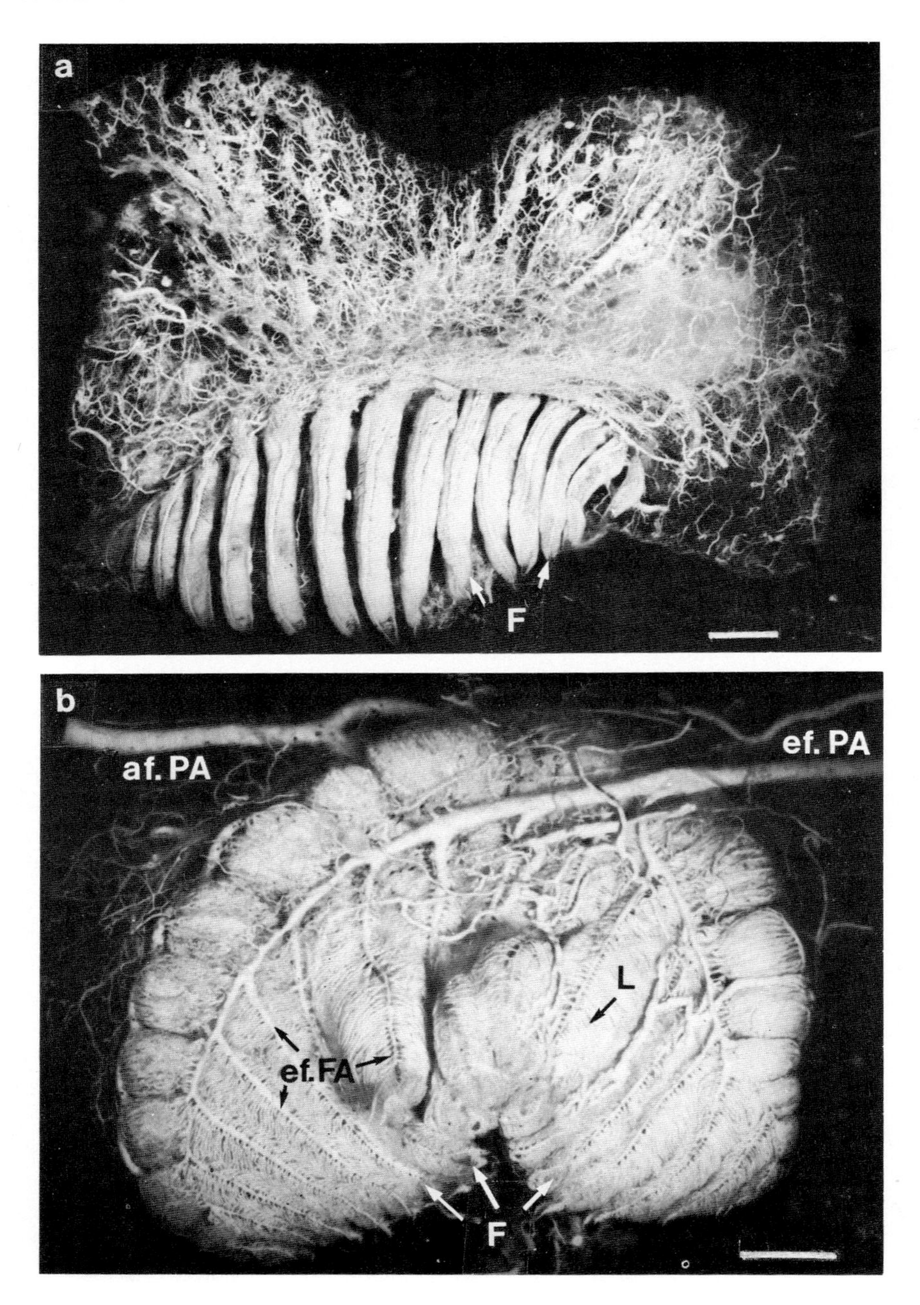
a
F
b
af. PA
ef. PA
L
ef. FA
F

The blood flow within this system is collected by a vein. The lamellae are made up of a capillary network that was described for the first time by Riess (1881) and is analogous to the branchial network.

3. The Different Patterns of Pseudobranch Organization

Different types of pseudobranch organization have been observed according to the fresh- or saltwater origin of the fish, but it has been noticed that forms, sizes, and external structures vary widely within orders, families, and species without any obvious reason, whereas vascular organization and innervation keep a similar organization throughout (Granel, 1927; Kryzanovsky, 1934) (Table I).

It has been noted long ago, however, that the pseudobranch is a much more impressive organ in Percoidei and in the great majority of marine fish. In Cyprinoidei and Salmonidei, which include a large number of freshwater fish, the pseudobranch is of smaller size. Among the 31 orders, 91 suborders, and 238 teleostean families, all species have pseudobranchs except the following: *Gymnarchus* and *Cobitis* (all of both genera), some species belonging to the anguilliform order, and some species belonging to the suborder of Siluroidei.

A useful classification of pseudobranchs is based on whether or not they are covered. In many freshwater teleosts the pseudobranch adheres to the operculum and is covered entirely by the opercular epithelium, which is sometimes so thick that the organ becomes invisible (Cyprinides). Beneath this epithelium filaments as well as lamellae are fused to each other, so that this type is often referred to as "glandular pseudobranch." This situation of course exists because the pseudobranch lamellae have no contact at all with the external medium. In the perch, the pseudobranch is almost entirely covered except the tips of filaments, which are free and bear separate lamellae.

In many saltwater fish various arrangements are seen, including the completely free pseudobranch (*Gobius niger*) and the entirely covered type (*Gadus*). In the free type, pseudobranchs are covered by a thin monolayered epithelium independently surrounding filaments and lamellae. In such an arrangement filaments and lamellae are completely individualized as they are in the gill, being only attached by their base. As a result, lamellae are in contact with the external medium. Another type of arrangement consists of filaments soldered onto the operculum by one side while the other side lies in contact with the external medium. This arrangement occurs in *Morone labrax* and *Serranus scriba*, for instance. Lamellae might be completely free (*M. labrax*) or partly fused as in *Spondyliosoma cantharus*. In both cases

Table I

Pseudobranch Sizes of Some Freshwater (FW) and Saltwater (SW) Fish

	Body length (cm)	Pseudobranch length (cm)	Maximum filament length (cm)	Number of filaments	Number of lamellae/filaments
Perca fluviatilis L. (FW)	10	0.8	0.5	25	60
Sander lucioperca L. (FW)	15	1	0.8	30	—
Salmo gairdneri G. (FW)	30	1.3	0.5	15	70
Spondyliosoma cantharus L. (SW)	10	0.4	0.3	15	60
Serranus scriba L. (SW)	15	0.7	0.25	20	40
Gobius niger L. (SW)	6	0.4	0.15	7	15

they adhere to the operculum. It is interesting to note that when lamellae are partly fused they remain free on the side of their afferent circulation (Dunel and Laurent, 1973a).

These various patterns have important consequences concerning structure and cellular components of the pseudobranchial epithelium. This distinguishes two kinds of epithelia, with respect to their situation regarding the external medium.

4. Ultrastructure of the Pseudobranchial Epithelium

Lamellae are vascularized by pillar capillaries similar to those of the gill. Pillar cells have flanges that line the capillary like a fenestrated endothelium (pillar cells are considered by some workers to be different from endothelial cells; see Chapter 2, Volume XA, this series). Pillar cells and their flanges are separated from the epithelium by a basal lamina. In covered pseudobranchs, where filaments and lamellae are fused, the epithelial origin is not obvious as it is in free pseudobranchs. However, developmental studies have clearly shown that this epithelium is derived from the opercular epithelium (Dunel, 1975). The pseudobranch development starts just after hatching by formation of a cellular excrescence. This excrescence comprises a central area from which originate the blood vessels of the filaments and lamellae and a peripheral area separated from the latter by the basal lamina. The peripheral area gives rise to the epithelium of the lamellae, which appear in their turn from the filament 6 days after hatching.

The epithelium of the pseudobranch consists of several cellular components but is mainly characterized by the presence of a typical cellular component called the pseudobranchial cells. This cell type, which has some characteristics in common with the chloride cells, has no obligatory relationships with the external medium as is shown by its occurrence in covered pseudobranchs.

All pseudobranchs have typical pseudobranchial cells except (as far as is known) species belonging to the genera *Xenocara*, *Fiereaster*, and *Ophidion* (Vialli, 1926). Pseudobranchial cells have never been observed within the gill of any species (Dunel and Laurent, 1973a).

Pseudobranchial cells form a jointing layer of hexagonal cells resting on the basal lamina, so that they are very closely associated with the blood

---

Fig 3. The epithelium of the pseudobranch, *Crenilabrus*. (a) Basal part of a chloride cell in a marine fish showing the tubular organization in the vascular side of the cell. (b) Pseudobranchial cell showing the distribution of the tubules, which are closely associated with the numerous mitochondria. PC, pillar cell; cl, capillary lumen. Bars = 1 μm. (adapted from Dunel and Laurent, 1973a.)

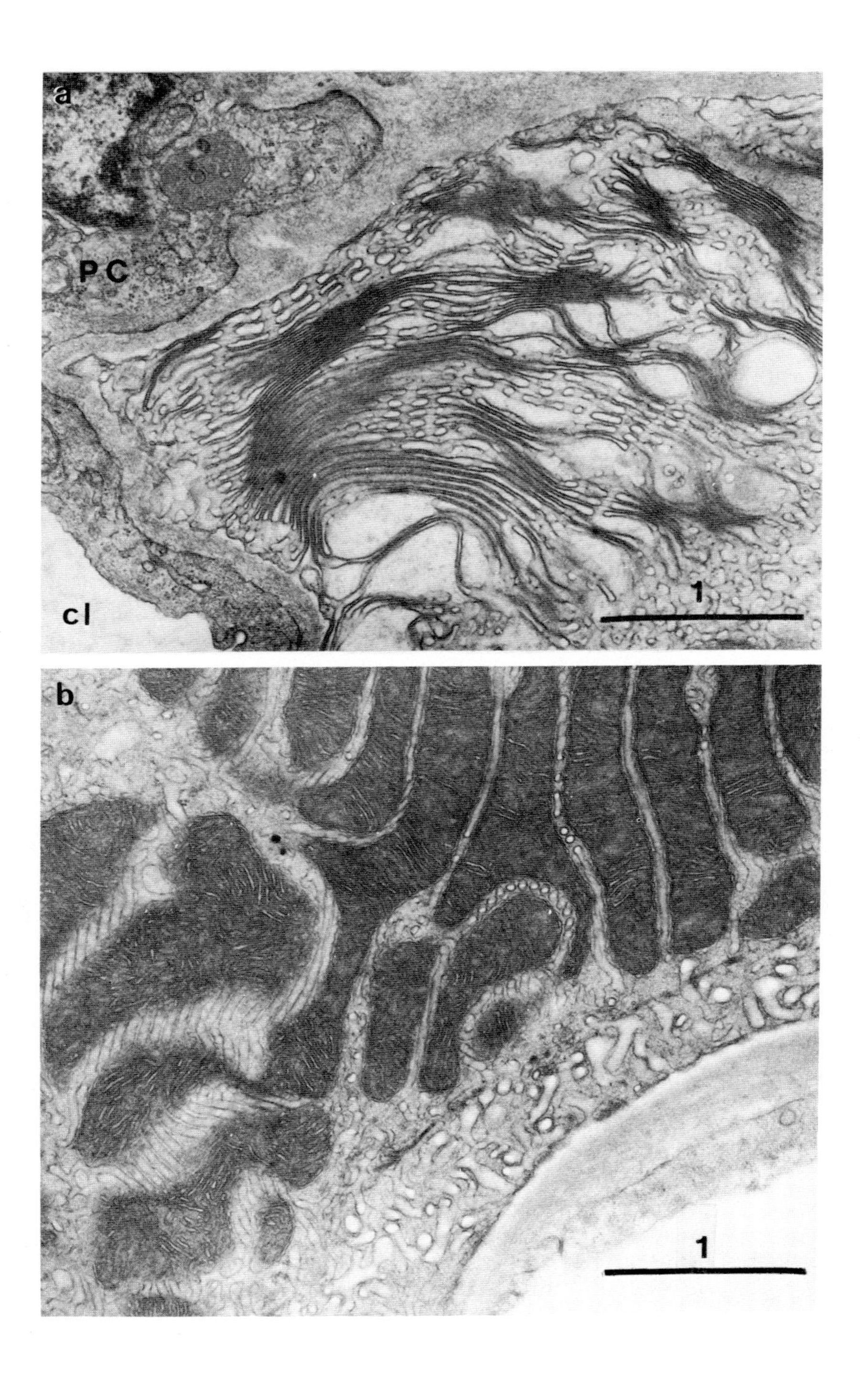
a
PC
cl
1
b
1

compartment. These cells are 10–15 μm wide and 6–10 μm high. They have constant characteristics in all species of teleosts from either marine or freshwater milieus (Fig. 6b) (Copeland and Dalton, 1959; Dunel and Laurent, 1971, 1973a).

These different cell parts may be distinguished according to an internal–external polarity:

1. On the *vascular or basal side*, the basal membrane invaginates inside the cell forming numerous and very long tubules penetrating and filling the first $\frac{2}{3}$ of the cell but never reaching the opposite membrane. Tubules, after initial rectilinear parallel courses (>0.5 μm), branch abundantly in the basal region, characterized by numerous skeleton fibrils (Fig. 3b).

2. In the *middle portion*, tubules emerging from the branching area again take linear and parallel courses. They have a constant diameter (400–500 Å) and are closely associated with the mitochondria. Mitochondria are large and appear in thin section to be arranged parallel to each other. Stereological analysis has shown that they actually form single branched organelles (A. Rambourg, personal communication).

The distance between the external membrane of the mitochondria and the tubules, which often run parallel to one another, does not exceed a few

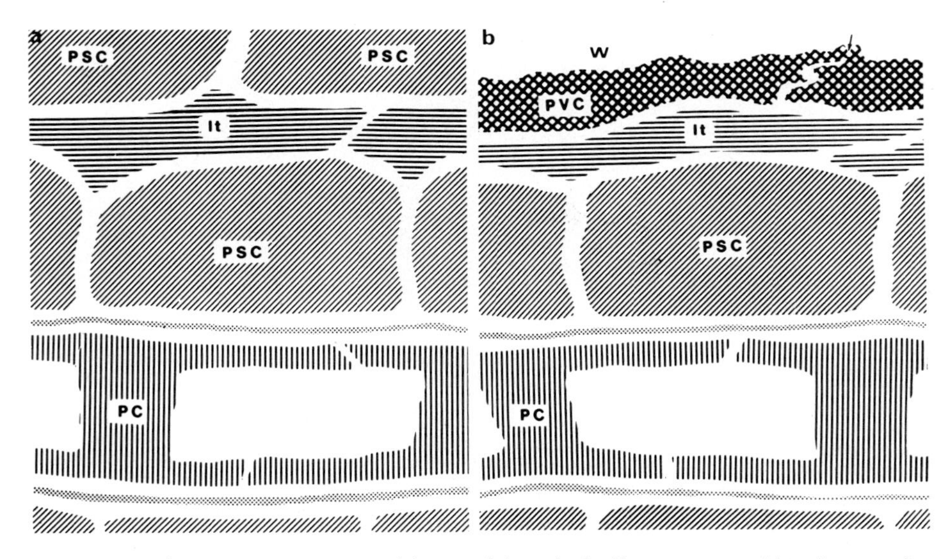

Fig. 4. Schematic representation of the pseudobranchial cell environment. (a) In the case of fused lamellae, a layer of lacunar tissue (lt) common to two lamellae is interposed between adjacent pseudobranchial cells (PSC). (b) In the case of free lamellae, a layer of lacunar tissue (lt) is still interposed between the pseudobranchial cells (PSC) and the pavement cells (PVC), which cover the free lamellae. In each case the pseudobranchial cells line the pillar capillary. PC, pillar cell. This diagram illustrates the micrograph in Fig. 5.

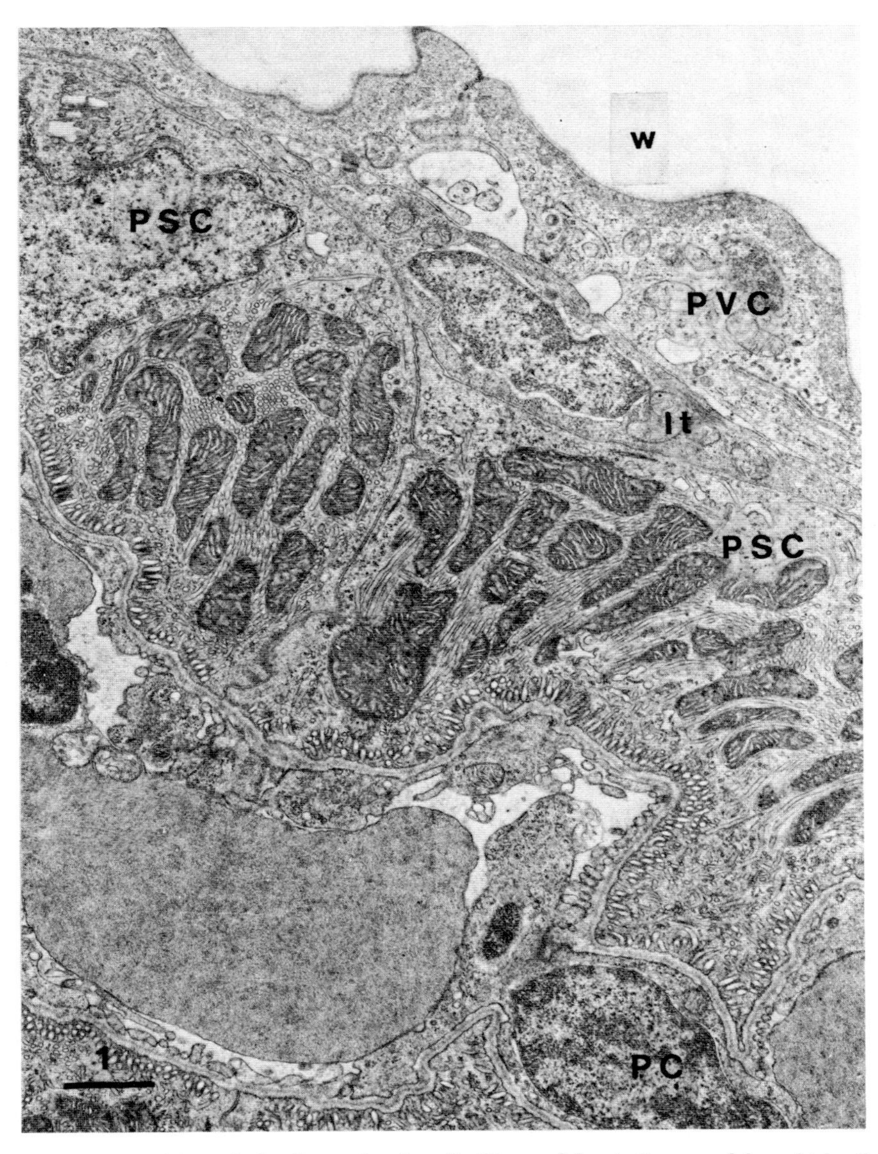

Fig. 5. Pseudobranchial cells in a free lamella (*Morone labrax*). Two pseudobranchial cells (PSC) line the capillary. A pillar cell is visible (PC). On the apical side lacunar tissue (lt) is interposed between the pseudobranchial cells and the pavement cell (PVC). W, external milieu. Bar = 1 μm.

angstroms. Often tubules come back toward the basal membrane after having passed around the mitochondria.

3. The *apical portion* does not contain any tubules or mitochondria. Here the rough endoplasmic reticulum is abundant. The apical membrane is very osmiophilic.

The environment of the pseudobranchial cell is schematized in Fig. 4a and b. In case of fused lamellae (Fig. 4a), a layer of lacunar tissue is interposed between adjacent layers of pseudobranchial cells. This lacunar tissue layer is common to two contiguous lamellae and is made up of very dark osmiophilic cells having long processes intermingled with those of neighboring cells and leaving large spaces between them. Lacunar cells are arranged in two layers; they are rich in tonofibrils and are attached to each other by desmosomes. In the case of free lamellae (Figs. 4b and 5), there is still a sheet of lacunar tissue covered by a thin layer of pavement cells separating the whole from the external medium. Electron-microscopic observations reveal that the tight part of the junctions linking neighboring pavement cells is long, suggesting the existence of several strands (Claude and Goodenough, 1973).

In some part of the pseudobranch with free lamellae, the epithelium consists of typical chloride cells in place of pseudobranchial cells. In the pseudobranch with partly fused lamellae, chloride cells are located in the free part of the lamellae and pseudobranchial cells in the fused part. It is of interest to mention that chloride cells are always located, as in the gill, on the afferent side of the pillar capillary, whereas pseudobranchial cells are located on the efferent side. In partly fused lamellae, the free part of the lamellae is on the afferent side of the pseudobranch. It is unknown why such constant localization occurs, except that chloride cells should have functional relationships with the external medium.

In the pseudobranch, chloride cells exhibit the characteristic differences of fresh- and saltwater species. This is in contrast with pseudobranchial cells, which have morphological characteristics remarkably constant regardless of the medium the fish inhabits. For instance, the tip of the pseudobranchial lamellae in the perch usually supports some free lamellae containing chloride cells typical of freshwater fish, whereas saltwater fish have pseudobranchial chloride cells equipped with an accessory cell (Fig. 6a). Furthermore,

---

Fig. 6. Pseudobranchial epithelium in a marine teleost. Bars = 1 μm. (a) A pseudobranchial chloride cell (CLC) (*Mullus surmuletus*) equipped with an accessory cell (AC) sending processes (arrows) in the apical cytoplasm of the chloride cell. PVC, pavement cell; W, external milieu (water). (Adapted from Dunel and Laurent, 1973a.) (b) Pseudobranchial cells (PSC) in fused lamellae (*Spondyliosoma cantharus*). Two contiguous pseudobranchial cells are separated by a layer of lacunar tissue (lt) where nervous fibers (nf) are observed. Note the arrangement of the tubules in the vascular side. PC, pillar cell. (Adapted from Laurent, 1974.)

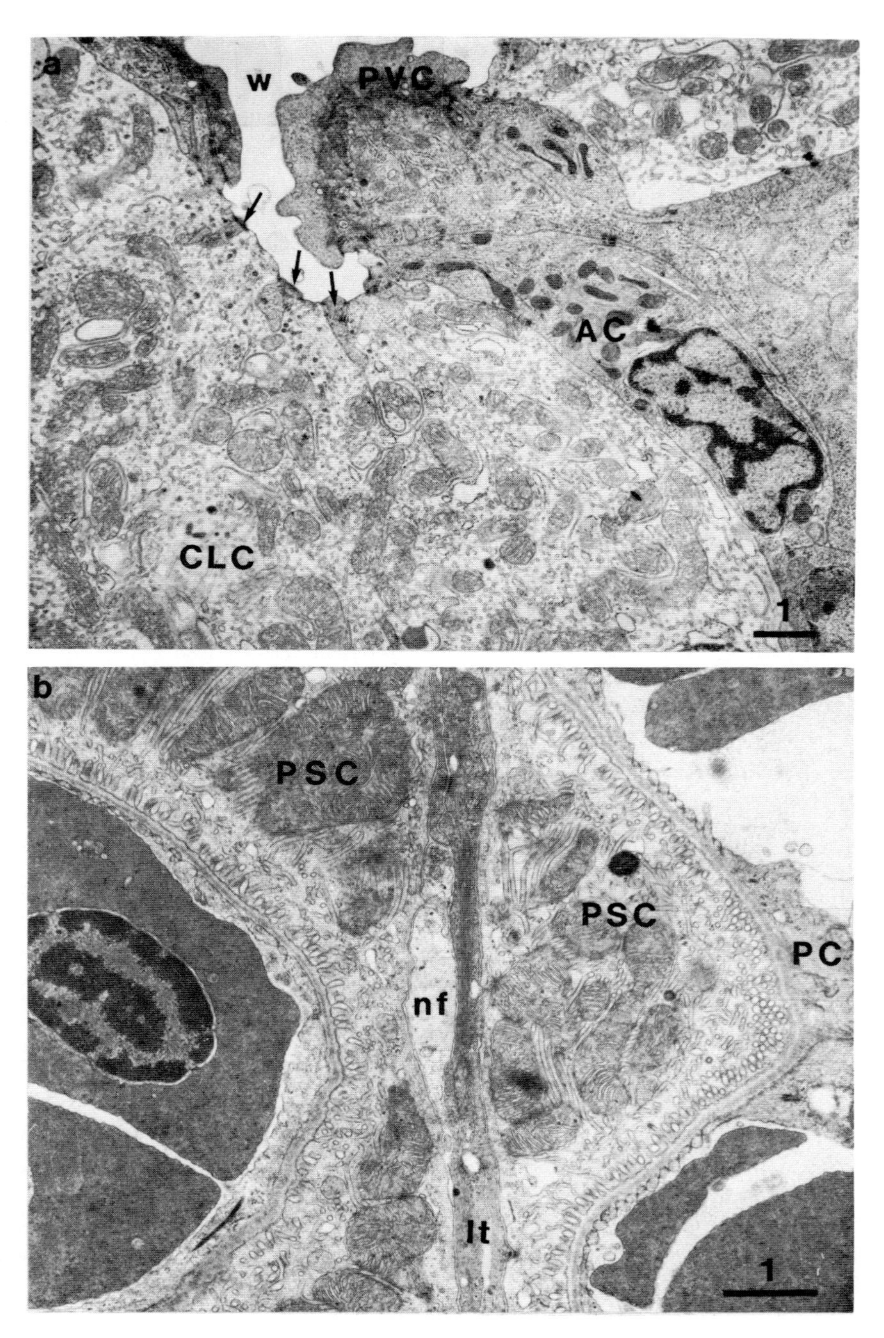
a
w
PVC
AC
CLC
1
b
PSC
PSC
PC
nf
lt
1

Table II

Comparison of Pseudobranchial and Chloride Cell Characteristics

Characteristic	Chloride cell	Pseudobranchial cell
Mitochondria	++ Small ellipsoid	+++ Long filamentous
Tubules	Branched, irregular diameter, forming a meshwork	Branched, then rectilinear tight relationships with mitochondria, very regular in diameter, periodic
Apical vesicles	Very numerous	None
Apical pit	Yes	No relationships with external medium
Innervation	None	Presence of numerous amyelinated profiles close to apical membrane
Vascular compartment	Pillar capillary afferent side	Pillar capillary afferent and efferent side

accessory cells have been demonstrated for the first time in saltwater pseudobranchial chloride cells (Dunel and Laurent, 1973a) and their presence demonstrated later in all saltwater chloride cells including gill (Dunel, 1975; see also Laurent and Dunel, 1980). Table II summarizes the distinguishing characteristics of pseudobranchial and chloride cell types.

5. Innervation of the Teleost Pseudobranch

According to Stork (1932), Laurent and Dunel (1966), and Dunel (1975), three different origins have been found for the teleost pseudobranch innervation: (1) a facial nerve origin as in *Sander lucioperca*, (2) a glossopharyngeal nerve origin as in *Micropterus dolomei*, and (3) a mixed origin as in *Salmo gairdneri*, *Tinca tinca*, and *Perca fluviatilis*. In the latter case indeed a branch emanating from the facial nerve (Jacobson's anastomosis) connects the glossopharyngeal branch just before its entrance into the pseudobranch.

In any case the pseudobranchial nerve gives rise to a plexus located in the base of the organ around and along the efferent pseudobranchial artery. This plexus consists of myelinated and nonmyelinated nerve fibers and includes a number of neurons. Ultrastructural studies reveal numerous axodendritic and axosomatic synapses.

From this first basal plexus emanate as many secondary plexus as there are filaments. These plexus are also located close to the efferent artery and obviously emerge from the latter; they have the same components but are even richer in axodendritic relationships. Some profiles are particularly rich in mitochondria. These observations indicate extensive neural convergence (Fig. 7).

Ultimately, amyelinated bundles separate from the secondary plexus and enter the lamellae via the lacunar tissue. These terminal plexus include numerous small bipolar neurons giving rise to a system of dendrites within the lacunar tissue separating contiguous lamellae (Fig. 8).

Several types of terminals were found innervating the different part of the pseudobranch in teleosts:

1. Numerous amyelinated fibers within the adventitia of the pseudobranchial efferent artery, vesiculated nerve endings, are seen close to the vascular smooth muscle fibers, which suggests a vascular control of the efferent artery, as in the gill (Laurent and Dunel, 1980).
2. In some species, mostly marine, a characteristic population of neurons is located in the efferent artery wall itself. These neurons show a polarization at their dendritic side toward the lumen of the vessel (Fig. 9). This organization suggests that they form a sensory system.
3. The pseudobranchial afferent artery displays arborescent endings after neurofibrillar methods of staining (Fig. 10), emanating from large-sized myelinated fibers. Ultrastructural studies confirm the existence of profiles that are not vesiculated but contain neurofilaments and mitochondria. It is concluded from characteristics and from experimental sections of the nerve followed by degeneration (see later) that these endings are probably mechano- or baroreceptors.
4. By far the most important innervation, in terms of richness, is that made up of the interlamellar plexus and its small bipolar neurons (Laurent and Dunel, 1965, 1966; Barets *et al.*, 1970; Dunel, 1975). In the case of fused lamellae, each plexus is common to two contiguous lamellae, so that the total number of fibers and neurons in each pseudobranch is enormous. Such an innervation consequently should be functionally important. Within each interlamellar space within the lacunae of the interlamellar tissue, numerous terminal or preterminal amyelinated profiles form the dendritic expansions of the small bipolar neurons. Axons of these neurons converge toward the axis of the filament and its plexus. There, axons converge on large multipolar associated neurons and connect them via axodendritic or axosomatic synapses. Then, amyelinated or small-sized myelinated axons of these multipolar neurons go back toward the central nervous system (CNS) via the glossopharyngeal nerve. Of course this innervation is not affected by nerve section.

In the case of pseudobranchs with free lamellae, nerve profiles are still present with the same characteristics in subepithelial position, between the pseudobranchial cells and the covering epithelium.

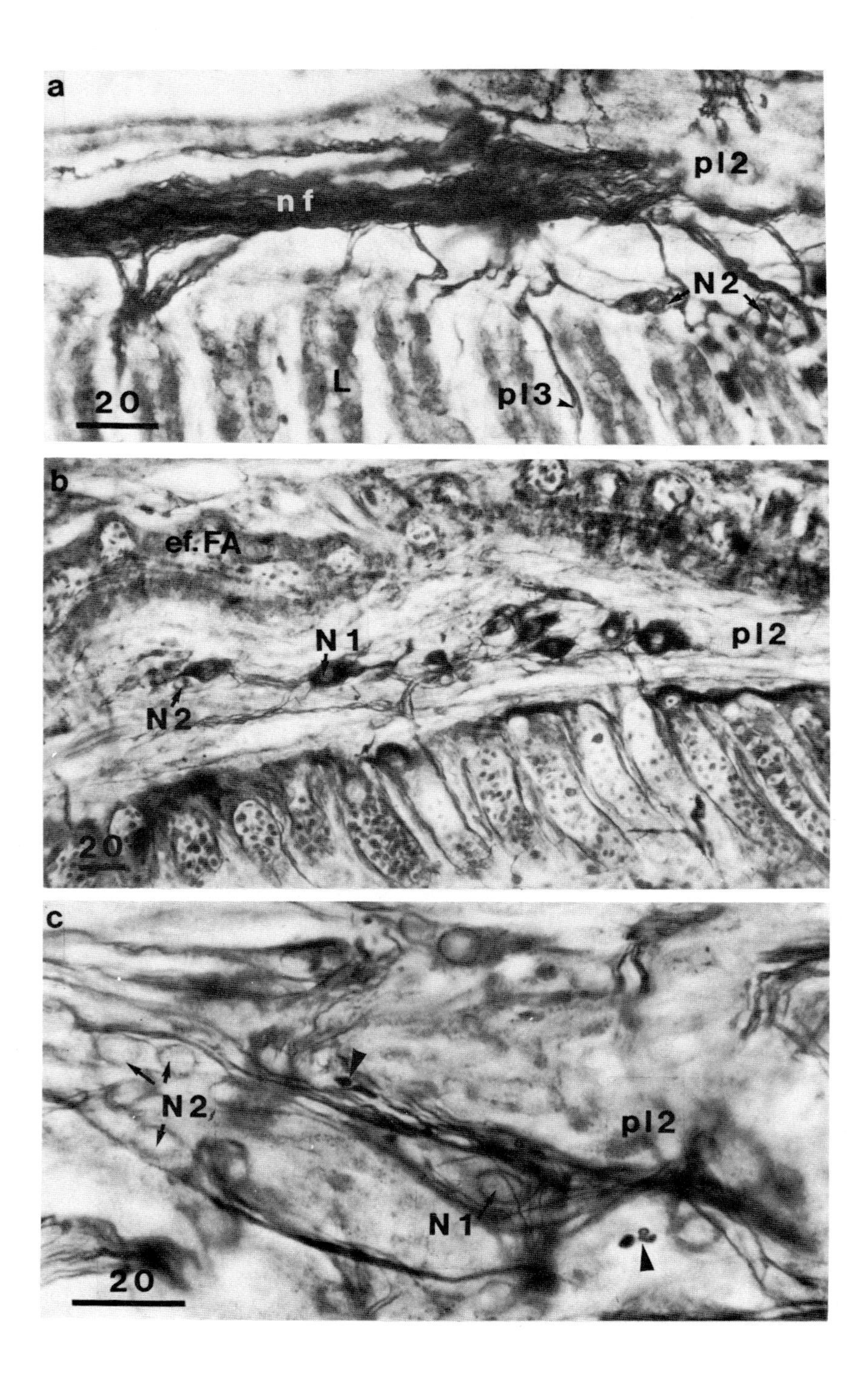
a
nf
pl2
N2
L
20
pl3
b
ef. FA
N1
pl2
N2
20
c
N2
pl2
N1
20

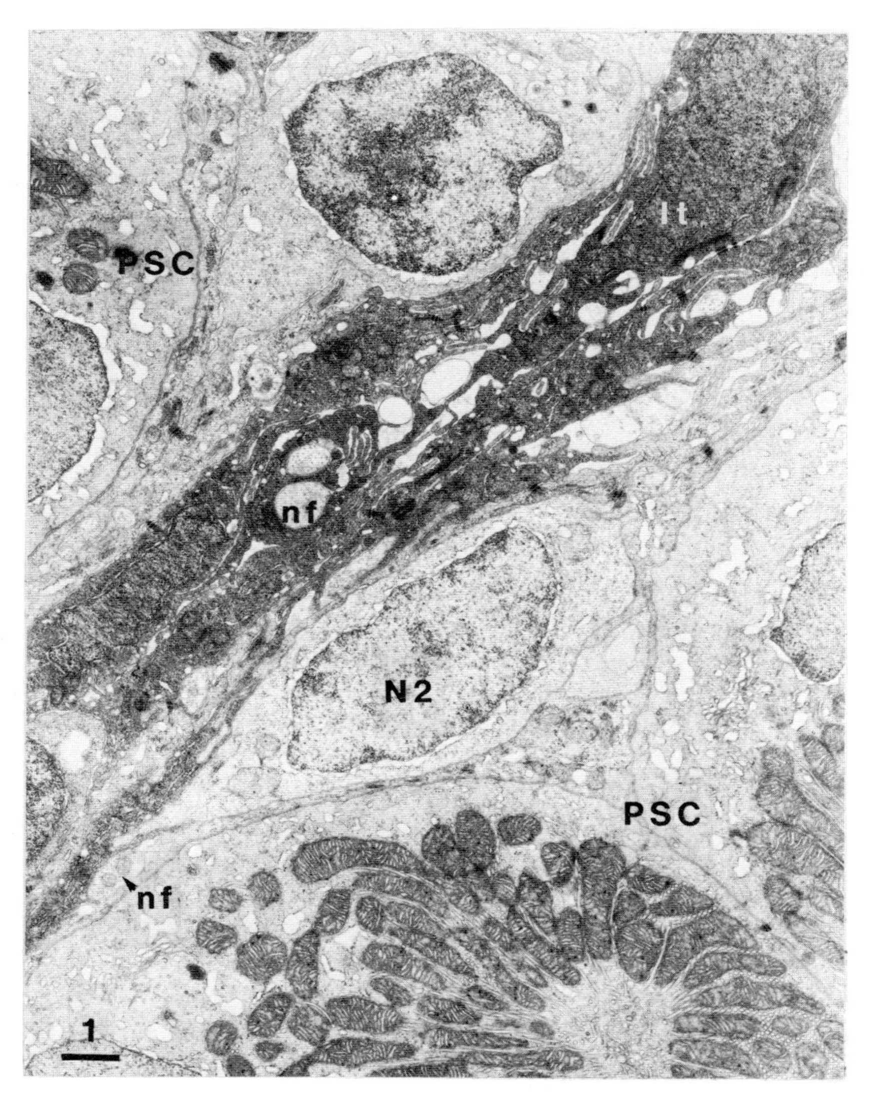

Fig. 8. A bipolar neuron (N2) in the interlamellar space between two contiguous lamellae. Nerve fibers (nf) are surrounded by the lacunar tissue (lt) located between adjacent pseudobranchial cells (PSC) of *Perca fluviatilis*. Bar = 1 μm. (Adapted from Laurent, 1974.)

Fig. 7. Innervation of teleost pseudobranch. Method of Cajal. (a) and (c) *Micropterus dolomei;* (b) *Sander lucioperca*. The secondary plexus (pl2) is located within the axis of the filaments and very close to the efferent artery (ef. FA). This plexus includes numerous neurons of two types: large multipolar neurons (N1) and small bipolar neurons (N2). Note that the fibers running between the lamellae (L) and constituting the interlamellar plexus (pl3) converge toward the secondary plexus (pl2). Nerve endings are visible in the plexus (arrowhead). Bars = 20 μm.

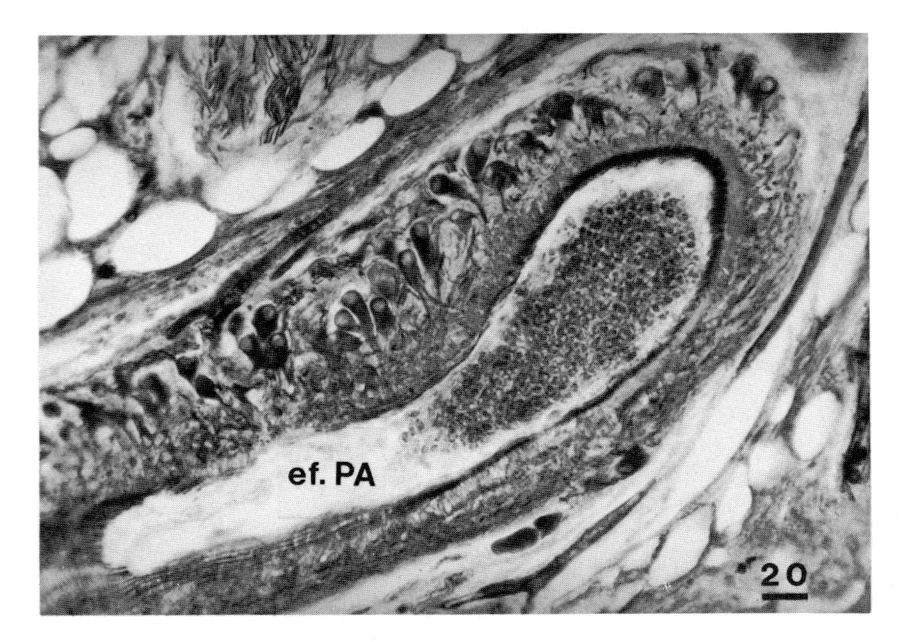

Fig. 9. Neurons located within the efferent pseudobranchial artery wall (ef. PA) in *Spondyliosoma cantharus*. Note the polarization of their dendrites toward the lumen of the vessel. Bar = 20 μm.

In addition to the interlamellar profiles, which are poor in organelles, some nerve endings containing small clear vesicles are also seen within the lacunar tissue. In the perch, zones of characteristic synaptic contacts occur between these endings and some pseudobranchial cells, suggesting the existence of an afferent control of these cells.

It is well known that a nerve fiber that is separated from its neuronal body degenerates after a certain period (this interval depends on temperature and size of the fiber). In the black bass, all nerve fibers more than 7μm in diameter degenerate after a transection of the glossopharyngeal nerve below its ganglion. This result clearly demonstrates that large nerve fibers, including myelinated fibers from 7 to 13 μm are afferent. A comparison carried out in the trout between the right pseudobranchial nerve left intact and the sectioned left nerve (40 days previously) leads to similar conclusions: all myelinated fibers ranging from 4 to 10 μm degenerate. However, this nerve section brought about the disappearance of the arborescent endings within the wall of the ipsilateral pseudobranchial afferent artery. These observations suggest that these endings are sensory and that their neuronal

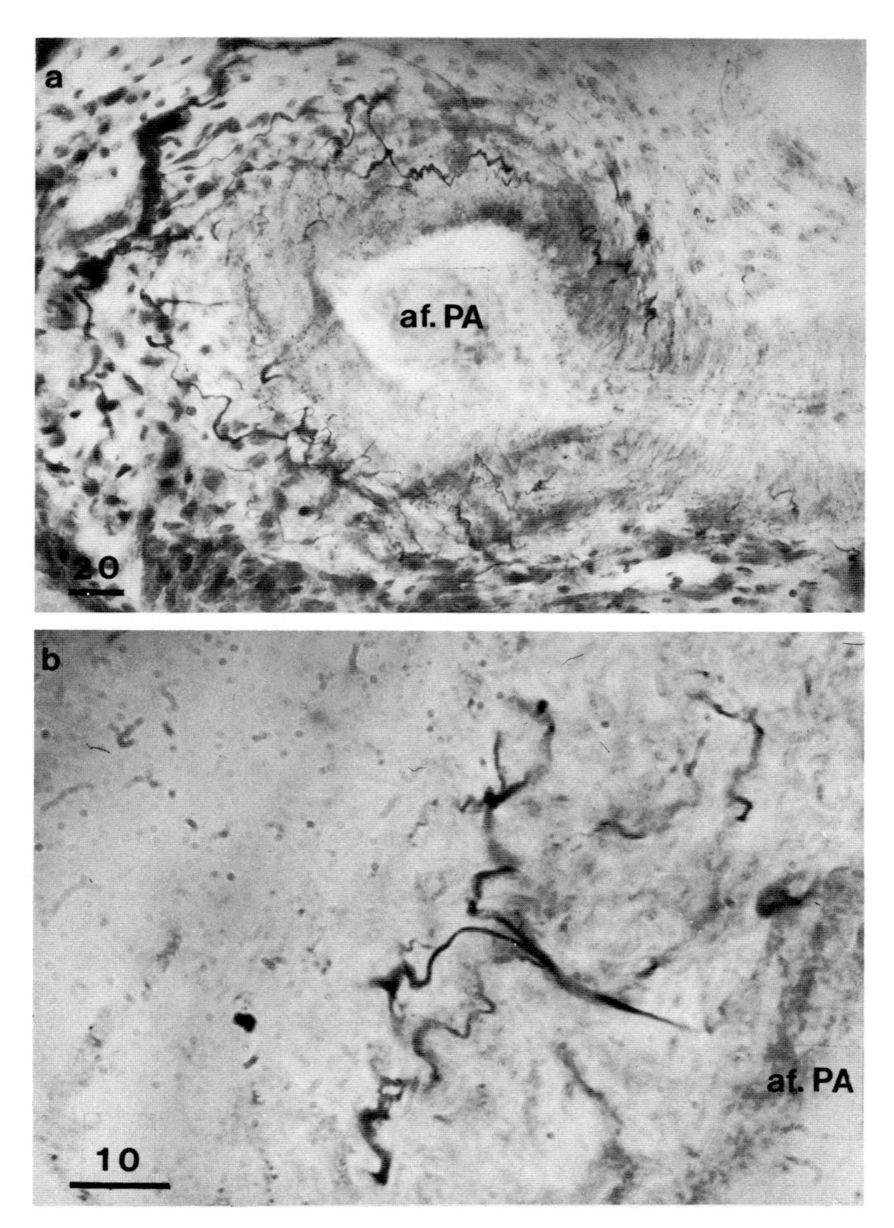

Fig. 10. Innervation of the afferent pseudobranchial artery (af. PA). (a) Numerous endings are observed within the adventitia. Bar = 20 μm. (Adapted from Laurent, 1976.) (b) Detail of an arboriform ending (*Tinca tinca*). Bar = 10 μm. (Adapted from Laurent and Dunel, 1964.)

body is located somewhere between the section and the CNS, obviously within the glossopharyngeal nerve ganglion.

Histological studies on the black bass and on the trout have shown that the glossopharyngeal branch to the pseudobranch comprises many smaller myelinated fibers. This group of fibers separates in two subgroups. In the first subgroup, nerve fibers degenerate after glossopharyngeal transection, suggesting that those fibers also have a neuronal body that is more centrally located than is the section. The second subgroup contains fibers that remain unaltered. The cell bodies of these fibers are located within the pseudobranch itself and represent about 80% of the total group. Cell bodies obviously are part of the different plexus described previously, and fibers constitute the centripetal axons of the multi- and bipolar intramural neurons. Indeed examination of these plexus after nerve section reveals that, in contrast with the arborescent endings of the pseudobranchial afferent artery, most of the plexus and particularly the interlamellar ones are not altered (Dunel, 1975).

Amyelinated fibers are also present within the nerve to the pseudobranch in rather small numbers, and their diameters range between 0.1 and 1 μm.

Degeneration experiments have therefore produced clear evidence for the existence of two afferent systems: a system with the largest fibers connecting the arterial mechanoreceptors (or baroreceptors) and a system with smaller fibers consisting of axons emerging from the pseudobranchial plexus.

B. The Elasmobranch Pseudobranch

The pseudobranch is a very minute organ in elasmobranchs, located on the mandibular arch as in teleosts, but shifted forward within the spiracle (spiracular pseudobranch). The spiracle is a channel connecting the pharyngeal cavity to the external medium. In a 30-cm *Torpedo* it measures 16 mm in length and 7 mm in width. The spiracle participates in gill ventilation (Couvreur, 1902). The spiracular pseudobranch lies on the anterior spiracular wall and consists of seven to eight filaments oriented parallel to the water stream. The middle filaments are 4 mm in length and the side filaments only 2 mm. The thickness of the filaments does not exceed 1 mm. Above the pseudobranch an erectile organ has the capability to close the spiracle.

1. Origin of the Vascularization

The blood supply is arterialized, as has been shown already for teleosts (Müller, 1839; Hyrtl, 1858; Parker, 1881; Dohrn, 1885; Virchow, 1890; Allis,

1909; Stork, 1932; Kryzanovsky, 1934). The pseudobranchial afferent artery originates from the hyoidean efferent artery.

Another artery, the so-called efferent pseudobranchial artery, is a tributary of the anterior carotid.

In elasmobranchs that do not have any pseudobranch, these two vessels (afferent and efferent) connect directly (*Callorhynchus*, Holocephalei) or through a simple rete (*Selache maxima*). This rete has been described in *Torpedo* by Carazzi (1905) as constituting a structure more complex than what exists in most elasmobranchs. Because of a connection between the cephalic and the ophthalmic circles, in contrast with teleosts, it has been suggested (Delage, 1974; Dunel, 1975) that these vessels do not function as afferent–efferent arteries. They both contribute to the blood supply of the pseudobranch, which is finally drained into the venous system (Fig. 11b). In *Torpedo marmorata*, for instance, the existence of a venous outflow has been shown by casting methods.

2. Distribution of the Vasculature

Previous studies (Vialli, 1924; Granel, 1924) have already demonstrated the existence within the elasmobranch pseudobranch of the basic gill vascular arrangement described in elasmobranchs.

Subsequent studies (Delage, 1974; Dunel, 1975; Laurent, 1976) have demonstrated in *Torpedo marmorata* that the pseudobranchial "afferent" artery forms a branched network at the base of this organ. From there single vessels penetrate the filaments and supply a mass of erectile tissue located at the base of each filament. From these cavernous bodies the blood flows into a few lamellae. These lamellae are not really functional and are reduced to a marginal vessel. The blood is finally collected by the so-called efferent artery. It is worth noting that this artery communicates, in turn, via sphincter-like structures, with the parenchymal sinuses of the filament (Fig. 12).

Finally the blood is drained, via a network of pseudobranchial veins, into the jugular vein. This arrangement is similar to that observed in elasmobranch gill: the sphincter-like structure corresponds with arteriovenous anastomoses and the lacunar pseudobranch tissue with central venous sinuses.

3. Organization of the Filaments

a. The Parenchyma and Its Innervation. The filamental parenchyma is a very complex system of sinuses lined by a thin endothelium. Large clear cells, each with a round nucleus and long processes, are located beneath this endothelium. They are thus separated from the blood by 0.1 μm. In addition to abundant glycogen granules, the cells contain dense-cored vesicles of

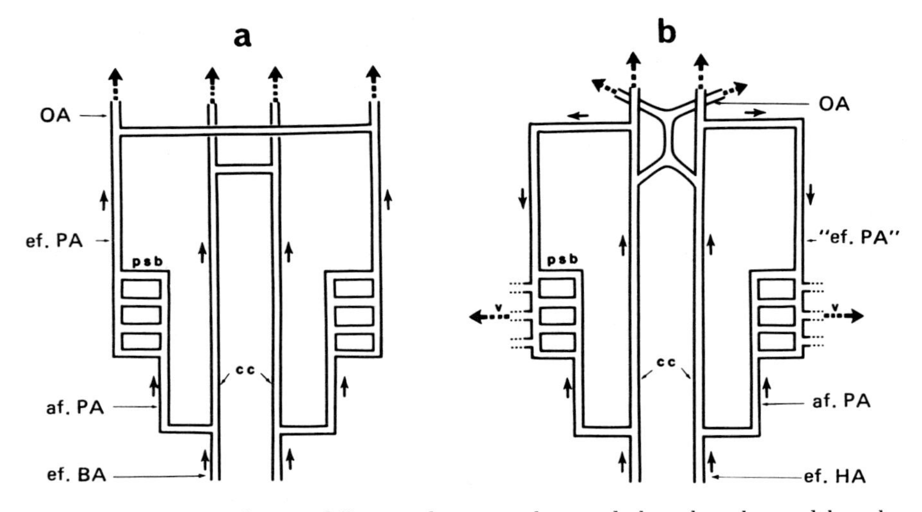

Fig. 11. Diagram showing differences between teleost and elasmobranch pseudobranch vascular organization. (a) Teleost pseudobranch receives blood from efferent branchial artery (ef. BA) via the afferent pseudobranchial artery (af. PA). Efferent pseudobranchial artery (ef. PA) drains off blood toward the ophthalmic artery (OA). Cephalic circle (cc) and ophthalmic circle are separated. psb, pseudobranch. (b) In elasmobranchs the cephalic and ophthalmic circles are interconnected, so afferent and efferent pseudobranchial arteries (af. PA, "ef. PA") both contribute to the blood supply of the pseudobranch. Blood is finally drained off by a venous system (v). ef. HA, efferent hyoidean artery. (Adapted from Laurent, 1976.)

1000 Å. Cellular processes make synaptic contacts with the dendrites of large neurons, which are another component of this lacunar tissue. The synaptic junctions show high presynaptic membrane density associated with dense-cored vesicles and a few small clear vesicles. Neurons are multipolar, grouped in small clusters, and often linked to each other by gap (electric) junctions. They send nonmyelinated axons of 1.5 μm, which join a bundle of nerve fibers running along the filament in the middle part of the lacunar tissue. These bundles after jointing connect to the brain via the palatine branch of the facial nerve. Degeneration experiments show that the filament bundles consist of myelinated fibers that degenerate after the facial nerve section, whereas the amyelinated fibers of 1.5 μm do not (Delage, 1974; Dunel, 1975).

Fig. 12. A filament of *Torpedo marmorata* pseudobranch. A multilayered epithelium (ep) surrounds the filament. (The external milieu, not visible here, will be on the top of the micrograph.) Note in the parenchyma the system of sinuses (s), two neurons (N), and nerve fibers (nf). Note also the sphincter-like structure (★) that connects the so-called efferent artery (ef. A) to the sinuses. Bar = 20 μm. In the inset, a higher magnification shows two large cells (arrows) located beneath the endothelium of the sinus. Bar = 10 μm.

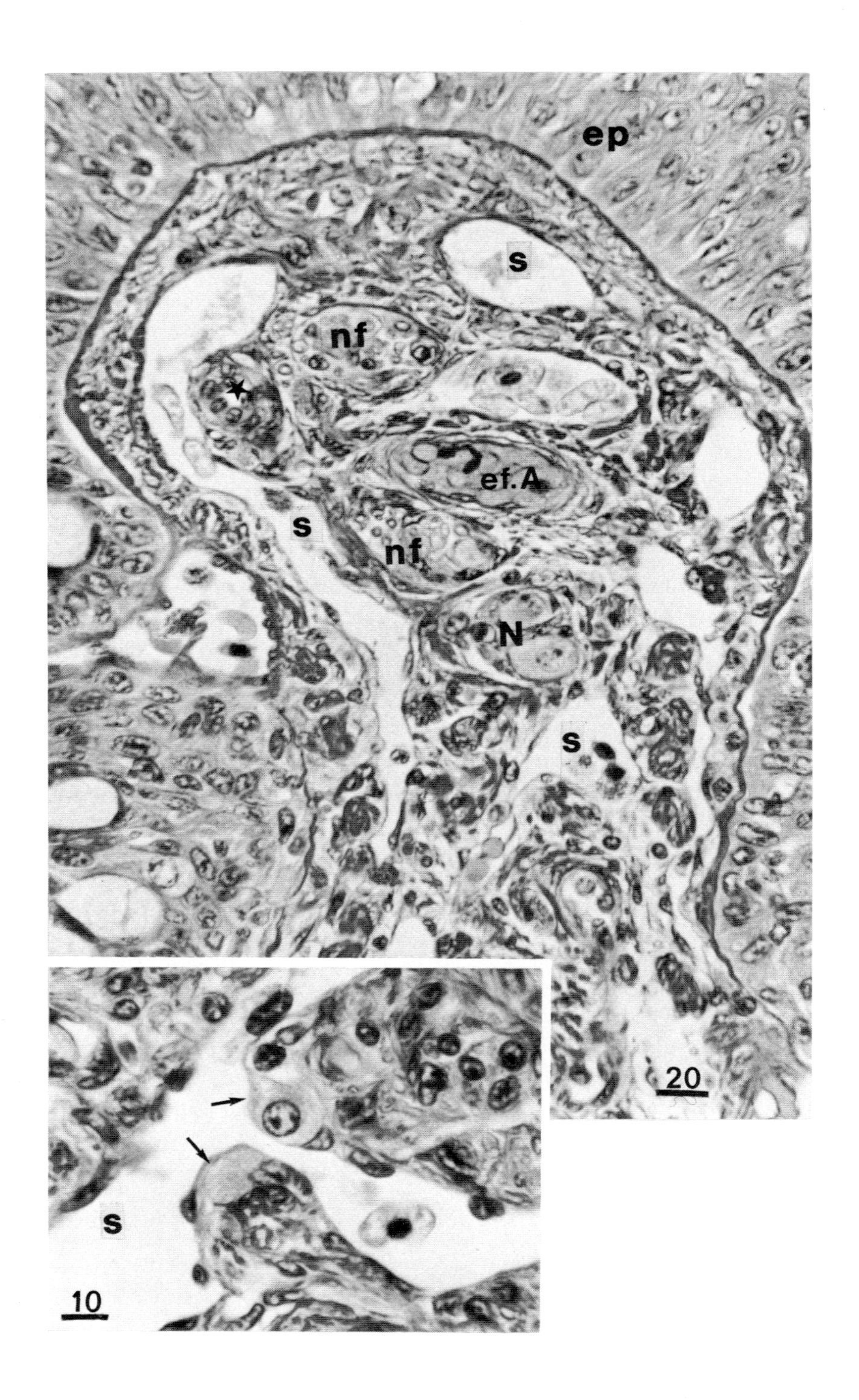
ep
s
nf
ef. A
s
nf
N
s
20
s
10

Thus, as shown in Fig. 12, on a strict morphological basis the innervation of the lacunar tissue of the filament is formed by primary sensory cells located very close to the blood, sending processes to the associated neurons, whose axons run to the central nervous system via the facial nerve.

b. The Epithelium and Its Innervation. A thick epithelium surrounds the filament. It is stratified and consists of five to six layers of cells. The superficial layers are squamous and contain large mucous cells. A germinative zone rests on a thick basal lamina. Epithelial cells are rich in electron-opaque granules of various sizes (up to 4000 Å in diameter). Abundant microfibrils are also present and form with desmosomes a dense endocellular skeleton. The most external layer of cells is characterized by a highly visible terminal web. Numerous clear exocytotic vesicles are seen associated with the external membrane, and transitional forms suggest a descent from endocellular large osmiophilic vesicles.

In addition to those components, long columnar cells are resting on the basal lamina of the epithelium. They have a large nucleus, glycogen granules, and are mostly characterized by cytoplasmic dense-cored vesicles of 800 Å. These vesicles are more concentrated at the base of the cell, close to the basal lamina. The cells contain numerous osmiophilic lysozomes. In addition, the cells display a formaldehyde-induced green fluorescence. The apex of the cells reach the epithelial surface where they develop microvilli and form, with neighboring epithelial cell, a characteristic luminal pit. At the bottom of these cells, some nonmyelinated nerve fibers originating from the palatine branch of the facial nerve cross the basal lamina and penetrate the epithelium deeply. These fibers show enlargements of several microns that are rich in glycogen granules, dark mitochondria, and vesicles of various sizes, some of them being granulated. They are probably related to the preceding epithelial cells, but no characteristic synapses have been seen yet. However taking all these characteristics into account, it seem reasonable to consider them neuroepithelial cells reminiscent of the components of the neuroepithelial bodies of the vertebrate airways. Similar structures have also been seen in the elasmobranch gill filament epithelium (Dunel-Erb *et al.*, 1982).

In conclusion, the pseudobranch in elasmobranchs and particularly in *Torpedo marmorata* is mostly a sensory organ that has no other significant

Fig. 13. (a) Pseudobranch (psb) and hyoidean hemibranch (hg) of *Lepisosteus osseus*. Microfil preparation. Bar = 1 mm. The pseudobranch receives blood from efferent hyoidean artery (ef.HA), which becomes the afferent pseudobranchial artery (af.PA). Note that the external structure is the same for pseudobranch and hemibranch. (b) Pseudobranch of *Amia*. Note the globular shape of the organ, which has the same organization as teleost pseudobranch (see Fig. 2b for comparison). Bar = 1 mm.

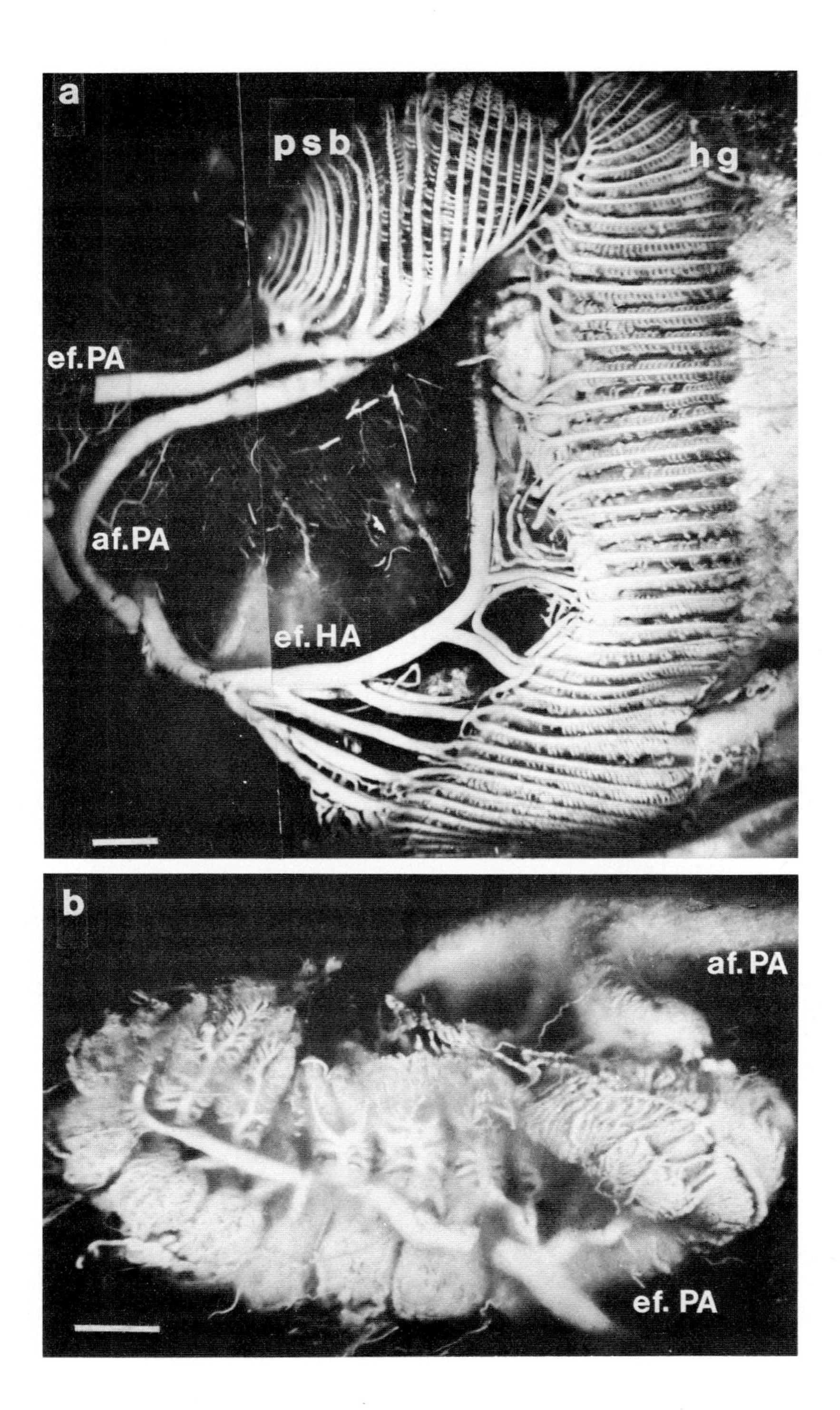
a
psb
hg
ef.PA
af.PA
ef.HA
b
af. PA
ef. PA

structure from a functional point of view than the parenchymal epitheloid cells and neuroepithelial cells.

C. The Pseudobranch in Lower Groups of Fish

A. Chondrostei

In *Acipenser* the pseudobranch is located in the spiracle near its pharyngeal opening. It consists of 10 to 15 filaments supporting thin lamellae. The filament adheres to the spiracle wall by one of its sides, corresponding with the afferent circulation. As in the *Acipenser* gill, there is a piece of cavernous tissue supplied by the afferent artery and running all the way along the filament. The lacunae of this tissue communicate with the lamellae. The lamellae in the pseudobranch and in the gill have similar structures, in that they do not include typical pseudobranch cells (Vialli, 1925; Granel, 1927).

In Polyodon it is interesting to mention that the cavernous tissue does not exist as such, but the afferent arteries have a spongy wall that might serve the same purpose.

Polypterus larvae and adults have no pseudobranch.

2. Holostei

The garpike (*Lepisosteus osseus*) has a pseudobranch and a hyoidean hemibranch as well (Fig. 13a). The pseudobranch develops in the larva close to the spiracular invagination, but the spiracle is never pierced in this animal. It secondarily migrates toward a subopercular position close to the hyoidean hemibranch (Granel, 1927). The pseudobranch is supplied from a branch of the hyoidean efferent artery and a ventral branch of the first-arch efferent artery. The efferent pseudobranchial artery connects the internal carotid and the ophthalmic artery. Filaments are free and bear numerous lamellae. As in *Acipenser*, afferent arteries have a spongy wall but no cavernous tissue. Lamellae have pillar cells but no pseudobranchial cells (Vialli, 1926). Thus, the *Lepisosteus* organization is very similar in gill and in pseudobranch, as it is in *Acipenser* (Granel, 1927; Vialli, 1926).

The bowfin (*Amia calva*) has a pseudobranch that is very similar to that of teleosts (Fig. 13b) and that contrasts with *Lepisosteus* (Vialli, 1925). Here the pseudobranch still has spatial relationships with the spiracle, but because of the small size of the spiracle, the bowfin pseudobranch develops inside the spiracle wall. This arrangement leads eventually to a covered, so-called glandular pseudobranch. The resemblance with teleosts is achieved by the structure of the lamellae, which include in their epithelium typical pseudobranchial cells (S. Dunel-Erb, unpublished).

The ultrastructural organization of nonteleost pseudobranchs, elasmobranchs excepted, is still almost unknown. Three main problems deserve attention:

1. The innervation is provided in *Acipenser*, *Lepisosteus*, and *Amia* by a branch of the glossopharyngeal nerve. The intrinsic distribution is still unknown.
2. In those three species afferent and efferent arteries have the same (or about the same) hemodynamic pressure (Stork, 1932). Therefore, a venous system, as in elasmobranchs, has to drain off the blood from the pseudobranch. If this were not the case, significant blood flow within the pseudobranch would be impossible.
3. The absence of pseudobranch cells in nonteleost fish with the exception of *Amia* presents a functional problem concerning these lower fish.

III. FUNCTIONS OF THE PSEUDOBRANCH

The function of the pseudobranch is still subject to speculation and controversy. Since its discovery it has successively or even simultaneously been considered a salt-regulating, a respiratory, a glandular, or a sensory organ. In addition, the possibility of the pseudobranch acting as a vascular rete mirabile functionally associated with vision has also received a great deal of attention.

A. The Concept of a Respiratory Pseudobranch

The respiratory role of the pseudobranch, at least in adult forms, has been questioned ever since it was established that its blood supply originates from efferent gill arteries, that is, that the incoming blood is already oxygenated before it enters the pseudobranch (Hyrtl, 1838). This point of view was subsequently confirmed by experiments (Parry and Holliday, 1960). Even after this observation, however, some authors still believed that the pseudobranch should have a respiratory function during the larval stage. In his important monograph "Die Pseudobranchie," Kryzanovsky (1934) claimed that this organ could play an important respiratory role during the period of time between the disappearance of the gas-exchanging vitelline membrane and the onset of efficient gas exchanges within the gills following their full development. The question of larval pseudobranch development has been studied in some detail at an ultrastructural level (Dunel, 1975). These studies led to conclusions having important functional implications.

In ancient elasmobranchs, which may be close to an ancestor of the jawed vertebrates, the mandibular hemibranch or pseudobranch should have shared a respiratory function with the other gill arches, the mandibular arch being an aortic arch similar to others (Romer and Parsons, 1970). In modern elasmobranchs, where metameric aortic arches are greatly modified from the primitive pattern, the development of pseudobranch and gill is simultaneous. The pseudobranch consists, even in larvae, of too small a number of lamellae to have any significant role in respiration at any time. However, during its early stage of development the organ is vascularized by tributaries of the ventral aorta (venous blood), and the regression of the gill-type anatomy of the pseudobranch is patent during the subsequent phases.

In nonteleost lower groups such as *Acipenser* and *Polyodon* (Chondrostei), and *Lepisosteus* and *Amia* (Holostei), the pseudobranch development is late compared to that of the gills. The larva is very active, the gill already functioning and the yolk resorbed. Thus, the respiratory function is unlikely, at least in these living modern species of lower groups.

In teleosts a similar conclusion can be drawn except in anguilliforms. Larvae of Murenoidei and Congroidei have a pseudobranch that disappears with adulthood. In leptocephalus the pseudobranch, which is really the first arch to develop, consists of a small number of filaments bearing long lamellae. This hemibranch has the structure of a gas exchanger and does not have any pseudobranchial cells (Dunel, 1975). In its early stages of development, leptocephalus has a vascular pattern typical of gill. Thus, the pseudobranch is supplied by some tributaries of the mandibular and hyoidean branches of the ventral aorta (Grassi, 1914). Later, an anastomosis develops between the afferent pseudobranchial artery and the efferent artery of the first arch, and replaces the primitive afferent vasculature. For some reason the pseudobranch disappears. It appears from these anatomic results that the pseudobranch of leptocephalus has the capability to function as a respiratory organ at least for a short period of time.

In *Salmo irideus*, in no period of the larval development, has the pseudobranch the capability of exchanging gas (Dunel, 1975; Dunel and Laurent, 1973b,c). In this species the pseudobranchial lamellae develop from the beginning on a pattern of fusion. They cannot have any respiratory function. However, the pseudobranch's connection with the ventral aorta regresses 1 week after hatching, even before the gill is really operating by itself (Dunel, 1975).

In *Morone labrax* (Percoidei), whose pseudobranch has free filaments and lamellae, pseudobranchial lamellae develop at the same time as gill lamellae. In this species, which has been carefully studied (Dunel, 1975), chloride and pseudobranchial cells invade the pseudobranch from the beginning of its development.

Thus, except maybe in anguilliforms, the larval as well as the adult pseudobranch is basically prevented from fulfilling a significant respiratory function. In teleosts, furthermore, the early development of pseudobranchial cells is another factor precluding an initial respiratory phase.

B. The Concept of the Pseudobranch as a Gland

The pseudobranch, particularly the glandular pseudobranch of some cyprinids, has long been suspected of having an endocrine function. A glandular pseudobranch, as already mentioned, corresponds to the typically buried or covered organ—that is, having no contact with the external medium. Granel (1927) thought that the acidophilic granulations seen with the light microscope within the pseudobranchial cells might constitute evidence in favor of an endocrine function. This viewpoint was accepted (Vialli, 1929). The first experimental basis for an endocrine function was presented by Leiner (1937, 1938), who postulated that a respiratory substance released from the pseudobranchial cells is essential to the respiratory transport mechanism in the thick nonvascularized retina of certain species. This respiratory pigment consisted of carbonic anhydrase (ca). The pseudobranch is indeed very rich in this enzyme (Leiner, 1940; Sobotka and Kann, 1941; Maetz, 1956; Laurent *et al.*, 1969). It has also been suggested that the pseudobranch gland may have some functional relationships with the swim bladder (Copeland, 1951; Fänge, 1950). Copeland and Dalton (1959), in their ultrastructural description of the pseudobranchial cells in *Fundulus*, suggested that the close association of endoplasmic reticulum with mitochondria was related to the production of ca or to some other unknown activity.

A series of careful experiments carried out by Maetz (1956) on two species, *Perca fluviatilis* and *Serranus scriba*, revealed that the pseudobranch does not secrete any ca destined for the eyes, since their tissues contained the enzyme in a very constant amount whether or not pseudobranchs were in place or were experimentally removed. Comparisons between incoming and outgoing blood concentrations do not show any significant difference. In addition, the ablation of 99% of the pseudobranch does not alter either vision or the retinal structure. However, ca inhibition results in blindness in these two species, a result that clearly indicates the role of this enzyme probably via pH regulation of the retina and the choroid gland (Maetz, 1956).

Parry and Holliday (1960) have claimed, however, that injection of pseudobranch extract to a pseudobranchectomized fish produced local temporary paling in contrast with the normally black coloration of those fish. These experiments indicated an endocrine role of the pseudobranch related to the control of skin chromatophores.

The relationships between the pseudobranch and the choroid rete have been investigated (Wittenberg and Haedrich, 1974). This rete is a vascular countercurrent organ located behind the retina and partly responsible for a high oxygen partial pressure in the retina. This rete is absent in all lower groups of fish including cyclostomes, elasmobranchs, chondrosteans, and holosteans with the exception of *Amia calva*. All species possessing a rete have a pseudobranch with pseudobranchial cells, and because the arterial blood comes to the choroid rete via the pseudobranchial vasculature, it has been suggested by Wittenberg and Haedrich that some blood modification occurs within the pseudobranch, allowing it to concentrate oxygen without concentrating carbon dioxide. Unilateral pseudobranchectomy significantly lowers retinal P_{O_2} (Fairbanks *et al.*, 1969). The oxygen concentration mechanism is suppressed by ca inhibitors (Fairbanks *et al.*, 1974). These results suggest that ca might be located on the luminal surface of the rete endothelium and that the pseudobranch might be a source of this enzyme. In contrast, degeneration of the pseudobranchial cells during pathological pseudobranch hypertrophy does not affect retinal P_{O_2} significantly. It is worth noting that such degeneration does not affect the arterial blood supply to the choroidal gland (Hoffert and Fromm, 1970), contrary to experimental pseudobranchectomy.

In conclusion, there is no convincing evidence in favor of a functional association of the pseudobranch with vision. The anatomic relationships between the pseudobranch and the choroid rete are well established but are of only indirect functional consequence, because any interruption of the blood supply to the pseudobranch affects the choroid rete. The role of the pseudobranch in secreting ca is still questionable.

C. The Pseudobranch: An Osmoregulatory Effector

The concept of a functional relationship between pseudobranch activity and osmoregulation has been already disproved (Parry and Holliday, 1960). The question was raised because of the belief that chloride and pseudobranchial cells have the same structure and function (Copeland and Dalton, 1959; Kessel and Beams 1962). Subsequently a distinction between these two cells was reported independently by Harb and Copeland (1969) in *Paralichtys lethostigma* and Dunel and Laurent (1971) in a series of marine fish (see also Dunel and Laurent, 1973a).

It is now well established that some species have a pseudobranch that is equipped with chloride cells and that consequently is involved in osmoregulatory processes (Dunel and Laurent, 1973a).

It is probably incorrect to consider that the blood supply of the pseudobranch from the efferent gill artery precludes any salt-regulating function. Indeed it has been observed that euryhaline fish transferred from fresh water to seawater display the characteristic occurrence of accessory cells beside the chloride cells in both gill and pseudobranch. The pseudobranch, like the opercular epithelium and skin, is a favorable site for chloride cell functioning, even if the blood reaching it has already been cleared across the gill epithelium. This indicates that the accessory cell development is triggered by some remote factors (hormonal).

(Na^+,K^+)-ATPase and some other enzymes have been studied in the pseudobranch. Pseudobranchs that have free lamellae have shown a high (Na^+,K^+)-ATPase activity in addition to phosphoglucomutase, monoamine oxidase, cytochrome oxidase, and Mg^{2+}-ATPase (Dendy *et al.*, 1973a). Because such pseudobranchs have both types of cells (pseudobranchial and chloride), the origin of an increased activity in response to a salt load is uncertain, although cytochemical observations indicate a plasma membrane localization (Dendy *et al.*, 1973b). In chloride cells, (Na^+,K^+)-ATPase activity is located on the basolateral membrane (Utida *et al.*, 1971), whereas the same enzyme is located on the apical membrane of pseudobranchial cells (Laurent *et al.*, 1968). In addition, it has been shown that pseudobranchial (Na^+,K^+)-ATPase differs from the same enzyme in chloride cells by its optimum pH: 6.4 and 7.4 in pseudobranchial and chloride cells, respectively. Moreover, only the chloride cell (Na^+,K^+)-ATPase activity increases with the salinity; pseudobranch activity is independent of any medium composition (Bonnet *et al.*, 1973).

Thus, there is no evidence in favor of a pseudobranchial cell involvement in osmoregulation. However, pseudobranchial cells, because of their similarity to gill chloride cells, may participate in this function.

D. The Pseudobranch: A Sensory Organ

1. Centripetal Nervous Activity

The discovery of a huge, partly afferent innervation within the teleost pseudobranch (Laurent and Dunel, 1964, 1966; Laurent, 1974; Dunel, 1975) has prompted investigation of centripetal nervous activity (Laurent and Rouzeau, 1972; Laurent, 1967).

Interoceptors were stimulated by various conditions imposed on an anesthetized and artificially ventilated fish (in situ experiments) or by changing the composition of saline solutions perfusing an isolated pseudobranch–nerve preparation (in vitro experiments). The overall afferent electrical activity has been differentiated into two types characterized by amplitude and form of the spikes.

a. Isolated Perfused Pseudobranch. A type A activity has a wide amplitude (50–200 μV) compared to type B activity, which appears as an irregular complex wave of less than 50 μV. These activities differ in their propagation speed and the effect of temperature, and by clear-cut differences in sensitivity to adequate stimuli:

1. Hydrostatic pressure has been seen to have a strong positive effect on type A activity but has left type B activity unaffected. The pattern of relationships between perfusion pressure and activity suggests that actual baroreceptor endings are stimulated. The sensitivity of the different vascular areas tested with a glass rod was maximum close to the afferent arteries, which is in accordance with the endings distribution (Laurent and Dunel, 1966).
2. Activity A, insensitive to osmotic pressure, is raised by increase in NaCl concentration in the perfusate. The increment is 3% per mEq Na^+ above 125 mEq.
3. Activity B is affected by hypoxic saline perfusion, an effect that is rapidly settled and still more rapidly canceled out when returning to normoxic conditions. The increment of activity is greater than 1% per torr P_{O_2} below 100 torr.
4. Activity B, insensitive to Na^+, is affected by osmotic pressure per se. Interactions with P_{O_2} have been observed.
5. pH affects activity B but not type A, and interactions with P_{O_2} are likely because P_{O_2} effects were more pronounced at pH 7.80 than at pH 7.32.
6. P_{CO_2} per se stimulates activity B, and carbonic anhydrase inhibition provokes a submaximal increase in activity B accompanied by a desensitization with respect to pH and P_{O_2}. This enzyme, which is found in large quantities within the pseudobranch (Leiner, 1937; Maetz, 1956), has been reported to be located in association with the tubules (Laurent *et al.*, 1969).

Thus, activity type A likely takes origin from baroreceptor endings, and results suggest that receptors sensitive to Na^+ are different, although they are connected to the same types of fibers.

In contrast, activity type B originates from chemoreceptor endings, which are affected by various stimuli: O_2, CO_2, pH, and osmotic pressure.

b. "In Situ" Pseudobranch. Activity type B recorded from an anesthetized, artificially ventilated trout is much more affected by the lowering of water P_{O_2} than could be expected from "in vitro" results: an increase of 500% in the frequency of a multifiber recording was obtained when the arterial oxygen pressure was reduced from normal to 60 torr. In contrast, background activity B in situ is lower at normoxic P_{O_2} than in vitro.

In addition to this activity, a large mechanoreceptor activity is also recorded from the same nerve. These receptors, which were located under the epithelium covering the pseudobranch, respond to abrupt changes in the gill chamber hydrostatic pressure.

Some efforts were directed toward understanding the origin and mechanism of activity B. It has been suggested, from consideration of fiber size versus the pattern of electrical activity, that activity B is initiated from the dendritic pole of the small bipolar neurons, a component of the interlamellar plexus (see Section II). On a strict morphological basis, it has already been concluded that only pseudobranchial cells have specific relationships with these dendrites. However, these possible functional connections are not synaptic but are formed by the extracellular medium bathing the naked amyelinated endings. It is clear that the nature of the extracellular environment is a function of the metabolic conditions imposed on the pseudobranchial cells (i.e., P_{O_2}, pH, P_{CO_2}, and P_{osm}). The development of the mitochondrial apparatus, the membrane differentiation into tubules, and the enzymatic equipment of these cells (Laurent *et al.*, 1968, 1969) appear to be concerned with specialization as a "test cell." The large variations in the membrane potentials of the pseudobranchial cells associated with P_{O_2} or pH changes constitute an argument in favor of such a mechanism and indicate an ionic redistribution between the intra- and extracellular milieus (Laurent, 1969). A supplementary argument is provided by the diphasic action of 2,4-dinitrophenol, which after a phase of stimulation suppresses activity B. An intense stimulation of activity B is also brought about by cyanide (Laurent and Rouzeau, 1972).

2. Physiological Role

Afferent type A fibers from teleost pseudobranch convey baroreceptor messages toward the central nervous system. The particular role played by the pseudobranchial receptors located within the afferent arteries in the regulation of the blood pressure in fish is still unknown and has not been differentiated from other baroreflexogenic areas. A simple test consisting of compressing the pseudobranchial arteries of a lightly anesthetized trout with a glass rod provokes a cardiac slowing or inhibition (Laurent, 1967).

It was early reported that the ventilatory and cardiac responses to hypoxia are unaffected by bilateral cranial section of the glossopharyngeal nerve in teleosts (Hughes and Shelton, 1962; Saunders and Sutterlin, 1971) and in elasmobranchs (Satchell, 1961). No attempt has been made in exercised fish to investigate the role of the glossopharyngeal nerve (see Jones and Randall, 1978) The effect of deafferentation of the pseudobranch has been tested in the resting trout. No significant effect on either amplitude or rate of respiration and cardiac responses to hypoxia has been observed (Randall and Jones, 1973; Bamford, 1974), whereas it has been found that bilateral ligation of gill arch I caused a marked reduction in P_{aO_2} (Davis, 1971), but no significant difference in P_{aO_2} has been observed between intact and denervated fish (Randall and Jones, 1973) although a decrease of oxygen consumption followed the pseudobranch deafferentation (Randall and Jones, 1973; P. Laurent, unpublished results). More recently it has been shown that cardiac responses to hypoxic water inhalation recorded from rainbow trout can be attributed to the stimulation of oxygen-sensitive chemoreceptors located in the anterodorsal region of the first branchial arch. For instance, irrigation of anterior arches only leads to a bradycardia in a hypoxic fish (Daxboeck and Holeton, 1978). The role played by the pseudobranch is not clear from these experiments because, in hypoxia, irrigation of the pseudobranch alone with aerated water failed to elicit any change in heart rate. To understand this discrepancy it might be supposed, in Daxboeck and Holeton's experiments, that aerated water works across the gill and pseudobranch epithelium, rendering the blood normoxic. This is not possible if only the covered pseudobranch is irrigated because of the thickness of the covering. However, Daxboeck and Holeton thought that the bradycardia observed is initiated through superficial (external) chemoreceptors.

The role of the pseudobranch as an osmoreceptor has been examined (Perry and Heming, 1981). The denervation of the rainbow trout pseudobranch did not affect internal ionic or acid–base equilibrium in freshwater fish. During transfer from fresh water into seawater and back again, the pseudobranch denervation did not affect plasma changes in $[Cl^-]$, $[Na^+]$, pH, total CO_2, or P_{CO_2}.

The data just reported apparently rule out any possibility for pseudobranchial receptors to represent a reflexogenic area involved in hypoxic regulation, in spite of electrophysiological data. Several possible reasons for such a discrepancy can be advanced:

1. Since some observations suggest that there are chemoreceptors located within the gill and that sections of the branchial nerve are unable to suppress the hypoxic responses, it can be concluded that other reflexogenic areas are involved. It is highly probable that ventilatory and cardiac responses do not arise only from one source but are actually made up of several different simultaneous or successively acting components (Laurent, 1974).
2. The experimental conditions are not suitable and/or the methods used are not accurate enough, so that the part played by a CNS control (Bamford, 1974) overcomes all other responses including those of pseudobranchial receptors. However, no attempt has yet been made to determine whether peripheral receptors are involved in some conditions other than

hypoxia, such as exercise, which involves several types of regulatory adjustments (see Jones and Randall, 1978).

IV. CONCLUDING REMARKS

The teleost pseudobranchs, like those of the elasmobranch *Torpedo*, are mainly characterized by a rich innervation that is doubtless afferent and should support a sensorial vegetative function.

Whereas no physiological data are yet available concerning elasmobranch pseudobranchs (elasmobranchs need wider morphological investigations), the results obtained from teleosts confirm the occurrence of centripetal nerve impulses traveling within the glossopharyngeal nerve. Factors affecting this electrical activity are related to the perfusate composition. These results indicate that adequate stimuli should come from the blood.

The existence of mitochondria-rich cells, so far considered to be typical of pseudobranchs, is another characteristic of the teleost pseudobranch (including also the holostean *Amia*). Another set of facts indicate that some relationships should exist between pseudobranchial cells and sensory innervations. More investigations are needed in lower groups of fish, where no pseudobranchial cells are present, to determine whether sensory innervation is also present even in the absence of those cells, as it is in *Torpedo*.

Curiously, no consistent report of established function(s) is yet available, with the exception of the baroreceptor endings, which presumably operate in concert with nonpseudobranchial ones.

Among other proposed functions (e.g., vision, secretion of hormones, osmoregulation), none are supported by unequivocal evidence, including the nonrespiratory function of pseudobranchs in larval forms (except anguilliforms?).

Finally, as stated by Granel (1927) in his important monograph, "La pseudobranchie n'est pas un organe rudimentaire, c'est-à-dire un simple reste sans fonction d'un organe ayant rempli autrefois son rôle, c'est un organe qui a changé de fonction ou a développé des fonctions accessoires et qui persiste à cause de cela."*

These functions remain uncertain, but the development of our knowledge during the last decade leads us to conclude that they should be at least partly sensory.

*The pseudobranch is not a rudimentary organ, that is, a mere nonfunctional remnant of an organ that performed its function at another time; it is an organ that has changed function or developed accessory functions and that persists because of that.

REFERENCES

Allis, E. P. (1909). The pseudobranchial and carotid arteries in the Gnathostome fishes. *Zool. Jahrb., Abt. Anat. Ontog. Tiere* **27,** 103–134.

Allis, E. P. (1912). The pseudobranchial and carotid arteries in *Esox*, *Salmo* and *Gadus*, together with a description of the arteries in the adult Amia. *Anat. Anz.* **41,** 113–142.

Bamford, O. S. (1974). Oxygen reception in the rainbow trout (*Salmo gairdneri*). *Comp. Biochem. Physiol. A.* **48A,** 69–76.

Barets, A., Dunel, S., and Laurent, P. (1970). Infrastructure du plexus nerveux terminal de la pseudobranchie d'un Téléostéen, *Sander lucioperca*. *J. Microsc.* (*Paris*) **9,** 619–624.

Bonnet, C. H., Bastide, F., and Laurent, P. (1973). Etude de l'ATPase ($Na^+ + K^+$) de la pseudobranchie de Téléostéens marins et d'eau douce. *J. Physiol.* (*Paris*) **67,** 247A.

Broussonet, P. M. A. (1785). Mémoire pour servir à l'histoire de la respiration des Poissons. *Mem. Acad. R. Sci.* (*Paris*) 174–196.

Carazzi, D. (1905). Sul sistema arterioso di Selache maxime e di altri squalidi (*Acanthias vulgaris, Mustelus vulgaris, Scyllium catulus, S. canicula, Squatina vulgaris*). *Anat. Anz.* **26,** 63–96, 124–134.

Claude, P., and Goodenough, D. A. (1973). Fracture faces of zonulae occludentes from "tight" and "leaky" epithelia. *J. Cell Biol.* **58,** 390–400.

Copeland, D. E. (1951). The function of the pseudobranch gland in Teleosts. *Am. J. Physiol.* **167,** 775.

Copeland, D. E., and Dalton, A. J. (1959). An association between mitochondria and the endoplasmic reticulum in cells of the pseudobranch gland of a Teleost. *J. Biophys. Biochem. Cytol.* **5,** 393–396.

Couvreur, E. (1902). Sur le mécanisme respiratoire de la Torpille. *C. R. Seances Soc. Biol. Ses. Fil.* **54,** 1252–1253.

Davis, J. C. (1971). Circulatory and ventilatory responses of rainbow trout (*Salmo gairdneri*) to artificial manipulation of gill surface area. *J. Fish. Res. Board Can.* **28,** 1609–1614.

Daxboeck, C., and Holeton, G. F. (1978). Oxygen receptors in the rainbow trout, *Salmo gairdneri*. *Can. J. Zool.* **56,** 1254–1259.

Delage, J. P. (1974). Recherches sur la structure de la pseudobranchie de la Torpille (*Torpedo marmorata* (L.). Thèse, Université de Bordeaux.

Dendy, L. A., Philpott, C. W., and Deter, R. L. (1973a). Localization of Na^+, K^+-ATPase and other enzymes in Teleost pseudobranch. I. Biochemical characterization of subcellular fractions. *J. Cell Biol.* **57,** 675–688.

Dendy, L. A., Philpott, C. W., and Deter, R. L. (1973b). Localization of Na^+, K^+-ATPase and other enzymes in Teleost pseudobranch. II. Morphological characterization of intact pseudobranch, subcellular fractions and plasma membrane substructure. *J. Cell Biol.* **57,** 689–703.

Dohrn, A. (1885). Studien zur Urgeschichte der Wirbeltierkörpers. VII. Entstehung und Differenzierung des Zungenbein - und Kiefer - Apparates der Selachier. *Mitt. Zool. Stn. Neapel* **6,** 1–48.

Dunel, S. (1975). Contribution à l'étude structurale et ultrastructurale de la pseudobranchie et de son innervation chez les Téléostéens. Ph.D. Thesis, Strasbourg.

Dunel, S., and Laurent, P. (1971). Ultrastructure comparée de la pseudobranchie des Téléostéens marins et d'eau douce. *J. Microsc.* (Paris) **11,** 48.

Dunel, S., and Laurent, P. (1973a). Ultrastructure comparée de la pseudobranchie chez les Téléostéens marins et d'eau douce. I. L'épithelium pseudobranchial. *J. Microsc.* (*Paris*) **16,** 53–74.

Dunel, S., and Laurent, P. (1973b). Morphogenèse comparée de la pseudobranchie et de la branchie chez un Téléostéen d'eau douce (*Salmo gairdneri*). *J. Microsc.* (*Paris*) **17**, 46a.

Dunel, S., and Laurent, P. (1973c). Histogenèse comparée des épitheliums pseudobranchial et branchial chez les Téléostéens d'eau douce. *J. Microsc.* (*Paris*) **17**, 47a.

Dunel-Erb, S., Bailly, Y., and Laurent, P. (1982). Neuroepithelial cells in fish gill primary lamellae. *J. Appl. Physiol.: Respir., Environ. Exercise Physiol.* **53**, 1342–1353.

Fairbanks, M. B., Hoffert, J. R., and Fromm, P. O. (1969). The dependence of the oxygen concentrating mechanism of the Teleost eye (*Salmo gairdneri*) on the enzyme carbonic anhydrase. *J. Gen. Physiol.* **54**, 203–211.

Fairbanks, M. B., Hoffert, J. R., and Fromm, P. O. (1974). Short circuiting of the ocular oxygen concentrating mechanism in the Teleost *Salmo gairdneri* using carbonic anhydrase inhibitors. *J. Gen. Physiol.* **64**, 263–273.

Fänge, R. (1950). Carbonic anhydrase and gas secretion in the swimbladder of fishes. *Int. Physiol. Congr., 18th, 1950* pp. 192–193.

Granel, F. (1924). Sur la branchie de l'évent (pseudobranchie) des Sélaciens. *C. R. Hebd. Seances Acad. Sci.* **178**, 2003–2005.

Granel, F. (1927). La pseudobranchie des Poissons. *Arch. Anat. Microsc.* **23**, 175–317.

Grassi, G. B. (1914). Funzione respiratoria delle cosidetta pseudobranchie dei Teleostei e altre particulari intorno ad esse. *Bios* **2** (1). (quoted from Granel, 1927)

Harb, J. M., and Copeland, D. E. (1969). Fine structure of the pseudobranch of the flounder *Paralichtys lethostigma*. A description of a chloride-type cell and a pseudobranch-type cell. *Z. Zellforsch. Microsk. Anat.* **101**, 167–174.

Hoffert, J. R., and Fromm, P. O. (1970). Quantitative aspects of glucose catabolism by rainbow and lake Trout ocular tissues including alterations resulting from various pathological conditions. *Exp. Eye Res.* **10**, 263–272.

Hughes, G. M., and Shelton, G. (1962). Respiratory mechanisms and their nervous control in fish. *Adv. Comp. Physiol. Biochem.* **1**, 275–369.

Hyrtl, J. (1838). Beobachtungen aus dem Gebiete der vergleichenden Gefässlehere. II. Uber den Bau der Kiemen der Fische. *Med. Jahrb.* **15**, 232–248.

Hyrtl, J. (1858). Das arterielle Gefässystem der Rochen. *Denkschr. Akad. Wiss.* (*Wien*) **15**, 1–36.

Jones, D. R., and Randall, D. J. (1978). The circulatory and respiratory systems during exercise. *In* "Fish Physiology" (W. S. Hoar and D. J. Randall, eds.), Vol. 7. pp. 425–492. Academic Press, New York.

Kessel, R. G., and Beams, H. W. (1962). Electron microscope studies on the gill filaments of *Fundulus heteroclitus* from sea water and fresh water with special reference to the ultrastructural organization of the "chloride cell." *J. Ultrastruct. Res.* **6**, 77–87.

Kryzanovsky, S. G. (1934). Die Pseudobranchie. Morphologie und biologische Bedeutung. *Zool. Jahrb., Abt. Anat. Ontog. Tiere* **58**, 171–238.

Laurent, P. (1967). La pseudobranchie des Téléostéens: Preuves électrophysiologiques de ses fonctions chémoreceptrice et baroréceptrice. C. R. *Hebd. Seances Acad. Sci.* **264**, 1879–1882.

Laurent, P. (1969). Action du pH et de la Po2 sur le potentiel de membrane des cellules de l'épithélium récepteur dans la pseudobranchie d'un poisson Téléostéen. *Rev. Can. Biol.* **28**, 149–155.

Laurent, P. (1974). Pseudobranchial receptors in Teleosts. *In* "Handbook of Sensory Physiology" (A. Fessard, ed.), Vol. 3, Part 3, Chapter 8, pp. 279–295. Springer-Verlag, Berlin and New York.

Laurent, P. (1976). Arterial chemoreceptive structures in Fish. *In* "Morphology and Mechanisms of Chemoreceptors" (Paintal, ed.), pp. 275–281.

Laurent, P., and Dunel, S. (1964). L'innervation de la pseudobranchie chez la Tanche. *C. R. Hebd. Seances Acad. Sci.* **258,** 6230–6233.

Laurent, P., and Dunel, S. (1966). Recherches sur l'innervation de la pseudobranchie des Téléostéens. *Arch. Anat. Microsc. Morphol. Exp.* **55,** 633–656.

Laurent, P., and Dunel, S. (1980). Morphology of gill epithelia in fish. *Am. J. Physiol.* **238,** R147–R159.

Laurent, P., and Rouzeau, J. (1972). Afferent neural activity from pseudobranch of Teleosts. Effects of Po2, pH, osmotic pressure and Na^+ ions. *Respir. Physiol.* **14,** 307–331.

Laurent, P., Dunel, S., and Barets, A. (1968). Tentative de localisation histochimique d'une ATPase Na^+-K^+ au niveau de l'épithelium pseudobranchial des Téléostéens. *Histochemie* **14,** 308–313.

Laurent, P., Dunel, S., and Barets, A. (1969). Localisation histochimique de l'anhydrase carbonique au niveau des chemorécepteurs artériels des Mammifères, des Batraciens et des Poissons. *Histochemie* **17,** 99–107.

Leiner, H. (1937). Die Kohlensäureanhydrase im Körper der Syngnathiden und die Bedeutung der Pseudobranchie der Knochenfische. *Verh. Dtsch. Zool. Ges.* **17,** 136–149.

Leiner, M. (1938). Die Augenkiemendrüse (Pseudobranchie) der Knochenfische. Experimentelle Untersuchungen über ihre physiologische Bedeutung. *Z. Vergl. Physiol.* **26,** 416–466.

Leiner, M. (1940). Das Atmungsferment Kohlensäure anhydrase im Tierkörper. *Naturwissenschaften* **28,** 165–171.

Maetz, J. (1956). Le rôle biologique de l'anhydrase carbonique chez quelques Téléostéens. *Bull. Biol. Fr. Belg., Suppl.* **40,** 1–129.

Müller, J. (1839). Vergleichende Anatomie der Myxinoiden. III. Uber das Gefässystem. *Abh. Dtsch. Akad. Wiss. Berlin* pp. 175–303.

Parker, T. J. (1881). On the venous system of the skate (*Raja nasuta*). *Trans. N. Z. Inst.* **13,** 413–418.

Parry, G., and Holliday, F. G. T. (1960). An experimental analysis of the function of the pseudobranch in Teleosts. *J. Exp. Biol.* **37,** 344–353.

Perry, S. F., and Heming, T. A. (1981). Blood ionic and acid-base status in rainbow trout (*Salmo gairdneri*) following rapid transfer from freshwater to seawater: Effect of pseudobranch denervation. *Can. J. Zool.* **59,** 1126–1132.

Randall, D. J., and Jones, D. R. (1973). The effect of deafferentation of the pseudobranch on the respiratory response to hypoxia and hyperoxia in the trout (*Salmo gairdneri*). *Respir. Physiol.* **17,** 291–301.

Riess, J. A. (1881). Der Bau der Kiemenblätter bei den Knochenfischen. *Arch. Naturgesch.* **47,** 518–550.

Romer, A. S., and Parsons, T. S. (1970). "The Vertebrate Body." Saunders, Philadelphia, Pennsylvania.

Satchell, G. H. (1961). The response of the dogfish to anoxia. *J. Exp. Biol.* **38,** 531–543.

Saunders, R. L., and Sutterlin, A. H. (1971). Cardiac and respiratory responses to hypoxia in the sea raven, *Hemitripterus americanus*, and an investigation of possible control mechanisms. *J. Fish. Res. Board Can.* **28,** 491–503.

Sobotka, H., and Kann, S. (1941). Carbonic anhydrase in fishes and Invertebrates. *J. Cell. Comp. Physiol.* **17,** 341–348.

Stork, H. A. (1932). Zur Homologierfrage der Teleostier-pseudobranchie. *Zool. Jahrb. Abt. Anat. Ontog. Tiere* **55,** 505–554.

Utida, S., Kamiya, M., and Shirai, N. (1971). Relationship between the activity of Na^+ -K^+ activated adenosinetriphosphatase and the number of chloride cells in Eel gills with special reference to sea-water adaptation. *Comp. Biochem. Physiol.* **38,** 443–447.

Vialli, M. (1924). La pseudobranchia dello spiracolo nei Ganoidi e nei Selachi. *Natura* **15.** (quoted from Granel, 1927)

Vialli, M. (1925). La pseudobranchia di *Amia calva* e di *Lepidosteus osseus*. *Atti Soc. Ital. Sci. Nat. Mus. Civ. Stor. Nat. Milano* **64.** (quoted from Granel, 1927)

Vialli, M. (1926). La pseudobranchie dei pesci. *Arch. Anat.* **23,** 50–117.

Vialli, M. (1929). Pseudobranchie spiracolare e pseudobranchie ioidea nei Teleostei. *Pubbl. Stn. Zool. Napoli* **9,** 443–456.

Virchow, H. (1890). Uber die Spritzlochkiemen der Selachier. *Arch. Anat. Physiol., Physiol. Abt.* 177–182.

Wittenberg, J. B., and Haedrich, R. L. (1974). The choroid rete mirabile of the fish eye. II. Distribution and relation to the pseudobranch and to the swimbladder rete mirabile. *Biol. Bull. (Woods Hole, Mass.)* **146,** 137–156.

10

PERFUSION METHODS FOR THE STUDY OF GILL PHYSIOLOGY

S.F. PERRY
Department of Biology
University of Ottawa
Ottawa, Ontario

P. S. DAVIE
Department of Physiology and Anatomy
Massey University
Palmerston North, New Zealand

C. DAXBOECK
Pacific Gamefish Foundation
Kailua-Kona, Hawaii

A. G. ELLIS and D. G. SMITH
Department of Zoology
University of Melbourne
Parkville, Victoria, Australia

*During the preparation of this chapter S. F. Perry was supported by NSERC and Eastburn postdoctoral fellowships; P. S. Davie was supported by a Killam postdoctoral scholarship; C. Daxboeck wishes to acknowledge the support of the Pacific Gamefish Foundation.

ISBN 0-12-350432-5

I. INTRODUCTION

Since Krakow (1913) examined the effects of some vasoactive agents on isolated, perfused pike gills, a large number of perfused gill preparations have been developed to study gill physiology. Although these studies have contributed significantly to our understanding of gill functions, interpretation of the data often is limited by the experimental methods. In general, isolated or in situ perfused gills allow the investigator to control or measure many variables concerning gill function. Control of perfusate composition, pressure and flow, and the exclusion of postbranchial circulation, together with increased accuracy of measurements, are the principal advantages. Against these advantages, the complex structure of the gills and their range of functions often make conclusions equivocal. In addition, perfusion and isolation invariably introduce new variables such as the effects of anesthetic, unnatural and mechanical strains and stresses, and the gradual death of the tissues. Many of the earlier investigators recognized both the advantages and the disadvantages of their preparations and critically discussed their perfusion methods (see Bateman and Keys, 1932a,b). To some extent the value of this earlier work has not been recognized.

The primary criterion by which to judge a preparation is its ability to reproduce the physiological behavior of the tissue as it occurs in intact animals. This depends largely on the function under investigation, and on the availability and accuracy of information about the function in vivo. Unfortunately, technical difficulties in obtaining information about gill function in vivo place limits on the data that are available for such comparisons. In fact, this is often the reason for developing a perfused preparation. In these cases the preparation becomes descriptive rather than experimental, and great caution should be exercised in using perfused preparations to establish such baseline data.

A. Types of Preparations

Perfused gill preparations fall into four main groups: isolated branchial arches, branchial baskets including all arches, heads, and whole bodies. Isolated branchial arches have been used by a number of investigators (Richards and Fromm, 1969; Sutterlin and Saunders, 1969; Rankin and Maetz, 1971; Shuttleworth, 1972, 1978; Capra and Satchell, 1974; Bergman *et al.*, 1974; Smith, 1977; Farrell *et al.*, 1979; Jackson and Fromm, 1981; Perry *et al.*, 1982; Stagg and Shuttleworth, 1982; Ellis and Smith, 1983). The preparation usually consists of an excised branchial arch perfused via the afferent branchial artery, while outflowing perfusate is collected from a catheter in the efferent branchial artery (Fig. 1). In order to maintain satisfactory outflow, ligatures often are placed around the whole arch at the level of the catheters, thereby occluding recurrent vessels to some degree. Isolated arches usually are suspended in a bathing medium by the catheters (Fig. 1).

Perfused branchial baskets were favored by earlier workers (Krakow, 1913; Ostlund and Fange, 1962; Reite, 1969) and will not be dealt with in detail. They involve cannulation of the bulbus arteriosus or ventral aorta for infusion of perfusate and cutting the arches near the roof of the mouth (Fig. 2). These preparations often were perfused *in situ*, allowing artificial ventila-

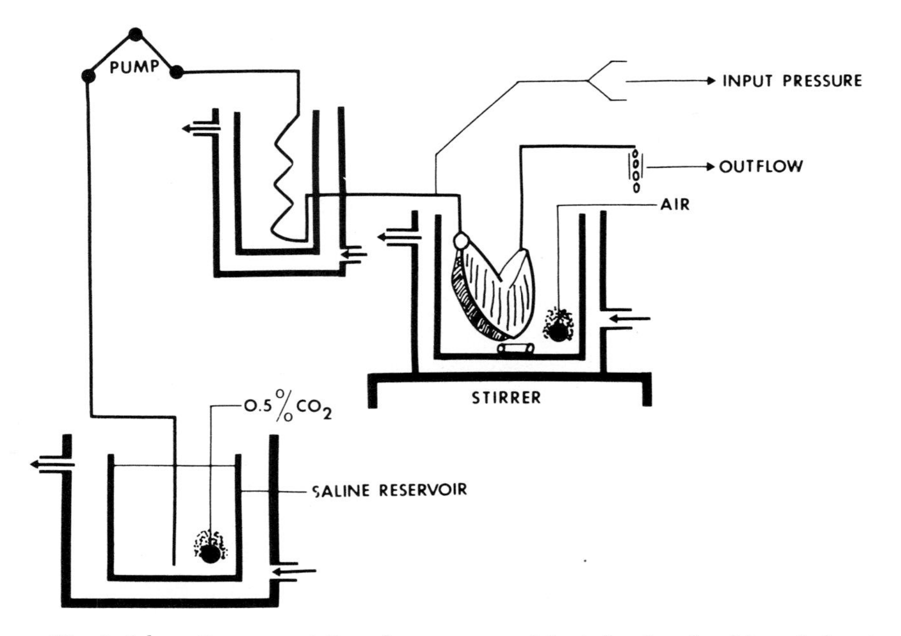

Fig. 1. Schematic representation of apparatus used for isolated perfused branchial arch preparation. (From Perry, 1981.)

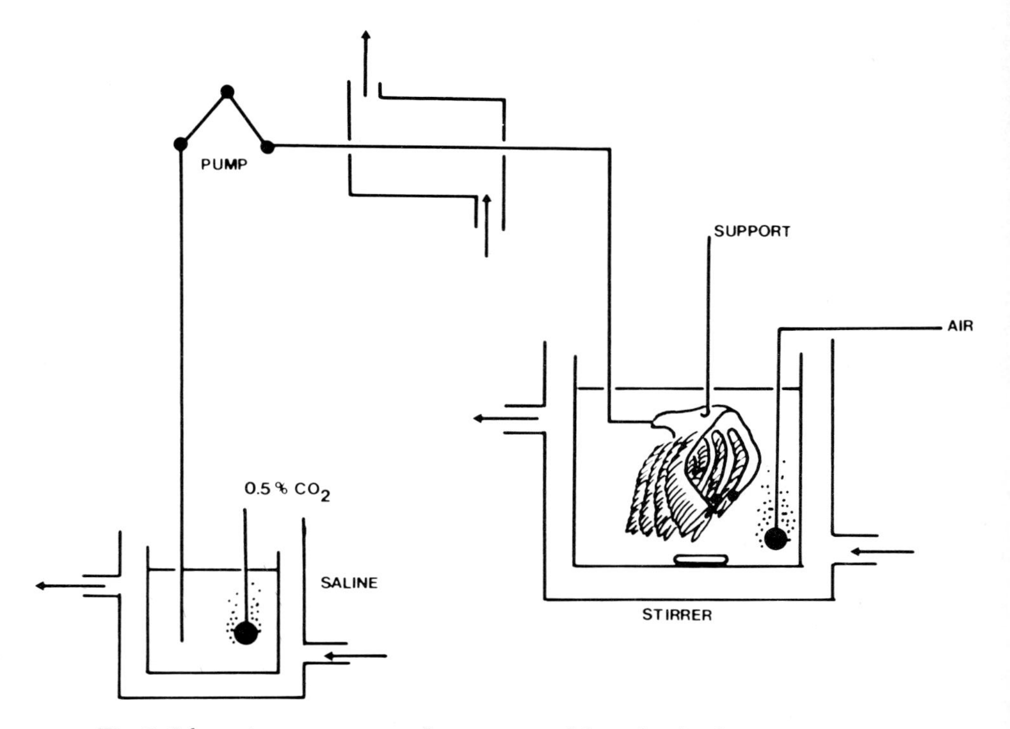

Fig. 2. Schematic representation of apparatus used for isolated, saline-perfused branchial basket preparation. (From Perry, 1981.)

tion and normal position to be maintained. Outflowing saline issued freely from the cut ends of the efferent branchial arteries and/or the dorsal aorta.

Perfused heads have been most widely used for gill physiology experiments (Keys, 1931; Keys and Bateman, 1932; Bateman and Keys, 1932a,b; Kirschner, 1969; Wood, 1974; Payan and Matty, 1975; Part and Svanberg, 1981; Perry, 1981; Perry and Daxboeck, 1984; Perry *et al.*, 1982, 1983, 1984a,b,c,d; Daxboeck and Davie, 1982; Davie and Daxboeck, 1982; Pettersson and Johansen, 1982; Pettersson, 1983). Perfused heads have input cannulae in the ventral aorta or bulbus arteriosus and output cannulae in the dorsal aorta at or about the level of the caudal operculum (Figs. 3 and 4). Occasionally, the heart has been used to pump the perfusate in "heart–gill preparations" (Keys, 1931). Many head preparations maintain a dorsal aortic pressure at near normal levels (Daxboeck and Davie, 1982). Ventilation is artificially maintained, although some head preparations spontaneously ventilate (Kawasaki, 1980).

Whole-body preparations essentially are identical to head preparations with the omission of the occlusive dorsal aortic catheter (Forster, 1976a,b). The occlusive dorsal aortic catheter can be replaced by a nonocclusive can-

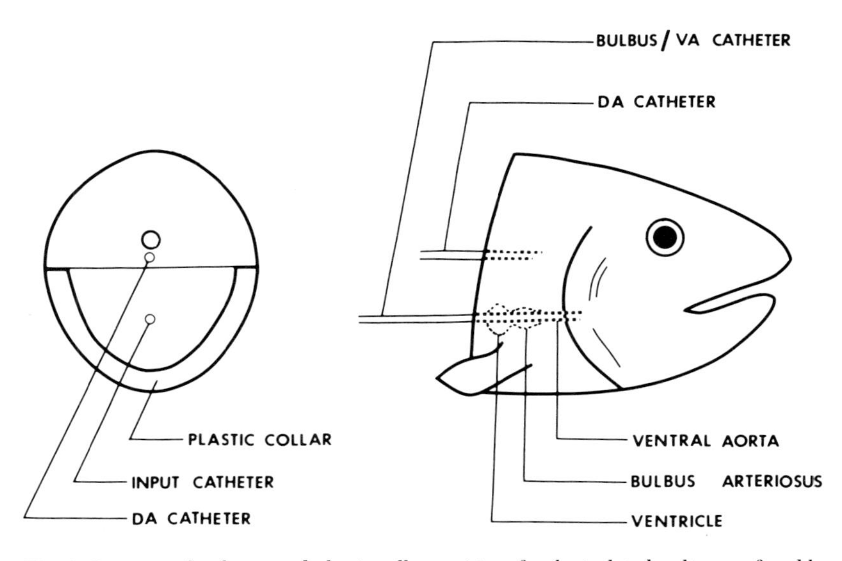

Fig. 3. Diagram of catheter and plastic collar positions for the isolated, saline-perfused head preparation. DA, dorsal aorta; VA, ventral aorta. (From Perry, 1981.)

nula for sampling and pressure measurements. Figure 5 illustrates a whole-fish preparation in which a nonocclusive cannula has been inserted into the dorsal aorta for postbranchial sampling (Wood *et al.*, 1978; Haswell *et al.*, 1978; Perry *et al.*, 1982). Ventilation usually is artificially maintained, but the use of blood as a perfusate produces a spontaneously ventilating perfused fish preparation (Davie *et al.*, 1982; Metcalfe and Butler, 1982).

Isolated branchial arches exclude tissues such as central nervous system, buccal and opercular epithelium, and pseudobranch, and represent truly isolated gills. The cost of isolation is impairment of recurrent circulation, abnormal ventilation, and mechanical stresses. They have been used most successfully in studies of vascular pharmacology rather than ion and gas exchange. Perfused heads have been used widely in hemodynamic and ion exchange investigations. Perfused heads with physiological dorsal aortic pressures have been useful in elucidating the hemodynamics and control of branchial circulation, because under these conditions, normal head and recurrent pressures are maintained. Perfused whole bodies sacrifice the ability to control dorsal aortic pressure for the additional information about systemic events and their effects on gill functions.

B. Perfusion Methods

Perfusion fluid has been infused through gill preparations by a great variety of methods. Constant-pressure perfusion has been employed most

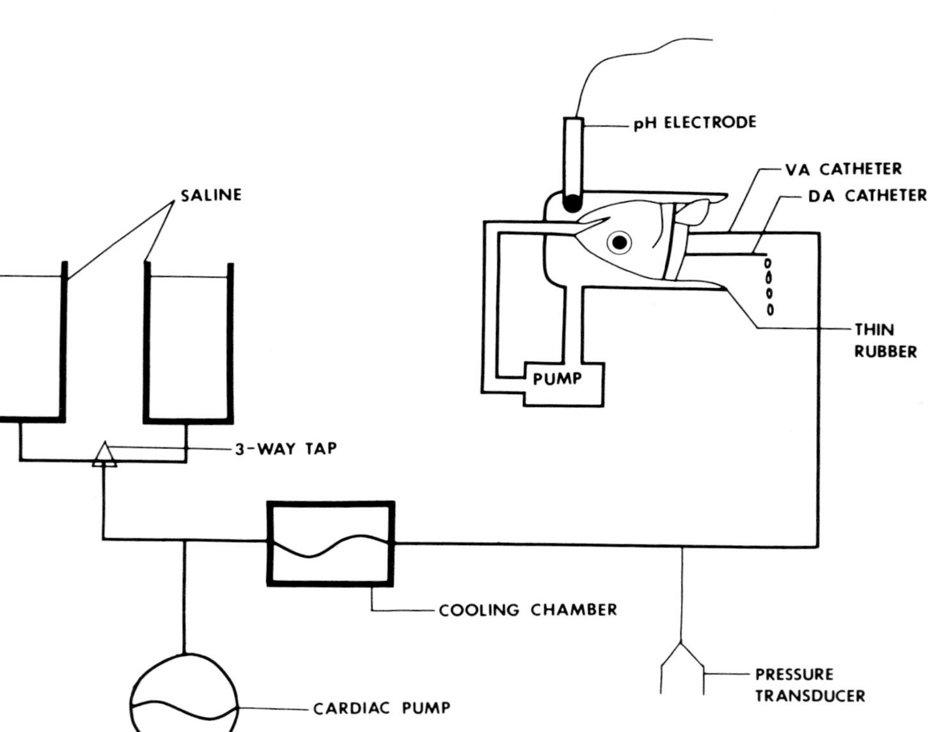

Fig. 4. Schematic representation of apparatus used for isolated, saline-perfused head preparation. Saline is pumped from reservoirs into the head via a ventral aortic (VA) catheter. Postgill saline is collected via a catheter implanted in the dorsal aorta (DA). (From Perry, 1981.)

commonly. Simply, this method involves a reservoir of perfusate suspended at a height above the preparation such that the resistance to flow of the delivery system and gill vascular beds is overcome and flow commences. Changes in inflow and outflow rates can be recorded, which, together with the pressure difference across the gills, allows calculation of gill vascular resistance (R_g; see Payan and Matty, 1975). Variations of this technique include the pressure differential–flow profile where a column of perfusate is allowed to fall as the perfusate flows through the gills (see Wood, 1974). Constant mean pressure with a superimposed pulse pressure method has been used by Bateman and Keys (1932a). Constant-flow perfusion by pulsatile (see Bergman *et al.*, 1974) or nonpulsatile method (see Shuttleworth, 1972) has also been employed. At constant flow rate, changes in pressure differential across the gills indicate alterations in gill vascular resistance. The value of pulsatile perfusion was first demonstrated by the

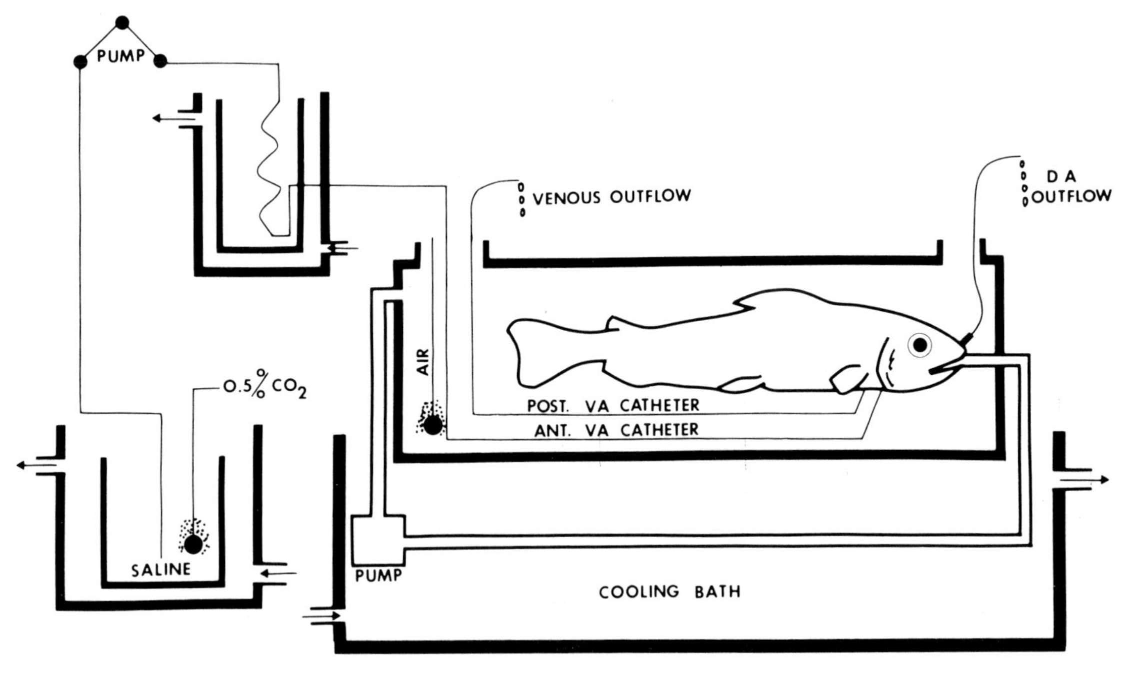

Fig. 5. Schematic representation of apparatus used for totally saline-perfused fish. Saline is pumped from a reservoir through a cooling chamber and into the fish via a ventral aortic (VA) catheter. Postgill saline is collected from a catheter implanted in the dorsal aorta (DA). (From Perry, 1981.)

"heart–gill preparations" of Keys (1931) and Keys and Bateman (1932), and many authors have since recommended pulsatile over nonpulsatile perfusion (Bergman *et al.*, 1974; Part and Svanberg, 1981; Daxboeck and Davie, 1982). Regulation of pulse pressure magnitude and the waveform may have beneficial effects on gill preparation performance and longevity (Metcalfe, 1981).

Except in spontaneously ventilating preparations (Davie *et al.*, 1982; Metcalfe and Butler, 1982), water flow over the gills always has been maintained by continuous flow. The action of ventilatory movements on gill perfusion has not been studied in detail, but Randall and Smith (1967) have shown a degree of synchrony between respiratory and cardiac cycles. Their results suggest a possible interaction between pressure pulses in the water and blood that augments recurrent flow.

In this chapter we summarize perfusion techniques commonly used in gill physiology and provide criteria for evaluation of each preparation type. Recommendations appear at the end of each section regarding the preparations we consider most suitable for different areas of investigation.

II. SALINES AND GAS MIXTURES

An important aspect of perfused gill preparations is correct simulation of in vivo conditions. Although it is optimal to perfuse gills with blood, often this is not feasible. Alternatively, investigators have resorted to perfusing isolated gill preparations with salines closely resembling the chemical composition of blood. Tables I and II display the chemical composition of some "typical" salines used in perfusion of freshwater and saltwater fish. It is apparent that considerable variations exist in the chemical composition of salines used in gill perfusion studies. Whereas many of the differences are insignificant, we are of the opinion that there are certain saline components that are more critical than others and should be carefully controlled. For example, the concentration of Ca^{2+} is usually kept low to prevent vasoconstriction of the gill vasculature. The pH of the saline is an important factor determining the viability of a perfused preparation. Blood pH values from live fish generally range between 7.6 and 8.1. Two components that contribute greatly to the final pH are the bicarbonate and the sodium phosphate concentrations. The HCO_3^- concentration in fish blood normally varies between 6 and 14 m*M*. It is clear that the salines of Rankin and Maetz (1971) and the original salines of Payan and Matty (1975) contain excessive amounts of HCO_3^-, giving rise to abnormally high pH values. Similarly, abnormally low pH values are obtained when monobasic sodium phosphate (NaH_2PO_4) is used. Thus, we recommend the use of dibasic sodium phosphate (Na_2HPO_4).

Rankin and Maetz (1971) added 2% polyvinylpyrrolidone (PVP; average

Table I

Chemical Composition of Two Different Physiological Salines Used for Saline Perfusion Studies of Freshwater Fish

	Cortland (Wolf, 1963)		Payan and Matty (1975)	
	g $liter^{-1}$	Molarity (m*M*)	g $liter^{-1}$	Molarity (m*M*)
NaCl	7.25	124.1	6.59	112.8
KCl	0.38	5.1	0.31	4.2
Na_2HPO_4[a]	0.41	2.9	0.14	1.0
$MgSO_4$	0.24	1.9	0.14	1.2
$CaCl_2$	0.16	1.4	0.14	1.3
$NaHCO_3$	1.00	11.9	1.10	13.1[c]
$(NH_4)_2SO_4$	—	—	0.02	0.1
KH_2PO_4	—	—	0.05	0.4
Glucose	1.00	5.6	1.00	5.6
PVP[b]	40.00	4%	—	—
Dextran	—	—	30.00	3%
Heparin	10,000 USP Units $liter^{-1}$		5000 USP Units $liter^{-1}$	

[a]The original Cortland saline used NaH_2PO_4. We have found that substituting Na_2HPO_4 results in a more realistic pH (approximately 8.0 at 0 P_{CO_2}).

[b]Polyvinylpyrrolidone, not included in the original Cortland saline. It was added first by Kirschner (1969) as a colloid osmotic filler.

[c]The saline of Payan and Matty (1975) has been modified to 13.1 m*M* $NaHCO_3$ from the original 26 m*M* $NaHCO_3$ (Perry *et al.*, 1983).

MW 40,000) to their saline in an attempt to approximate the colloid osmotic pressure of blood. Normally, 2–4% PVP, dextran (MW 60,000–80,000), or plasma protein (Jackson and Fromm, 1980; Davie, 1981) is added to saline in an attempt to prevent edema. Indeed, Wood (1974) has shown that addition of 4% PVP slows the increase in the branchial vascular resistance, normally observed in the totally saline-perfused rainbow trout. Although it is likely that addition of a colloid osmotic filler does enhance the viability of perfused preparations, its role in the prevention of edema is less clear. The phenomenon of edema formation in saline-perfused gills is discussed more fully in Section III. Another important feature of saline preparation is the addition of heparin (anticoagulant). This is especially vital initially, when one is clearing the gills of blood cells. Normally, heparin is injected intraperitoneally or directly into the circulatory system of the animal prior to preparation, as well as added to the saline.

Rankin and Maetz (1971) observed that the use of nonfiltered saline in a

Table II

Chemical Composition of Two Different Physiological Salines Used for Saline Perfusion Studies of Saltwater Fish

	Rankin and Maetz (1971)		Shuttleworth *et al.* (1974)	
	g liter^{-1}	Molarity (m*M*)	g liter^{-1}	Molarity (m*M*)
NaCl	8.36	143.1	10.00	171.1
KCl	0.25	3.4	0.30	4.0
NaH_2PO_4	—	—	0.40	3.3
$NaHPO_4 \cdot 12H_2O$	0.50	1.6	—	—
$MgSO_4$	—	—	0.15	1.2
$MgSO_4 \cdot 7H_2O$	1.01	4.1	—	—
$CaCl_2$	0.14	1.3	0.17	1.5
$NaHCO_3$	2.20	26.2	1.26	15.0
$(NH_4)_2SO_4$	0.05	0.4	—	—
KH_2PO_4	0.04	0.3	—	—
Glucose	1.00	5.6	10.00[a]	55.6
PVP	20.00	2%	—	—
Heparin	—	—	2,500 USP Units liter^{-1}	

[a]This concentration of glucose is approximately 10 times greater than that measured in resting rainbow trout (T. P. Momssen, personal communication).

constant-pressure perfused gill arch resulted in a rapid decline of flow rates that could be prevented by the addition of 10^{-5} *M* adrenaline or noradrenaline. When the saline was filtered through a 0.22-μm Millipore filter before use, constant high flow rates could be maintained in the absence of catecholamines. It is now common practice to filter all saline (0.22- or 0.45-μm pore size) prior to use in a perfused gill preparation.

It is interesting and unfortunate that while most investigators pay great attention to the chemical composition of saline, little or no effort is put into obtaining correct gas equilibration. Normally, blood entering the gills of live, resting fish contains levels of oxygen and carbon dioxide of 20 to 50 and 2–4 torr, respectively. Thus, in order to simulate *in vivo* conditions, one must equilibrate the perfusate with gas mixtures in this range. It is apparent that most researchers have used gas mixtures well outside of the physiological range or no gas mixtures at all. Primarily hyperoxic and hypercapnic salines have been used in the past [95% O_2–5% CO_2: Rankin and Maetz (1971), Payan and Matty (1975), Smith (1977), Farrell *et al.* (1979); 97% O_2–3% CO_2: Wahlqvist (1980), Pettersson and Nilsson (1979); 99% O_2–1% CO_2: Goldstein *et al.* (1982)]. Perry and Daxboeck (1984) have examined the effects of hypoxic, hyperoxic, and hypercapnic salines in an isolated, perfused head preparation (Fig. 6). It is clear that elevation of the partial pressure of CO_2 to 5% (37.5 torr) from a normal level causes a significant increase in afferent perfusion pressure during constant-flow perfusion (Fig. 6a). This most likely reflects an increase in branchial vascular resistance caused by local vasoconstriction. The phenomenon is due to the elevated P_{CO_2} and not to the accompanying decrease in saline pH. When P_{CO_2} is held constant and pH lowered to similar levels, no increase in perfusion pressure is observed (Fig. 6b). Although hypoxic or hyperoxic conditions in the saline caused no apparent changes in perfusion pressure (Fig. 6c), this does not prove that

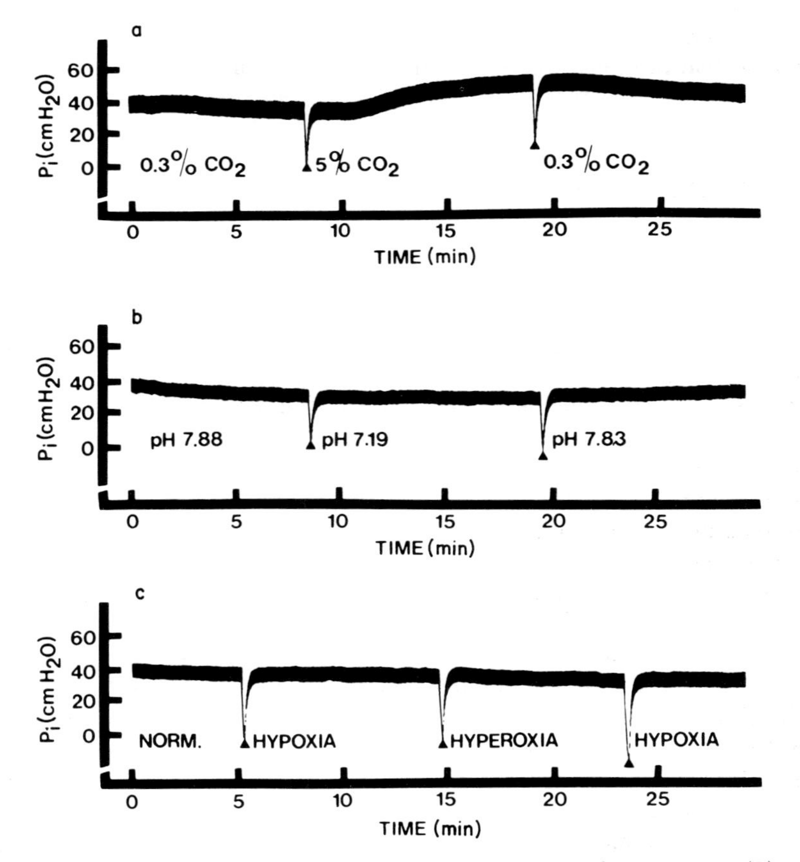

Fig. 6. Effect of saline pH and gas composition on afferent perfusion pressure (P_i) in the isolated, saline-perfused trout head preparation. (a) Effect of changing from 0.3% CO_2 (2.25 torr, pH 7.9) to 5% CO_2 (37.5 torr, pH 7.1). (b) Effect of decreased pH with constant 0.3% CO_2. (c) Effect of changing O_2 content: normoxia = 35 torr, hypoxia = 5 torr, hyperoxia = 350 torr. (From Perry and Daxboeck, 1984.)

perfusing with hyperoxic saline has no deleterious effects on the preparation. Indeed it has been demonstrated that elevated levels of O_2 inhibit CO_2 excretion in a totally perfused trout preparation (Haswell *et al.*, 1978). These authors attribute the inhibition to a decrease in the functional surface area of the gill, caused by hyperoxia. This explanation is supported by the fact that Wood and Jackson (1980) were able to induce hypercapnic acidosis in rainbow trout maintained in hyperoxic water.

It must be stressed that we are not recommending total use of saline in perfused gill preparations. Clearly, the greater O_2-carrying capacity, buffering capacity, and viscosity of blood make it a superior perfusion fluid, and thus it should be used whenever possible. However, often this is not feasible or practical, and one must resort to using saline. In these instances it is essential that the chemical composition, pH, and gas equilibration closely resemble that of blood entering the gills of live, intact fish.

III. EDEMA FORMATION IN PERFUSED GILL PREPARATIONS

The lamellar blood channels of fish gills are perfused with blood under arterial pressure and in that sense bear a close resemblance to the capillaries of a glomerular kidney. Therefore, in fish gills, as in renal glomeruli, ultrafiltration will occur unless the plasma colloid osmotic pressure greatly exceeds that of the tissues, or unless the permeability of the pillar cells that form the exchange channels is low. The protein permeability of systemic capillaries in fish is high (Hargens *et al.*, 1974), so there is a strong possibility of ultrafiltration or exudation of fluid from the gill capillaries. Unless this exudate is effectively cleared from the system, its accumulation may impair diffusional exchange by increasing the thickness of the blood–water barrier (Hughes, 1966).

Lymph clearance in mammals has long been known to be aided by the mechanical interaction between muscular skeletal elements and lymphatic valves (Starling, 1896), or by the pumping action of the arterial pulse (Hering, 1867) even in isolated vascular beds (DeLanger, 1958; Wilkins *et al.*, 1962). Most studies of diffusional exchange in fish gills have been performed on isolated branchial arches or head preparations perfused at constant pressure and in the absence of normal breathing movements (Tables III and IV). Despite the likely possibility of edema formation in such preparations, few authors have commented on structural changes in gill preparations after extended perfusion with any one of a number of physiological saline solutions (Tables III and IV). When Ringer-perfused gills have been examined histologically, edema often has been revealed as an enlargement of the

Table III

Summary of Reports of Colloid-Free Ringer Perfusions

Preparation type[a]	Perfusion type[b]	Species studied	Perfusate[c]	Purpose of investigation	Reference
IBB	CP	*Esox* spp.	R	Vascular control	Krakow (1913)
IH	H	*Anguilla* spp.	R	Ion exchange	Keys (1931)
VA V	PIST	*Anguilla vulgaris*	R	Vascular control and ion exchange	Keys and Bateman (1932)
VA	PIST	*A. vulgaris*	R	Gas exchange	Bateman and Keys (1932)
IH	CP	*Salmo gairdneri*	R	Ion exchange	Schiffman (1961)
IBB	CP	*Anguilla anguilla, Gadus collarias, Zoarces viviparis, Labrus berggytta, Squalus acanthias*	SW	Vascular control	Ostlund and Fange (1962)
IA	CP	*S. gairdneri*	R	Vascular control	Richards and Fromm (1969)
IH	CP	*S. gairdneri*	C	Ion exchange	Randall *et al.* (1972)
IA	CP	*Anguilla dieffenbachii*	R	Ion exchange	Shuttleworth (1972); Shuttleworth and Freeman (1974)
IA	CP	*Platychthys flesus*	R	Ion exchange	Shuttleworth *et al.* (1974)
IA	CP	*Anguilla rostrata*	R	Ion exchange	Fenwick and So (1974)
IA	PERI	*S. gairdneri*	R	Vascular control	Bergman *et al.* (1974)
IA	PERI	*S. gairdneri, Salmo trutta*	C	Vascular control	Smith (1977)
IA	CP	*A. rostrata*	R	Ion exchange	So and Fenwick (1977)
VA	CP	*S. gairdneri*	C	Gas exchange	Haswell and Randall (1978)
IA	PERI	*P. flesus*	R	Ion exchange	Shuttleworth (1978)
IA	PERI	*Ophidon elongatus*	CM	Vascular control	Farrell *et al.* (1979)
IA/½BB	CP	*Gadus morhua*	R	Vascular control	Pettersson and Nilsson (1979)
IH	CP	*Cyprinus carpio*	R	Respiratory rhythm	Kawasaki (1980)
IA	CP	*G. morhua*	R	Vascular control	Wahlqvist (1980)
IA/IH	PERI	*G. morhua*	R	Gas exchange	Johansen and Pettersson (1981)
IH	PERI	*G. morhua*	R	Gas exchange	Pettersson and Johansen (1982)
½BB	PERI	*G. morhua*	R	Vascular control	Nilsson and Pettersson (1981)

[a] IA, isolated arch; IH, isolated head; VA, ventral aortic perfusion where head not isolated or isolation not specified; IBB, isolated branchial basket; ½BB, perfusion of only one side of branchial basket.

[b] CP, constant pressure; PERI, peristaltic pump; PIST, piston pump.

[c] R, Ringer solution with composition as specified composition; C, Cortland Ringer (Wolf, 1963); CM, modified Cortland Ringer solution; SW, seawater diluted 1:2 vol/vol with distilled water.

Table IV

Summary of Reports of Colloidal Ringer Fish Gill Perfusions: Type and Concentration of Artificial Colloid[a]

Preparation type	Perfusion type	Species studied	Perfusate	Colloid and concentration	Purpose of investigation	Reference
VA	PERI	*Anguilla anguilla*	R	4% PVP	Ion exchange and vascular control	Kirschner (1969)
IA	CP	*A. anguilla*	R	2% PVP	Vascular control	Rankin and Maetz (1971)
VA	CP	*Salmo gairdneri*	C	4% PVP	Vascular control	Wood (1974)
VA	PERI	*S. gairdneri*	C	4% PVP	Vascular control	Wood (1975)
IH	CP	*S. gairdneri*	R	3% Dextran	Ion exchange	Payan and Matty (1975); Payan *et al.* (1975); Girard and Payan (1977a,b); Payan (1978)
IH	CP/PERI	*S. gairdneri*	R	3% Dextran	Ion exchange	Girard (1976)
VA	PERI	*A. anguilla*	R	2% PVP	Vascular control	Forster (1976a,b)
IH	CP	*S. gairdneri*	R	3% Dextran	Vascular control	Girard and Payan (1976); Payan and Girard (1977)
IH	CP	*S. gairdneri*	R	3% Dextran	Water exchange	Isaia *et al.* (1978, 1979)
VA	CP	*Salmo* spp.	C	4% PVP	Gas exchange	Haswell *et al.* (1978)
VA	PIST	*S. gairdneri*	C	4% PVP	Gas exchange	Wood *et al.* (1978)
IA	CP	*S. gairdneri*	C	4% PVP	Vascular control	Booth (1979)
IH	PIST	*S. gairdneri*	R	2% PVP	Vascular control	Colin *et al.* (1979)
IH	CP/PIST	*Myoxocephalus octodecimspinosus*	R	3% PVP	Vascular control	Claiborne and Evans (1980)
IA	CP	*A. anguilla, Salmo trutta*	R	2% PVP	Vascular control	Bolis and Rankin (1980)
IA	PERI/PIST	*S. gairdneri*	R	4% PVP	Ion exchange	Part and Svanberg (1981)
IH	PIST	*S. gairdneri*	C	4% PVP	Gas exchange	Part *et al.* (1982a)
IH	PIST	*S. gairdneri*	R	4% PVP	Vascular control	Part *et al.* (1982b)
IH	PIST	*S. gairdneri*	C	4% PVP	Gas and ion exchange	van der Putte and Part (1982)
IA	PERI	*S. gairdneri*	C	4% PVP	Gas exchange	Perry *et al.* (1982)
IH	PIST	*S. gairdneri*	R	4% PVP	Ion exchange	Perry *et al.* (1983); Perry *et al.* (1984a)
IH	PIST	*S. gairdneri*	R	4% PVP	Gas exchange	Perry *et al.* (1984b); Perry and Daxboeck (1984)

[a]Abbreviations as in Table III.

subepithelial interstitial space (Richards and Fromm, 1969; Part *et al.*, 1982a); the degree of enlargement correlates well with the impairment of O_2 transfer (Part *et al.*, 1982a).

In this section we will discuss the problem of edema formation in perfused gill preparations. Primarily we will focus on the study of Ellis and Smith (1983), in which scanning electron microscopy (SEM) was used to assess the damage caused by Ringer perfusion of isolated eel branchial arches (holobranchs). The use of an alternative perfusate that would enable exchange studies to be performed on undamaged gills will also be discussed.

The degree of edema in isolated eel arches has been qualitatively assessed by estimation of the degree of expansion of the subepithelial interstitial spaces. Four categories were recognizable and are defined in Fig. 8.

The vascular anatomy of the gill filament and lamellae of *Anguilla australis* is generally similar to that reported for *Anguilla anguilla* by Laurent and Dunel (1976). The subepithelial interstitial spaces of the filaments and lamellae in "unperfused" tissue have the appearance of small slits or gaps occurring beneath or within the epithelium (Fig. 7). No systematic differences are revealed between "unperfused" gills and gills that had been "briefly perfused" (5-min saline perfusion) (Figs. 9 and 10). Perfusion of isolated holobranchs with eel Ringer resulted in rapidly developing edema, involving widespread uplifting of lamellar and occasionally filamental epithelial layers due to gross inflammation of the subepithelial interstitial spaces (Fig. 12). Severe edema resulted from perfusion for less than 30 min, with perfusion variables within physiological limits.

Hydrodynamic Factors

The occurrence of edema was independent of input pressure (P_i), or pulse pressure (ΔP), with severe edema occurring over a range of P_i from about 40 to 80 cm H_2O. No systematic difference in the degree of edema was observed between gills subjected to pulsatile, as opposed to nonpulsatile perfusion, whether constancy of pressure was achieved by use of a pressure head or by inclusion of a Windkessel (air chamber) in the pump outflow line. These results indicate that the arterial pulse cannot, by itself, clear the subepithelial spaces of edematous fluid, and contrast with the results of Daxboeck and Davie (1982), which showed that EtOH clearance from subepithelial spaces was enhanced by pulsatility.

Severe edema occurred over the entire range of perfusion rates used (1.0–1.4 ml min^{-1} $arch^{-1}$ kg^{-1}). Outflow rates (F_o) were extremely variable, and there was no apparent correlation between F_o and the degree of edema. Controlled variation of efferent pressure (0.0–16 cm H_2O) did not prevent the development of severe edema.

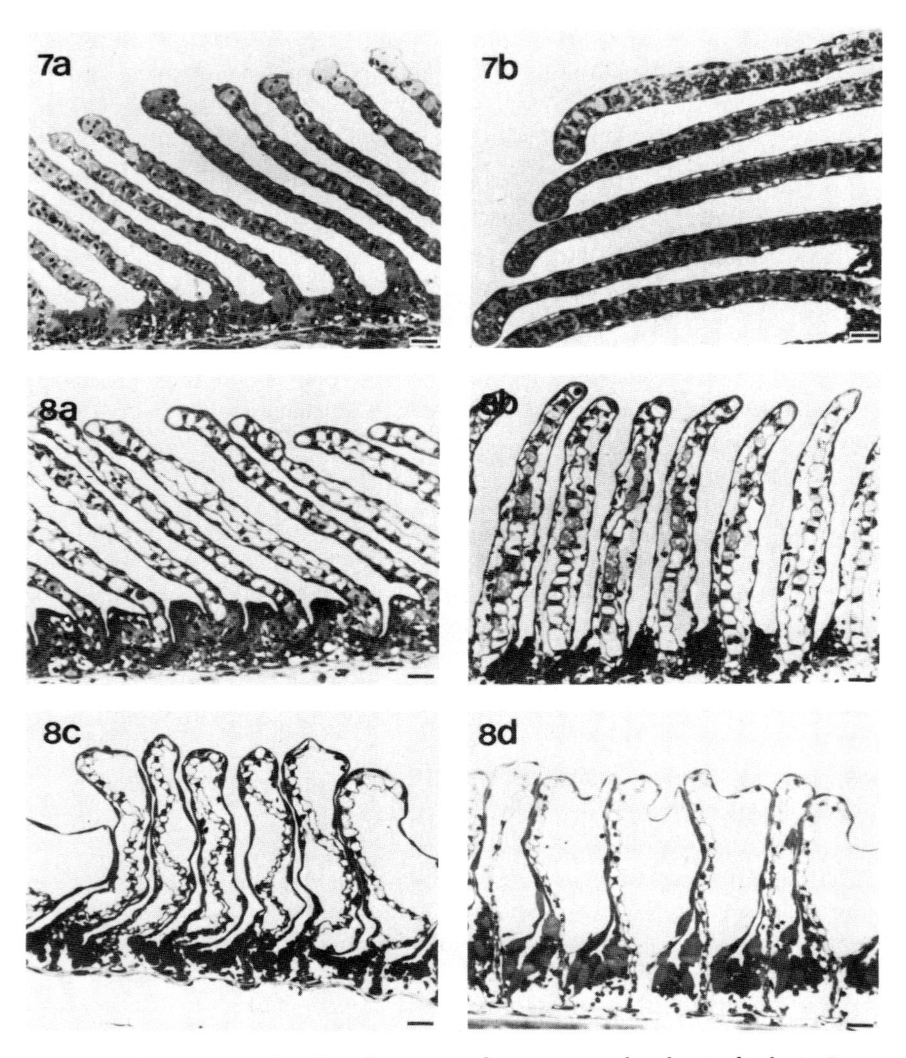

Fig. 7. Photomicrographs of lamellae sectioned transverse to their longitudinal axis. Bars = 20 μm. (a) Unperfused tissue, (b) blood-perfused tissue. (From Ellis and Smith, 1983.)

Fig. 8. Photomicrographs of lamellae sectioned transverse to their longitudinal axis, displaying the four recognizable categories of edema. Bars = 20 μm. (a) Mild edema: lamellar subepithelial spaces slightly enlarged but not more than the width of the blood lacunae, lamellar epithelium attached for most of its length. (b) Moderate edema: lamellar subepithelial spaces enlarged, space broken in places by persistent attachments of epithelium connecting to the pillar cell system, interstitial spaces as wide or wider than blood lacunae. (c) Severe edema: lamellar subepithelial interstitial space continuous from filament epithelium to margin of

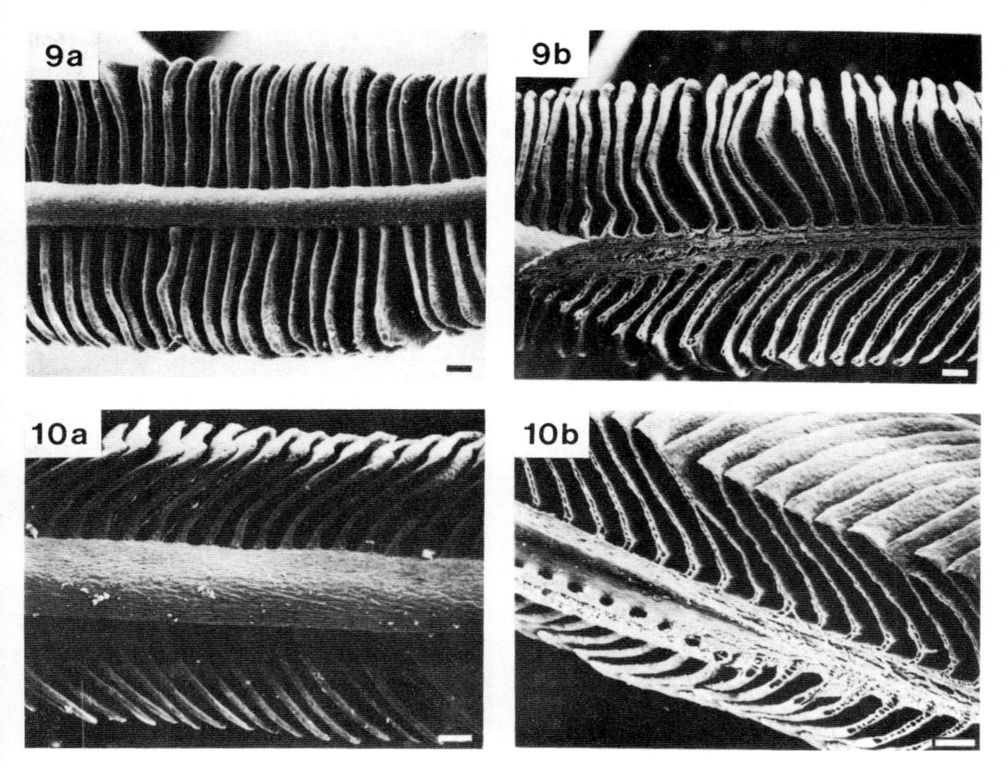

Fig. 9. Scanning electron micrographs of critical point-dried tissue that was not perfused. Bars = 50 μm. (a) Buccal view, (b) razor section. (From Ellis and Smith, 1983.)

Fig. 10. Scanning electron micrographs of critical point-dried tissue that had been briefly perfused. Bars = 50 μm. (a) Buccal view, (b) razor section. (From Ellis and Smith, 1983.)

When the osmotic pressure difference between the perfusate and the bathing medium was reduced to zero, by perfusing and bathing the gill with eel Ringer, edema persisted. This confirmed that the fluid was arising by exudation or ultrafiltration. Even if artificial colloid osmotic substances were added to the perfusate, moderate to severe edema developed within 15 min. The colloid substitutes, at the concentrations used, are listed along with the extent of edema observed, in Table V. Relatively high concentration (8%) of PVP reduced, but did not prevent edema over a 15-min perfusion. Pro-

---

lamella, lamellar height reduced, interlamellar water space reduced, filamental subepithelial interstitial spaces enlarged. (d) Very severe edema: lamellar subepithelial space continuous with filamental subepithelial space, both grossly enlarged; filament epithelium detached, interlamellar water space absent. (From Ellis and Smith, 1983.)

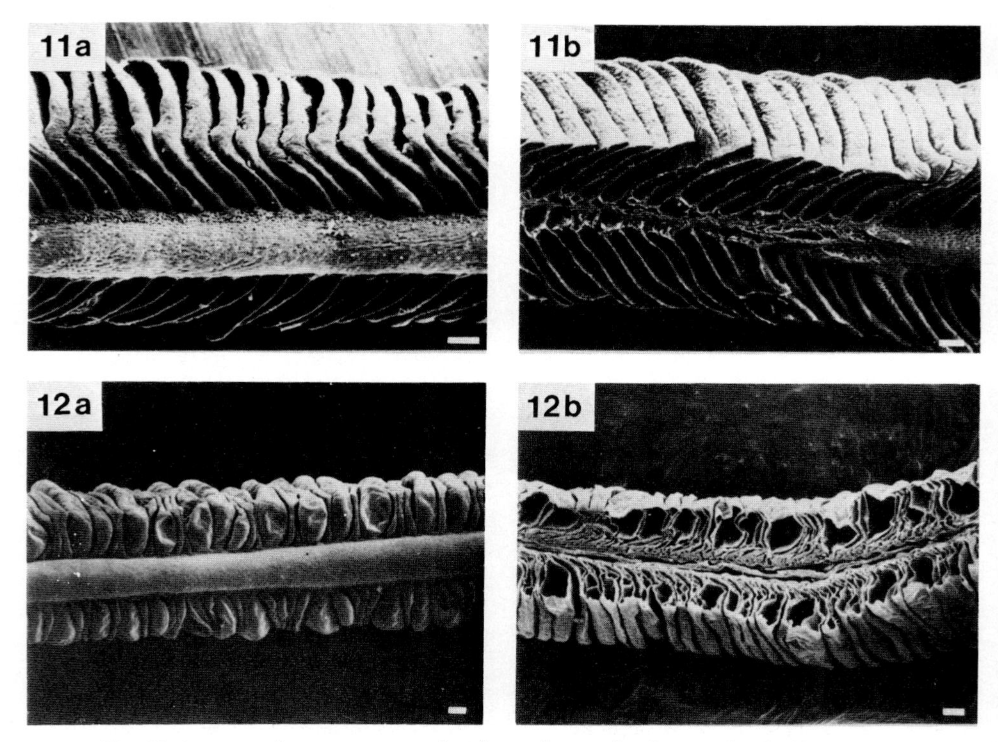

Fig. 11. Scanning electron micrographs of critical point-dried tissue that had been perfused with mammalian plasma. Bars = 50 μm. (a) Buccal view, (b) razor section. (From Ellis and Smith, 1983.)

Fig. 12. Scanning electron micrographs of critical point-dried tissue that had been perfused with eel Ringer. Bars = 50 μm. (a) Buccal view, (b) razor section. (From Ellis and Smith, 1983.)

longed perfusion resulted in severe edema; moreover, the high viscosity of an 8% PVP–Ringer solution made it unsuitable as a perfusate.

Perfusion with fresh mammalian whole blood prevented edema for perfusion periods up to 30 min. Mild edema developed when perfusions were continued for longer than 30 min (Fig. 7b). Perfusions with mammalian whole blood diluted with eel Ringer indicated that the extent of edema was proportional to the degree of dilution. Perfusion with fresh bovine calf plasma prevented edema for periods of up to 30 min duration (Fig. 11). However, if the calf plasma was more than 5 days old, severe edema resulted (Fig. 13).

The study of Ellis and Smith (1983) has shown that edema is an inevitable consequence of Ringer perfusion in eel gills, which greatly increases the blood–water diffusion barrier. As shown in Tables III and IV, there have

Table V

Artificial Colloid Substitutes, Concentrations Used, and Resulting Degree of Edema[a,b]

Artificial colloid substitute	Concentration	Degree of edema[f]
Dextran[c]	2%	++to+++
Dextran	3%	++to+++
Dextran + Protein[d]	2% 0.2%	++to+++
PVP[e]	1%	++ (n = 1)
PVP	2%	++to+++ (n = 1)
PVP	4%	+to+++
PVP	8%	++to+++

[a]From Ellis and Smith (1983).
[b]Perfusion time in each case was 15 min.
[c]Dextran T.70, MW 70,000.
[d]Bovine serum albumin, MW 69,000.
[e]Polyvinylpyrrolidone, MW 30,000–40,000.
[f]+ Mild; ++ moderate; +++ severe; ++++ very severe.

been many studies of "Ringer-perfused gills." In only a few of the studies listed have the gills been examined histologically after perfusion (Richards and Fromm, 1969; Payan and Matty, 1975; Part and Svanberg, 1981; Part *et al.*, 1982a; van der Putte and Part, 1982). Except for the study of Richards and Fromm (1969), histological examination of the gills revealed no structural abnormalities following saline perfusion under the experimental conditions (10^{-5} to 10^{-6} M adrenaline in the perfusate). In addition, a recent histological comparison among gills from unperfused, saline-perfused, and blood-perfused trout heads has shown no morphological differences (Fig. 14). Perhaps eel gills are especially vulnerable to edema formation: Keys and Hill (1933) showed that the colloid osmotic pressure of eel serum (25–28 cm H_2O) is comparable with that of mammals (25–33 cm H_2O). This is a result of retention of extensive fat stores in all body fluid compartments. Starvation reduces the serum osmotic pressure to typical teleostean values in about a week (Keys and Hill, 1933).

The source of accumulating fluid in eel gills is the perfusate. By analogy with the situation in glomerular kidneys, net filtration pressure is given by the difference between summed osmotic and hydrostatic pressures in perfusate and tissue fluid. Thus, the addition of colloid substitutes to the perfusate

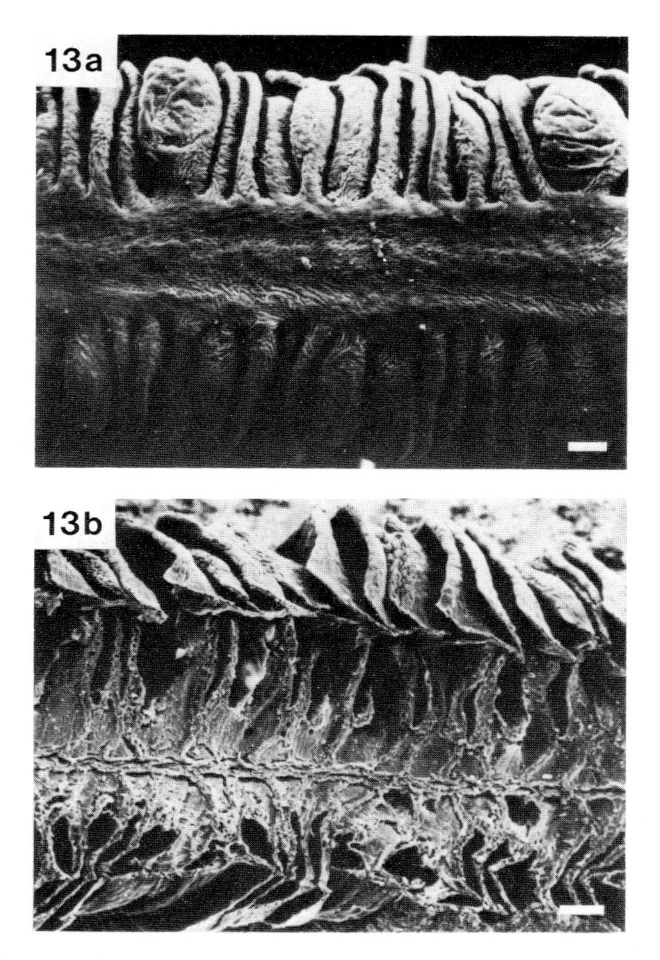

Fig. 13. Scanning electron micrographs of critical point-dried tissue that had been perfused with mammalian plasma that was more than 5 days old. Bars = 50 μm. (a) Buccal view, (b) razor section. (From Ellis and Smith, 1983.)

Fig. 14. Scanning electron micrographs of critical point-dried unperfused trout gills and gills that had been perfused for 30 min. (a) Unperfused, (b) blood perfusion, (c) saline perfusion. Magnification ×160. (From D. J. Lauren, S. F. Perry, and C. E. Booth, unpublished observations.)

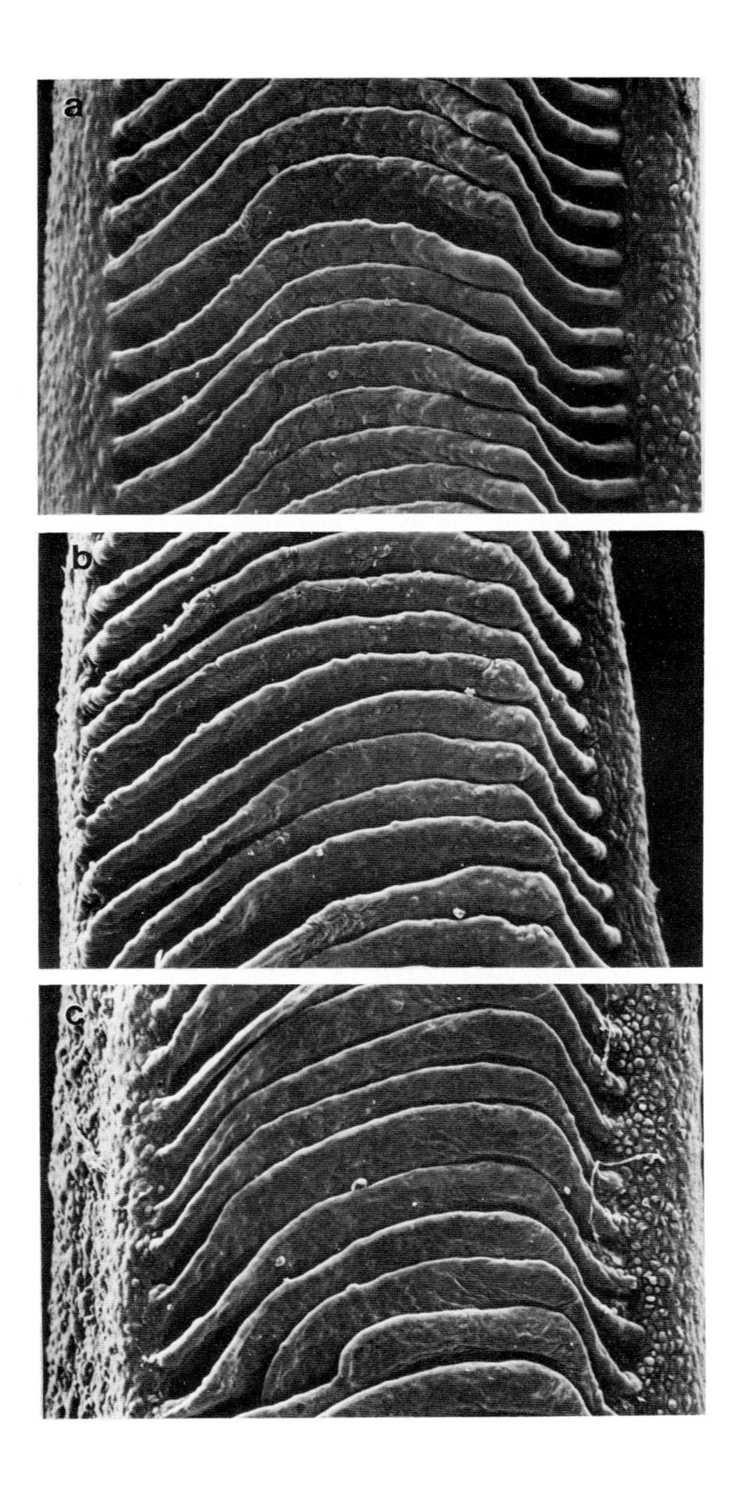
a
b
c

should reduce filtration, and hence the degree of edema. Clearly this is not the case; only by perfusing with unworkably viscous solutions of 8% PVP in eel Ringer is edema reduced even a little. The failure of nonprotein colloids to prevent edema during studies of perfused extremities in mammals has been noted long ago for rabbits (Krogh and Harrop, 1921; Schlesser and Freed, 1942) and frogs (Drinker, 1927; Danielli, 1940). Moreover, it has been shown that, in perfused cat hindlimbs, dextrans do not exert their full osmotic pressure unless at least 0.2% protein is present in the perfusate (Dow and Hamilton, 1963). They proposed that the functional dimensions, and hence their filtration coefficients, were reduced by a layer of adsorbed plasma protein. This explanation may be particularly relevant to the situation in fish; Hargens *et al.* (1974) have demonstrated that systemic capillaries in *Gadus morhua* and *Pleuronectes platessa* are exceptionally leaky with respect to proteins, and the pillar cell pores of *Anguilla australis* are large enough to allow dextran of MW 170,000 to pass easily into the extravascular subepithelial space (D. J. Randall and B. J. Gannon, unpublished observations). It may be for this reason that bovine serum albumin was ineffective in preventing edema in perfused eel gills.

It is interesting that O_2 transfer is improved greatly in mammalian plasma versus saline-perfused eel gills. This phenomenon will be discussed in detail in Section VI.

The results of Ellis and Smith (1983) suggest the possibility that mammalian plasma, with a colloid osmotic pressure similar to that of the eel, and with ready availability, may be an ideal perfusate for studies of diffusional exchange in eel gills. Its use in other teleosts would probably require dilution to match the lower osmotic pressure seen there.

Although not all saline-perfused gill preparations display extensive edema formation as perfused eel arches do, we do recommend that histological examination of the gills be performed routinely as a test of gill viability.

IV. ION EXCHANGE

Experiments designed to study branchial ion transport have been conducted, primarily on intact animals. In recent years, however, investigators have resorted to employing various types of perfused gill preparations in an attempt to define the mechanisms of piscine ion transport more carefully.

In this section we will discuss the various types of perfused gill preparations used to investigate ionic transport (including the opercular epithelium). The advantages and disadvantages, as well as problems of interpretation, will be evaluated for each preparation. Before discussing perfused preparations, it is necessary to review studies performed in vivo, so as to explain the need for perfused gill preparations.

A. Whole-Animal Studies

The advantages of using intact animals for ion exchange studies are that correct ventilation and perfusion of the gill are achieved, and that neural and hormonal inputs are present. The disadvantages include the inability to modify with precision the chemical composition of blood entering the gills, the difficulty in separating primary from secondary effects (e.g., cardiovascular), and the impossibility of determining whether transport is across the basolateral or apical membranes. Furthermore, the effect of stress in whole-animal studies is difficult to assess, and although most investigators do all that is possible to prevent stress, one is never sure.

Although studies using intact animals have contributed greatly to our knowledge of ion exchange mechanisms in fish, the limitations just outlined have made it necessary to resort to perfused preparations.

B. Perfused Gill Preparations

If a perfused gill preparation is to be useful in studying ion exchange, the following criteria or requirements should be met:

1. In the absence of ventilation, the gills must be irrigated in a fashion that will ensure adequate water flow over the lamellae.
2. The gills must be perfused with pulsatile flow in the physiological range so as to mimic the action of the heart.
3. There must be adequate stirring or mixing of the external medium to avoid formation of boundary layers or dead spaces.
4. The correct geometry of the gills must be maintained.
5. Long periods of branchial ischemia should be avoided.
6. Surgery should be as rapid as possible, and anesthesia should be avoided if possible.
7. The volume of the external medium should be kept small, so that slight changes in ion concentration can be detected easily.
8. Leakage of perfusate into the bathing medium must be kept within physiological limits.
9. The chemical and gas composition, and the pH of the perfusate should be similar to venous blood of the species being examined.

Various types of perfused preparations have been used over the years in studies of branchial ion exchange (see review by Evans *et al.*, 1982). These include the early "heart–gill preparations" (Keys, 1931; Keys and Bateman, 1932), the isolated, perfused head preparation (Payan and Matty, 1975;

Payan *et al.*, 1975; Girard, 1976; Girard and Payan, 1977a,b; Payan, 1978; Claiborne and Evans, 1980, 1981; Perry, 1981; Goldstein *et al.*, 1982; Perry *et al.*, 1983, 1984a,d), the isolated, perfused branchial arch preparation (Shuttleworth, 1972, 1978; Shuttleworth and Freeman, 1974; Shuttleworth *et al.*, 1974; Farmer and Evans, 1981; Stagg and Shuttleworth, 1982), and the isolated opercular epithelium (Karnaky *et al.*, 1976; Degnan *et al.*, 1977; Zadunaisky, 1979; Degnan and Zadunaisky, 1980a,b; Mayer-Gostan and Maetz, 1980).

1. The "Heart–Gill Preparation"

The original "heart–gill preparation" (Keys, 1931; Keys and Bateman, 1932) was prepared by perfusing the hepatic vein of *Anguilla vulgaris* with Ringer solution that was pumped via the intact heart into the gill vasculature. This preparation was limited in the same manner as are intact animals, because the direct effects of vasoactive substances on the gill vasculature could not be separated from accompanying effects on the heart. A modification of this technique (Kirschner, 1969), in which Ringer is pumped into the ventral aorta by a cardiac pump, overcame this problem. Unfortunately, this preparation is subject to rapid deterioration and is not often used for ion transport studies.

2. The Isolated, Perfused Head Preparation

One of the most widely used preparations for the study of ion exchange is the isolated, saline-perfused head preparation first described by Payan and Matty (1975). The perfused head is prepared as follows. After heparinizing the animal, the unanesthetized fish is decapitated by cutting behind the opercular openings. The head is placed onto an operating table and the gills irrigated with water. Immediately the pericardium is cut and the ventricle severed to prevent air from being pumped into the gills. Viscera are removed from the body cavity and a catheter inserted into the bulbus arteriosus through the severed ventricle and tied in place (Fig. 3). The gills are cleared of blood by perfusing with filtered (0.22- or 0.45-μm pore size) saline (Payan and Matty, 1975; see Section II) at a constant pressure of 50 cm H_2O (Perry *et al.*, 1983). A tightly fitting semicircular plastic collar is placed inside the abdominal cavity and sutured into position, thus making the body wall rigid. Finally, a catheter is inserted into the dorsal aorta for collection of postgill saline. The surgery can be completed in approximately 8 min. Next, the head is placed inside a cylindrical plastic container and held in place by a thin rubber membrane (balloon or condom) that prevents leakage of the recirculated external medium contained inside (Fig. 4). The gills are perfused either at constant pressure (Payan and Matty, 1975) or constant flow

(Perry *et al.*, 1983) from reservoirs containing saline. Preferably the saline is equilibrated with gas tensions approximating venous blood (i.e., 0.3% CO_2, 2.25 torr; 5.0% O_2, 40 torr; pH 7.9–8.1; Perry and Daxboeck, 1984). Typically, input pressure is monitored via a T junction in the input catheter connected to a pressure transducer and displayed on a chart recorder.

The isolated, saline-perfused head preparation meets many of the criteria necessary for a useful perfused gill preparation. The surgery is rapid, no anesthetic is used, long periods of ischemia are avoided, the gills are well irrigated and perfused, and the volume of the external medium can be very small (e.g., 150 ml; Perry *et al.*, 1983). The major criticism of this preparation is that, dorsal aortic pressure normally is kept at zero to prevent perfusate from leaking into the external environment. Certainly this must affect the pattern of flow through the gills. This lack of back pressure may also contribute to the rapid deterioration often observed in this preparation. Girard (1976) found that gill resistance increased five-fold within 30 min, whereas Wood (1974) demonstrated that relatively linear and stable pressure versus flow relationships were possible if postbranchial efferent pressures were maintained by a column of irrigation solutions. Recently, it has been shown that raising dorsal aortic pressure to 10 to 15 cm H_2O does not cause leaks in saline-perfused trout heads (Perry *et al.*, 1984c,d) but does lead to a significant decrease in the ratio of dorsal aortic to arteriovenous flow.

The problem of hemodynamic deterioration can be partially overcome by the addition of 10^{-7} *M* adrenalin or noradrenaline to the perfusate or by irrigating the gills with hyperoxic water (Perry and Daxboeck, 1984). It is interesting that hemodynamic deterioration may be limited to the trout head. It has been shown recently that isolated head preparations of the sculpin and dogfish "pup" can maintain relatively constant gill resistances for 3 to 8 hr (Claiborne and Evans, 1980; Evans and Claiborne, 1982). Using the trout head, it is obvious that experiments of this duration are impossible under normal physiological conditions. If the isolated, saline-perfused trout head is to be used, we urge that experiments last no longer than 60 min (30 min is preferable) and that input pressure be monitored continuously, acting as an indicator of gill viability.

A great advantage of the perfused head preparation is that efferent flow can be partitioned into its arterial and venous components. This is possible because the fluid leaving the lamellae in the gill may return via efferent filamental and branchial arteries to the dorsal aorta or be shuttled through anastomoses between the efferent filamental artery and the filamental venous sinus (efferent arteriovenous anastomoses; see Chapter 2, Volume XA, this series) to the venous circulation. Because of this partitioning of efferent flow, it is possible to separate ionic fluxes into a respiratory epithelium component (flux across respiratory cells) and a nonrespiratory epithelium component (flux across chloride cells). In this manner, Girard and Payan (1977a) demonstrated that in fresh water, Na^+ and Cl^- influxes occur across the respiratory epithelium, whereas in seawater the major portion of salt flux occurs across chloride cells (Girard and Payan, 1977b). Unfortunately, under "normal" conditions, the saline-perfused freshwater trout head displays rates of Na^+ and Cl^- influxes well below *in vivo* levels. Only when 10^{-5} *M* adrenaline is added to the perfusate does the head display in vivo rates of Na^+ influx, while Cl^- influx is unaffected (Girard and Payan, 1980). According to Payan (1978), the stimulatory effect of adrenaline on Na^+ uptake is specific in nature and not due to accompanying hemodynamic alterations. Investigations of Na^+ uptake using the perfused head now are routinely performed using adrenaline in the perfusate. Experiments involving Cl^- uptake have been limited because of the inability of this preparation to absorb Cl^- from the external medium. However, it has been shown by Perry *et al.*(1983) that branchial Cl^- uptake in the perfused trout head also is subject to adrenergic control; it is enhanced by α-receptor stimulation and inhibited by β-receptor stimulation (see Fig. 15). Rates of Cl^- influx approximating in vivo values are attainable by adding 10^{-6} *M* noradrenaline to the saline.

One of the most common problems in the analysis of gill perfusion studies is separating specific from nonspecific (e.g., hemodynamic) effects and is most prevalent in studies employing vasoactive substances. For this reason it is essential that perfusion pressure be monitored continuously so as to give an indication of gill hemodynamics. For example, Perry *et al.* (1984a) observed that Cl^- influx in the perfused trout head was stimulated by the addition of HCO_3^- to HCO_3^--free saline (Fig. 16) and was associated with a large reduction of input pressure (Fig. 17). Subsequent analysis indicated that stimulation of Cl^- uptake by perfusate HCO_3^- was not a specific effect (e.g., increased entry of HCO_3^- into the gill epithelium) but was due to accompanying hemodynamic changes.

Although use of the perfused head has been extremely helpful in elucidating mechanisms of branchial ion exchange, there are serious problems that must be overcome if its use is to be continued. These are its inability to display in vivo rates of Na^+ and Cl^- uptake and its rapid deterioration in the absence of catecholamines, as well as presence of leaks, in some instances, when physiological dorsal aortic back pressure is applied. All of these phenomena may reflect the use of saline as the perfusion medium; it is not unlikely that the O_2-carrying capacity of the saline contributes to tissue necrosis. As discussed in Section III, edema formation may be a consequence of saline perfusion, and indeed it has been observed in saline-perfused trout heads in the absence of adrenaline (Part *et al.*, 1982a). An additional problem may be the difference in viscosity between blood and saline.

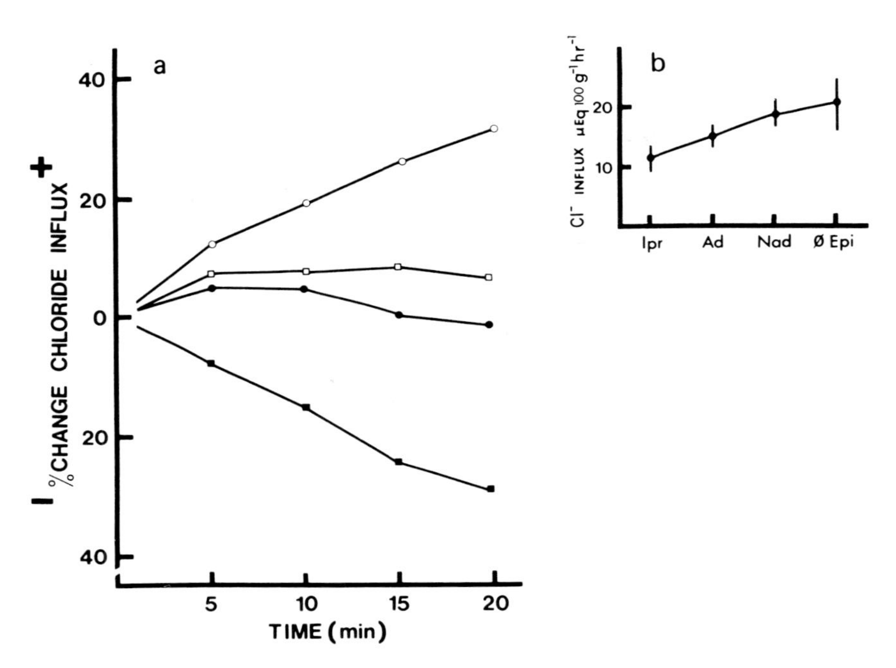

Fig. 15. Effect of various catecholamines on branchial chloride uptake in the isolated, saline-perfused head of rainbow trout. (a) Percentage change during 20 min perfusion. (○----○) phenylephrine (∅Eph; n = 5); (□----□) noradrenaline (Nad; n = 7); (●----●) adrenaline (Ad; n = 5); (■----■) isoprenaline (Ipr; n = 7). (b) Absolute chloride influx using same four catecholamines (all concentrations 10^{-5} M). (From Perry *et al.*, 1983.)

Perry *et al.* (1984d) reported increased stability of Na^+ uptake and ammonia excretion during blood perfusion so it is possible that the use of blood or plasma as the perfusate may solve some of these problems and lead to a perfused head preparation more representative of the intact animal.

3. The Isolated, Perfused Branchial Arch Preparation

Isolated, saline-perfused branchial arches (holobranchs) also have been used to study ion exchange in fish, although to a lesser extent than the perfused head preparation. Although methods may vary slightly, isolated arches are typically prepared by cannulation of the ventral aorta or bulbus arteriosus, allowing perfusion of the anesthetized animal. Once the gills are cleared of blood, the branchial basket is removed and individual arches dissected free. Afferent and efferent arch arteries are cannulated, and the holobranch is either suspended vertically (Rankin and Maetz, 1971; Shut-

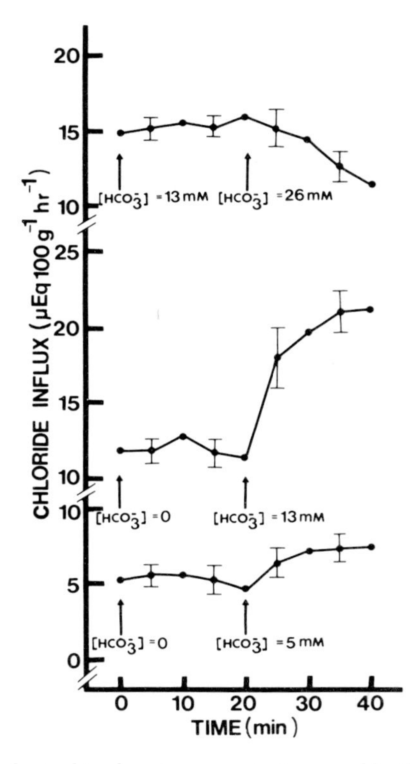

Fig. 16. Effect of perfusate bicarbonate concentration on chloride influx in the isolated, saline-perfused head of rainbow trout. For each curve shown, $n = 4$. (From Perry *et al.*, 1984a.)

tleworth, 1972) or placed horizontally (Farrell *et al.*, 1979) in a constant-temperature bath (see Fig. 2). Normally the external bath is aerated or agitated to allow proper mixing of the external medium. Perfusion is accomplished either by a constant-pressure head (Rankin and Maetz, 1971), or by pulsatile (Bergman *et al.*, 1974; Farrell *et al.*, 1979; Farmer and Evans, 1981; Perry *et al.*, 1982) or nonpulsatile (Shuttleworth, 1972; Shuttleworth *et al.*, 1974; Fenwick and So, 1974) constant flow.

According to the criteria we have listed for a successful perfused preparation, the isolated arch appears to be less desirable for studies of ion exchange than the perfused head preparation.

1. A major problem is the difficulty in providing adequate irrigation (ventilation). Even vigorous stirring of the irrigation bath apparently does not mimic the irrigation patterns found in the intact animal. The result is the formation of boundary layers in the fluid in direct contact with the gill

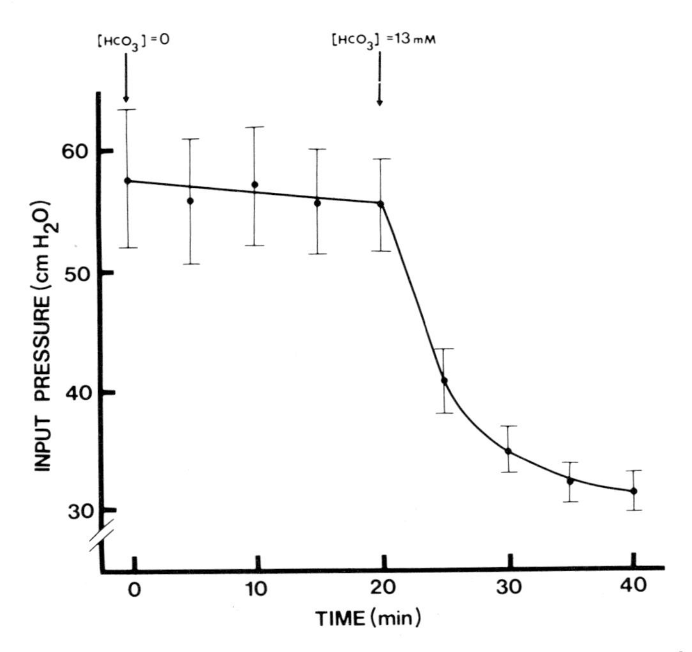

Fig. 17. Effect of perfusate bicarbonate concentration on input pressure in the isolated, saline-perfused head of rainbow trout ($n = 5$). (From Perry *et al.*, 1984a.)

epithelium, which may not allow it to reach the active site of ion transport between the lamellae (Maetz, 1971). This may be the reason investigators using this preparation report ion flux rates much lower than in vivo values (Shuttleworth and Freeman, 1974; Farmer and Evans, 1981). This problem of inadequate irrigation has led to the practice of aerating the external medium with high-oxygen mixtures, which may have deleterious effects on branchial hemodynamics.

2. The catheterization of afferent and efferent branchial arch arteries often is difficult (especially in rainbow trout) and time-consuming, which leads to extended periods of branchial ischemia.

3. The procedures involved in the preparation make it necessary that anesthetic be used. The most commonly used anesthetic, MS 222, is known to have profound effects on gill electrolyte composition (Houston *et al.*, 1971) and may affect ion-transporting properties of the gill epithelium.

4. Leakage of perfusate into the external medium is extremely difficult to prevent and leads to inaccuracies in flux measurements as well as a constantly changing external environment (Richards and Fromm, 1970). Perry (1981) noted that efferent flow in saline-perfused trout holobranchs is, at

best, 75% of afferent flow. Similarly, Farrell *et al.* (1979) reported that efferent flow ranged between 30 and 88% of afferent flow in saline-perfused arches of the lingcod (*Ophiodon elongatus*). Leakage can be diminished to a certain extent by reducing efferent perfusion pressure. However, even at zero efferent pressure, leaks still occur as a result of venous drainage from the arch. The only nonleaky gill arch preparation appears to be the eel (*Anguilla dieffenbachii*) gill preparation of Shuttleworth (1972). This may be a result of the low flow rates used (10 μl / min^{-1}): assuming that 70% of the filaments on each arch are perfused at approximately one-eighth of *in vivo* cardiac output, then, for rainbow trout, the correct flow would be 1.5 ml min^{-1} kg^{-1}, on the basis of cardiac output calculations of Kiceniuk and Jones (1977).

5. Other problems encountered using this preparation are mucus accumulation on gill filaments and the difficulty in maintaining correct gill geometry after excising individual arches from the branchial basket.

Even considering the limitations just listed, the perfused arch has provided valuable information unavailable with other techniques. This has been primarily a result of the ease in which transepithelial potential (TEP) can be measured. Stable TEPs for periods of at least 8 hr have been reported (Shuttleworth, 1978). In an important study, Shuttleworth *et al.* (1974) demonstrated that addition of ouabain abolished the TEP across the flounder gill bathed and perfused with Ringer solution, indicating electrogenic salt extrusion coupled to (Na^+,K^+)-activated ATPase. Unfortunately, perfusion pressures were not monitored, and the results of Farmer and Evans (1981) indicate considerable vasosensitivity of the isolated gill preparation to ouabain. It is possible that inhibition of ion transport by ouabain may in fact be due to local vasoconstriction. Once again, the importance of monitoring vascular pressures, especially during pharmacological manipulations, cannot be overstated.

4. The Isolated Opercular Epithelium

Although not a perfused gill preparation, the isolated opercular epithelium has contributed much to our knowledge of gill ion transport and for that reason is discussed here. The opercular epithelium lining the inside of the gill chamber of the teleost, *Fundulus heteroclitus*, contains an abundance of chloride cells (Burns and Copeland, 1950; Karnaky and Kinter, 1977; Karnaky *et al.*, 1977), which are morphologically identical to the chloride cells of the gill epithelium. Chloride-rich tissues have also subsequently been located in the operculum of *Fundulus grandis* (Krasny and Evans, 1980) and *Sarotherodon mossambicus* (Foskett *et al.*, 1979), as well as the jaw epithelium of *Gillichthys mirabilis* (Marshall and Bern, 1980).

Normally, results obtained using the opercular epithelium are extrapolated to include branchial chloride cells. Strictly speaking, this procedure is not valid but is accepted given the similarity in morphological and biochemical analyses of branchial and opercular chloride cells.

The preparation of the isolated opercular epithelium is described in detail by Degnan *et al.* (1977) and involves techniques first described by Ussing and Zerahn (1951). Briefly, the epithelium is exposed by removing the gills and the branchiostegal rays of the operculum. Using a dissecting microscope, the epithelium is freed from the underlying bony operculum using microforceps, and mounted between two halves of a Lucite chamber (see Degnan *et al.*, 1977).

The isolated, short-circuited opercular epithelium obviates many of the technical problems involved in intact fish and perfused gill preparations, and can provide some of the needed biophysical data required for a better understanding of fish ionic regulation. Short-circuit techniques have never been applied to the gill because of the complex histology of this tissue. Other advantages are that problems of perfusion of the gill vasculature and the presence of boundary or unstirred layers are eliminated. Unfortunately, lack of "normal" perfusion in the opercular epithelium also leads to limitations. For example, the cells of the epithelium must obtain their O_2 requirement from saline superperfusing the preparation. For this reason, researchers have resorted to equilibrating the saline, bathing the opercular epithelium with high-O_2 mixtures (e.g., 95% O_2–5% CO_2, pH 7.1; Degnan and Zadunaisky, 1980a). Although it is unclear what effect these hyperoxic–hypercapnic salines might have on the opercular epithelium, clearly these conditions are far removed from the situation in vivo.

C. Conclusions and Recommendations

For ion exchange studies involving flux measurements, we recommend use of the isolated, perfused head preparation. It is advantageous because of its "nonleakiness," ease of preparation, and the small external volume required. Investigators using this preparation should attempt to use blood or plasma as the perfusion medium instead of saline, as well as levels of catecholamines that are more representative of in vivo values (e.g., 10^{-7}–10^{-8} M adrenaline or noradrenaline). We also suggest that input pressure and O_2 uptake be monitored continually as indicators of gill viability.

For ion exchange studies involving TEP measurements, the isolated arch is recommended because of its stability (at least in eel and flounder).

V. HEMODYNAMICS

The gill vascular bed is made up of a high-pressure respiratory circulation, possibly with arterio-arterial anastomoses, and a low-pressure arteriovenous or recurrent circulation (see Chapter 2, Volume XA, this series). Most studies of gill hemodynamics have concentrated on the former (Wood, 1974; Payan and Girard, 1977), often with the partial occlusion of the latter circulation (Rankin and Maetz, 1971; Farrell *et al.*, 1979). The importance of maintaining both components of the gill circulation has been recognized by many workers (Kirschner, 1969; Smith, 1977), although little quantitative information about the recurrent circulation is available (see Daxboeck and Davie, 1982). Gill circulation is subjected to fluctuations in mean pressure, pulse pressure, heart rate, and stroke volume (Stevens and Randall, 1967a,b; Jones and Randall, 1978). These changes cause passive alterations in resistance to blood flow through gills. Descriptions of physical factors that affect gill resistance have been provided by a number of investigators (Wood, 1974; Farrell *et al.*, 1979). Although many changes in gill hemodynamics observed in intact fish are partly due to passive changes of the gill circulation resulting from altered flow and/or pressure, active alterations caused by the autonomic nervous system and/or circulating catecholamines have received much more attention (Krakow, 1913; Keys and Bateman, 1932; Rankin and Maetz, 1971; Wood, 1974, 1975; Smith, 1977). Normally, nervous and/or hormonal activity maintains branchial vascular tone. It is against this tonic activity that physical factors effect their changes.

An understanding of the vascular connections in the gill is essential to a discussion of how best to investigate gill hemodynamics. A detailed description of gill circulation is given in Chapter 2, Volume XA, this series.

A. Preparations for Study of Gill Hemodynamics

The study of gill hemodynamics requires a preparation that displays stable pressures and flows with time of perfusion. Many preparations show progressive increases in resistance to flow through the arterio-arterial circulation during perfusion, which makes baseline values difficult to establish and is indicative of progressive deterioration of the circulatory pathways through the gills.

1. Progressive Deterioration: The Problem and Its Solution

The pressure difference between ventral and dorsal aortas often increases with time during constant-flow perfusion, or the flow from the dorsal aorta decreases with time during constant-pressure perfusion (Rankin and Maetz, 1971; Wood, 1974; Payan and Matty, 1975; Daxboeck and Davie, 1982). Studies that report at least a partial solution to this problem show a number of common methodological features, and some research offers suggestions as to why some techniques produce more stable preparations than others.

An important feature in attaining a stable preparation is proper preparation of saline and use of physiological gas tensions (see Section II). Addition of blood proteins or use of blood as the perfusate significantly improves the stability of hemodynamic performance in some preparations (Saunders and Sutterlin, 1971; Davie *et al.*, 1982; Metcalfe and Butler, 1982; Ellis and Smith, 1983). Pulsatile perfusion also has been qualitatively assessed as superior to nonpulsatile perfusion (Kirschner, 1969; Bergman *et al.*, 1974; Part and Svanberg, 1981; Davie and Daxboeck, 1982; see also Section III). Payan and co-workers have overcome the problem of progressive deterioration by perfusing with saline containing 10^{-5} *M* adrenaline and using only the first hour of perfusion for experimentation. This expedient method avoids the problem rather than solving it and is inconvenient if responses to catecholamines are under investigation. Other workers suggest a period of at least 60 min of perfusion before experimentation to allow recovery from the most acute effects of preparing the tissues for perfusion (Jackson and Fromm, 1981; Davie *et al.*, 1982).

Curiously, eel gill preparations appear to be relatively free from progressive deterioration (Rankin and Maetz, 1971; Shuttleworth, 1972, 1978). This may be because the recurrent circulation is derived from both afferent and efferent filamental arteries (Laurent and Dunel, 1976) and may offer a true "shunt" when edema at the respiratory barrier increases resistance to blood flow through lamellae.

If the respiratory barrier is so sensitive to inappropriate pulse characteristics, then one might expect to observe the physiological consequences of increased barrier thickness (e.g., low O_2 uptake) in perfused preparations. Indeed, many investigators have been unable to show significant oxygen movement across saline-perfused gills (Wood *et al.*, 1978; Daxboeck and Davie, 1982). Only by maintenance of normal respiratory barrier structure by carefully controlled perfusion conditions (Part *et al.*, 1982a; Pettersson and Johanssen, 1982; Pettersson, 1983; Perry and Daxboeck, 1984; Perry *et al.*, 1984c; Ellis and Smith, 1983) has significant oxygen uptake been reported.

2. Spontaneously Ventilating Blood-Perfused Trout Preparation

The spontaneously ventilating blood-perfused trout preparation (SVBPTP; Davie *et al.*, 1982) allows independent adjustments to be made to many of the perfusion parameters, while gas and ion exchange as well as dorsal and ventral aortic pressures can be monitored. In essence, this preparation consists of a fish with an artificial heart and a large extracorporeal blood reservoir.

A detailed discussion of the preparation of the spontaneously ventilating blood-perfused trout is given by Davie *et al.* (1982). Briefly, the procedure involves anesthetizing a fish and inserting catheters into the bulbus arteriosus in the orthograde (input catheter, IC) and retrograde (venous return catheter, VRC) directions (see Fig. 18). Additionally, the dorsal aorta and buccal cavity are cannulated. The duration of the operation averages 45 min. Interruptions of perfusion with either saline or blood last less than 1 min.

Following completion of surgery, the fish is transferred to a holding box (Fig. 19) and blood perfusion started. Once placed into the box and perfused with blood, the fish regains equilibrium and resumes normal breathing movements within 30 min. Blood perfusion is accomplished by means of a

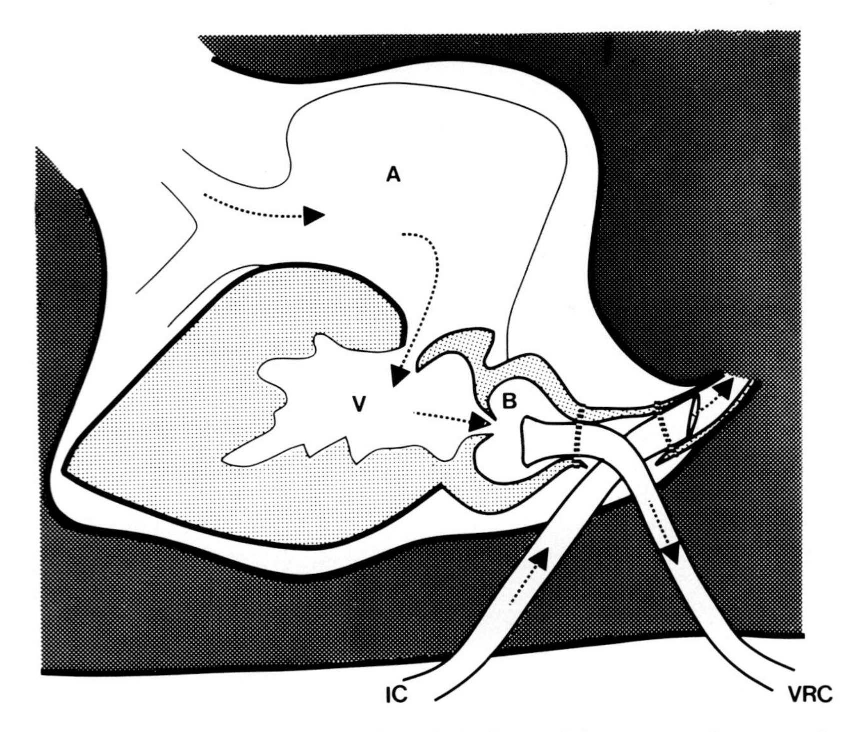

Fig. 18. Schematic sagittal section through the heart of the trout, to illustrate catheter positions used for blood perfusion experiments. Top of the page is dorsal, right is cranial. A, atrium; B, bulbus arteriosus; IC, input catheter into ventral aorta, leading to gills; V, ventricle; VRC, venous return catheter, flow aided by ventricular contractions. Arrows within lumina of tubes and heart chambers indicate direction of blood flow. (From Davie *et al.*, 1982.)

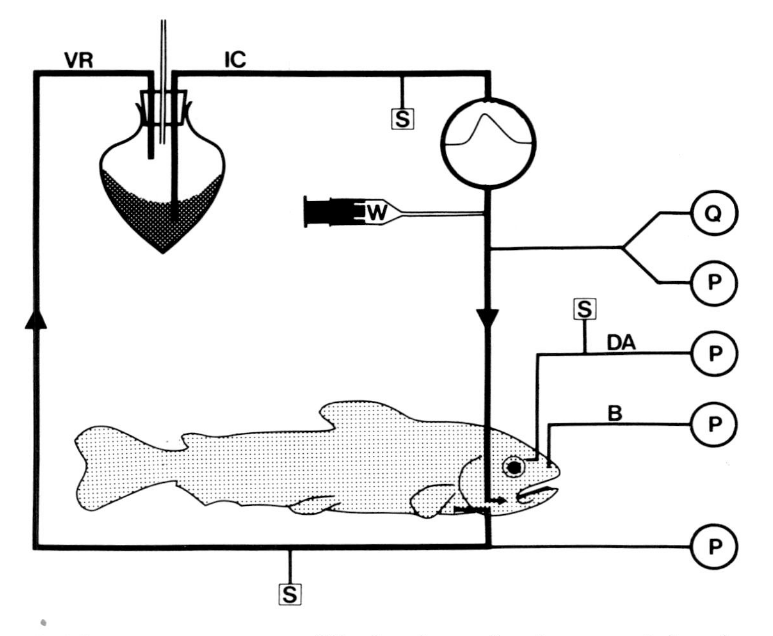

Fig. 19. Schematic representation of blood perfusion of rainbow trout. B, buccal cannula; DA, dorsal aortic cannula; IC, input catheter; Q, flow record displayed on chart recorder; P, pressure record displayed on chart recorder; S, sampling sites; VR, venous return to tonometer flasks; W, Windkessel (air chamber) to adjust pulse pressure. (From Davie *et al.*, 1982.)

cardiac pump that draws blood, equilibrated with venous gas tensions, from extracorporeal reservoirs (see Davie *et al.*, 1982).

a. Evaluation of the Performance of the SVBPTP. The SVBPTP represents significant improvements over other perfused preparations used for quantitative study of the hemodynamics of the respiratory circulation of gills. Most importantly, blood perfusion keeps the animal alive. In addition, the fish display behavior similar to that of intact animals: they maintain equilibrium, show visual tracking, can swim, and exhibit bradycardia when exposed to hypoxic water (Fig. 20). Preparations displayed rates of gas exchange similar to in vivo values (see Section VI). The large extracorporeal blood volume allows repeated sampling without depletion of the total circulating blood volume in the fish.

In addition to these qualitative advantages, the perfusion technique allows independent variation of perfusion flow rate, by either frequency or stroke volume, and pulse pressure. The effects of these perfusion parameters on ventral and dorsal aortic pressures can be recorded, permitting a com-

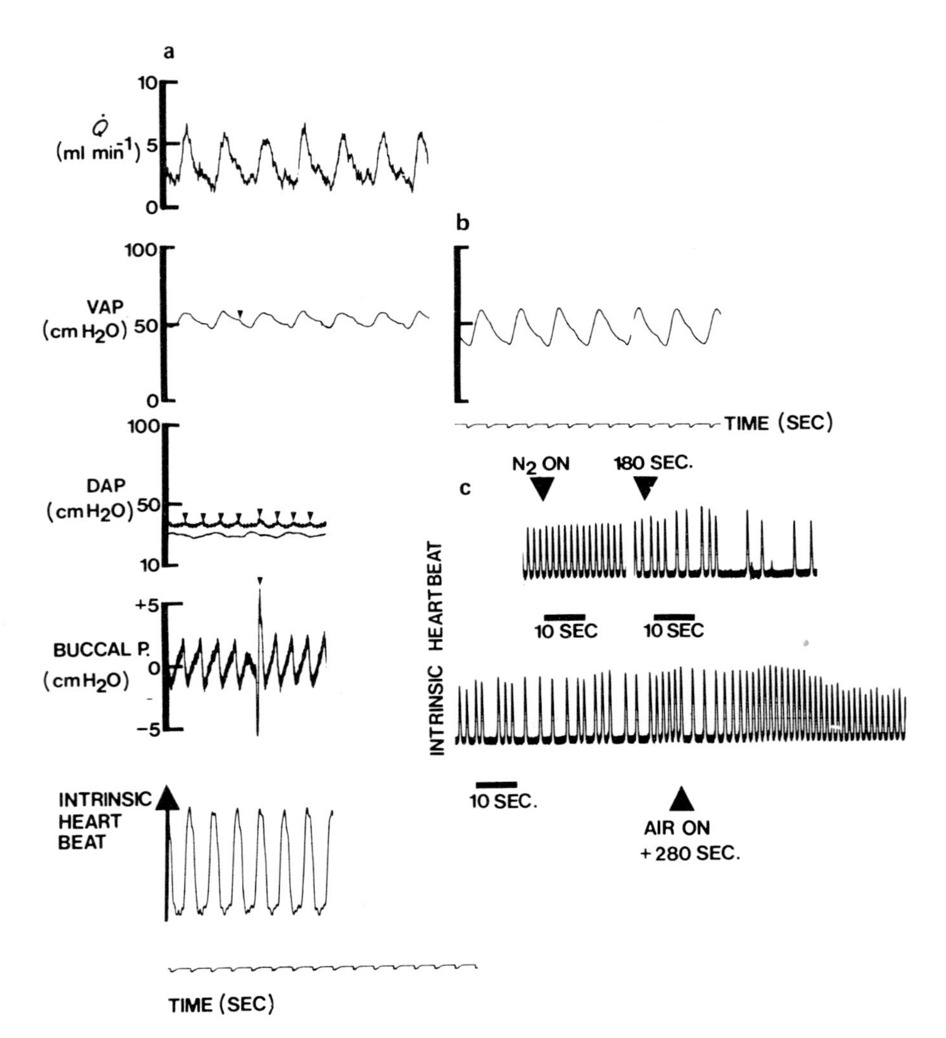

Fig. 20. Record of pressure and flows from spontaneously ventilating blood-perfused rainbow trout. (a) Simultaneous records of flows and pressures from a preparation that displayed cardiorespiratory interaction. Arrowheads show interactions of respiratory movements on the pressure and flow traces. $\dot{Q}$, perfusion flow rate; VAP, ventral aortic pressure; DAP, dorsal aortic pressure; Buccal P., buccal pressure record of ventilatory movements. (b) Record of VAP during increased pulse pressure. (c) Record of intrinsic heart activity during exposure to hypoxic water showing the "on" response (bradycardia) after 150 sec in hypoxic water and the "off" response (posthypoxic tachycardia) after resumption of normoxic water flow. (From Davie *et al.*, 1982.)

plete analysis of the changes in vascular resistance across the arterio-arterial circulation of the gills.

In our quantitative assessment of the hemodynamic performance of this preparation we have chosen to use sets of *in vivo* data where as many variables as possible have been measured simultaneously.

b. Flow, Pressures, and Vascular Resistances. Mean perfusion flow rate was 16.2 ml min^{-1} kg^{-1}. Cardiac output has been measured directly in resting rainbow trout by Wood and Shelton (1980a), who reported values of 36.7 ml min^{-1} kg^{-1}. Cardiac outputs of resting rainbow trout, estimated by the Fick principle, are 21.6 ml min^{-1} kg^{-1} (Stevens and Randall, 1967b) and 17.6 ml min^{-1} kg^{-1} (Kiceniuk and Jones, 1977). However, as stated previously, stroke volumes calculated from all available in vivo data for trout are similar (0.46 ml kg^{-1}). Consequently, the difference in flow appears to arise from heart rate differences.

Blood flow measured directly in other fish species range between 6 and 21 ml min^{-1} kg^{-1} (*Ophiodon elongatus* 6.0, Stevens *et al.*, 1972; *Anguilla australis* 11.3, Davie and Forster, 1980; *Gadus morhua* 20.8, Jones *et al.*, 1974). From these limited data on cardiac output from resting intact trout and other species, it is apparent that flow in blood-perfused fish approximates in vivo values.

c. Dorsal Aortic Pressure (DAP). Flow rate initially was adjusted to achieve a DAP of 40 cm H_2O. During the life of a preparation, DAP decreased to give a mean DAP of 34.8 cm H_2O. In resting intact trout, DAPs measured range from 34 to 42 cm H_2O (34.5, Wood and Shelton, 1980a; 35.8; Stevens and Randall, 1967a,b; 41.2, Kiceniuk and Jones, 1977). Other species in which DAPs have been measured in vivo have yielded similar values (see Helgason and Nilsson, 1973; Stevens *et al.*, 1972; Davie and Forster, 1980). Thus DAP in blood-perfused fish compares well with published in vivo data.

Davie *et al.* (1982) chose initially to alter flow to achieve a DAP of 40 cm H_2O, because fish, like most other vertebrates, probably regulate pressure by altering flow and vascular resistance (Wahlqvist and Nilsson, 1977; Smith, 1978; Wood and Shelton, 1980b). Because DAP records from trout are relatively common, and because the dorsal aorta lies between the two main vascular resistance sites, it seems a logical point around which to set cardiovascular parameters.

d. Ventral Aortic (Input) Pressure (VAP). Mean VAP in blood-perfused fish was 58.8 cm H_2O. This value appears to be slightly high for trout, indicating a high flow rate ($\dot{Q}$) or high gill vascular resistance. Other species in which VAPs have been measured (*Gadus morhua* 58, Helgason and Nilsson, 1973; *Ophiodon elongatus* 59.2, Stevens *et al.*, 1972; *Anguilla australis* 52.4, Davie and Forster, 1980) give values closer to those measured in blood-perfused fish. If fish baroreceptors are located in the ventral aorta or gill vasculature (Ristori, 1970; Ristori and Desseaux, 1970), rather than the postbranchial vasculature, then perhaps VAP would be a better variable to set flow against. This could be done using a double-bore input catheter—one bore for inflowing perfusate, the other to measure ventral aortic pressure—thus obviating the need to correct for input catheter resistance. It is apparent, nonetheless, that VAPs in blood-perfused fish are similar to values from in vivo studies.

e. Branchial Vascular Resistance (R_g). The pressure difference across the gills divided by the flow rate gives the vascular resistance of the arterio-arterial gill vessels. Mean R_g of blood-perfused fish was 14.2 cm H_2O ml^{-1} min^{-1} 100 g^{-1} (Table VI). In vivo R_g values typically are around 6 cm H_2O ml^{-1} min^{-1} 100 g^{-1} (6.02, Kiceniuk and Jones, 1977; 5.3, Stevens and Randall, 1967b; 3.4, Wood and Shelton, 1980a). The differences arise from higher VAPs rather than lower DAPs or higher flow in blood-perfused fish. Either blood-perfused fish have significant branchial vascular tone, or in vivo measurements have been taken from fish in which there was little branchial vascular tone. Pharmacological experiments designed to elicit decreases in

Table VI

Effects of Cardiac Output (Q), Stroke Volume (SV), Pulse Pressure (PP), and Cardiac Frequency (f) on Gill Resistance Change (R_g%) in Resting Blood-Perfused Trout[a]

Treatment[b]	Number of animals	R_g% (from normal)
(1) SV+; PP+; f0; $\dot{Q}$+	7	−22.91 ±9.09
(2) SV+; PP+; f−; $\dot{Q}$0 (bradycardia)	8	−25.22 ±8.98
(3) SV−; PP−; f0; $\dot{Q}$−	7	−35.38 ±17.96
(4) SV−; PP−; f+; $\dot{Q}$0 (tachycardia)	8	+4.48 ±18.63
(5) SV0; PP+; f0; $\dot{Q}$0	8	+15.37 ±11.16
(6) SV0; PP−; f0; $\dot{Q}$0	3	+14.99 ±15.89

[a]From Davie *et al.* (1982).

[b]0, No change from normal; + increase from normal; − decrease from normal. Means ± SE.

R_g in intact fish (e.g., injection of isoprenaline, Wood and Shelton, 1980a) showed only small effects, indicating low vascular tone in the gills of these animals. Wood and Shelton (1980a) did not measure both flow and VAP in the same animals, which makes their values more difficult to compare. In vivo R_g measurements generally have been taken from fish of greater weight than those used for blood perfusion. Wood (1974) and Payan and Matty (1975) have shown inverse relationships between weight and baseline vascular resistance in isolated saline-perfused preparations. The large size of fish used for in vivo measurements may partially explain the lower R_g values calculated from *in vivo* measurements.

Branchial vascular resistance values from isolated, saline-perfused gills or heads of trout are higher than those measured in vivo or those from blood-perfused preparations at equivalent pulsatile flow and DAP (14.4 cm H_2O ml^{-1}, Davie and Daxboeck, 1982). Wood *et al.* (1978) perfused whole trout with saline; using pulsatile flow of ~17 ml min^{-1} kg^{-1} and DAP maintained at 40 cm H_2O by a tourniquet to simulate systemic resistance, they measured R_g values of 11.8 cm H_2O ml^{-1} min^{-1} 100 g^{-1}. This low value for saline-perfused preparations very closely approaches those measured *in vivo* and may indicate little branchial vascular tone as in the systemic circulation.

f. Hemodynamic Responses of the SVBPTP to Changes in Perfusion Parameters. The effects of changes in flow by stroke volume and frequency, and pulse pressure on gill vascular resistance are summarized in Table VI. Simulation of perfusion conditions during mild exercise by increasing $\dot{Q}$ 1.5 times resulted in a significant fall in R_g. Because this set of perfusion conditions involves changes in flow, stroke volume, and pulse pressure, further experiments were conducted to examine which component or combination of components had the most marked effect on R_g. Increased pulse pressure, flow, or frequency alone caused smaller changes in R_g than simulated exercise. Increases in stroke volume and pulse pressure at a lower frequency, thereby maintaining $\dot{Q}$, were effective in lowering R_g. This suggests that R_g is decreased by a lowering of mean ventral aortic pressure. Higher peak systolic pressures may augment decreases in R_g because more vessels open when critical closing pressures are overcome in more distal lamellae on gill filaments (Farrell *et al.*, 1979).

Response of this preparation to 10^{-6} *M* adrenaline in blood was as predicted from saline-perfused preparations (Wood, 1974). Gill vascular resistance decreased by 40%, and systemic vascular resistance increased by 50%. Although dorsal aortic pressure was elevated, it was responsible for less than half the observed reduction in R_g (see Wood, 1974).

The SVBPTP is superior to many other preparations used to study hemodynamics of the respiratory circulation in gills. Also, it offers some clues as to why fish increase cardiac output by stroke volume rather than frequency

(Kiceniuk and Jones, 1977), why hypoxia is associated with bradycardia and maintained cardiac output (Holeton and Randall, 1967), and why the bulbus arteriosus and ventral aorta show adrenergic control of tone that modulates the size and shape of the pressure pulse (Klaverkamp and Dyer, 1974; Capra and Satchell, 1977).

Naturally, it also has disadvantages: it is time-consuming and expensive in terms of fish, it does not allow measurements of the dynamics of the recurrent circulation, and gill tissues are not directly visible. Its maximum lifetime has not yet been determined, because Davie *et al.* (1982) terminated their experiments voluntarily, following 6 to 9 hr perfusion. Similarly, Metcalfe and Butler (1982) have reported that extracorporeal bypass in dogfish can maintain viability of the fish for as long as 21 hr, after which time the experiment was terminated.

3. Other Preparations Used to Study Gill Hemodynamics

Apart from branchial baskets perfused in situ (Reite, 1969) or as isolated preparations (Krakow, 1913), two other types of preparations have been used to study gill hemodynamics. These are isolated, perfused head preparations (see Wood, 1974; Payan and Matty, 1975) and isolated, perfused branchial arches (see Rankin and Maetz, 1971; Shuttleworth, 1972, 1978). These preparations have made significant contributions to our understanding of gill physiology and have provided some information about the hemodynamics of the recurrent circulation.

Many preparations have been used to study control of blood flow to the recurrent circulation. Cholinergic stimulation constricts contractile tissue surrounding the proximal part of the efferent filamental artery, increasing flow of perfusate into recurrent channels (Smith, 1977). α-Adrenergic stimulation constricts arteriovenous anastomoses between efferent filamental arteries and filamental venous sinuses, decreasing flow into recurrent vessels (Payan and Girard, 1977). Concomitantly, β-adrenergic stimulation serves to decrease vascular resistance of the arterio-arterial respiratory circulation by mechanisms that involve the lamellar arterioles and perhaps lamellar pillar cells (Rankin and Maetz, 1971; Payan and Girard, 1977; Claiborne and Evans, 1980). Estimates of the ratio of arterio-arterial flow to recurrent flow from isolated head preparations are confounded by the inability to collect only recurrent outflow. All preparations thus far have collected both recurrent gill perfusate and perfusate from the general head circulation. Thus the range of ratios of arterio-arterial to recurrent flow vary between 1:0.5 to 1:5 (Payan and Girard, 1977; Colin and Leray, 1977; Ristori and Laurent, 1977; Claiborne and Evans, 1980; Daxboeck and Davie, 1982). In addition to collection of a mixed venous outflow, most preparations fail to maintain physiological efferent pressure. Consequently the driving pressure for the recurrent circulation is likely to approximate zero. Despite low efferent pressures, venous return from perfused heads is significant. Clearly, either general head circulation is large, or extravascular fluid from gill tissues makes a larger contribution to recurrent fluid than has previously been assumed.

Pulse pressure in the recurrent circulation might be derived from pulsations in adjacent arteries rather than through the narrow arterio-venous anastomoses. Daxboeck and Davie (1982) suggest that pulsatility augments flow from the recurrent circulation. Pulsatility also may act on extravascular fluid in lamellar walls forcing it into the filamental venous sinuses through the numerous pocket valves. Similarly, small pressures generated during ventilation might act in concert with arterial pulses during synchronization of mouth closing with systole.

The hemodynamics of recurrent circulation is poorly understood. Perfused whole fish and isolated heads are difficult preparations for quantitative study of recurrent hemodynamics. Isolated gill arches however, may offer better preparations for research into this area of gill hemodynamics.

Isolated, Perfused Branchial Arches. Perfusion of isolated branchial arches has led to significant qualitative and quantitative contributions to hemodynamics (see Farrell *et al.*, 1979). It is difficult to compare data from perfused arches with those from live, intact fish. Exclusion of a portion of the arch from perfusion by ligatures, unusual mechanical strains on arches suspended in stirred baths, and the absence of an intact central nervous system (Kirschner, 1969) may further complicate the problem of progressive deterioration. In addition, the first arch of teleosts is innervated differently from the others (Smith and Jones, 1978) and may receive less water flow in situ (Pauling, 1968). As a consequence of these restrictions, isolated arches have been used to answer "yes–no" questions rather than "how much" questions. Research using isolated arches to study quantitative aspects of gill hemodynamics show results that are similar to other preparations, but absolute values of pressures and flows are more difficult to compare.

Where isolated arches might prove useful is in the investigation of hemodynamics of the recurrent circulation. Perhaps with improved salines and better controlled perfusion parameters, the performance of isolated arches might improve (Jackson and Fromm, 1981; Holbert *et al.*, 1979; Ellis and Smith, 1983). Additionally, the practice of occluding the recurrent circulation by ligation of the whole arch might be disadvantageous (Holbert *et al.*, 1979). Arches from large fish may have branchial veins large enough to cannulate. Such preparations might provide data on flows from the recurrent circulation separate from venous drainage of the head region and help to elucidate the precise function of the recurrent blood circulation.

B. Conclusions and Recommendations

All perfused gill preparations are hemodynamic experiments whether the experimental objectives are to investigate gas exchange, ionic regulation, or hemodynamics per se. Each investigator must be aware of the hemodynamic effects of the perfusion regime used on the function under investigation. For investigations of arterio-arterial gill hemodynamics, SVBPPs are recommended. We also believe that the arterio-venous circulation operates well in these preparations, but it is not accessible for sampling or measurement.

Necessary requirements for any preparation used to study gill hemodynamics are as follows, in order of priority:

1. Maintain physiological flow rates as well as afferent and efferent pressures of both arterio-arterial and arteriovenous circulations.
2. Maintain physiological perfusion parameters (frequency, stroke volume, pulse pressures); in particular the shape of the pressure pulse should mimic that of the heart.
3. Use blood proteins in the saline if blood is not available.
4. Maintain gill geometry and the normal relationship to physiological ventilatory water flow.

VI. GAS TRANSFER

The transfer of respiratory gases across the gills of fish has received a great deal of investigative interest over the years (see Randall, 1970; Piiper and Scheid, 1975; Jones and Randall, 1978; Randall *et al.*, 1982). Many of these studies have concentrated on obtaining *in vivo* measurements of blood and water gas tensions, from which subsequent predictions of unmeasured variables were made. In this sense, these experiments tend to be limited for the detailed study of gill gas transfer, because of the rather restricted number of variables that can be measured simultaneously. Technical difficulties often preclude the acquisition of direct measurements of gill perfusion and ventilation rates, in concert with blood and water gas tensions, thereby making it difficult to obtain direct *in vivo* rates of gas transfer across the gills. Many of these variables, however, have been measured as individual points, in numerous independent experiments. A further complication in using intact fish is that the degree of control that any investigator has over individual variables is most usually restricted to manipulating inspired water tension and its volume flow over the gills. These limitations have led researchers to use a variety of perfused preparations (see Section I), which have been useful not only in clarifying some aspects of gas exchange, but also for ion transfer and the nature of gill vascular perfusion patterns.

In the following section we will examine in detail the pros and cons of various perfused preparations with respect to their suitability for describing the mechanisms of gas transfer across gills *in vivo*, and discuss these in light of the possible problems inherent in using intact animals in studies of oxygen uptake and carbon dioxide excretion.

A. Oxygen Uptake

Metabolic rates, as measured by oxygen uptake in vivo, and oxygen transfer across the gills of fishes have been studied extensively (Randall *et al.*, 1967; Fry, 1971a; Randall, 1970; Brett, 1972; Jones and Randall, 1978). In an effort to understand the mechanisms of branchial oxygen transfer more thoroughly, various perfused preparations have been employed. However, not all have been successful for the purposes intended (see, in particular, Wood *et al.*, 1978; Daxboeck and Davie, 1982), in that no significant degree of O_2 uptake was measurable across the gills. In these preparations (see Fig. 5), problems with the length of time for completion of surgery and the stress of the operations themselves, the low oxygen content of the saline solutions, and the serious implications of edema formation (see Section III) on the thickness of the oxygen diffusion barrier may have complicated oxygen transfer studies.

To reduce operating time an isolated head preparation, developed by Payan and Matty (1975), has been used to examine specifically the detailed mechanisms of oxygen transfer across gills (Part *et al.*, 1982a; Pettersson and Johansen, 1982; Pettersson, 1983; Perry and Daxboeck, 1984; Perry *et al.*, 1984b). Most of these investigators have used catecholamines in the perfusate of their preparations and have looked for changes in the diffusion–perfusion limitations for O_2 uptake. As stated earlier, however (Sections IV and V), one criticism of these isolated head preparations is that they lack any dorsal aortic back pressure (exceptions are the studies of Pettersson and Johansen, 1982 and Perry *et al.*, 1984c,d) and therefore are subject to the consequences of this omission on branchial vascular resistance and flow patterns (see Wood, 1974; Farrell *et al.*, 1979; Daxboeck and Davie, 1982). Despite having some representative systemic pressures in some nonisolated preparations of trout heads, they still did not yield significant O_2 uptake rates, and therefore these pressures would not appear to ameliorate the problems inherent in these particular instances (see Wood *et al.*, 1978; Daxboeck and Davie, 1982). In addition, these preparations, as well as the isolated head (Payan and Matty, 1975), may suffer from progressive deterioration with time, as indicated by continual increase in gill vascular resistance.

Improving the Viability of the Isolated Head Preparation

If one is to investigate O_2 uptake ($\dot{M}_{O_2}$) in isolated, saline-perfused head preparations, and assuming the recommendations for saline composition are followed (see Section II), then it may be necessary to increase the oxygen content of the external medium so as to meet the demands of gill tissue metabolism because of the low O_2-carrying capacity of the perfusate (see Johansen and Pettersson, 1981; Daxboeck *et al.*, 1982). In this way, net O_2 uptake across the gills is more easily measurable, and the preparation remains stable longer than if normoxic water is used to irrigate the gills (Perry and Daxboeck, 1984; see Fig. 21). However, the addition of moderate levels of any catecholamines, known to be present and vasoactive in fishes (e.g., adrenaline or noradrenaline, 10^{-7} *M*), is also recommended as a general procedure in order to increase the viability of the preparations. Note in particular in Table VII that this level of noradrenaline is useful in maintain-

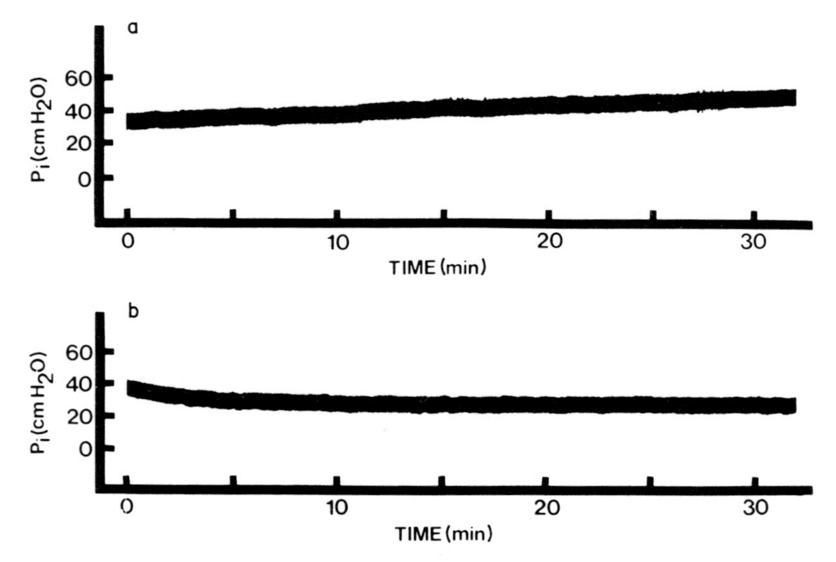

Fig. 21. Afferent perfusion pressure (P_i) in the isolated, saline-perfused head of rainbow trout irrigated with (a) normoxic water (P_{O_2} = 140 torr) or (b) hyperoxic water (P_{O_2} = 250 torr). (From Perry and Daxboeck, 1984.)

Table VII

Effect of External Hyperoxia on Oxygen Transfer, Flow Distribution, and Perfusion Pressure (P_i) in the Isolated, Saline-Perfused Head of Rainbow Trout[a,b]

Time (min)	$\dot{Q}$ (ml min^{-1})	$\dot{Q}_{DA}$ (ml min^{-1})	$\dot{Q}_{AV}$ (ml min^{-1})	$P_{A_{O_2}}$ (torr)	$P_{A_{O_2}} - P_{V_{O_2}}$ (torr)	$\dot{M}_{O_2}$ (μM min^{-1} 100 g^{-1})	P_i (cm H_2O)
Normoxic controls (10^{-7} M noradrenaline, $n = 6$)							
0	3.92 ± 0.06	3.45 ± 0.04	0.47 ± 0.07	102.8 ± 4.8	65.7 ± 3.8	0.23 ± 0.02	34.0 ± 2.2
10	3.85 ± 0.07	3.49 ± 0.07	0.36 ± 0.04	101.5 ± 5.2	63.7 ± 4.4	0.22 ± 0.02	32.5 ± 2.0
20	3.84 ± 0.07	3.45 ± 0.07	0.39 ± 0.05	99.0 ± 5.9	61.5 ± 5.0	0.21 ± 0.02	34.0 ± 2.2
Hyperoxic controls (no hormone, $n = 9$)							
0	3.83 ± 0.16	3.41 ± 0.16	0.42 ± 0.10	192.3 ± 8.0	152.7 ± 8.1	0.48 ± 0.03	31.7 ± 1.5
5	3.87 ± 0.08	3.38 ± 0.17	0.49 ± 0.10	193.6 ± 8.2	153.6 ± 8.4	0.48 ± 0.04	29.9 ± 1.5
10	3.91 ± 0.05	3.40 ± 0.17	0.51 ± 0.12	190.7 ± 9.9	148.4 ± 10.1	0.48 ± 0.04	30.1 ± 1.7
15	3.91 ± 0.06	3.38 ± 0.18	0.53 ± 0.10	183.0 ± 12.1	141.3 ± 12.6	0.45 ± 0.05	30.6 ± 2.0
20	3.89 ± 0.06	3.34 ± 0.19	0.55 ± 0.11	177.7 ± 13.6	137.2 ± 14.3	0.44 ± 0.05	31.3 ± 2.2

[a] Data from Perry and Daxboeck (1984).

[b] Means ± SE.

Table VIII

Effect of Perfusion Flow Rate ($\dot{Q}$) on Oxygen Transfer in the Isolated, Saline-Perfused Head of Rainbow Trout[a,b]

Flow condition[c]	Time (min)	$\dot{Q}$ (ml min^{-1})	$\dot{Q}_{DA}$ (ml min^{-1})	$P_{A_{O_2}}$ (torr)	$P_{A_{O_2}} - P_{V_{O_2}}$ (torr)	$\dot{M}_{O_2}$ (μ*M* min^{-1} 100 g^{-1})	P_i (cm H_2O)
Normal $\dot{Q}$	0	3.84 ± 0.06	3.34 ± 0.08	199.1 ± 6.3	160.0 ± 6.3	0.50 ± 0.06	39.0 ± 1.7
High $\dot{Q}$	7.5	5.59 ± 0.10	4.81 ± 0.20	174.0 ± 6.0	131.8 ± 5.3	0.59 ± 0.05	41.0 ± 2.5
Subnormal $\dot{Q}$	15.0	2.20 ± 0.10	1.91 ± 0.10	184.3 ± 9.6	143.0 ± 8.7	0.26 ± 0.04	29.2 ± 1.6

[a] Data from Perry *et al.* (1984).
[b] $n = 6$; means ± SE. See also Fig. 22.
[c] See text for details.

ing constant input pressure (P_i) and significant rates of $\dot{M}_{O_2}$, even in normoxic water. The extraction effectiveness for oxygen in these instances is quite adequate (70–80%).

If these preparations are to be used to test specific effects of catecholamines on O_2 transfer, one should be aware that any effects recorded may be difficult to interpret, since they could be due to hemodynamic changes in the flow distribution pattern within the branchial vasculature (see Sections IV and V), as well as changes in gill tissue permeability (see Chapter 1, this volume). Nonetheless, various investigators have examined the effects of catecholamines or perfusion flow rate changes on the perfusion–diffusion limitations to oxygen transfer in isolated head preparations (Part *et al.*, 1982a; Pettersson and Johansen, 1982; Pettersson, 1983; Perry *et al.*, 1984b). Simply by increasing the perfusate flow rate by 1.5 times normal, the transfer of O_2 across the gills, as measured by arterial–venous (input–output) O_2 differences, is decreased (Fig. 22). When flow is then decreased to a level below normal, the arterial–venous difference is increased from that recorded at the high flow rate (Table VIII and Fig. 22). Clearly,

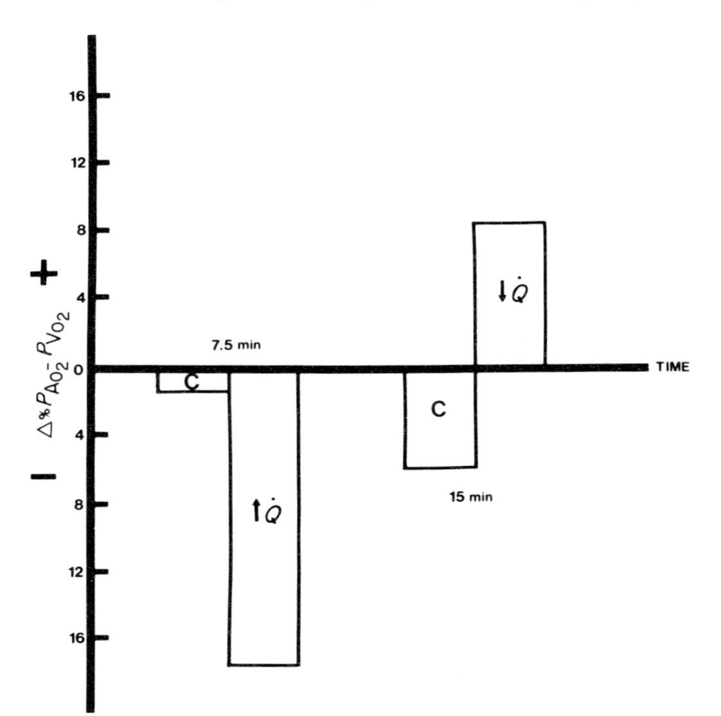

Fig. 22. Effect of perfusion flow rate ($\dot{Q}$) on oxygen transfer in the isolated, saline-perfused head of rainbow trout (n = 6). C, control; Δ% $P_{A_{O_2}} - P_{V_{O_2}}$, = percentage change of the P_{O_2} difference across the gills. (From Perry *et al.*, 1984b.)

this indicates that some diffusion limitations for O_2 transfer exist in the perfused gill circulation, and as such, conflicts with the conclusion reached by Part (1982) that there are no diffusion limitations. However, the data must be examined more closely if we want to make the preceding assertions with certainty. In the preparations of Part (1982), 10^{-6} *M* adrenaline was added to the perfusate. It is clear from Tables IX and X and Fig. 23 that there is a dose-dependent response to adrenaline in the isolated, saline-perfused head, as measured by its stimulatory effect on O_2 transfer. This is evident not only with "normal" perfusion flow rate, but also with increased perfusion. Therefore, it would appear that by using 10^{-6} *M* adrenaline, the preparation is rendered completely perfusion limited and the gills allow maximal transfer of O_2 into the perfusate at all flow rates used.

For studies of oxygen transfer, it may also be beneficial to increase the oxygen content of the irrigation water or the perfusion media (other than blood). A simple method is to use higher levels of O_2 than that in ordinary air to equilibrate the water. C. Daxboeck and S. F. Perry (unpublished data) have tried free mammalian hemoglobin in Cortland saline solution as an alternate method of increasing the carrying capacity of the perfusate, in the absence of a readily available source of large quantities of fish blood. In these experiments the perfusate O_2 content was increased dramatically from that of simple saline, but there were such serious and irreversible vascular effects (e.g., increased gill vascular resistance to flow to the point of rupture) that this avenue of approach is of no use. Another way in which the O_2-carrying capacity of the perfusate can be increased is by the use of fluorocarbons, which have been used as a breathing medium in liquid-filled mammalian lung experiments (Matthews and Kylstra, 1978). These may be useful in perfused gills as well, but to date their vascular effects in fish gills have not been tested. A third possibility is the use of whole blood, collected from donor fish, as the perfusion medium (Perry *et al.*, 1984c). However, the volume of blood needed in isolated head preparations might make this method impractical in some circumstances. Alternatively, other sources of whole blood or plasma (e.g., mammalian) that are more readily available in large quantities have been used in isolated gill arch studies (see Section III).

In a study of gas transfer across eel gills (*Anguilla australis;* Ellis and Smith, 1983), it has been shown that O_2 transfer is greatly improved with mammalian plasma-perfused gills alone, indicated by a maximum extraction ratio of O_2 of 72%, a ratio far higher than that achieved in saline-perfused trout arches. It appears that by preventing edema formation and hence the increase in blood–water barrier thickness, plasma perfusion may be useful for studies of diffusional O_2 exchange across fish gills. Problems do arise, nonetheless, with the use of isolated branchial arches for studying O_2 transfer because of inadequate duplication of *in vivo* ventilatory water flow dy-

Table IX

Effect of Adrenaline on Oxygen Transfer, Flow Distribution, and Perfusion Pressure (P_i) at Normal Perfusion Flow Rate ($\dot{Q}$) in the Isolated, Saline-Perfused Trout Head[a]

Time (min)	$\dot{Q}_{tot}$ (ml min^{-1})	$\dot{Q}_{DA}$ (ml min^{-1})	$\dot{Q}_{AV}$ (ml min^{-1})	P_{AO_2} (torr)	$P_{AO_2}-P_{VO_2}$ (torr)	$\dot{M}_{O_2}$ (μ*M* min^{-1} 100 g^{-1})	P_i (cm H_2O)
Control (*n* = 9)							
0	3.83 ± 0.10	3.41 ± 0.16	0.42 ± 0.10	192.3 ± 8.0	152.7 ± 8.1	0.48 ± 0.03	31.7 ± 1.5
5	3.87 ± 0.08	3.38 ± 0.17	0.49 ± 0.10	193.6 ± 8.2	153.6 ± 8.4	0.48 ± 0.04	29.9 ± 1.5
10	3.91 ± 0.05	3.40 ± 0.17	0.51 ± 0.12	190.7 ± 9.9	148.4 ± 10.1	0.48 ± 0.04	30.1 ± 1.7
15	3.91 ± 0.06	3.38 ± 0.18	0.53 ± 0.10	183.0 ± 12.1	141.3 ± 12.6	0.45 ± 0.05	30.6 ± 2.0
20	3.89 ± 0.06	3.34 ± 0.19	0.55 ± 0.11	177.7 ± 13.6	137.2 ± 14.3	0.44 ± 0.05	31.3 ± 2.2
10^{-7} *M* Adrenaline (*n* = 6)							
Before adrenaline	3.87 ± 0.04	3.51 ± 0.08	0.36 ± 0.08	191.8 ± 3.5	145.2 ± 3.8	0.59 ± 0.03	36.4 ± 3.7
10	3.87 ± 0.06	3.52 ± 0.08	0.35 ± 0.07	194.8 ± 4.7	152.5 ± 4.6	0.62 ± 0.05	33.3 ± 3.2
15	3.84 ± 0.04	3.46 ± 0.05	0.38 ± 0.06	189.8 ± 6.2	147.7 ± 6.3	0.60 ± 0.05	32.8 ± 3.3
20	3.85 ± 0.04	3.49 ± 0.07	0.36 ± 0.06	186.8 ± 7.5	144.3 ± 7.6	0.58 ± 0.05	32.6 ± 3.3
10^{-6} *M* Adrenaline (*n* = 6)							
Before adrenaline	3.80 ± 0.04	2.77 ± 0.20	1.03 ± 0.19	166.3 ± 5.7	128.5 ± 5.6	0.44 ± 0.05	39.5 ± 2.4
10	3.84 ± 0.07	3.07 ± 0.18	0.77 ± 0.19	175.5 ± 4.6	138.2 ± 4.0	0.48 ± 0.05	38.3 ± 3.3
15	3.86 ± 0.04	3.16 ± 0.16	0.70 ± 0.17	181.3 ± 4.8	144.3 ± 3.0	0.49 ± 0.04	35.4 ± 3.2
20	3.85 ± 0.06	3.15 ± 0.14	0.70 ± 0.14	179.0 ± 5.3	144.2 ± 2.6	0.49 ± 0.14	34.6 ± 3.0

[a]Data from Perry *et al.* (1984).

Table X

Effect of Adrenaline on Oxygen Transfer, Flow Distribution, and Perfusion Pressure (P_i) at Increased Perfusion Flow Rate ($\dot{Q}$) in the Isolated, Saline-Perfused Trout Head[a,b]

Time (min)	$\dot{Q}$ (ml min^{-1})	$\dot{Q}_{DA}$ (ml min^{-1})	$\dot{Q}_{AV}$ (ml min^{-1})	$P_{A_{O_2}}$ (torr)	$P_{A_{O_2}} - P_{V_{O_2}}$ (torr)	$\dot{M}_{O_2}$ (μM min^{-1} 100 g^{-1})	P_i (cm H_2O)
Control ($n = 5$)							
0	5.77 ± 0.10	5.03 ± 0.21	0.74 ± 0.18	183.6 ± 10.6	146.2 ± 10.9	0.68 ± 0.05	21.8 ± 2.1
10	5.81 ± 0.11	4.84 ± 0.23	0.97 ± 0.19	167.0 ± 16.8	129.8 ± 16.8	0.61 ± 0.08	21.2 ± 2.4
15	5.87 ± 0.13	4.88 ± 0.23	0.99 ± 0.19	155.4 ± 17.2	117.6 ± 16.1	0.56 ± 0.08	21.2 ± 2.2
20	5.84 ± 0.11	4.77 ± 0.24	1.07 ± 0.20	144.0 ± 17.0	106.2 ± 15.5	0.50 ± 0.08	21.2 ± 2.3
10^{-7} *M* Adrenaline ($n = 5$)							
Before adrenaline	5.62 ± 0.13	4.94 ± 0.20	0.68 ± 0.19	184.2 ± 8.5	143.8 ± 8.0	0.78 ± 0.04	35.0 ± 3.3
10	5.58 ± 0.12	4.89 ± 0.24	0.69 ± 0.22	183.2 ± 10.3	141.6 ± 9.9	0.71 ± 0.05	35.6 ± 2.9
15	5.53 ± 0.10	4.83 ± 0.23	0.70 ± 0.21	182.6 ± 11.2	141.8 ± 10.7	0.71 ± 0.05	35.6 ± 3.4
20	5.50 ± 0.12	4.97 ± 0.16	0.53 ± 0.17	180.8 ± 7.8	140.4 ± 7.4	0.70 ± 0.03	35.8 ± 3.7
10^{-6} *M* Adrenaline ($n = 6$)							
Before adrenaline	5.82 ± 0.12	4.53 ± 0.25	1.29 ± 0.35	180.5 ± 5.7	140.2 ± 6.6	0.77 ± 0.09	22.6 ± 2.9
10	5.92 ± 0.09	5.21 ± 0.19	0.71 ± 0.16	195.7 ± 7.7	156.8 ± 7.8	0.88 ± 0.10	18.9 ± 2.0
15	5.93 ± 0.12	5.21 ± 0.18	0.72 ± 0.16	200.2 ± 6.5	161.5 ± 6.2	0.81 ± 0.10	18.7 ± 2.1
20	5.91 ± 0.11	5.13 ± 0.20	0.78 ± 0.17	202.8 ± 5.6	163.0 ± 6.2	0.91 ± 0.10	18.6 ± 2.0

[a] Data from Perry *et al.* (1984).
[b] Means ± SE.

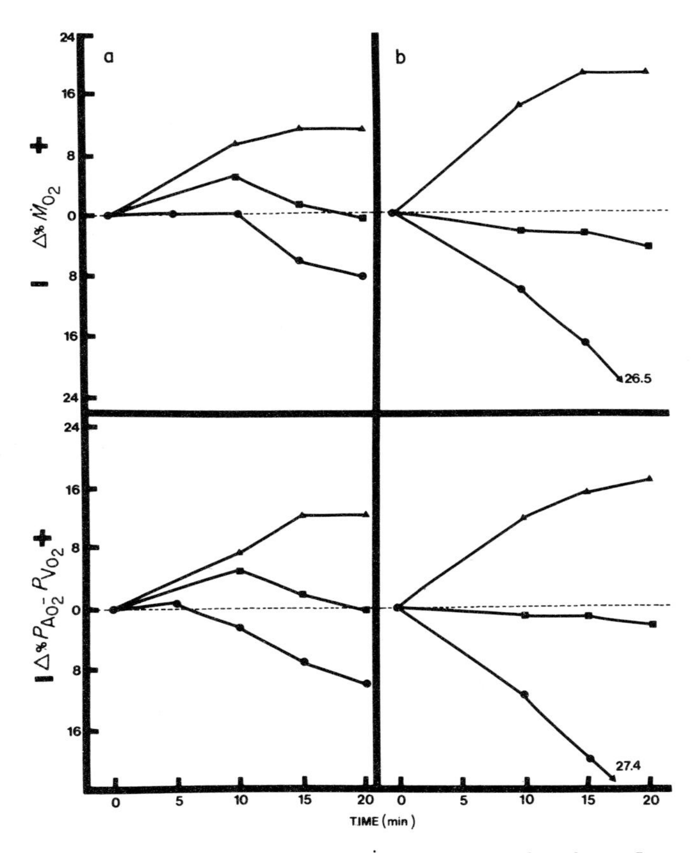

Fig. 23. Effect of adrenaline on O_2 transfer ($\dot{M}_{O_2}$) at (a) normal perfusion flow rate (4.0 ml min^{-1}) and (b) increased perfusion flow rate (5.8 ml min^{-1}) in the isolated, saline-perfused head of rainbow trout. $\Delta\%\ P_{A_{O_2}} - P_{V_{O_2}}$, percentage change of the P_{O_2} difference across the gills; $\Delta\%\dot{M}_{O_2}$, percentage change of oxygen uptake; (●----●), control (no adrenaline, $n = 9$); (■----■), 10^{-7} *M* adrenaline ($n = 6$); (▲----▲) 10^{-6} *M* adrenaline ($n = 6$). (From Perry *et al.*, 1984b.)

namics and gill arch orientation, leading to considerable diffusional dead space. Also, when dealing with single arches, one must consider that less than one-eighth of the total gill surface area of the fish is being used, and that of this, some of the circulation is ligated because of the nature of the preparation procedure (see Farrell *et al.*, 1979). Furthermore, as a result of the length of time taken to get each arch ready for experimentation, a considerable amount of mucus production by the gills will contribute even more dead space, further increasing diffusion limitations.

In an attempt to reconcile all of the foregoing criticisms of the various

Table XI

Summary of Variables for the Normal Resting State of the Spontaneously Ventilating Blood-Perfused Rainbow Trout at 7°C[a,b]

	Pressure (cm H_2O)		Cardiac pump		$\dot{Q}$ (ml min^{-1} 100 g^{-1})	Hct (%)	Acid–base balance		P_{O_2} (mm Hg)	C_{O_2} (mM)	P_{CO_2} (mm Hg)	C_{CO_2} (mM)	Plasma osmolarity (m Osmol)	[Cl−] (mM)
	Mean	Pulse	f (bpm)	SV (ml)			[H+] (nM)	pH						
							Input							
X	58.8	10.7	40	0.125	1.619	10.3	17.66	7.76	24.9	0.90	3.36	10.28	275.4	11.9
±SE	2.0	0.2	0	0.003	0.025	0.2	0.40		1.1	0.05	0.15	0.23	0.62	0.70
n	45	45	45	45	45	45	44		44	44	44	45	5 (pooled)	5
							Dorsal aorta							
X	34.8	2.05	—	—	—	8.80	19.72	7.72	103.4	1.58	3.66	9.02	265.7	115.8
±SE	0.95	0.13	—	—	—	0.2	0.71		2.4	0.08	0.16	0.20	0.30	0.60
n	45	45	—	—	—	42	44		44	44	44	45	5 (pooled)	5
							Venous return							
X	—	—	—	—	—	9.4	17.55	7.76	13.0	0.42	3.65	10.01	273.8	108.2
±SE	—	—	—	—	—	0.6	0.83		0.5	0.04	0.18	0.20	0.69	1.60
n	—	—	—	—	—	39	42		41	39	41	42	5 (pooled)	5

Breathing rate (f_g) no. min^{-1} = 69.4 ± 0.96 (n = 42).
Intrinsic heart rate, no. min^{-1} = 48.5 ± 0.99 (n = 37).

[a] Data from Davie *et al.* (1982).
[b] Experiments involved 15 fish: 306.2 ± 9.3g. n = number of observations.

types of isolated preparations, a spontaneously ventilating blood-perfused trout preparation has been developed to study, among other things, oxygen transfer across gills (Davie *et al.*, 1982; see also Fig. 19). This particular preparation and ones like it (see Metcalfe and Butler, 1982) are perfused by whole blood (usually from donor fish) via a cardiac action simulator pump, have in vivo systemic pressures and resistances to blood flow, can irrigate their gills by spontaneous rhythmic breathing motion, and have their full complement of eight gill arches intact. Its advantages for the study of gill physiology have been detailed elsewhere (see Section V). Tables XI and XII illustrate the ability of the SVBPTP to display in vivo values of blood gas tensions as well as oxygen uptake and carbon dioxide excretion. However, the disadvantages are that the surgery is lengthy and quite involved, and when the preparation is ready for use, it is likely to be highly stressed and therefore to contain high levels of circulating catecholamines. A viable alternative for short-term experiments is the blood-perfused trout head preparation (Perry *et al.*, 1984c).

B. Carbon Dioxide Excretion

Unlike oxygen uptake studies, very few studies of CO_2 excretion in fish have involved perfused preparations. This is because at physiological P_{CO_2}

Table XII

Summary of the Vascular Resistances to Flow (R_g), and Gas Transfer across the Branchial (Input–Dorsal Aorta) and Systemic (Dorsal Aorta–Venous Return) Circulation in the Resting State of Spontaneously Ventilating Blood-Perfused Rainbow Trout, at 7°C[a,b]

	R_g (cm H_2O ml^{-1} min^{-1} 100 g^{-1})	M_{gO_2} (μmol min^{-1} 100 g^{-1})	M_{gCO_2} (μmol min^{-1} 100 g^{-1})	RE_g (M_{gCO_2}/M_{gO_2}
X	14.23	1.17	2.05	1.85
±SE	1.14	0.08	0.15	0.12
n	44	44	44	43
	R_s	M_{sO_2}	M_{sCO_2}	RQ_s
X	19.21	1.97	1.63	0.83
±SE	0.88	0.11	0.09	0.09
n	44	39	44	39

[a] Data from Davie *et al.* (1982).

[b] Experiments involved 15 fish: 306.2 ± 9.3 g. n = number of observations.

[c] R_g, R_s: vascular resistance to flow across branchial and systemic circulation, respectively. M_{gO_2}, M_{gCO_2}, M_{sO_2}, M_{sCO_2}: gas transfer (O_2 and CO_2) across branchial and systemic circulation, respectively.

and pH, and in the absence of red blood cells, CO_2 excretion is not measurable in perfused preparations (Perry *et al.*, 1982). Only when perfusate P_{CO_2} is elevated to 1 or 2% do saline-perfused preparations display significant CO_2 excretion (Haswell and Randall, 1978; Haswell *et al.*, 1978; Perry *et al.*, 1982; see Table XIII).

It is now apparent that in order to examine CO_2 excretion in perfused preparations, blood must be used as the perfusion medium. In only two studies of CO_2 excretion (Perry *et al.*, 1982, 1984c) has blood been employed as a perfusate. Perry *et al.* (1982) noted a positive linear relationship between CO_2 excretion and hematocrit. These data, together with the results of in vitro (Cameron, 1978; Obaid *et al.*, 1979; Heming and Randall, 1982)

Table XIII

Effects of Perfusate P_{CO_2} and Bicarbonate (HCO_3^-) Concentration on CO_2 Excretion ($\dot{M}_{CO_2}$) of Various Saline-Perfused Preparations[a]

Treatment	Afferent saline or input	Efferent saline or dorsal aortic	Δ	Δ (%)
Isolated, saline-perfused trout holobranchs				
Normal P_{CO_2}				
C_{CO_2} (m*M*)	8.71	8.88	0.17	2.0
P_{CO_2} (mm Hg)	4.01	4.76	0.75	18.7
High P_{CO_2}				
C_{CO_2} (m*M*)	10.82	10.50	−0.32[b]	−2.6
P_{CO_2} (mm Hg)	15.40	13.14	−2.26[b]	−14.7
Totally saline-perfused trout				
Normal P_{CO_2}				
C_{CO_2} (m*M*)	10.57	11.61	1.04	9.8
P_{CO_2} (mm Hg)	3.14	4.80	1.66	52.9
High P_{CO_2}				
C_{CO_2} (m*M*)	11.00	10.46	−0.54[b]	−4.9
P_{CO_2} (mm Hg)	9.16	9.88	0.72[b]	7.9
Totally saline-perfused coho salmon (marine)				
Normal HCO_3^-				
C_{CO_2} (m*M*)	11.86	11.72	−0.14	−1.2
High HCO_3^-				
C_{CO_2} (m*M*)	33.15	33.30	0.15	0.5

[a]From Perry *et al.* (1982).
[b]Significantly different from normal.

and *in vivo* (Wood *et al.*, 1981) studies, clearly show the involvement of fish erythrocytes in CO_2 excretion (see Chapter 5, Volume XA, this series for a detailed discussion of CO_2 excretion in fish).

Earlier studies concerning the patterns of CO_2 elimination across fish gills have been conducted on intact fish. *In vivo* studies are limited and difficult to interpret for the following reasons:

1. For technical reasons, it is difficult to measure inspired and expired water CO_2 tensions accurately. An alternative is to measure arterial and venous CO_2 contents so as to give an estimate of CO_2 excretion (unless cardiac output is known, $\dot{M}_{CO_2}$ can only be estimated).
2. Occasionally it is necessary to examine the effects of drugs or to alter blood chemistry. This is not easily accomplished *in vivo* for reasons discussed earlier (Section IV).
3. Results can be misinterpreted easily because of effects of any particular treatment on the cardiovascular system. For example, it is known that severe anemia is associated with increased cardiac output (Cameron and Davis, 1970; Wood *et al.*, 1979; Wood and Shelton, 1980b). This response, by increasing the delivery of physically dissolved CO_2 to the gills, would tend to maintain net CO_2 excretion rates near normal under these conditions, and thereby mask any effect of decreased red blood cell number on this excretion rate. This may explain the results of Haswell and Randall (1978) of severely anemic fish still maintaining CO_2 excretion.

C. Conclusions and Recommendations

The primary problems associated with saline-perfused preparations for studying branchial gas transfer are edema formation, low O_2-carrying capacity, and low buffering capacity of the perfusate. While these problems may be partially overcome by performing short-term experiments, by adding catecholamines, or by using hyperoxic water, we advise that future studies employ blood as the perfusion medium. Clearly this is a necessity when examining CO_2 excretion. The isolated head is easily and quickly prepared and may be best suited for short-term (1 hr or less) experiments. We also recommend that physiological dorsal aortic pressure be used.

For long-term experiments, in which most cardiorespiratory variables can be controlled and manipulated, a spontaneously ventilating blood-perfused preparation is ideal. With care and further refinement of techniques, this type of preparation is to be highly encouraged for its use in studying oxygen uptake and carbon dioxide excretion in fish.

VII. GENERAL CONCLUSIONS AND RECOMMENDATIONS

In this chapter we have discussed the use of perfused preparations in studies of gill physiology. We have stressed the importance of simulating in vivo conditions in terms of perfusate composition, irrigation and perfusion flow patterns, and gill geometry. It is clear that no one preparation is capable of meeting all of the criteria necessary to be considered ideal for investigations of all three of the major areas of gill physiology we have discussed: ion exchange, hemodynamics, and gas transfer. The type of preparation selected will depend primarily on the nature of the investigation. Here is an assessment of the three most commonly used perfused preparations.

A. Perfused Branchial Arches

The disadvantages of using this preparation include (1) its extreme leakiness. (2) problems of irrigation leading to diffusional dead spaces and boundary layers, (3) lengthy operating procedures leading to extended periods of branchial ischemia, (4) excessive mucus accumulation, (5) disturbance of the normal gill geometry by removal from the branchial basket, and (6) the capacity to perfuse less than one-eighth of the total gill surface area of the fish. In our opinion, these disadvantages make it undesirable as a tool for studying gill physiology. Clearly it is unacceptable for gas transfer experiments, and its worthiness in studies of ion exchange is apparently species dependent. For example, the perfused eel or flounder arch is vastly superior to the perfused trout arch. Perhaps these particular preparations should be limited to studies of transepithelial potential, which remain stable for as long as 8 hr. Although perfused branchial arches have been useful in hemodynamic studies, these can be performed equally well in perfused heads or totally perfused preparations. Where isolated arches still might prove useful is in the investigation of the hemodynamics of the recurrent circulation.

B. Isolated Perfused Heads

The major disadvantages of this preparation include (1) its progressive deterioration with time, making it suitable only for short-term experiments, and (2) low flux rates of ions and O_2 in the absence of catecholamines. The primary advantages include (1) simple and rapid surgery with no need of anesthetic, (2) non-leakiness with or without dorsal aortic pressure, and (3) partitioning of efferent flow into arterial and venous components. This preparation probably is best suited for ion and gas transfer studies of short duration (no longer than 60 min). Future studies using the isolated head should incorporate the use of plasma or blood as the perfusion medium, if possible. This procedure may solve the problems of edema formation and low O_2-carrying capacity. Also, we suggest imposing physiological levels of dorsal aortic pressure. We do not advocate using high levels of catecholamines in the perfusate, as practiced earlier, but do recommend using 10^{-7} or 10^{-8} *M* adrenaline routinely, to help solve the problems of progressive deterioration.

C. Totally Perfused Preparations

Totally perfused preparations are well suited for hemodynamic studies, because the integrity of the gill vasculature is kept intact. The continued use of totally saline-perfused fish is to be discouraged, however, because of the structural damage saline perfusion is known to cause. Instead, plasma (fish or mammalian) or whole-blood (fish only) perfusion is recommended. By using whole blood it is possible to create a spontaneously ventilating preparation. The spontaneously ventilating blood-perfused preparation is the closest approximation of in vivo conditions that has been achieved in a perfused preparation. The disadvantages are that (1) it is highly stressed and (2) it is both difficult and time-consuming to prepare. Despite these disadvantages, its continued use is highly encouraged.

Although gill perfusion studies have been very important in elucidating mechanisms of piscine ion regulation, gas transfer, and hemodynamics, clearly they cannot replace whole-animal investigations. If a study can be performed equally well in vivo, then this is the obvious choice for investigation. When it does become necessary to employ a perfused preparation, one must use extreme caution in interpreting results and not be overly hasty in assuming that similar mechanisms operate under normal conditions in intact fish.

ACKNOWLEDGMENTS

We wish to thank D. J. Randall and R. Boutilier for critically reading the manuscript. We are grateful to D. Lauren and C. Booth for providing assistance in histological preparation of perfused trout gills.

REFERENCES

Bateman, J. B., and Keys, A. (1932a). Chloride and vapour pressure relations in the secretory activity of the gills of the eel. *J. Physiol. (London)* **75**, 226–240.

Bateman, J. B., and Keys, A. (1932b). Respiration of isolated gill tissue of the eel. *J. Physiol. (London)* **77**(3), 271.

Bergman, H. L., Olson, K. R., and Fromm, P. O. (1974). The effects of vasoactive agents on the functional surface area of isolated perfused gills of rainbow trout. *J. Comp. Physiol.* **94,** 267–286.

Bolis, L., and Rankin, R. C. (1980). Interactions between vascular actions of detergents and catecholamines in perfused gills of European eel, *Anguilla anguilla* L. and brown trout, *Salmo trutta* L. *J. Fish Biol.* **16,** 61–73.

Brett, J. R. (1972). The metabolic demand for oxygen in fish, particularly salmonids, and a comparison with other vertebrates. *Respir. Physiol.* **14,** 151.

Burns, J., and Copeland, P. E. (1950). Chloride excretion in the head region of *Fundulus heteroclitus. Biol. Bull. (Woods Hole, Mass.)* **99,** 381–385.

Cameron, J. N. (1978). Chloride shift in fish blood. *J. Exp. Zool.* **206,** 285–289.

Cameron, J. N., and Davis, J. C. (1970). Gas exchange in rainbow trout with varying blood oxygen capacity. *J. Fish. Res. Board Can.* **27,** 1069–1085.

Capra, M. F., and Satchell, G. H. (1974). Beta adrenergic dilatory responses in isolated saline-perfused arteries of an elasmobranch fish, *Squalus acanthias. Experientia* **30,** 927–928.

Capra, M. F., and Satchell, G. M. (1977). The adrenergic responses of isolated saline-perfused prebranchial arteries and gills of the elasmobranch *Squalus acanthias. Gen. Pharmacol.* **8,** 67–71.

Claiborne, J. B., and Evans, D. H. (1980). The isolated, perfused head of the marine telost fish, *Myoxocephalus octodecimspinosus:* Hemodynamic effects of epinephrine. *J. Comp. Physiol.* **138,** 79–85.

Claiborne, J. B., and Evans, D. H. (1981). The effect of perfusion and irrigation flow rate variations on NaCl efflux from the isolated, perfused head of the marine teleost, *Myoxocephalus octodecimspinosus. Mar. Biol. Lett.* **2,** 123–130.

Colin, D., and Leray, C. (1977). Réponses Hemodynamiques de la branchie à l'adenosine chez le truitte (*Salmo gairdneri*). Etude sur tête perfuseé. *C. Hebd. Seances R. Acad. Sci.* **284,** 1191–1194.

Colin, D., Kirsch, R., and Leray, C. (1979). Hemodynamic effects of adenosine on gills of the trout (*Salmo gairdneri*). *J. Comp. Physiol.* **130,** 325–330.

Danielli, J. F. (1940). Capillary permeability and edema in the perfused frog. *J. Physiol. (London)* **98,** 109–129.

Davie, P. S. (1981). Vascular resistance responses of an eel tail preparation. Alpha constriction and beta dilation. *J. Exp. Biol.* **90,** 65–84.

Davie, P. S., and Daxboeck, C. (1982). Effect of pulse pressure on fluid exchange between blood and tissues in trout gills. *Can. J. Zool.* **60,** 1000–1006.

Davie, P. S., and Forster, M. E. (1980). Cardiovascular responses to swimming in eels. *Comp. Biochem. Physiol.* **A 67A,** 367–373.

Davie, P. S., Daxboeck, C., Perry, S. F., and Randall, D. J. (1982). Gas transfer in a spontaneously ventilating, blood-perfused trout preparation. *J. Exp. Biol.* **101,** 17–34.

Daxboeck, C., and Davie, P. S. (1982). Effects of pulsatile perfusion on flow distribution within an isolated, saline-perfused trout head preparation. *Can. J. Zool.* **60,** 994–999.

Daxboeck, C., Davie, P. S., Perry, S. F., and Randall, D. J. (1982). Oxygen uptake in a spontaneously ventilating, blood-perfused trout preparation. *J. Exp. Biol.* **101,** 35–46.

Degnan, K. J., and Zadunaisky, J. A. (1980a). Ionic contributions to the potential and current across the opercular epithelium. *Am. J. Physiol.* **238,** 231–239.

Degnan, K. J., and Zadunaisky, J. A. (1980b). Passive sodium movements across the opercular epithelium: The paracellular shunt pathway and ionic conductance. *J. Membr. Biol.* **55,** 175–185.

Degnan, K. J., Karnaky, K. J., and Zadunaisky, J. A. (1977). Active chloride transport in the *in vitro* opercular skin of a teleost (*Fundulus heteroclitus*), a gill-like epithelium rich in chloride cells. *J. Physiol. (London)* **271**, 155–191.

DeLanger, C. D. (1958). Blood pressure, pulse pressure and tissue flow. *Cardiologia* **33**, 249.

Dow, P., and Hamilton, W. F., eds. (1963). "The Handbook of Physiology," Sec. 2, Vol. II, pp. 994–995. Am. Physiol. Soc., Washington, D.C.

Drinker, C. K. (1927). The permeability and diameter of the capillaries in the web of the brown frog (R. *temporaria*) when perfused with solutions containing pituitary extract and horse serum. *J. Physiol. (London)* **63**, 249–269.

Ellis, A. G., and Smith, D. G. (1983). Oedema formation and impaired O_2 transfer in Ringer-perfused gills of the eel, *Anguilla australis. J. Exp. Zool.* **227**, 371–380.

Evans, D. H., and Claiborne, J. B. (1982). Hemodynamics of the isolated, perfused head of the dogfish "pup." *Bull. Mt. Desert Isl. Biol. Lab.* **219**, 395–397.

Evans, D. H., Claiborne, J. B., Farmer, L., Mallery, C., and Krasny, E. J. (1982). Fish gill ionic transport: methods and models. *Biol. Bull.* **163**, 108–130.

Farmer, L. L., and Evans, D. H. (1981). Chloride extrusion in the isolated perfused teleost gill. *J. Comp. Physiol.* **141**, 471–476.

Farrell, A. P., Daxboeck, C., and Randall, D. J. (1979). The effect of input pressure and flow on the pattern and resistance to flow in the isolated perfused gill of a teleost fish. *J. Comp. Physiol.* **133**, 233–240.

Fenwick, J. C., and So, Y. P. (1974). A perfusion study of the effect of stanniectomy on the net influx of calcium-45 across an isolated eel gill. *J. Exp. Zool.* **188**, 125–131.

Forster, M. E. (1976a). Effects of catecholamines on the heart and on branchial and peripheral resistance of the eel (*Anguilla anguilla*). *Comp. Biochem. Physiol.* **55**, 27–32.

Forster, M. E. (1976b). Effects of adrenergic blocking drugs on the cardiovascular system of the eel, *A. anguilla. Comp. Biochem. Physiol.* **55**, 33–36.

Foskett, J. K., Turner, T., Logsdon, C., and Bern, H. (1979). Electrical correlates of chloride cell development in subopercular membrane of the tilapia *Sarotherodon mossambicus* transferred to seawater. *Am. Zool.* **19**, 995.

Fry, F. E. J. (1971). The effect of environmental factors on physiology. *In* "Fish Physiology" (W. S. Hoar and D. J. Randall, eds.), Vol. 6, pp. 1–98. Academic Press, New York.

Girard, J. P. (1976). Salt excretion by the perfused head of trout adapted to seawater and its inhibition by adrenaline. *J. Comp. Physiol.* **111**, 77–91.

Girard, J. P., and Payan, P. (1976). Effect of epinephrine on vascular space of gills and head of rainbow trout. *Am. J. Physiol.* **230**, 1555–1560.

Girard, J. P., and Payan, P. (1977a). Kinetic analysis and partitioning of sodium and chloride influxes across the gills of salt water adapted trout. *J. Physiol. (London)* **267**, 519–536.

Girard, J. P., and Payan, P. (1977b). Kinetic analysis of sodium and chloride influxes across the gills of the trout in freshwater. *J. Physiol. (London)* **273**, 195–209.

Girard, J. P., and Payan, P. (1980). Ion exchange through respiratory and chloride cells in freshwater and seawater adapted teleosteans. *Am. J. Physiol.* **238**, 260–268.

Goldstein, L., Claiborne, J. B., and Evans, D. H. (1982). Ammonia excretion by the gills of two marine teleost fish: The importance of NH_4^+ permeance. *J. Exp. Zool.* **219**, 395–397.

Hargens, A. R., Millard, P. W., and Johansen, K. (1974). High capillary permeability in fishes. *Comp. Biochem Physiol. A* **48A**, 675–680.

Haswell, M. S., and Randall, D. J. (1978). The pattern of carbon dioxide excretion in the rainbow trout, *Salmo gairdneri. J. Exp. Biol.* **72**, 17–22.

Haswell, M. S., Perry, S. F., and Randall, D. J. (1978). The effect of perfusate oxygen levels on CO_2 excretion in the perfused gill. *J. Exp. Zool.* **205**, 309–314.

Helgason, S., and Nilsson, S. (1973). Drug effects on post branchial blood pressure and heart rate in a free swimming marine teleost, *Gadus morhua*. *Acta Physiol. Scand.* **88,** 533–540.

Heming, T. A., and Randall, D. J. (1982). Fish erythrocytes are bicarbonate permeable: Problems with determining carbonic anhydrase activity using the modified boat technique. *J. Exp. Zool.* **219,** 125–128.

Hering, E. (1867). Uber den Bau der Birbeltierleber. *Arch. Mikrosk. Anat.* **3,** 88.

Holbert, P. W., Boland, E. J., and Olsen, K. R. (1979). The effect of epinephrine and acetylcholine on the distribution of red cells within the gills of the channel catfish (*Ictalurus punctatus*). *J. Exp. Biol.* **79,** 135–146.

Holeton, G., and Randall, D. J. (1967). The effect of hypoxia upon the partial pressure of gases in the blood and water afferent and efferent to the gills of rainbow trout. *J. Exp. Biol.* **46,** 317–327.

Houston, A. H., Madden, J. A., Woods, R. J., and Miles, H. M. (1971). Variations in the blood and tissue chemistry of brook trout (*Salvelinus fontinalis*), subsequent to handling, anesthesia and surgery. *J. Fish. Res. Board Can.* **28,** 635–642.

Hughes, G. M. (1966). The dimensions of fish gills in relation to their function. *J. Exp. Biol.* **45,** 177–195.

Isaia, J., Girard, J. P., and Payan, P. (1978). Kinetic study of gill epithelium permeability to water diffusion in the freshwater trout, *Salmo gairdneri*. *J. Membr. Biol.* **41,** 337–347.

Isaia, J., Payan, P., and Girard, J. P. (1979). A study of the water permeability of the gills of freshwater and seawater adapted trout (*Salmo gairdneri*): Mode of action of epinephrine. *Physiol. Zool.* **52,** 269–279.

Jackson, W. F., and Fromm, P. O. (1980). Effect of acute acid stress on isolated perfused gills of rainbow trout. *Comp. Biochem. Physiol.* **67,** 141–145.

Jackson, W. F., and Fromm, P. O. (1981). Factors affecting H_2O transfer capacity of isolated perfused trout gills. *Am. J. Physiol.* **240,** 235–245.

Johansen, K., and Pettersson, K. (1981). Gill O_2 consumption in a teleost fish, *Gadus morhua*. *Respir. Physiol.* **44,** 277–284.

Jones, D. R., and Randall, D. J. (1978). The respiratory and circulatory systems during exercise. *In* "Fish Physiology" (W. S. Hoar and D. J. Randall, eds.), Vol. 7, pp. 425–501. Academic Press, New York.

Jones, D. R., Langille, B. L., Randall, D. J., and Shelton, G. (1974). Blood flow in the dorsal and ventral aortae of the cod (*Gadus morhua*). *Am. J. Physiol.* **266,** 90–95.

Karnaky, K. J., Jr., and Kinter, W. B. (1977). Killifish opercular skin: A flat epithelium with a high density of chloride cells. *J. Exp. Zool.* **199,** 355–364.

Karnaky, K. J., Jr., Kinter, L. B., Kinter, W. B., and Stirling, C. E. (1976). Teleost chloride cell. II. Autoradiographic localization of Na^+ - K^+ - ATPase in killifish *Fundulus heteroclitis* adapted to low and high salinity environments. *J. Cell Biol.* **70,** 157–177.

Karnaky, K. J., Jr., Degnan, K. J., and Zadunaisky, J. A. (1977). Chloride transport across isolated opercular epithelium of killifish. A membrane rich in chloride cells. *Science* **195,** 203–205.

Kawasaki, R. (1980). Maintenance of respiratory rythmn-generation by vascular perfusion with physiological saline in the isolated head of the carp. *Jpn. J. Physiol.* **30,** 575–589.

Keys, A. B. (1931). The heart-gill preparation of the eel and its perfusion for the study of a natural membrane *in situ*. *Z. Vergl. Physiol.* **15,** 352–363.

Keys, A., and Bateman, J. B. (1932). Branchial responses to adrenaline and pitressin in the eel. *Biol. Bull.* (*Woods Hale, Mass.*) **63,** 327–336.

Keys, A., and Hill, R. M. (1933). The osmotic pressure of the colloids of fish sera. *J. Exp. Biol.* **11,** 28–34.

Kiceniuk, K. W., and Jones, D. R. (1977). The oxygen transport system in trout (*Salmo gairdneri*) during exercise. *J. Exp. Biol.* **69,** 247–261.

Kirschner, L. B. (1969). Ventral aortic pressure and sodium fluxes in perfused eel gills. *Am. J. Physiol.* **217,** 596–604.

Klaverkamp. J. F., and Dyer, D. C. (1974). Autonomic receptors in isolated rainbow trout vasculature. *Eur. J. Pharmacol.* **28,** 25–34.

Krakow, N. P. (1913). Uber die Wirkung von Giften auf die Gefasse isolierter Fischkiemen. *Pfluegers Arch. Gesamte Physiol. Menschen Tiere* **151,** 583–603.

Krasny, E. J., and Evans, D. H. (1980). Effects of catecholamines on active Cl^- secretion by the opercular epithelium of *Fundulus grandis. Physiologist* **23,** 63.

Krogh, A., and Harrop, G. A. (1921). On the substance responsible for capillary tonus. *J. Physiol. (London)* **54,** 125.

Laurent, P., and Dunel S. (1976). Functional organization of the teleost gill. I. Blood pathways. *Acta Zool. (Stockholm)* **57,** 189–209.

Maetz, J. (1971). Fish gills: Mechanisms of salt transfer in freshwater and seawater. *Philos. Trans. R. Soc. London, Ser. B* **262,** 209–251.

Marshall, W. S., and Bern, H. (1980). Ion transport across the isolated skin of the teleost, *Gillichthys mirabilis. In* "Epithelial Transport in the Lower Vertebrates" (B. Lahlou, ed.), pp. 337–350. Cambridge Univ. Press, London and New York.

Matthews, W. H., and Kylstra, J. A. (1978). Investigation of a new breathing liquid. *Underwater Physiol. Proc. Symp. Underwater Physiol., 6th 1975.*

Mayer-Gostan, N., and Maetz, J. (1980). Ionic exchange in the opercular membrane of *Fundulus heteroclitus* adapted to seawater. *In* "Epithelial Transport in the Lower Vertebrates" (B. Lahlou, ed.), pp. 233–248. Cambridge Univ. Press, London and New York.

Metcalfe, J. D. (1981). Branchial blood flow in the dogfish, *Scyliorhinus canicula.* Ph.D. Thesis, University of Birmingham, U.K.

Metcalfe, J. D., and Butler, P. (1982). Differences between directly measured and calculated values for cardiac output in the dogfish: A criticism of the Fick Method. *J. Exp. Biol.* **99,** 255–268.

Nilsson, S., and Pettersson, K. (1981). Sympathetic nervous activity of the Atlantic cod, *Gadus Morhua. J. Comp. Physiol.* **144,** 157–163.

Obaid, A. L., Critz, A. M., and Crandall, E. D. (1979). Kinetics of HCO_3^-/Cl^- exchanges in dogfish erythrocytes. *Am. J. Physiol.* **237,** 132–128.

Ostlund, E., and Fange, R. (1962). Vasodilation by adrenaline noradrenaline, and the effects of some other substances on perfused fish gills. *Comp. Biochem. Physiol.* **5,** 307–309.

Part, P., and Svanberg, O. (1981). Uptake of cadmium in perfused rainbow trout (*Salmo gairdneri*) gills. *Can. J. Fish. Aquat. Sci.* **38,** 917–924.

Part, P., Tuurala, H., and Soivio, A. (1982a). Oxygen transfer, gill resistance and structural changes in rainbow trout (*Salmo gairdneri*) gills perfused with vasoactive agents. *Comp. Biochem. Physiol. C* **71C,** 7–13.

Part, P., Kiessling, A., and Ring, O. (1982b). Adrenaline increase vascular resistance in perfused rainbow trout gills. *Comp. Biochem. Physiol. C* **72C,** 107–108.

Part, P. (1982). Symposium on gas exchange, gas transport and acid-base regulation in lower vertebrates. Max-Plank Institut fur Experimentelle Medizin, 1982. Abstract.

Pauling, J. E. (1968). A method of estimating the relative volumes of water flowing over different gills of a freshwater fish. *J. Exp. Biol.* **48,** 533–544.

Payan, P. (1978). A study of the Na^+/NH_4^+ exchange across the gill of the perfused head of the trout (*Salmo gairdneri*). *J. Comp. Physiol.* **124,** 181–188.

Payan, P., and Girard, J. P. (1977). Adrenergic receptors regulating patterns of blood flow through the gills of trout. *Am. J. Physiol.* **232,** 18–23.

Payan, P., and Matty, A. J. (1975). The characteristics of ammonia excretion by an isolated perfused head of trout (*Salmo gairdneri*): Effect of temperature and CO_2-free Ringer. *J. Comp. Physiol.* **96,** 167–184.

Payan, P., Matty, A. J., and Maetz, J. (1975). A study of the sodium pump in the perfused head preparation of the trout, *Salmo gairdneri,* in freshwater. *J. Comp. Physiol.* **104,** 33–48.

Perry, S. F. (1981). Carbon dioxide excretion and acid-base regulation in the rainbow trout (*Salmo gairdneri*): Involvement of the branchial epithelium and red blood cell. Ph.D. Thesis, University of British Columbia, Canada.

Perry, S. F., and Daxboeck, C. (1984). The isolated, saline-perfused trout head preparation: It's suitability for the study of oxygen transfer across the gill. (submitted for publication).

Perry, S. F., Davie, P. S. Daxboeck, C., and Randall, D. J. (1982). A comparison of CO_2 excretion in a spontaneously ventilating, blood-perfused trout preparation and saline-perfused gill preparations: Contribution of the branchial epithelium and red blood cell. *J. Exp. Biol.* **101,** 57–6O.

Perry, S. F., Payan, P., and Girard, J. P. (1983). Adrenergic control of branchial chloride transport in the isolated, saline-perfused head of the freshwater trout (*Salmo gairdneri*). *J. Comp. Physiol.* (in press).

Perry, S. F., Payan, P., and Girard, J. P. (1984a). The origin of HCO_3^- for gill Cl^-/HCO_3^- exchange is plasma CO_2, not plasma HCO_3^-. (submitted for publication).

Perry, S. F., Daxboeck, C., and Dobson, G. P. (1984b). The effect of perfusion flow rate and adrenergic stimulation on oxygen transfer in the isolated, saline-perfused head of rainbow trout (*Salmo gairdneri*). (submitted for publication).

Perry, S. F., Booth, C., and McDonald, D. G. (1984c). A comparison between saline and blood-perfusion in the isolated head of rainbow trout (*Salmo gairdneri*): I. Gas transfer, acid-base balance and haemodynamics. (submitted for publication).

Perry, S. F., Booth, C., and McDonald, D. G. (1984d). A comparison between saline and blood-perfusion in the isolated head of rainbow trout (*Salmo gairdneri*): II. Ionic exchange. (Submitted for publication).

Perry, S. F. Lauren, D., and Booth, C. (1984e). Lack of branchial edema in perfused heads of rainbow trout. (submitted for publication).

Pettersson, K. (1983). Adrenergic control of oxygen transfer in perfused gills of the cod, *Gadus morhua. J. Exp. Biol.* **102,** 327–335.

Pettersson, K., and Johansen, K. (1982). Hypoxic vasoconstriction and the effects of adrenaline on gas exchange efficiency in fish gills. *J. Exp. Biol.* **97,** 263–272.

Pettersson, K., and Nilsson, S. (1979). Nervous control of the branchial vascular resistance of the Atlantic cod, *Gadus morhua. J. Comp. Physiol.* **129,** 179–183.

Piiper, J., and Scheid, P. (1975). Gas transport efficacy of gills, lungs and skin: Theory and experimental data. *Respir. Physiol.* **23,** 209–221.

Randall, D. J. (1970). Gas exchange in fishes. *In* "Fish Physiology" (W. S. Hoar and D. J. Randall, eds.), Vol. 4, p. 253. Academic Press, New York.

Randall, D. J., and Smith, L. S. (1967). The effect of environmental factors on circulation and respiration in teleost fish. *Hydrobiologia* **29,** 113–124.

Randall, D. J., Holeton, G. F., and Stevens, E. D. (1967). The exchange of O_2 and CO_2 across the gills of rainbow trout. *J. Exp. Biol.* **46,** 339–348.

Randall, D. J., Baumgarten, D., and Malyusz, M. (1972). The relationship between gas and ion transfer across the gills of fishes. *Comp. Biochem. Physiol.* **41,** 629–637.

Randall, D. J., Perry, S. F., and Heming, T. A. (1982). Gas transfer and acid-base regulation in salmonids. *Comp. Biochem. Physiol. B* **73B,** 93–103.

Rankin, J. C., and Maetz, J. (1971). A perfused gill preparation: Vascular action of neurohypophysial hormones and cathecholamines. *J. Endocrinol.* **51,** 621–635.

Reite, O. B. (1969). The evolution of vascular smooth muscle responses to histamine and 5-hydroxytryptamine. I. Occurence of stimulatory actions in fish. *Acta Physiol. Scand.* **75,** 221–239.

Richards, B. D., and Fromm, P. O. (1969). Patterns of blood flow through filaments and lamellae of isolated perfused rainbow trout (*Salmo gairdneri*) gills. *Comp. Biochem. Physiol.* **29,** 1063–1070.

Richards, B. D., and Fromm, P. O. (1970). Sodium uptake by isolated perfused gills of rainbow trout (*Salmo gairdneri*). *Comp. Biochem. Physiol.* **33,** 303–310.

Ristori, M. T. (1970). Reflèxe de barosensibilité chez un posson téléostéen (*Cyprinus* carpio). *C. R. Seances Soc. Biol. Ses Fil.* **164,** 1512–1516.

Ristori, M. T., and Desseaux, G. (1970). Sur l'existence d'un gradient de sensibilité dans les recepteurs branchiaux de *Cyprinus carpio* L. *C. R. Seances Soc. Biol. Ses Fil.* **164,** 1517–1519.

Ristori, M. T., and Laurent, P. (1977). Action de l'hypoxie sur le systeme vasculaire branchial de la tête perfusée de truite. *C. R. Hebd. Seances Acad. Sci.* **171,** 809–813.

Saunders, R. L., and Sutterlin, A. M. (1971). Cardiac and respiratory responses to hypoxia in the sea raven, *Hemitripterus americanus*, and an investigation of possible control mechanisms. *J. Fish. Res. Board Can.* **28,** 491–503.

Schiffman, R. H. (1961). A perfusion study of the movement of strontium across the gills of rainbow trout (*Salmo gairdneri*). *Biol. Bull.* (*Woods Hole, Mass.*) **120,** 110–117.

Schlesser, I. H., and Freed, S. C. (1942). The effect of peptone on capillary permeability and its neutralization by adrenal cortical extract. *Am. J. Physiol.* **137,** 426–430.

Shuttleworth, T. J. (1972). A new isolated perfused gill preparation for the study of ionic regulation in teleosts. *Comp. Biochem. Physiol. A.* **43A,** 59–64.

Shuttleworth, T. J. (1978). The effect of adrenaline on potentials in the isolated gills of the flounder (*Platichthys flesus*). *J. Comp. Physiol.* **124,** 129–136.

Shuttleworth, T. J., and Freeman, R. F. H. (1974). Factors affecting the net fluxes of ions in the isolated perfused gills of freshwater *Anguilla dieffenbachii. J. Comp. Physiol.* **94,** 297–307.

Shuttleworth, T. J., Potts, W. T. W., and Harris, J. N. (1974). Bioelectric potentials in the gills of the flounder, *Platichthys flesus. J. Comp. Physiol.* **94,** 321–329.

Smith, D. G. (1977). Sites of cholinergic vasoconstriction in trout gills. *Am. J. Physiol.* **233,** 222–229.

Smith, D. G. (1978). Neural regulation of blood pressure in rainbow trout (*Salmo gairdneri*). *Can. J. Zool.* **56,** 1678–1683.

Smith, F. M., and Jones, D. R. (1978). Localisation of receptors causing hypoxic bradycardia in trout (*Salmo gairdneri*). *Can. J. Zool.* **56,** 1620–1625.

So, Y. P., and Fenwick, J. C. (1977). Relationship between net ^{45}Ca influx across a perfused isolated eel gill and the development of post-stanniectomy hypercalcemia. *J. Exp. Zool.* **200,** 259–264.

Stagg, R. M. and Shuttleworth, T. J. (1982). The effects of copper on ionic regulation by the gills of the seawater-adapted flounder *Platichthys flesus* L. *J. Comp. Physiol.* **149,** 83–90.

Starling, E. H. (1896). On the absorption of fluid from the connective tissue spaces *J. Physiol.* (*London*) **19,** 312–326.

Stevens, E. D., and Randall, D. J. (1967a). Changes in blood pressure, heart rate and breathing rate during moderate swimming activity in rainbow trout. *J. Exp. Biol.* **46,** 307–315.

Stevens, E. D., and Randall, D. J. (1967b). Changes of gas concentration in blood and water during moderate swimming activity in rainbow trout. *J. Exp. Biol.* **46,** 329–337.

Stevens, E. D., Bennion, G. R., Randall, D. J., and Shelton, G. (1972). Factors affecting arterial pressures and blood flow from the heart in intact, unrestrained Lingcod, *Ophiodon elongatus. Comp. Biochem. Physiol.* **43,** 681–695.

Sutterlin, A. M., and Saunders, R. L. (1969). Proprioceptors in the gills of teleosts. *Can. J. Zool.* **47,** 1209–1212.

Ussing, H., and Zerahn, K. (1951). Active transport of sodium as the source of electric current in the short-circuited isolated frog skin. *Acta Physiol. Scand.* **3,** 110–127.
van der Putte, I., and Part, P. (1982). Oxygen and chromium transfer in perfused gills of rainbow trout (*Salmo gairdneri*) exposed to hexavalent chromium at two different pH levels. *In* "Aquatic Toxicology" (L. J. Weber, ed.), Vol. 2, pp. 31–45. Raven Press, New York.
Wahlqvist, I. (1980). Effects of catecholamines on isolated systemic and branchial vascular beds of the cod, *Gadus morhua. J. Comp. Physiol.* **137,** 139–143.
Wahlqvist, I., and Nilsson, S. (1977). Teh role of sympathetic fibres and circulating catecholamines in controlling the blood pressure and heart rate in the cod, *Gadus morhua. Comp. Biochem. Physiol. C* **57C,** 65–67.
Wilkins, H., Regelson, W., and Hoffmeister, F. S. (1962). The physiological importance of pulsatile blood flow. *N. Engl. J. Med.* **256,** 443.
Wolf, K. (1963). Physiological salines for freshwater teleosts. *Prog. Fish-Cult.* **25,** 135–140.
Wood, C. M. (1974). A critical examination of the physical and adrenergic factors affecting blood flow through the gills of the rainbow trout. *J. Exp. Biol.* **60,** 241–265.
Wood, C. M. (1975). A pharmacological analysis of the adrenergic and cholinergic mechanisms regulating branchial vascular resistance in rainbow trout (*Salmo garidneri*). Can. J. Zool. **53,** 1569–1577.
Wood, C. M., and Jackson, E. B. (1980). Blood acid-base regulation during environmental hyperoxia in the rainbow trout (*Salmo gairdneri*). *Respir. Physiol.* **42,** 351–372.
Wood, C. M., and Shelton, G. (1980a). Cardiovascular dynamics and adrenergic responses of the rainbow trout *in vivo. J. Exp. Biol.* **83,** 247–270.
Wood, C. M., and Shelton, G. (1980b). The reflex control of heart rate and cardiac output in the rainbow trout: Interactive influences of hypoxia, haemorrage, and systemic vascular tone. *J. Exp. Biol.* **83,** 271–284.
Wood, C. M., McMahon, B. R., and McDonald, D. G. (1978). Oxygen exchange and vascular resistance in the totally perfused rainbow trout. *Am. J. Physiol.* **234,** 201–208.
Wood, C. M., McMahon, B. R., and McDonald, D. G. (1979). Respiratory gas exchange in the resting starry flounder, *Platichthys stellatus:* A comparison with other teleosts. *J. Exp. Biol.* **78,** 167–179.
Wood, C. M., McDonald, D. G., and McMahon, B. R. (1981). The influence of experimental anaemia on blood acid-base regulation *in vivo* and *in vitro* in the starry flounder (*Platichthys stellatus*) and the rainbow trout (*Salmo gairdneri*). *J. Exp. Biol.* **96,** 221–237.
Zadunaisky, J. A. (1979). Characteristics of Cl^- secretion in some non-intestinal epithelia. *In* "Mechanisms of Intestinal Secretion" (H. J. Binder, ed.), pp. 53–64. Alan R. Liss, Inc., New York.

AUTHOR INDEX

Numbers in italics refer to the pages on which the complete references are listed.

N

SYSTEMATIC INDEX

Note: Names listed are those used by the authors of the various chapters. No attempt has been made to provide the current nomenclature where taxonomic changes have occurred. Boldface letters refer to Parts A and B of Volume X.

A

B

E

F

G

H

U

X

Z

SUBJECT INDEX

Note: Boldface **A** refers to entries in Volume XA; **B** refers to entries in Volume XB.

A

B

C

D

E

F

G

H

I

J

K

L

O